AF388993

Cocoa, Chocolate and Human Health

Cocoa, Chocolate and Human Health

Special Issue Editors

Sabine Ellinger
Benno F. Zimmermann

MDPI • Basel • Beijing • Wuhan • Barcelona • Belgrade • Manchester • Tokyo • Cluj • Tianjin

Special Issue Editors

Sabine Ellinger
Niederrhein University of Applied Sciences
Germany

Benno F. Zimmermann
University of Bonn
Germany

Editorial Office
MDPI
St. Alban-Anlage 66
4052 Basel, Switzerland

This is a reprint of articles from the Special Issue published online in the open access journal *Nutrients* (ISSN 2072-6643) (available at: https://www.mdpi.com/journal/nutrients/special_issues/cocoa_health).

For citation purposes, cite each article independently as indicated on the article page online and as indicated below:

LastName, A.A.; LastName, B.B.; LastName, C.C. Article Title. *Journal Name* **Year**, *Article Number*, Page Range.

ISBN 978-3-03928-588-4 (Pbk)
ISBN 978-3-03928-589-1 (PDF)

Cover image courtesy of Cacao Collectors GmbH & Co. KG, Westenriederstraße 15, 80331 München, Germany.

Contents

About the Special Issue Editors

Sabine Ellinger has been professor in Nutrition Science at Niederrhein University of Applied Sciences, in Mönchengladbach, Germany, since 2012. She is associated with the Faculty of Agriculture, University of Bonn, as a private lecturer, where she also completed her Ph.D. She has received several research fellowships. Her main research interests are the cardiometabolic effects of flavonoid-rich foods, with focus on cocoa. Further interests include the impact of nutrition on wound healing. She has supervised many M.S. and Ph.D. students in these areas at University of Bonn and at Niederrhein University of Applied Sciences. She has published more than 30 peer-reviewed journal articles, which have amassed over 2000 citations.

Benno F. Zimmermann, Dr., is a researcher and lecturer at Bonn University, Germany; after earning a Ph.D. (Dr. rer. nat.) at Bonn University under the supervision of Rudolf Galensa, he gathered professional experience in a commercial setting as head of R&D at Institut Kurz, Köln, Germany. His main research interests are the analysis of flavonoids and other polyphenols in food and biological fluids and LC-MS development in general. He has published more than 45 peer-reviewed journal articles that have amassed over 750 citations.

 *nutrients*

Editorial

Cocoa, Chocolate, and Human Health

Benno F. Zimmermann [1] and Sabine Ellinger [2,*]

[1] Department of Nutritional and Food Sciences – Food Sciences, University of Bonn, Endenicher Allee 11-13, 53115 Bonn, Germany; benno.zimmermann@uni-bonn.de

[2] Faculty of Nutrition, Food, and Hospitality Sciences, Niederrhein, University of Applied Sciences, Rheydter Str. 277 , 41065 Mönchengladbach, Germany

* Correspondence: Sabine.Ellinger@hs-niederrhein.de

Received: 26 February 2020; Accepted: 2 March 2020; Published: 5 March 2020

Cocoa has been used as a ceremonial and hedonistic food for thousands of years in the tropical parts of America and for hundreds of years in the western world. In the last decades, health-related aspects of cocoa have come into the focus of research. This Special Issue entitled "Cocoa, Chocolate, and Human Health" presents the most recent findings on cocoa and health in 14 peer-reviewed articles, including nine original contributions and five reviews from cocoa experts around the world.

Polyphenols of all kinds have attracted great attention because of their possible beneficial effects [1,2], but one has to keep in mind the enormous variety within the group of polyphenolic compounds. Cocoa and cocoa-containing food such as chocolate are particularly rich in flavan-3-ols, i.e., mainly epicatechin and its close relatives, the proanthocyanidins [3]. Bioavailability and metabolism of the native flavanols and the process-derived flavanols in cocoa and cocoa products are the subjects of three contributions. Stereoisomers of (–)-epicatechin that are generated during roasting and alkalization are less bioavailable than the native (–)-epicatechin [4]. Phase-II-conjugates of epicatechin and metabolites without an intact flavanol core like phenolic acids are found in plasma and urine [5]. Further metabolites such as valerolactones are formed by the gut microbiome by the degradation of non-absorbed flavanols. These microbial metabolites are present in human plasma at roughly five times higher concentrations than epicatechin conjugates [6]. The impact of all these metabolites for health has not yet been completely elucidated. Nevertheless, current data suggest possible evidence that these microbial metabolites might also be relevant for human health [7].

Besides flavanols, theobromine and other methylxanthines, peptides [8], and volatile aroma compounds [9] might also affect human health, e.g., theobromine, which seems to improve memory in rats [10]. Many studies, being intervention studies or epidemiological observations, do not focus on single compounds, but on cocoa as such; in some cases, enriched in polyphenols. In these studies, an observed effect can hardly be ascribed to a single constituent but proves the effectiveness of cocoa as a functional food.

In this Special Issue, a positive influence of cocoa on hearing problems, exercise performance, and metabolic syndrome is discussed with mixed results. Hearing loss was found to be inversely associated with chocolate consumption in a middle-aged subgroup, but tinnitus did not depend on chocolate consumption [11]. In a review of thirteen clinical trials with athletes, a reduction of exercise-induced oxidative stress was found. However, regarding exercise performance and recovery, inconsistent results in literature did not allow a clear conclusion to be drawn [12].

There is evidence that cocoa flavanols may modulate some risk factors related to the metabolic syndrome, such as hypertension and disorders in glucose and lipid metabolism [13,14]. Several cardiometabolic parameters in type 2 diabetics were not affected by a flavanol-rich cocoa powder as simultaneous treatment with potent pharmaceuticals such as oral antidiabetic and antihypertensive drugs might have exhausted the effect of cocoa [15]. Also, the cocoa bean shell as a by-product of cocoa production contains valuable phytochemicals and can be used as an ingredient for functional food [16].

Three chapters focus on technical processes affecting cocoa components. During ripening and post-harvest processing, such as fermentation, drying, and roasting of cocoa beans, chemical changes occur to a lesser or greater extent, which concern almost all compounds. This has been known for many years regarding the flavanols, but proteins [17] and volatile compounds [9] are also formed or decomposed. Chocolate, the most popular cocoa product, shows remarkable losses in polyphenols and vitamin E during 18 months of storage. This is accompanied by changes in sensory profiles, while the flavor still remains acceptable [18].

Food and food supplements containing cocoa, enriched cocoa, or cocoa extracts are available for the costumers. The hoped-for effects are up to now only partly covered by scientific evidence. However, we have to bear in mind that manufacturers of such products do not only want to make the world a better place, but they also have financial interests.

As guest editors, we would like to acknowledge all the authors for their valuable contributions and the reviewers for their constructive remarks. Special thanks to the publishing team of *Nutrients* at MDPI for their professional support in all aspects of this Special Issue.

Funding: This research received no external funding.

Conflicts of Interest: The authors declare no conflict of interest.

References

1. Williamson, G. The role of polyphenols in modern nutrition. *Nutr. Bull.* **2017**, *42*, 226–235.
2. Grosso, G. Effects of Polyphenol-Rich Foods on Human Health. *Nutrients* **2018**, *10*, 1089.
3. Damm, I.; Enger, E.; Chrubasik-Hausmann, S.; Schieber, A.; Zimmermann, B.F. Fast and comprehensive analysis of secondary metabolites in cocoa products using ultra high-performance liquid chromatography directly after pressurized liquid extraction. *J. Sep. Sci.* **2016**, *39*, 3113–3122.
4. Ellinger, S.; Reusch, A.; Henckes, L.; Ritter, C.; Zimmermann, B.F.; Ellinger, J.; Galensa, R.; Stehle, P.; Helfrich, H.-P. Low Plasma Appearance of (+)-Catechin and (−)-Catechin Compared with Epicatechin after Consumption of Beverages Prepared from Nonalkalized or Alkalized Cocoa—A Randomized, Double-Blind Trial. *Nutrients* **2020**, *12*, 231.
5. Mayorga-Gross, A.L.; Esquivel, P. Impact of Cocoa Products Intake on Plasma and Urine Metabolites: A Review of Targeted and Non-Targeted Studies in Humans. *Nutrients* **2019**, *11*, 1163.
6. Gómez-Juaristi, M.; Sarriá, B.; Martínez-López, S.; Bravo-Clemente, L.; Mateos, R. Flavanol Bioavailability in Two Cocoa Products with Different Phenolic Content. A Comparative Study in Humans. *Nutrients* **2019**, *11*, 1441.
7. Márquez Campos, E.; Stehle, P.; Simon, M.C. Microbial Metabolites of Flavan-3-Ols and Their Biological Activity. *Nutrients* **2019**, *11*, 2260.
8. Marseglia, A.; Dellafiora, L.; Prandi, B.; Lolli, V.; Sforza, S.; Cozzini, P.; Tedeschi, T.; Galaverna, G.; Caligiani, A. Simulated Gastrointestinal Digestion of Cocoa: Detection of Resistant Peptides and in Silico/in Vitro Prediction of Their Ace Inhibitory Activity. *Nutrients* **2019**, *11*, 985.
9. Mota-Gutierrez, J.; Barbosa-Pereira, L.; Ferrocino, I.; Cocolin, L. Traceability of Functional Volatile Compounds Generated on Inoculated Cocoa Fermentation and Its Potential Health Benefits. *Nutrients* **2019**, *11*, 884.
10. Islam, R.; Matsuzaki, K.; Sumiyoshi, E.; Hossain, M.E.; Hashimoto, M.; Katakura, M.; Sugimoto, N.; Shido, O. Theobromine Improves Working Memory by Activating the CaMKII/CREB/BDNF Pathway in Rats. *Nutrients* **2019**, *11*, 888.
11. Lee, S.-Y.; Jung, G.; Jang, M.-J.; Suh, M.-W.; Lee, J.; Oh, S.-H.; Park, M. Association of Chocolate Consumption with Hearing Loss and Tinnitus in Middle-Aged People Based on the Korean National Health and Nutrition Examination Survey 2012–2013. *Nutrients* **2019**, *11*, 746–749.
12. Massaro, M.; Scoditti, E.; Carluccio, M.; Kaltsatou, A.; Cicchella, A. Effect of Cocoa Products and Its Polyphenolic Constituents on Exercise Performance and Exercise-Induced Muscle Damage and Inflammation: A Review of Clinical Trials. *Nutrients* **2019**, *11*, 1471.

13. Rynarzewski, J.; Dicks, L.; Zimmermann, B.; Stoffel-Wagner, B.; Ludwig, N.; Helfrich, H.-P.; Ellinger, S. Impact of a Usual Serving Size of Flavanol-Rich Cocoa Powder Ingested with a Diabetic-Suitable Meal on Postprandial Cardiometabolic Parameters in Type 2 Diabetics—A Randomized, Placebo-Controlled, Double-Blind Crossover Study. *Nutrients* **2019**, *11*, 417.
14. Jaramillo, F.M.E. Cocoa Flavanols: Natural Agents with Attenuating Effects on Metabolic Syndrome Risk Factors. *Nutrients* **2019**, *11*, 751.
15. Dicks, L.; Kirch, N.; Gronwald, D.; Wernken, K.; Zimmermann, B.; Helfrich, H.-P.; Ellinger, S. Regular Intake of a Usual Serving Size of Flavanol-Rich Cocoa Powder Does Not Affect Cardiometabolic Parameters in Stably Treated Patients with Type 2 Diabetes and Hypertension—A Double-Blinded, Randomized, Placebo-Controlled Trial. *Nutrients* **2018**, *10*, 1435.
16. Rojo-Poveda, O.; Barbosa-Pereira, L.; Mateus-Reguengo, L.; Bertolino, M.; Stévigny, C.; Zeppa, G. Effects of Particle Size and Extraction Methods on Cocoa Bean Shell Functional Beverage. *Nutrients* **2019**, *11*, 867.
17. Rawel, H.; Huschek, G.; Sagu, S.; Homann, T. Cocoa Bean Proteins—Characterization, Changes and Modifications due to Ripening and Post-Harvest Processing. *Nutrients* **2019**, *11*, 428.
18. Roda, A.; Lambri, M. Changes in Antioxidants and Sensory Properties of Italian Chocolates and Related Ingredients under Controlled Conditions During an Eighteen-Month Storage Period. *Nutrients* **2019**, *11*, 2719–2721.

 *nutrients*

Review

Impact of Cocoa Products Intake on Plasma and Urine Metabolites: A Review of Targeted and Non-Targeted Studies in Humans

Ana Lucía Mayorga-Gross [1,*] and Patricia Esquivel [2]

1 Centro Nacional de Ciencia y Tecnología de Alimentos, Universidad de Costa Rica, San Pedro 11501-2060, Costa Rica
2 Escuela de Tecnología de Alimentos, Universidad de Costa Rica, San Pedro 11501-2060, Costa Rica; patricia.esquivel@ucr.ac.cr
* Correspondence: analucia.mayorga@ucr.ac.cr

Received: 13 February 2019; Accepted: 25 April 2019; Published: 24 May 2019

Abstract: Cocoa is continuously drawing attention due to growing scientific evidence suggesting its effects on health. Flavanols and methylxanthines are some of the most important bioactive compounds present in cocoa. Other important bioactives, such as phenolic acids and lactones, are derived from microbial metabolism. The identification of the metabolites produced after cocoa intake is a first step to understand the overall effect on human health. In general, after cocoa intake, methylxanthines show high absorption and elimination efficiencies. Catechins are transformed mainly into sulfate and glucuronide conjugates. Metabolism of procyanidins is highly influenced by the polymerization degree, which hinders their absorption. The polymerization degree over three units leads to biotransformation by the colonic microbiota, resulting in valerolactones and phenolic acids, with higher excretion times. Long term intervention studies, as well as untargeted metabolomic approaches, are scarce. Contradictory results have been reported concerning matrix effects and health impact, and there are still scientific gaps that have to be addressed to understand the influence of cocoa intake on health. This review addresses different cocoa clinical studies, summarizes the different methodologies employed as well as the metabolites that have been identified in plasma and urine after cocoa intake.

Keywords: cocoa; chocolate; metabolites; biomarkers; metabolomics; urine; plasma; procyanidins; methylxanthines; polyphenols

1. Introduction

1.1. Theobroma cacao L. composition

Cocoa (*Theobroma cacao* L.) is a tree from the Malvaceae family. Its seeds are covered by a sweet and sour mucilage which contains approximately 11% of sugars, mainly sucrose, and an acidic environment with a pH of about 3.5–3.8. Citric acid is the main organic acid present in the pulp. Others such as oxalic, phosphoric, malic and tartaric acid are also present [1–6].

Fat is the main component of cocoa beans: it accounts to almost 50% of the cotyledon dry weight, where 98% corresponds to neutral lipids, and 2% to polar lipids, mainly phospholipids and glycolipids. It has been reported that the major cocoa fatty acids are palmitic, stearic and oleic [7,8].

Proteins represent 17–20% of the dried bean [9,10]. Vicilin-like globulins and albumins are the two most abundant proteins and are key factors in quality development during fermentation. Concerning the amino acid profile, glutamic acid and aspartic acid showed the highest contents while cysteine showed the lowest ones, in both fermented and non-fermented cocoa derived products [11].

From a functional point of view, cocoa has been considered an important source of different bioactive compounds. The more important groups of compounds present in cacao are flavonoids, mainly flavan-3-ols, phenolic acids, methylxanthines, peptides, *N*-phenylpropenoyl-L-amino acids, and stilbenes [5,12–20].

Polyphenols represent approximately 13% of the dried unfermented cocoa beans [21], where proanthocyanidins can reach 58% of the total polyphenol content, followed by catechins and anthocyanins with 37% and 4%, respectively [22]. Proanthocyanidins are usually (epi) catechin-based, i.e., procyanidins, and the main anthocyanins are cyanidin-3-galactoside and cyanidin-3-arabinoside [21,23,24].

The main methylxanthines are theobromine, caffeine, theophylline, and 7-methylxanthine. Theobromine is the most concentrated methylxanthine, and it could be present in concentration levels of 1–2% (dry basis) in cocoa seeds; caffeine also has an important contribution from 0–2% (dry basis) [5].

Additionally, *N*-phenylpropenoyl-L-amino acids have been identified in cocoa and described as polyphenol-amino acid conjugates, some of which have been related to antioxidant mechanisms, inducers of mitochondrial activity and inhibitors of pathogen adhesion in stomach tissues [20].

Cocoa and cocoa derived products suffer significant changes throughout the processing, and this has to be taken into account when studying the effects of cocoa intake on human health.

Traditionally, cocoa undergoes different processing steps after harvesting, some of which are fermentation, drying, roasting [6], and size reduction steps.

During fermentation and drying cocoa beans have extensive proteolysis, oxidation, and polymerization reactions, among others, resulting in a decrease on total flavonoids and the emergence of new metabolites [5,25–28].

Also, during roasting, high temperatures lead to glucose and sugar depletion, and to the formation of pyrazines, pyrroles, quinoxalines, pyrones, phenylalk-2-enals, lactones, diketopiperazines, and derivatives of phenylalanine, furanones and others. Compounds, such as triacylglycerols, alcohols, esters, and acids, do not go through significant changes [5]. On the other hand, flavanols and procyanidins can be affected by different temperature dependent reactions, as for example epimerization [3,29–31].

Roasted beans [3,5] are the starting point for chocolate and cocoa powder production, which are the cocoa derived products most commonly consumed. In both cases, formulation steps are needed, which increase the diversity of compounds present in these matrices and their interactions.

1.2. Cocoa and Health: General Aspects

A variety of health benefits have been associated to the intake of cocoa and its derived products, many of them attributed to the intake of polyphenols, particularly flavonoids [32–34]: improvement in insulin resistance by lowering serum insulin and in flow-mediated dilatation [35], as well as improvement in blood pressure, maintenance of normal endothelium-dependent vasodilation, vascular and platelet function, increased cerebral blood flow, potential cancer prevention, and anti-inflammatory and antioxidant activity [36–40]. Nonetheless, contradictory evidence has been reported indicating that more research is still needed [41–46].

It has been proposed that cocoa polyphenols can also have an influence on the gut microbiota, promoting the development of microorganisms as *Lactobacillus* spp. and *Bifidobacterium* spp. Microbiota can metabolize polyphenols into different bioactives, as valerolactones, phenylpropionic and phenylacetic derivatives some of which could activate anti-inflammatory pathways [33,47,48].

Other bioactive compounds, like methylxanthines, have shown possible beneficial effects. For example, caffeine may improve exercise performance, but more research is needed to understand the mechanisms behind this statement [49].

To understand the effects of cocoa on human health, it is fundamental to know which metabolites are produced after cocoa intake, as this can help understand the absorption, distribution, metabolism, and excretion mechanisms [47,50]. The development of different types of clinical trials contributes to this knowledge; and, in relation to cocoa, there is a particular need to promote long term research.

This review collects the results of different studies that traced several metabolites in urine and plasma after the acute or chronic cocoa intake.

2. Methodological Considerations

Data collection was done by searching individual and compound keywords in different databases, such as: cocoa, chocolate, humans, randomized, acute, health, untargeted, non-targeted, metabolomics, urine, plasma, pharmacokinetics, polyphenols, (epi) catechin, procyanidin, methylxanthines, and theobromine. The main databases used were: Google Scholar, ScienceDirect, ACS Publications, and Springer Link. The searches were done in between August 2016 and February 2019 and included studies from 1999 to February 2019.

3. Results

Tables 1 and 2 summarize several studies that analyze the effects of the intake of different cocoa derivatives, such as chocolate (Table 1) and cocoa powder drinks or extracts (Table 2) in human plasma and urine samples using a targeted methodology. Meanwhile Table 3 summarizes non-targeted researches, including chocolate or cocoa powder drink intake studies, as well as one cross sectional study.

The methodological aspects are detailed, including the main objective of the studies, the experimental design applied [51–54], information regarding the subjects, matrices and dose employed, the biological samples studied, the analytical instrumental technique used to measure the metabolites, and the statistics applied to process data.

A total of 34 studies were reviewed, 30 of them used a targeted approach, and 4 of them were untargeted. The clinical studies employed different cocoa derived matrices. Nine of these studies included chocolate intake intervention, 24 used cocoa powders or cocoa extracts beverages, and 2 of them had a cross-sectional experimental design which did not apply dietary interventions [55,56].

Most researches followed acute single dose interventions, nonetheless only 2 of them evaluated the effect of a daily cocoa-derived products intake after 4 weeks [57,58]. Additionally, 10 studies included a non-cocoa-derived control intake in their experimental design.

Regarding the type of samples analyzed, most studies (n) investigated plasma ($n = 8$), urine ($n = 14$) or both plasma and urine ($n = 12$). Additionally, one study analyzed saliva [59] and another studied feces [60].

The number of volunteers (n) that finalized clinical studies was varied. In five studies only 5 volunteers participated, in ten studies $5 < n \leq 10$ were part of the clinical trial, eight studies included $10 < n \leq 20$ and twelve studies recruited and selected volunteers in a range from 21 to 80 volunteers.

Most clinical studies investigated samples from healthy non-smokers. Only two trials obtained samples from smokers [61] and from volunteers with high cardiovascular risk [57]. Ten studies investigated only men, 18 studies men and women, and one study investigated children [56].

LC-UV/Vis (liquid chromatography couple to ultraviolet or visible detectors) and FLD (fluorescence detectors), and LC-MS (liquid chromatography couple to mass spectrometers) were the predominant analytical platforms reported. In particular, when using an untargeted approach, LC-MS was present in all studies; and one study also applied NMR (nuclear magnetic resonance) [20].

Further on, Tables 4–8 were used to classify the metabolites identified in urine and plasma, in different studies with different approaches.

Table 1. Summary of the main aspects of targeted cocoa studies, involving chocolate intake.

Objectives	Experimental Design	Subjects	Matrices Description and Dose	Biological Samples	Sampling period	Analytical Technique	Statistical Analysis	Year	Reference
Analyze plasma kinetics of epicatechin after dark chocolate intake.	Non-randomize cross-over. 1-week washout.	8 males. Average age and BMI of 40 years and 23.9 kg/m^2.	40 g and 80 g of dark chocolate with bread.	Plasma.	0, 1, 2, 3, 4, and 8 h.	HPLC-UV or fluorescence.	Student's and Wilcoxon tests.	1999	[62]
Evaluate changes in plasma epicatechin levels, antioxidant capacity and plasma lipid oxidation products after chocolate intake.	Randomized, cross-over. 1-week washout.	14 women and 6 men. From 20 to 56 years, with average BMI of 23.8 kg/m^2.	0, 27, 53, and 80 g of semi-sweet chocolate rich on procyanidins, and 47 g of bread.	Plasma.	0, 2, and 6 h.	HPLC coupled with a coulometric detector.	ANOVA with control for multiple measurements, and Tukey–Kramer test.	2000	[63]
Evaluate (-)-epicatechin and its metabolites in plasma and urine after cocoa and chocolate intervention.	Cross-over. 6-day washout.	5 males. Ages from 30 to 33 years and BMI from 20.4 to 23.9 kg/m^2.	Chocolate or cocoa.	Urine and plasma.	Plasma: Baseline, 1, 2, 4, 8, and 24 h.Urine: 0–8 h, 8–24 h.	HPLC and LC-MS negative mode.	Student's *t*-test.	2000	[64]
Quantify in urine, different phenolic acids formed in the colon after the consumption of chocolate.	Uncontrolled.	7 men and 4 women. Mean age of 24, mean height of 172 cm and mean weight of 67 kg.	80 g of chocolate.	Urine.	Baseline, 0–3, 3–6,6–9, 9–24, and 24–48 h.	GC-MS, HPLC-DAD, HPLC-ESI-MS in negative ionization mode.	ANOVA and Tukey test.	2003	[65]
Determine theobromine and caffeine in saliva, plasma and urine after dark chocolate intake.	Uncontrolled.	5 healthy volunteers with no dietary restrictions.	41 g of dark chocolate bars.	Urine, saliva, and plasma.	Baseline and 90 min.	UPLC-DAD triple quadrupole MS/MS.	No tests were applied. Data was expressed as averages with confidence intervals.	2010	[59]
Identify and quantify (-)-epicatechin conjugates in plasma and urine after chocolate intake.	Uncontrolled.	5 healthy volunteers. Average age of 23 years and average BMI of 22 kg/m^2.	100 g of 70% chocolate having 79 mg of (-)-epicatechin, 26 mg of (±)-catechin, and 49 mg of procyanidin B2.	Urine and plasma.	Plasma: Baseline, 15, 30, 45 min, and 1, 1.25, 1.5, 1.75, 2, 2.25, 2.5, 3, 4, 6, 8, 10, 14, 18, and 24 h. Urine: Baseline, 0–5 h, 5–10 h, 10–24 h, and 0–24 h.	UPLC-ESI-Quattro Micro API.	One-way ANOVA with Tukey test.	2012	[66]
Analyze the relation between chocolate consumption, glucose metabolism, and exercise performance.	Single blinded, randomized, cross-over.	16 male cyclists (from 4 to 20 h/week of training).	Cocoa enriched dark 70% chocolate and cocoa depleted control.	Plasma.	−10, 0, 15, 30, 45, 60, 90, and 120, 140, 180, 210, 240, 300, and 360 min.	HPLC-UV/Vis	Mixed model with F-test, Hidges-Lehmann, Wilcoxonsigned-rank and Student's *t*-test.	2014	[49]
Evaluate the effects of intake of milk chocolate, cocoa powder or dark chocolate consumption on uric acid crystallization.	Cross-over.	11 males and 9 females with ages from 22 to 65 years.	2 × 20 g of a cocoa derived product/day (milk chocolate, chocolate powder and dark chocolate).	Urine.	Baseline and urine obtained after 12 h overnight at the end of the intervention.	HPLC-UV/Vis.	ANOVA, Bonferroni and Wilcoxon signed-rank tests.	2018	[67]

Notes: The abbreviation BMI corresponds to body mass index, and BW to body weight.

Table 2. Summary of the main aspects of targeted cocoa studies, involving cocoa powder or cocoa extracts-based drinks intake.

Objectives	Experimental Design	Subjects	Matrices Description and Dose	Biological Samples	Sampling period	Analytical Technique	Statistical Analysis	Year	Reference
Determine the presence of specific procyanidins in human plasma after consumption of cocoa flavanol extract.	Uncontrolled.	3 men and 2 women. Age: 23–34 years, average body weight 70.5 kg.	0.375 g of cocoa extract/kg BW in 300 mL of water (average of 26.4 g of cocoa).	Plasma.	Baseline, 0.5, 2 and 6 h.	HPLC-coulometric electrochemical multiple-array detection; HPLC-MS.	Kruskal–Wallis one-way ANOVA and Tukey or Dunn tests.	2002	[68]
Evaluate the effect of a flavanol-rich cocoa beverage on the circulating pool of nitric oxide and of endothelial dysfunction.	Randomized, double-blinded, cross-over.	6 males and 5 females, with mean age of 31, and mean BMI of 21.8 kg/m^2. They smoked an average of 17 cigarettes/day.	High and low flavanol content cocoa drink in water.	Plasma.	Baseline and 2 h.	HPLC-FLD.	ANOVA, pairwise tests with Bonferroni correction for multiple comparisons.	2005	[61]
Measure different metabolites in urine after polyphenol-rich beverages intake, with a fast method.	Randomized, cross-over. 14-day washout.	5 women and 4 men, with ages in between 20–32 years and BMI in between 18.9–24.8 kg/m^2.	10 g of cocoa powder in 200 mL of water (other non-cocoa beverages were tested in the research), or hot water.	Urine.	0–24 h.	HPLC-ESI-MS/MS	Mann–Whitney U test.	2005	[69]
Developing a rapid and reproducible method for analysis of (-)-epicatechin metabolites in plasma and urine.	Randomized, cross-over.	2 women and 3 men in a range of 18–49 years.	250 mL of milk or 40 g of cocoa powder in 250 mL of milk.	Plasma and urine.	Plasma: 0 and 2 h Urine: 0 and 6 h.	HPLC coupled to an API-QQQ-MS.	Student's t-test.	2005	[70]
Evaluate if dietary flavanols and their metabolites can function as vasoactive mediators.	Randomized, double-blinded cross-over. Minimum 2-day washout.	16 males within 25–32 years and with a BMI of 19–23 kg/m^2.	High and low flavanol content cocoa powder with 300 mL of water.	Plasma.	Baseline, 1, 2, 3, 4, and 6 h.	HPLC-MS	ANOVA, pairwise tests with Bonferroni correction for multiple comparisons.	2006	[71]
Analyze (-)-epicatechin metabolites and total antioxidant activity after cocoa beverage intake.	Randomized and cross-over.	9 women and 12 men, within 18 and 50 years, and with a BMI of 19.1–27.7 kg/m^2.	40 g of cocoa powder with 250 mL of water and 250 mL of milk as control.	Urine.	Baseline, 0, 6, 12, and 24 h.	API-QQQ-MS/MS.	ANCOVA and Student's t-test; Pearson's correlation.	2007	[72]
Analyze the effect of milk in the bioavailability of (-)-epicatechin from a cocoa powder.	Randomized and cross-over. 1-week washout.	9 women, 12 men, within 18 and 50, and with a BMI of 19.1–27.7 kg/m^2.	250 mL of milk used as a control, 40 g of cocoa powder dissolved in 250 mL of water or milk.	Plasma.	Baseline, 2 and 6 h.	LC-MS.	ANCOVA.	2007	[73]
Determine the effect of milk protein addition on the uptake of cocoa polyphenols by analyzing metabolites in plasma samples, after a cocoa drink intervention.	Randomized, controlled, double blind, and cross-over. 1-week washout.	10 men and 14 women. Age in the range of 52–65 years with BMI in the range of 18–30 kg/m^2. 13 volunteers were taking medications or dietary supplements.	200 mL of dairy and non-dairy chocolate drink.	Plasma.	Baseline, 0.5, 1, 1.5, 2, 3, 4, 6, and 8 h.	Analytical: HPLC-fluorescence.	Paired t-test with Bonferroni correction for multiple tests.	2007	[74]

Table 2. *Cont.*

Objectives	Experimental Design	Subjects	Matrices Description and Dose	Biological Samples	Sampling period	Analytical Technique	Statistical Analysis	Year	Reference
Quantify and evaluate human metabolism of N-phenylpropenoyl-L-amino acids present in a cocoa drink.	Uncontrolled.	4 males and 4 females, from 24 to 30 years.	Cocoa powder-based beverage.	Urine.	Baseline, 1, 2, 3, 4, 6 and 8 h.	LC-MS/MS multiple reaction monitoring and NMR.	Not specified.	2008	[20]
Evaluate the impact of milk addition on the (-)-epicatechin metabolites after intake of a cocoa drink.	Randomized cross-over. 1-week washout.	9 women and 12 men with ages between 18–50 and BMI from 19.1 to 27.7 kg/m^2.	Cocoa beverage with 40 g of cocoa powder and: (a) 250 mL of milk; (b) 250 mL of water or (c) 250 mL of milk without cocoa powder.	Urine.	0–6, 6–12 and 12–24 h.	HPLC coupled to an API-QQQ-MS.	ANCOVA and Student's *t*-test.	2008	[75]
Develop an analytical method for determining cocoa metabolites in human and rat urine.	Uncontrolled.	9 women and 12 men, within 18 and 50 years and a mean BMI of 21.6 kg/m^2.	40 g of cocoa powder in 250 mL of water.	Urine.	Baseline and 24 h after intake.	SPE and LC-MS/MS.	Wilcoxon test.	2009	[76]
Determine the effect of milk on the bioavailability of cocoa flavan-3-ol metabolites by evaluating plasma and urine samples after cocoa powder intervention.	Controlled, cross-over. 4-week washout.	6 man and 3 women. Ages in the range of 20–43. Average BMI of 24.7 kg/m^2.	10 g of cocoa powder in 250 mL of milk or water with 1 g of paracetamol and 5 g of lactulose.	Urine and plasma.	Plasma: baseline 0.5, 1, 2, 3, 4, 6, 8, and 24 h. Urine: baseline, 0–2, 2–5, 5–8 and 8–24 h.	HPLC-PDA-MS^2.	2-factor repeated measures ANOVA and Student's *t*-test.	2009	[77]
Evaluate plasma and urine metabolites after cocoa powder intake in high cardiovascular risk subjects.	Randomized, controlled and cross-over.	High CVD risk patients: 19 men and 23 women. Average age of 69.7 years.	2 × 20 g of cocoa powder/day with 250 skimmed milk or only 500 mL/day of skimmed milk for 4 weeks.	Urine and plasma.	0–24 h.	LC-MS/MS.	Wilcoxon test.	2009	[57]
Evaluate the effect of milk on the urinary excretion of colonic microbial-derived phenolic acids after cocoa powder intake.	Randomized and cross-over.	12 men and 9 women. Age in the range of 18–50 years with BMI of 21.6 kg/m^2.	40 g of cocoa powder in 250 mL of water or milk.	Urine.	Baseline, 0–6, 6–12, and 12–24 h.	LC-MS/MS and LC-PAD.	Wilcoxon test for related samples.	2010	[78]
Study the stereochemical configuration of four different flavanols on their absorption, metabolism, and biological activity.	Randomized, double-blinded cross-over.	7 males within 18 and 35 years old, with average BMI of 24 kg/m^2.	1.5 mg/kg BW of (-)-epicatechin, (+)-epicatechin, (+)-catechin, (-)-catechin, with 0.5 g/kg BW of low-flavanol cocoa based dairy drink.	Urine and plasma.	Urine: collected over 24 h. Plasma:baseline, 2 and 4 h.	HPLC-UV/VisAnd fluorescence.	Two-way repeated measures ANOVA and Tukey's test.	2011	[79]

Table 2. *Cont.*

Objectives	Experimental Design	Subjects	Matrices Description and Dose	Biological Samples	Sampling period	Analytical Technique	Statistical Analysis	Year	Reference
Quantify different metabolites in plasma and urine, after the consumption of a cocoa drink with added flavanols and procyanidins.	Randomized, double-masked, cross-over.	12 males in between 18 and 35 years old, with average BMI of 24 kg/m^2.	0.48 g/kg BW of flavanol (F) and procyanidin (P) free mimetic cocoa drink powder reconstituted in 4 g/kg of BW of milk with 1% fat with (a) cocoa extract with monomeric F and P, (b) cocoa extract high in P or (c) (-)-epicatechin isolated from cocoa.	Urine and plasma.	Urine: 0–7 h and 7–24 h. Plasma: Baseline, 1, 2, and 4 h.	HPLC-UV and coulometric electrochemical detector, and HPLC-diode array detector.	Two-way repeated measures ANOVA and Tukey's test.	2012	[80]
Study the bioavailability of methylxanthines in two soluble cocoa products by evaluating plasma and urine samples the after intervention.	Randomized, controlled, single-blind, cross-over. 10-day washout.	3 males and 10 females. Ages in the range of 18–45 years. Average BMI of 22.5 kg/m^2 for men and 23.4 kg/m^2 for women.	15 g of the powder without enrichment and 25 g of methylxanthine-enriched powder, with 200 mL of semi-skimmed milk.	Urine and plasma.	Plasma: baseline, 0.5, 1, 2, 3, 4, 6, and 8 h. Urine: baseline, 0–4, 4–8, 8–12, and 12–24 h.	HPLC-DAD, HPLC-Q-ToF in positive ionization mode.	Mixed model.	2014	[81]
Analyze the bioavailability, metabolism and microbial breakdown of (-)-epicatechin, procyanidin B1, and a cocoa procyanidin fraction.	Randomized, double-blinded cross-over. 1-week washout.	7 healthy male volunteers with ages in between 24 and 31 years and with a BMI of 22 and 26 kg/m^2.	Pure (-)-epicatechin, pure procyanidin B1 and a purified cocoa polymeric procyanidin fraction.	Plasma, urine and feces.	Plasma: baseline, 1, 2, 4, 8, 24 and 48 h. Urine: 0–4, 4–8, 8–24 h, and 48 h. Feces: 0–24 h.	GC-MS/MS and HPLC-ESI-Q-MS.	Not specified.	2015	[60]
Validate and HPLC method for measuring theobromine in urine and evaluating theobromine urinary levels in children with different cocoa intake patterns.	Cross-sectional.	80 healthy children from 8–17 years: 26 did not consume cocoa derived products, 19 of them did not consumed cocoa powder but did consume 1 cocoa derived product daily, 12 children just consumed cocoa powder in breakfast and 23 consumed cocoa derived products > once a day.	No dietary intervention or recommendations.	Urine.	12 h in the day and 12 h in the night.	HPLC-UV.	Kruskal–Wallis and Wilcoxon signed-rank tests.	2015	[56]

Table 2. *Cont.*

Objectives	Experimental Design	Subjects	Matrices Description and Dose	Biological Samples	Sampling period	Analytical Technique	Statistical Analysis	Year	Reference
Compare absorption, metabolism, and excretion after a cocoa drink intake.	Randomized cross-over. 1-week washout.	40 males. Group 1 (young): 18–35 years and average BMI of 24 kg/m^2; Group 2 (elderly): 65–80 years and average BMI of 27 kg/m^2.	Fruit-flavored cocoa powder drinks with 5.3 mg or 10.7 mg of cocoa flavanols/kg BW.	Plasma and urine.	Plasma: 0, 0.5, 1, 1.5, 2, 3, 4, 5, 6, and 24-h. Urine: 0–2, 2–4, 4–6, 6–12, 12–24 h.	HPLC-FLD/UV and electrochemical detection, and HPLC-UV.	Two-way ANOVA with Bonferroni and Tukey tests.	2015	[82]
Evaluate the effects of milkand dark chocolate and cocoa powder intake on uric acid crystallization.	Cross-over.	11 males and 9 females with ages from 22 to 65 years.	2 × 20 g/day of a cocoa derived product: milk chocolate, chocolate powder and dark chocolate.	Urine.	Baseline and urine obtained after 12 h overnight at the end of the intervention.	HPLC-UV/Vis.	ANOVA, Bonferroni and Wilcoxon signed-rank test.	2018	[67]
Analyze the use of gVL-3′/4′-sulphate and gVL-3′/4′-O-glucuronide as biomarkers of flavan-3-ols intake.	Randomized, double-masked cross-over.	I. 8 males, from 25–60 years. II. 14 males, from 25–40 years.	I. 8 different nondairy drinks with flavan-3-ols (F) with 34.8 mg of (-)-epicatechin or the equivalent concentration of one of the following: (-)-epigallocatechin, (-)-epicatechin-3-O-gallate, (-)-epigallocatechin-3-O-ga (isolated from green tea), 1:1:1 of theaflavin-3-O-gallate, theaflavin-3′-O-gallate and theaflavin-3,3′-O-digallate, thearubigins, (isolated from black tea), procyanidin B2 isolated from cocoa, and a control without added F. II. Four different nondairy drinks with different amounts of flavan-3-ols ranging from 95 mg to 1424 mg per volunteer.	Urine.	I. 0–6 and 6–24 h. II. 0–4, 4–8, 8–12, and 12–24 h.	UPLC-MS.	ANOVA and Student's t-test.	2018	[83]

Table 2. *Cont.*

Objectives	Experimental Design	Subjects	Matrices Description and Dose	Biological Samples	Sampling period	Analytical Technique	Statistical Analysis	Year	Reference
Evaluate the effect of cocoa intake on lipid profiles and biomarkers of oxidative stress, and arachidonic acid/ eicosapentaenoic acid ratio.	Randomized, three-arm parallel group.	48 healthy subjects divided in three groups: low-cocoa (n = 16), middle-cocoa (n = 16), and high-cocoa group (n = 16). Average age and BMI were close to 44 years and 23 kg/m^2.	Cocoa capsules dosed to different groups: a) low-cocoa group: 1 g of cocoa (55 mg/day), middle-cocoa group: 2 g of cocoa (110 mg flavanols/day), and high-cocoa group: 4 g of cocoa (220 mg flavanols/day), for 4 weeks.	Plasma and urine.	0, 1, 2, and 4 (at the beginning of the study and after 4 weeks of intervention).	HPLC-FLD-UV	ANOVA and Tukey's test.	2018	[58]

Notes: The abbreviation BMI corresponds to body mass index, and BW to body weight.

Table 3. Summary of the main aspects of untargeted cocoa intervention studies.

Objectives	Experimental Design	Subjects	Matrices and Dose	Biological Samples	Sampling period	Analytical Technique	Statistical Analysis	Year	Reference
Study the chemical profile of plasma and urine after flavan-3-ol enriched dark chocolate, standard dark chocolate and white chocolate intake.	Randomized, controlled and cross-over. Minimum 2-wk washout.	16 males and 26 females with average BMI of 24.5 kg/m^2.	60 g of: flavan-3-ol (FLA) enriched dark chocolate, standard dark chocolate low in FLA and white chocolate with no FLA.	Plasma and urine.	Baseline, 2 and 6 h. (1)	1H NMR (600 MHz) and HPLC-ToF-MS.	PCA and PLS-DA. Kruskal–Wallis and Dunn tests.	2017, 2013	[84,85]
Evaluate the changes in humane urine metabolome after cocoa powder intake.	Randomized, controlled and cross-over. 1-week washout.	5 men and 5 women. Ages in the range of 18–50 and average BMI of 21.6 kg/m^2.	40 g of cocoa powder in 250 mL of water or milk and 250 mL of milk as a control.	Urine.	Baseline, 0–6, 6–12 and 12–24 h.	HPLC-Q-ToF. Positive ionization mode.	PCA, PLS-DA, (OSC) PLS and OSC-PLS-DA.	2009	[86]
Analyze the changes in urine metabolome after cocoa powder intervention.	Uncontrolled.	5 women and 5 men between 18–50 years with BMI 21.6 kg/m^2.	40 g of cocoa powder with 250 mL of milk.	Urine.	Baseline, 0–6, 6–12, and 12–24 h.	HPLC-Q-ToF-MS in positive mode.	PLS-DA, OSC-PLS-DA and two-way hierarchical clustering (HCA) applying Bonferroni correction.	2010	[87]
Apply an untargeted metabolomics approach to propose a model that can discriminate the urinary metabolome of regular cocoa product consumers and non-consumers, revealing dietary biomarkers, in a free-living population.	Randomized, controlled, parallel group, multicenter, and cross-sectional.	64 high risk (2) free-living subjects. 32 were classified as cocoa consumers (at least 3 servings/week of chocolate and/or cocoa powder) and 32 as non-cocoa consumers (0 g/day) (3).	No dietary interventions or recommendations.	Urine.	Baseline.	Analytical: HPLC-Q-ToF-MS. Positive and negative ionization modes.	OSC-PLS-DA, Mann–Whitney and Student's *t*-test.	2015	[55]

Notes: (1) Samples taken 2 h after intake were analyzed only by NMR, and not by LC-MS. (2) At least with type 2 diabetes mellitus or with at least three conventional cardiovascular risk factors. (3) Baseline data and urine samples were obtained from a PREDIMED study of 275 volunteers.

Table 4. Metabolites analyzed in plasma after different acute cocoa intake interventions, by targeted methodologies.

Chemical Category	Metabolite	Treatment 1			Treatment 2			Reference
		Type of Matrix	Concentration	Sampling Period (h)	Type of Matrix	Concentration	Sampling Period (h)	
Flavonoids and conjugates		N.R.	N.R.	N.R.	Dark chocolate	233 ± 60 nmol/L	3.2 ± 0.2	[66]
	(-)-epicatechin 3′-sulfate	Drink with 5.3 mg cocoa flavanols/kg body weight (young)	98 ± 12 nmol/L	1.8 ± 0.2	Drink with 5.3 mg cocoa flavanols/kg body weight (elderly)	125 ± 13 nmol/L	1.7 ± 0.2	[82]
		Drink with 10.7 mg cocoa flavanols/kg body weight (young)	251 ± 20 nmol/L	1.5 ± 0.1	Drink with 10.7 mg cocoa flavanols/kg body weight (elderly)	212 ± 13 nmol/L	1.7 ± 0.1	[82]
	(-)-epicatechin 4′-sulfate	N.R.	N.R.	N.R.	Dark chocolate	11 ± 3 nmol/L	3.5 ± 0.3	[66]
		N.R.	N.R.	N.R.	Dark chocolate	290 ± 49 nmol/L	3.2 ± 0.2	[66]
	(-)-epicatechin-3′-β-D -glucuronide	Drink with 5.3 mg cocoa flavanols/kg body weight (young)	309 ± 41 nmol/L	1.1 ± 0.1	Drink with 5.3 mg cocoa flavanols/kg body weight (elderly)	371 ± 34 nmol/L	1.3 ± 0.1	[82]
		Drink with 10.7 mg cocoa flavanols/kg body weight (young)	551 ± 67 nmol/L	1.2 ± 0.1	Drink with 10.7 mg cocoa flavanols/kg body weight (elderly)	645 ± 66 nmol/L	1.2 ± 0.1	[82]
	(-)-epicatechin-4′-β-D- glucuronide	N.R.	N.R.	N.R.	Dark chocolate	44 ± 11 nmol/L	3.4 ± 0.3	[66]
	(-)-epicatechin-7′-β-D-glu	N.R.	N.R.	N.R.	Dark chocolate	22 ± 6 nmol/L	12.8 ± 4.8	[66]
	(-)-epicatechin glucuronide	Cocoa powder with water	330 ± 156 nmol/L	2	Cocoa powder with milk	273 ± 138 nmol/L	2	[73]
	(4-hydroxyphenyl) acetic acid	N.R.	N.R.	N.R.	Cocoa polymeric procyanidin concentrate	120 ± 40 ng/mL	2 ± 1	[60]
	epicatechin-*O*-sulfate	Cocoa powder with water	83 ± 8 nmol/L	1.4 ± 0.2	Cocoa powder with milk	77 ± 14 nmol/L	1.3 ± 0.2	[77]
	3′-*O*-methyl-(-)-epicatechi 4′-sulfate	N.R.	N.R.	N.R.	Dark chocolate	49 ± 14 nmol/L	3.6 ± 0.3	[66]
	3′-*O*-methyl-(-)-epicatechin 7-sulfate	N.R.	N.R.	N.R.	Dark chocolate	40 ± 10 nmol/L	3.8 ± 0.2	[66]
		Drink with 5.3 mg cocoa flavanols/kg body weight (young)	76 ± 6 nmol/L	1.7 ± 0.2	Drink with 5.3 mg cocoa flavanols/kg body weight (elderly)	63 ± 6 nmol/L	1.8 ± 0.1	[82]
		Drink with 10.7 mg cocoa flavanols/kg body weight (young)	167 ± 19 nmol/L	1.7 ± 0.1	Drink with 10.7 mg cocoa flavanols/kg body weight (elderly)	109 ± 11 nmol/L	1.9 ± 0.2	[82]

Table 4. *Cont.*

Chemical Category	Metabolite	Treatment 1			Treatment 2			Reference
		Type of Matrix	Concentration	Sampling Period (h)	Type of Matrix	Concentration	Sampling Period (h)	
		N.R.	N.R.	N.R.	Dark chocolate	153 ± 43 nmol/L	3.8 ± 0.2	[66]
	3′-*O*-methyl-(-)-epicatechi-5-sulfate	Drink with 5.3 mg cocoa flavanols/kg body weight (young)	75 ± 7 nmol/L	1.4 ± 0.1	Drink with 5.3 mg cocoa flavanols/kg body weight (elderly)	72 ± 7 nmol/L	1.4 ± 0.1	[82]
		Drink with 10.7 mg cocoa flavanols/kg body weight (young)	176 ± 14 nmol/L	1.6 ± 0.1	Drink with 10.7 mg cocoa flavanols/kg body weight (elderly)	128 ± 11 nmol/L	1.8 ± 0.1	[82]
	4′-methyl-(-)-epicatechin	Low-flavanol cocoa drink	N.R.	N.R.	High-flavanol cocoa drink	N.R.	N.R.	[71]
	4′-*O*-methyl-(-)-epicatechi	Low-flavanol cocoa drink	25 ± 5 nmol/L	2	High-flavanol cocoa drink	41 ± 10 nmol/L	2	[61]
	4′-*O*-methyl-(-)-epicatechin 5-sulfate	N.R.	N.R.	N.R.	Dark chocolate	18 ± 6 nmol/L	3.8 ± 0.2	[66]
	4′-*O*-methyl-(-)-epicatechi 7-sulfate	N.R.	N.R.	N.R.	Dark chocolate	13 ± 4 nmol/L	3.8 ± 0.2	[66]
	4′-*O*-methyl-(-)-epicatechin-*O*-β-D-glucuronide	Low-flavanol cocoa drink	N.R.	N.R.	High-flavanol cocoa drink	N.R.	N.R.	[71]
		Low-flavanol cocoa drink	151 ± 46 nmol/L	2	High-flavanol cocoa drink	287 ± 58 nmol/L	2	[61]
Flavonoids and conjugates	Epicatechin-*O*-β-D-glucu Epicatechin-7-β-D-glucuronide	Low-flavanol cocoa drink	N.R.	N.R.	High-flavanol cocoa drink	N.R.	N.R.	[71]
		Low-flavanol cocoa drink	9 ± 3 nmol/L	2	High-flavanol cocoa drink	39 ± 13 nmol/L	2	[61]
		Chocolate drink milk-free	0.21 ± 0.2 μmol/L	1–1.5	Chocolate drink with milk	0.20 ± 0.02 μmol/L	0.5–1	[74]
		No cocoa extract	0.00 μmol/L	0	Cocoa extract in water	0.16 ± 0.03 μmol/L	0.5–2	[68]
	Catechin	N.R.	N.R.	N.R.	Cocoa based dairy drink	149 ± 18 nmol/L	2	[79]
		Low-flavanol cocoa drink	N.R.	N.R.	High-flavanol cocoa drink	N.R.	N.R.	[71]
		Low-flavanol cocoa drink	9 ± 3 nmol/L	2	High-flavanol cocoa drink	18 ± 3 nmol/L	2	[61]
		N.R.	N.R.	N.R.	Cocoa based dairy drink	889 ± 114 nmol/L	2	[79]
	Epicatechin	Chocolate drink milk-free	12.89 ± 0.95 μmol/L	1–2	Chocolate drink with milk	12.42 ± 0.97 μmol/L	0.5–1	[74]
		No cocoa extract	0.08 ± 0.46 μmol/L	0	Cocoa extract in water	5.96 ± 0.60 μmol/L	0.5–2	[68]
		Bread	19 ± 14 nmol/L	2	80 g of chocolate and 47 g of bread	355 ± 49 nmol/L	2	[63]
		Chocolate	0.15 ± 0.04 μmol/L	1–2	Cocoa	0.22 ± 0.06 μmol/L	0–1	[64]

Table 4. *Cont.*

Chemical Category	Metabolite	Treatment 1			Treatment 2			Reference
		Type of Matrix	Concentration	Sampling Period (h)	Type of Matrix	Concentration	Sampling Period (h)	
Flavonoids and conjugates	Epicatechin	40 g of chocolate	103 ± 29 ng/mL	2.00 ± 0.00	80 g of chocolate	196 ± 57 ng/mL	2.57 ± 0.50	[62]
		Milk	N.D.	2	Milk and cocoa powder	625.7 ± 198.3 nmol/L	2	[70]
		Low-flavanol cocoa drink	N.R.	N.R.	High-flavanol cocoa drink	N.R.	N.R.	[71]
		Low-flavanol cocoa drink	3 ± 0 nmol/L	2	High-flavanol cocoa drink	19 ± 6 nmol/L	2	[61]
		Low-cocoa group	578 ± 61 nmol/L	2	High-cocoa group	N.R.	N.R.	[58]
	Epicatechin glucuronide	Chocolate	0.78 ± 0.28 µmol/L	1–2	Cocoa	0.91 ± 0.21 µmol/L	1–2	[64]
	Epicatechin sulfate	Chocolate	1.11 ± 0.43 µmol/L	0–1	Cocoa	1.14 ± 0.21 µmol/L	1–2	[64]
	Epicatechin sulfoglucuronide	Chocolate	1.19 ± 0.44 µmol/L	0–1	Cocoa	1.28 ± 0.76 µmol/L	0–1	[64]
	Methylated epicatechin sulfate	Chocolate	0.95 ± 0.27 µmol/L	1–2	Cocoa	1.00 ± 0.34 µmol/L	1–2	[64]
	Methylated epicatechin sulfoglucuronide	Chocolate	0.71 ± 0.14 µmol/L	1–2	Cocoa	0.78 ± 0.51 µmol/L	4–8	[64]
	O-methyl-(epi)-catechin-O-sulfate	Cocoa powder with water	60 ± 8 nmol/L	1.0 ± 0.2	Cocoa powder with milk	50 ± 8 nmol/L	1.3 ± 0.2	[77]
	Procyanidin B2	No cocoa extract	Detected	0	Chocolate extract in water	41 ± 4 nmol/L	0.5–2	[68]
		N.R.	N.R.	N.R.	Cocoa dairy drink	4.0 ± 0.6 nmol/L	2	[80]
Methylxanthines	3-methylxanthine	Cocoa powder with milk	0.6 ± 1.4 µmol/L	3.4 ± 3.3	Cocoa powder enriched with methylxanthines with milk	1.8 ± 0.9 µmol/L	2.2 ± 2.5	[81]
	7-methylxanthine	Cocoa powder with milk	2.1 ± 1.4 µmol/L	3.3 ± 1.6	Cocoa powder enriched with methylxanthines with milk	4.6 ± 1.8 µmol/L	4.5 ± 1.5	[81]
	Caffeine	Cocoa powder with milk	2.1 ± 1.3 µmol/L	2.1 ± 2.0	Cocoa powder enriched with methylxanthines with milk	12.1 ± 2.6 µmol/L	2.0 ± 1.1	[81]
		No chocolate dose	<2.5–21.4 µmol/L	0	Dark chocolate	4.6–25.3 µmol/L	1.5	[59]
	Paraxanthine	Cocoa powder with milk	9.5 ± 1.3 µmol/L	4.5 ± 2.9	Cocoa powder enriched with methylxanthines with milk	12.0 ± 2.2 µmol/L	3.5 ± 2.5	[81]
	Theobromine	No chocolate dose	<2.5–7.1 µmol/L	0	Dark chocolate	43.2–67.5 µmol/L	1.5	[59]
		Cocoa powder with milk	15.8 ± 3.3 µmol/L	1.9 ± 1.0	Cocoa powder enriched with methylxanthines with milk	51.9 ± 13 µmol/L	2.2 ± 1.1	[81]

Table 4. *Cont.*

Chemical Category	Metabolite	Treatment 1			Treatment 2			Reference
		Type of Matrix	Concentration	Sampling Period (h)	Type of Matrix	Concentration	Sampling Period (h)	
Methylxanthin		Cocoa depleted control	N.R.	N.R.	Dark chocolate	70 µmol/L	3	[49]
		40 g of chocolate	6.365 ± 0.894 µg/mL	2.25 ± 0.88	80 g of chocolate	11.414 ± 1.190 µg/mL	3.25 ± 0.45	[62]
		Non-cocoa consumers	0.04–0.17 mg/kg	0–24	High cocoa consumers	0.33–0.66 mg/kg	0–24	[56]
	Theobromine	Baseline	2.3 ± 2.4 mg/L	0–12	Milk chocolate	7.6 ± 5.4 mg/L	0–12	[67]
		Baseline	2.3 ± 2.4 mg/L	0–12	Cocoa powder	19.3 ± 5.9 mg/L	0–12	[67]
		Baseline	2.3 ± 2.4 mg/L	0–12	Dark chocolate	30.6 ± 10.3 mg/L	0–12	[67]
	Theophylline	Cocoa powder with milk	11.5 ± 2.6 µmol/L	3.1 ± 3.2	Cocoa powder enriched with methylxanthines with milk	12.3 ± 4.3 µmol/L	1.9 ± 2.3	[81]
Phenolic acids	Ferulic acid	N.R.	N.R.	N.R.	Cocoa polymeric procyanidin concentrate	15 ± 2 ng/mL	3 ± 1	[60]

Notes: N.R. means non reported or non-cocoa matrix was used.

Table 5. Metabolites detected in plasma samples, that discriminated the baseline from treatments (enriched dark chocolate, dark chocolate and white chocolate), and times after intake where increase in measured signal was the highest, in an untargeted study [84].

Metabolite	Sampling Period (h)
β-hydroxybutyrate	6
Acetone	6
Acetoacetate	6
Aspartate	6
Lactate	2

Table 6. Metabolites analyzed in urine after acute cocoa interventions and studied with targeted methodologies.

Chemical Category	Metabolite	Treatment 1			Treatment 2			Reference
		Type of Matrix	Concentration	Sampling Period (h)	Type of Matrix	Concentration	Sampling Period (h)	
	(-)-catechin	N.R.	N.R.	N.R.	Dairy cocoa drink	4.0 ± 0.4 µmol	0–24	[79]
		N.R.	N.R.	N.R.	Dairy cocoa drink	13 ± 2 µmol	0–24	[79]
	(-)-epicatechin	N.R.	N.R.	N.R.	Non dairy cocoa drink	N.R.	0–24	[76]
		N.R.	N.R.	N.R.	Cocoa polymeric procyanidin concentrate	0.6 ± 0.2 ng/mg creatintine	0–4	[60]
Flavonoids and conjugates	(-)-epicatechin glucuronide	Non-dairy cocoa drink	194.95 µg/g creatinine	0–6	Dairy cocoa drink	112.79 µg/g creatinine	0–6	[75]
	(-)-epicatechin sulfate (isomer 1)	Non-dairy cocoa drink	48.83 µg/g creatinine	0–6	Dairy cocoa drink	30.86 µg/g creatinine	0–6	[72,75]
	(-)-epicatechin sulfate (isomer 2)	No-dairy cocoa drink	195.29 µg/g creatinine	6–12	Dairy cocoa drink	128.47 µg/g creatinine	6–12	[72,75]
	(-)-epicatechin sulfate (isomer 3)	Non-dairy cocoa drink	5.07 µg/g creatinine	0–6	Dairy cocoa drink	72.45 µg/g creatinine	0–6	[72,75]
	(-)-epicatechin-3'-sulfate	N.R.	N.R.	N.R.	Dark chocolate	5.80 ± 1.78 µmol	5–10	[66]
	(-)-epicatechin-3'-β-D-glucuronide	N.R.	N.R.	N.R.	Dark chocolate	8.74 ± 2.92 µmol	5–10	[66]
	(-)-epicatechin-4'-sulfate	N.R.	N.R.	N.R.	Dark chocolate	0.37 ± 0.13 µmol	5–10	[66]
	(-)-epicatechin-4'-β-D-glucuronide	N.R.	N.R.	N.R.	Dark chocolate	0.56 ± 0.14 µmol	5–10	[66]
	(-)-epicatechin-5-sulfate	N.R.	N.R.	N.R.	Dark chocolate	0.72 ± 0.36 µmol	5–10	[66]
	(-)-epicatechin-7-β-D-glucuronide	N.R.	N.R.	N.R.	Dark chocolate	4.59 ± 0.74 µmol	10–24	[66]
	(-)-epicatechin-O-glucuronide	Cocoa powder with water	405 ± 44 µmol/L	0–2	Cocoa powder with milk	136 ± 24 µmol/L	2–5	[77]
	(-)-epicatechin-O-sulfate (isomer 2)	Cocoa powder with water	1127 ± 196 µmol/L	0–2	Cocoa powder with milk	737 ± 118 µmol/L	2–5	[77]
	(+)-catechin	N.R.	N.R.	N.R.	Dairy cocoa drink	9.8 ± 1.4 µmol	0–24	[79]
	(+)-epicatechin	N.R.	N.R.	N.R.	Dairy cocoa drink	10 ± 1 µmol	0–24	[79]
Flavonoids and conjugates	Epicatechin	Hot water	N.D.	0–24	Cocoa powder with water	4.3 ± 4.2 µmol	0–24	[69]
	(epi) catechin-O-sulfate (isomer 1)	Cocoa powder with water	928 ± 110 µmol/L	0–2	Cocoa powder with milk	476 ± 75 µmol/L	0–2	[77]
	3'-O-methyl-(-)-epicatechin	N.R.	N.R.	N.R.	Cocoa polymeric procyanidin concentrate	0.2 ± 0.3 ng/mg creatinine	0–4	[60]
	3'-O-methyl-(-)-epicatechin 4'-sulfate	N.R.	N.R.	N.R.	Dark chocolate	1.27 ± 0.39 µmol	5–10	[66]
	3'-O-methyl-(-)-epicatechin 5-sulfate	N.R.	N.R.	N.R.	Dark chocolate	8.87 ± 3.05 µmol	5–10	[66]
	3'-O-methyl-(-)-epicatechin 7-sulfate	N.R.	N.R.	N.R.	Dark chocolate	1.55 ± 0.55 µmol	5–10	[66]
	3'-O-methyl-(-)-epicatechin-3'-β-D-glucuronide	N.R.	N.R.	N.R.	Dark chocolate	0.69 ± 0.12 µmol	5–10	[66]
	4'-O-methyl-(-)-epicatechin 5-sulfate	N.R.	N.R.	N.R.	Dark chocolate	0.92 ± 0.34 µmol	5–10	[66]
	4'-O-methyl-(-)-epicatechin 7-sulfate	N.R.	N.R.	N.R.	Dark chocolate	0.55 ± 0.29 µmol	5–10	[66]
	O-methyl-(epi)-catechin-O-sulfate	Cocoa powder with water	1146 ± 231 µmol/L	0–2	Cocoa powder with milk	823 ± 131 µmol/L	0–2	[77]
Flavonoids and conjugates	Naringenin	Hot water	0.1 ± 0.2 µmol	0–24	Cocoa powder with water	0.3 ± 0.4 µmol	0–24	[69]
	Procyanidin B2	N.R.	N.R.	N.R.	Non-dairy cocoa drink	N.R.	0–24	[76]

Table 6. *Cont.*

Chemical Category	Metabolite	Treatment 1			Treatment 2			Reference
		Type of Matrix	Concentration	Sampling Period (h)	Type of Matrix	Concentration	Sampling Period (h)	
Lignans	Enterodiol	N.R.	N.R.	N.R.	Non-dairy cocoa drink	N.R.	0–24	[76]
		Hot water	0.5 ± 1.1 µmol	0–24	Cocoa powder with water	0.2 ± 0.3 µmol	0–24	[69]
	Enterolactone	N.R.	N.R.	N.R.	Non-dairy cocoa drink	N.R.	0–24	[76]
		Hot water	3.8 ± 2.8 µmol	0–24	Cocoa powder with water	3.6 ± 2.9 µmol	0–24	[69]
Methylxanthines	1,3,7-trimethyluric acid	Cocoa powder with milk	0.7 ± 0.3 µmol/L	0–2	Cocoa powder with milk, enriched with methylxanthines	1.8 ± 0.6 µmol/L	9.7 ± 6.9	[81]
	1,3-dimethyluric acid	Cocoa powder with milk	1.3 ± 1.1 µmol/L	21.5 ± 6.0	Cocoa powder with milk, enriched with methylxanthines	3.4 ± 1.9 µmol/L	18.0 ± 7.5	[81]
	1,7-dimethyluric acid	Cocoa powder with milk	3.9 ± 2.7 µmol/L	20.0 ± 7.7	Cocoa powder with milk, enriched with methylxanthines	13.5 ± 7.8 µmol/L	12.7 ± 8.7	[81]
	1-methyluric acid	Cocoa powder with milk	9.2 ± 4.1 µmol/L	19.7 ± 8.2	Cocoa powder with milk, enriched with methylxanthines	45.2 ± 16.4 µmol/L	12.4 ± 7.5	[81]
	1-methylxanthine	Cocoa powder with milk	5.5 ± 2.7 µmol/L	21.2 ± 6.8	Cocoa powder with milk, enriched with methylxanthines	15.3 ± 7.3 µmol/L	13.7 ± 7.7	[81]
	3,7-dimethyluric acid	Cocoa powder with milk	2.7 ± 1.1 µmol/L	21.2 ± 6.8	Cocoa powder with milk, enriched with methylxanthines	7.3 ± 2.9 µmol/L	14.3 ± 8.6	[81]
Methylxanthines	3-methylxanthine	Cocoa powder with milk	34.8 ± 8.9 µmol/L	14.8 ± 9.1	Cocoa powder with milk, enriched with methylxanthines	98.1 ± 21.9 µmol/L	12.0 ± 7.4	[81]
	7-methylxanthine	Cocoa powder with milk	110.1 ± 40.1 µmol/L	16.3 ± 8.7	Cocoa powder with milk, enriched with methylxanthines	187.4 ± 82.1 µmol/L	12.0 ± 7.2	[81]
	Caffeine	Cocoa powder with milk	2.1 ± 0.7 µmol/L	14.5 ± 9.3	Cocoa powder with milk, enriched with methylxanthines	9.9 ± 3.5 µmol/L	7.7 ± 5.5	[81]
		Baseline	<2.5–23.4 µmol/L	0	Dark chocolate	4.2–24.7 µmol/L	1.5	[59]
	Paraxanthine	Cocoa powder with milk	3.5 ± 1.8 µmol/L	15.1 ± 9.0	Cocoa powder with milk, enriched with methylxanthines	10.7 ± 5.8 µmol/L	11.7 ± 7.5	[81]
	Theobromine	Cocoa powder with milk	50.4 ± 18.4 µmol/L	14.2 ± 9.6	Cocoa powder with milk, enriched with methylxanthines	177.4 ± 45.0 µmol/L	7.7 ± 5.5	[81]
		Baseline	2.5–94.9 µmol/L	0	Dark chocolate	131.9–449.4 µmol/L	1.5	[59]
	Theophylline	Baseline	1.0 ± 1.4 µmol/L	14.9 ± 8.9	Dark chocolate	2.5 ± 1.3 µmol/L	7.6 ± 5.8	[81]

Table 6. *Cont.*

Chemical Category	Metabolite	Treatment 1			Treatment 2			Reference
		Type of Matrix	Concentration	Sampling Period (h)	Type of Matrix	Concentration	Sampling Period (h)	
N-phenylpropenoyl-L-amino acids	N-[3′,4′-dihydroxy-(E)-cinnamoyl]-L-aspartic acid	Baseline	N.D.	0	Non-dairy cocoa drink	4.76 ± 4.01 µg	2	[20]
	N-[3′,4′-dihydroxy-(E)-cinnamoyl]-L-dopa	Baseline	N.D.	0	Non-dairy cocoa drink	1.25 ± 1.19 µg	2	[20]
	N-[3′,4′-dihydroxy-(E)-cinnamoyl]-L-glutamic acid	Baseline	N.D.	0	Non-dairy cocoa drink	N.D.	N.D.	[20]
	N-[3′,4′-dihydroxy-(E)-cinnamoyl]-L-tyrosine	Baseline	N.D.	0	Non-dairy cocoa drink	0.30 ± 0.39 µg	2	[20]
	N-[4′-hydroxy-(E)-cinnamoyl]-L-aspartic acid	Baseline	N.D.	0	Non-dairy cocoa drink	64.37 ± 20.22 µg	2	[20]
	N-[4′-hydroxy-(E)-cinnamoyl]-L-dopa	Baseline	N.D.	0	Non-dairy cocoa drink	0.85 ± 0.96 µg	2	[20]
	N-[4′-hydroxy-(E)-cinnamoyl]-L-glutar acid	Baseline	N.D.	0	Non-dairy cocoa drink	22.17 ± 12.57 µg	2	[20]
N-phenylpropenoyl-L-amino acids	N-[4′-hydroxy-(E)-cinnamoyl]-L-tryptophane	Baseline	N.D.	0	Non-dairy cocoa drink	<0.1 µg	1–4	[20]
	N-[4′-hydroxy-(E)-cinnamoyl]-L-tyrosine	Baseline	N.D.	0	Non-dairy cocoa drink	18.91 ± 5.53 µg	2	[20]
	N-[4′-hydroxy-3′-methoxy-(E)-cinnamoyl]-L-aspartic acid	Baseline	N.D.	0	Non-dairy cocoa drink	9.70 ± 2.93 µg	2	[20]
	N-[4′-hydroxy-3′-methoxy-(E)-cinnamoyl]-L-tyrosine	Baseline	N.D.	0	Non-dairy cocoa drink	1.91 ± 0.59 µg	2	[20]
	N-[cinnamoyl]-L-aspartic acid	Baseline	N.D.	0	Non-dairy cocoa drink	0.43 ± 0.33 µg	2	[20]
Phenolic acids and others	3-hydroxybenzoic acid	N.R.	N.R.	N.R.	Cocoa polymeric procyanidin concentrate	1 ± 1 ng/mg creatinine	0–4	[66]
		N.R.	N.R.	N.R.	Non-dairy cocoa drink	N.R.	0–24	[76]
	3,4-dihydroxyphenyl acetic acid	Baseline	15.8 ± 4.4 nmol/mg creatinine	−24–0	Chocolate	38.8 ± 12.3 nmol/mg creatinine	0–24	[65]
		Cocoa powder with water	1.60 ± 0.37 nmol/mg creatinine	6–12	Cocoa powder with milk	0.45 ± 0.10 nmol/mg creatinine	12–24	[78]
		N.R.	N.R.	N.R.	Cocoa polymeric procyanidin concentrate	10 ± 10 ng/mg creatinine	0–4	[60]
		N.R.	N.R.	N.R.	Non-dairy cocoa drink	N.R.	0–24	[76]

Table 6. *Cont.*

Chemical Category	Metabolite	Treatment 1			Treatment 2			Reference
		Type of Matrix	Concentration	Sampling Period (h)	Type of Matrix	Concentration	Sampling Period (h)	
	3,4-dihydroxyphenyl propionic acid	N.R.	N.R.	N.R.	Cocoa polymeric procyanidin concentrate	6 ± 5 ng/mg creatinine	0–4	[60]
		Baseline	3.1 ± 0.7 nmol/mg creatinine	−24–0	Chocolate	2.6 ± 0.6 nmol/mg creatinine	0–24	[65]
		Cocoa powder with water	16.84 ± 2.81 nmol/mg creatinine	0–6	Cocoa powder with milk	16.9 ± 4.0 nmol/mg creatinine	0–6	[78]
		N.R.	N.R.	N.R.	Non-dairy cocoa drink	N.R.	0–24	[76]
	3-hydroxyphenyl acetic acid	Cocoa powder with water	10.1 ± 3.42 nmol/mg creatinine	12–24	Cocoa powder with milk	7.36 ± 1.61 nmol/mg creatinine	12–24	[78]
Phenolic acids and others		N.R.	N.R.	N.R.	Non-dairy cocoa drink	N.R.	0–24	[76]
		N.R.	N.R.	N.R.	Cocoa polymeric procyanidin concentrate	30 ± 20 ng/mg creatinine	0–4	[60]
	3-hydroxyphenyl propanoic acid	N.R.	N.R.	N.R.	Cocoa polymeric procyanidin concentrate	10 ± 7 ng/mg creatinine	0–4	[60]
	3-methoxy-4-hydroxy-phenylacetic acid	Cocoa powder with water	11.30 ± 1.63 nmol/mg creatinine	6–12	Cocoa powder with milk	9.30 ± 0.74 nmol/mg creatinine	6–12	[78]
		N.R.	N.R.	N.R.	Non-dairy cocoa drink	N.R.	0–24	[76]
	4-hydroxy-5-(3′,4′-dihydroxyphenyl) valeric acid	N.R.	N.R.	N.R.	Cocoa polymeric procyanidin concentrate	0.2 ± 0.3 ng/mg creatinine	0–4	[60]
	4-hydroxybenzoic acid	N.R.	N.R.	N.R.	Cocoa polymeric procyanidin concentrate	16 ± 6 ng/mg creatinine	0–4	[60]
		N.R.	N.R.	N.R.	Non-dairy cocoa drink	N.R.	0–24	[76]
	4-hydroxyphenyl acetic acid	N.R.	N.R.	N.R.	Cocoa polymeric procyanidin concentrate	210 ± 50 ng/mg creatinine	0–4	[60]

Table 6. *Cont.*

Chemical Category	Metabolite	Treatment 1			Treatment 2			Reference
		Type of Matrix	Concentration	Sampling Period (h)	Type of Matrix	Concentration	Sampling Period (h)	
	4-hydroxyphenyl propanoic acid	N.R.	N.R.	N.R.	Cocoa polymeric procyanidin concentrate	0.02 ± 0.04 ng/mg creatinine	0–4	[60]
	4-O-methylgallic acid	N.R.	N.R.	N.R.	Non-dairy cocoa drink	N.R.	0–24	[76]
		Hot water	N.D.	0–24	Cocoa powder with water	0.6 ± 1.1 µmol	0–24	[69]
	5-(3′,4′-dihydroxyphenyl) valerolactone	N.R.	N.R.	N.R.	Cocoa polymeric procyanidin concentrate	2 ± 4 ng/mg creatinine	48	[60]
	Caffeic acid	N.R.	N.R.	N.R.	Cocoa polymeric procyanidin concentrate	1.2 ± 0.5 ng/mg creatinine	0–4	[60]
		N.R.	N.R.	N.R.	Non-dairy cocoa drink	N.R.	0–24	[76]
		Hot water	0.1 ± 0.2 µmol	0–24	Cocoa powder with water	0.4 ± 0.4 µmol	0–24	[69]
	Dihydroxyphenyl valeric acid	N.R.	N.R.	N.R.	Cocoa polymeric procyanidin concentrate	0.01 ng/mg creatintine	-	[60]
Phenolic acids and others	Ferulic acid	N.R.	N.R.	N.R.	Cocoa polymeric procyanidin concentrate	21 ± 4 ng/mg creatinine	0–4	[60]
		Baseline	131 ± 59 nmol/mg creatinine	−24–0	Chocolate	321 ± 99 nmol/mg creatinine	24–48	[65]
		N.R.	N.R.	N.R.	Non-dairy cocoa drink	N.R.	0–24	[76]
	Hippuric acid	Cocoa powder with water	193.16 ± 27.89 nmol/mg creatinine	0–6	Cocoa powder with milk	122.82 ± 18.83 nmol/mg creatinine	12–24	[78]
		N.R.	N.R.	N.R.	Non-dairy cocoa drink	N.R.	0–24	[76]
		Baseline	2943 ± 1923 nmol/mg creatinine	−24–0	Chocolate	974± 115 nmol/mg creatinine	0–24	[65]
	m-coumaric acid	N.R.	N.R.	N.R.	Cocoa polymeric procyanidin concentrate	1 ± 1 ng/mg creatinine	0–4	[60]
		Hot water	0.5 ± 0.9 µmol	0–24	Cocoa powder with water	1.4 ± 1.4 µmol	0–24	[69]

Table 6. *Cont.*

Chemical Category	Metabolite	Treatment 1			Treatment 2			Reference
		Type of Matrix	Concentration	Sampling Period (h)	Type of Matrix	Concentration	Sampling Period (h)	
Phenolic acids and others	*m*-coumaric acid	N.R.	N.R.	N.R.	Non-dairy cocoa drink	N.R.	0–24	[76]
	Methyl-5-(3′, 4′-dihydroxyphenyl)-valerolactone	N.R.	N.R.	N.R.	Cocoa polymeric procyanidin concentrate	0.5 ± 0.8 ng/mg creatinine	48	[60]
	m-hydroxybenzoic acid	Baseline	Non detected	−24–0	Chocolate	8.93 ± 3.9 nmol/mg creatinine	24–48	[65]
		Cocoa powder with water	0.56 ± 24 nmol/mg creatinine	6–12	Cocoa powder with milk	0.60 ± 0.28 nmol/mg creatinine	12–24	[78]
	m-hydroxyphenyl acetic acid	Baseline	21.2 ± 3.5 nmol/mg creatinine	−24–0	Chocolate	156 ± 54 nmol/mg creatinine	24–48	[65]
	m-hydroxyphenyl propionic acid	Baseline	5.4 ± 2.4 nmol/mg creatinine	−24–0	Chocolate	13.4 ± 4.1 nmol/mg creatinine	24–48	[65]
	p-coumaric acid	N.R.	N.R.	N.R.	Cocoa procyanidin concentrate	1 ± 1 ng/mg creatinine	48	[60]
		N.R.	N.R.	N.R.	Non-dairy cocoa drink	N.R.	0–24	[76]
	Phenylacetic acid	Cocoa powder with water	163.77 ± 22.10 nmol/mg creatinine	0–6	Cocoa powder with milk	167.72 ± 23.80 nmol/mg creatinine	0–6	[78]
		Baseline	70.1 ± 11.3 nmol/mg creatinine	−24–0	Chocolate	36.8 ± 8.1 nmol/mg creatinine	24–48	[65]
	p-hydroxybenzoic acid	Baseline	54.2 ± 8.1 nmol/mg creatinine	−24–0	Chocolate	41.2 ± 10.4 nmol/mg creatinine	24–48	[65]
		Cocoa powder with water	6.39 ± 0.66 nmol/mg creatinine	0–6	Cocoa powder with milk	3.01 ± 0.97 nmol/mg creatinine	12–24	[78]

Table 6. *Cont.*

Chemical Category	Metabolite	Treatment 1			Treatment 2			Reference
		Type of Matrix	Concentration	Sampling Period (h)	Type of Matrix	Concentration	Sampling Period (h)	
Phenolic acids and others		Cocoa powder with water	3.13 ± 0.60 nmol/mg creatinine	0–6	Cocoa powder with milk	1.76 ± 0.30 nmol/mg creatinine	6–12	[78]
	p-hydroxyhippuric acid	Baseline	71.1 ± 13.6 nmol/mg creatinine	−24–0	Chocolate	90.5 ± 21.3 nmol/mg creatinine	24–48	[65]
		N.R.	N.R.	N.R.	Non-dairy cocoa drink	N.R.	0–24	[76]
	Protocatechuic acid	Cocoa powder with water	11.07 ± 1.19 nmol/mg creatinine	0–6	Cocoa powder with milk	8.8 ± 2.2 nmol/mg creatinine	0–6	[78]
		N.R.	N.R.	N.R.	Non-dairy cocoa drink	N.R.	0–24	[76]
		N.R.	N.R.	N.R.	Cocoa polymeric procyanidin concentrate	40 ± 20 ng/mg creatinine	8–24	[60]
	Vanillic acid	Cocoa powder with water	5.96 ± 1.15 nmol/mg creatinine	0–6	Cocoa powder with milk	9.85 ± 1.27 nmol/mg creatinine	0–6	[78]
		Baseline	64.6 ± 25.5 nmol/mg creatinine	−24–0	Chocolate	228 ± 33 nmol/mg creatinine	0–24	[65]
		N.R.	N.R.	N.R.	Non-dairy cocoa drink	N.R.	0–24	[76]
		N.R.	N.R.	N.R.	Cocoa polymeric procyanidin concentrate	14 ± 4 ng/mg creatinine	0–4	[60]

Notes: N.R. refers to non-reported or non-cocoa matrix was used, and N.D. to non-detected.

Table 7. Metabolites analyzed in urine after acute cocoa intervention studied with non-targeted methodologies.

Period (h)	Purine Metabolites	Polyphenol Metabolites	Nicotinic Acid Metabolites	Amino Acid Metabolites	Others	Reference
Basel	*N*-methylguanine	-	-	-	-	[87]
0–6	3-methyluric acid 3-methylxanthine 3,7-dimethyluric acid 7-methyluric acid 7-methylxanthineAMMU (4) Caffeine Theobromine	3′-methoxy-4′-hydroxyphenyl valerolactone 3′-methoxy-4′-hydroxyphenyl valerolactone glucuronide 4-hydroxy-5-(3,4-dihydroxyphenyl)-valeric acid 5-(3′,4′-dihydroxyphenyl)-γ-valerolactone glucuronide Epicatechin-O—sulfate O-methyl epicatechin Vanillic acid Vanilloylglycine	Hydroxynicotinic acid Trigonelline	Tyrosine	3,5-diethyl-2-methyl pyrazine Cyclo(Ser-Tyr) Cyclo(Pro-Pro) Hydroxyacetophenone	[86]
	3-methylxanthine 7-methylxanthine Theobromine	Vanilloylglycine	-	-	-	[87]
	3-methyluric acid 3-methylxanthine 3,7- dimethyluric acid 7-methyluric acid 7-methylxanthine AMMU Caffeine Theobromine	3′-methoxy-4′-hydroxyphenyl valerolactone 3′-methoxy-4′-hydroxyphenyl valerolactone glucuronide 4-hydroxy-5-(3,4-dihydroxyphenyl)-valeric acid 5-(3′,4′-dihydroxyphenyl)–valerolactone sulfate 5-(3′,4′-dihydroxyphenyl)-valerolactone glucuronide Hippurate Epicatechin-O-sulfate Vanilloylglycine	Hydroxynicotinic acid N1-methylnicotinamide	Alanine Arginine Glycine Tyrosine Valine	2-hydroxyisobutyrate 3-hydroxyisobutyrate 3-hydroxyisovalerate 4-hydroxyphenyl acetate Creatinine Dimethylamine Lactate O-feruoylquinate Pyruvate	[84]
6–12	3-methyluric acid 3-methylxanthine 3,7-dimethyluric acid 7-methyluric acid 7-methylxanthine AMMU Caffeine Theobromine	3′-methoxy-4′-hydroxyphenyl valerolactone 3′-methoxy-4′-hydroxyphenyl valerolactone glucuronide 4-hydroxy-5-(3,4-dihydroxyphenyl)-valeric acid 5-(3′,4′-dihydroxyphenyl)-g-valerolactone glucuronide 5-(3′,4′-dihydroxyphenyl)-γ-valerolactone sulfate 5-(3′,4′-dihydroxyphenyl)-γ-valerolactone glucuronide	-	-	Cyclo(Ser-Tyr) 3,5-diethyl-2-methyl pyrazine	[86]
	3-methylxanthine 7-methylxanthine Theobromine	Dihydroxyphenyl valerolactone glucuronide	-	-	Furoylglycine	[87]
12–24	3-methyluric acid 3-methylxanthine 3,7-dimethyluric acid 7-methyluric acid 7-methylxanthine AMMU Theobromine	3′-methoxy-4′-hydroxyphenyl valerolactone 3′-methoxy-4′-hydroxyphenyl valerolactone glucuronide 4-hydroxy-5-(3,4-dihydroxyphenyl)-valeric acid 5-(3′,4′-dihydroxyphenyl)-g-valerolactone glucuronide 5-(3′,4′-dihydroxyphenyl)-γ-valerolactone glucuronide	-	-	Cyclo(Ser-Tyr)	[86]
	3-methylxanthine 7-methylxanthine Theobromine	-	-	Xanthurenic acid	-	[87]

Notes: (4) AMMU means 6-amino-5-[*N*-methylformylamino]-1-methyluracil.

Table 8. Metabolites analyzed in urine and plasma after acute, short-term, or regular cocoa intake interventions, studied with non-targeted and targeted methodologies.

Type of Metabolite	Metabolites in Urine		Metabolites in Plasma (Targeted)	
	Non-Targeted [55]	Targeted [57]	[57]	[58]
Polyphenols metabolites	(DHPV) 5-(3′,4′-dihydroxphenyl)-valerolactone sulfoglucuronide (HDHPVA) 4-hydroxy-5-(dihydroxyphenyl)-valeric acid glucuronide (HHMPVA) 4-hydroxy-5-(hydroxyl-methoxyphenyl)-valeric acid glucuronide (HHPVA) 4-hydroxy-5-(hydroxyphenyl)-valeric acid sulphate (HPV) hydroxyphenyl-valerolactone glucuronide (HPVA) 4-hydroxy-5-(phenyl)-valeric acidsulphate (MHPV) methoxyhydroxyphenyl valerolactone DHPV glucuronide (1) and (2) DHPV sulphate (epi) catechin glucuronide (epi) catechin sulphate HDHPVA HDHPVA sulphate HHPMVA sulphate HPV sulphate MHPV glucuronide Vanillic acid Vanillin sulphate	(-)-epicatechin 3-hydroxybenzoic acid 3-hydroxyhippuric acid 3-hydroxyphenyl acetic acid 3-hydroxyphenyl propionic acid 3-methoxy-4-hydroxyphenyl acetic acid 3,4-dihydroxyphenyl acetic acid 3,4-dihydroxyphenyl propionic acid4-hydroxybenzoic acid 4-hydroxyhippuric acid5-(3′,4′-dihydroxyphenyl)-γ-valerolactone 5-(3′-methoxy, 4′-hydroxyphenyl)-γ-valerolactone O-glucuronide 5-(3′-methoxy, 4′-hydroxyphenyl)-γ-valerolactone O-sulfate 5-(3′,4′-dihydroxyphenyl)-γ-valerolactone O-glucuronide isomers Caffeic acid Epicatechin-3′-O-glucuronide Epicatechin-7-O-glucuronide Epicatechin-O-glucuronide isomers Epicatechin-O-sulfate isomers Ferulic acidm-coumaric acidO-methyl-epicatechin-O-glucuronide isomersO-methyl-epicatechin-O-sulfate isomersp-coumaric acid Phenylacetic acid Protocatechuic acid Vanillic acid	3-hydroxyhippuric acid 3-hydroxyphenyl acetic acid 3-hydroxyphenyl propionic acid 3,4-dihydroxyphenyl acetic acid 3,4-dihydroxyphenyl propionic acid 4-hydroxybenzoic acid 4-hydroxyhippuric acid 5-(3′,4′-dihydroxyphenyl)-γ-valerolactone 5-(3′-methoxy, 4′-hydroxyphenyl)-γ-valerolactone O-glucuronide 5-(3′,4′-dihydroxyphenyl)-γ-valerolactone O-glucuronide (1) and (2) Caffeic acid Ferulic acid p-coumaric acid Phenylacetic acid Protocatechuic acid Vanillic acid	3′-O-methyl-(-)-epicatechin 4′-O-methyl-(-)-epicatechin Epicatechin
Purine metabolites	3-methyluric acid 3,7-dimethyluric acid 7-methylxanthine AMMU 1 and 2 (5) Theobromine Xanthine	-	-	-
Others	Aspartyl-phenylalanine Cyclo(aspartyl-phenylalanyl) Furoylglycine Methylglutaryl carnitine	-	-	-

Notes: (5) 1 and 2 correspond to isomers AMMU= 6-amino-5-[*N*-methylformylamino]-1-methyluracil.

In particular, studies that mentioned acute intake of cocoa are summarized in the following tables: (a) Tables 4 and 6 include information of different metabolites as concentrations measured in the biological samples and the corresponding sampling periods. The type of matrix analyzed is also indicated. Both tables summarize information of targeted studies, the first focused on plasma (Table 4) and the other in urine (Table 6) analysis; (b) Table 5 shows the time after intake of different chocolates, where the highest signals of the different metabolites were detected or expected in plasma samples, (c) Table 7 focuses on the main metabolites found in urine that discriminated between different periods of time after intake, with an untargeted approach.

On the other hand, discriminating metabolites in between regular/chronic consumers and non-regular/chronic consumers observed in urine and plasma are listed on Table 8. This covers both targeted and untargeted studies.

Metabolites were organized by means of the chemical group, mainly methylxanthines, flavonoids and conjugates, phenolic acids, and lignans. As for the data collected from untargeted studies, data organization was done in a way that respected as much as possible the authors' original classification. Additionally, in each category metabolites were organized alphabetically.

4. Discussion

4.1. Methylxanthines

Theobromine showed some of the higher concentrations in plasma and urine after cocoa intake in different human intervention trials (Tables 4 and 5). One study [81] showed that unmetabolized theobromine in urine was close to 1.5 times higher when consuming methylxanthine enriched powder compared to the non-enriched cocoa powder, which indicates that renal clearance could vary in function of exposure. Due to a possible saturation of the pathway, a lack of enzymes and/or cofactors to process the metabolite may have had affected [81,88].

There is evidence that suggests that excretion also depends on gender and age. After consumption of chocolate in a cross-over study, theobromine excretion showed by women almost doubled man's levels and was also higher for volunteers younger than 29 years, and lower compared to the ones older than 54 years [84].

A variety of methylxanthines metabolites were also observed in non-targeted studies (Table 7), which follow characteristic metabolic pathways [89]. Some of them have been proposed as possible biomarkers of regular cocoa intake as AMMU, 3-methyluric acid, 7-methylxanthine, 3-methylxanthine, theobromine, and 3,7-dimethyluric acid [55].

4.2. Polyphenols

Polyphenols such as (epi) catechin, and the corresponding glucuronides, sulfates, with or without methylations were the most frequently analyzed metabolites.

One example is a study that evaluated plasma and urine samples after drinking a cocoa beverage. The most abundant flavanol metabolites were (-)-epicatechin-3′-β-D-glucuronide, (-)-epicatechin-3′-sulfate, 3′-*O*-methyl-(-)-epicatechin-5-sulfate, and 3′-*O*-methyl-(-)-epicatechin-7-sulfate which represented close to 94% of the total flavanol metabolites that were measured in plasma [82].

This is consequence of the metabolic pathways that flavanols and phenolic acids follow. When absorbed, these metabolites are transferred to the liver or other tissues where phase I and phase II reactions occur. In phase I, cytochrome enzymes lead to hydroxylations, oxidations and reductions; meanwhile, phase II produces conjugates by action of glucuronosyl transferases, sulfotransferases, and catechol-*O*-methyl transferases [75,90–92].

In general terms these metabolites reached maximum concentrations in plasma in less than 4 hours after intake (Table 4). One exception was (-)-epicatechin-7′-β-D-glucuronide, that in one study required 12.8 ± 4.8 h after cocoa intake to reach maximal concentrations [66].

When these metabolites are present in plasma, they could exert particular effects. For example, one study analyzed the impact of chocolate intake on oxidative damage, and observed an increase in plasma antioxidant activity and plasma lipid oxidation products decrease as epicatechin plasma levels increased [63]. Additionally, elimination rates have an influence on the probabilities of accumulation in blood or tissues, and the subsequent effect on health [93].

In this matter, when studying urine samples, the required time to reach maximal concentration was found to be largely variable. While (epi) catechin-*O*-sulfate isomer required a time in between 0–2 h after cocoa powder in milk intake to reach the higher concentration [77], (-)-epicatechin-7-β-D-glucuronide required a time in between 10–24 h after dark chocolate consumption [66].

Influence of age on metabolism was analyzed in an intervention that evaluated two cocoa powder drinks. Non-significant differences or small differences were observed for the different pharmacokinetic parameters, in the renal clearance, total excreted metabolites, and apparent volume of distribution in two age groups: 18–35 years and 65–80 years. Some of the small differences observed were associated with changes in renal function due to normal aging processes [82].

Studies converge in that increasing polymerization degree ($n > 2$) of procyanidins hinders their absorption. Procyanidin B2 dimer has been detected in plasma after consuming a cocoa beverage, showing a maximum concentration in the period between 0.5–2 h (Table 5). In this same study, B5 dimer was not detected [68]. Similar results were observed in another study where volunteers drank samples with cocoa extracts that contained procyanidins from 2–10 units. In this study, B2 dimer was detected in plasma in approximately 80% of the volunteers, indicating maximum concentrations at 2 h after intake, and with concentrations close to the limit of detection. Additionally, B5 dimer was not detected in any of the plasma samples [80].

Polyphenol absorption and excretion can depend on the food matrix, but contradictory information exists. For example, excretion patterns of microbial phenolic acids metabolites measured in urine differed depending on the liquid vehicle of cocoa powder consumed: water or milk [78]. One randomized study compared the levels of (-)-epicatechin glucuronide in plasma after the intake of cocoa powder with milk or water, and reported no statistical differences in the results from each group [73]. Also, milk protein did not affect the bioavailability of polyphenols present in a chocolate drink [74]. In another study the excretion of different flavan-3-ol metabolites within 24 h after cocoa intake was close to 20% of the ingested dose if cocoa was provided as a water-based beverage, but only 10% if ingested as a milk-based cocoa drink [77]. When consuming two cocoa beverages, either prepared with water or with milk, the total excretion was not affected, but kinetics was different. In the same period of time after intake (-)-epicatechin glucuronide was the most concentrated metabolite when water was the vehicle—doubling the concentration of total excreted (-)-epicatechin sulfates-but when milk was used, glucuronides and total sulfates reached similar concentrations in urine [75]. No significant differences in excretion patterns were found when comparing a variety of metabolites in urine after consuming cocoa powder in milk or in water [86]. It is suggested that polyphenol absorption and metabolism might be affected because of the interactions between milk polypeptides and polyphenols, which can decrease the bioaccessibility [94,95]. In addition, polyphenol absorption could also be affected by carbohydrates, as it has been reported that flavanol uptake could increase when simultaneously consuming these macronutrients [50].

Flavanol absorption and metabolism has also been demonstrated to be dependent on the stereochemistry. Differences in absorption and metabolism were detected after the consumption of equal amounts of (-)-epicatechin, (-)-catechin, (+)-catechin, and (+)-epicatechin in a cocoa based milk drink [79].

Non-absorbed flavonoids and phenolic acids could follow metabolic pathways that are carried out by the colonic microbiota [66,90–92]. Some epicatechin metabolites can return to small intestine including non-absorbed epicatechin pass to the colon, which can account a total close to 70 % of the total ingested [96].

Microbial metabolism can lead to the production of phenolic acids, sulfated and glucuronidated valerolactones [57]. Some examples are ferulic acid, phenylacetic acid, hippuric acid (Table 6), 4-hydroxy-5-(dihydroxyphenyl)-valeric acid glucuronide, and methoxyhydroxyphenyl valerolactone (Tables 6 and 7).

In general terms, these types of metabolites require longer times to reach mean concentrations in urine and plasma than the metabolites following phase II metabolism (Tables 4 and 6, Tables 7 and 8). To illustrate, ferulic and *m*-hydroxyphenyl acetic acids showed maximum concentrations in between 24 and 48 h after chocolate intake, and vanillic acid was the only metabolite that returned to basal level concentrations in the 48 h of observation [65].

Research suggests that consumption of cocoa and derivatives could modulate human metabolism. A randomized controlled cross-over intervention trial analyzed the acute consumption of three types of chocolates in 42 healthy volunteers and concluded that several endogenous metabolites were affected (Tables 5 and 7). In particular, several amino acids, organic acids, creatinine, lactate, and N^1-methylnicotinamide decreased meanwhile pyruvate, tyrosine, and *p*-hydroxyphenylacetate increased after the intervention. Flavan-3-ols, methylxanthines and their metabolites were suggested to modulate endogenous and colonic microbial metabolism [84].

Studies with longer periods of intervention detected additional metabolome changes due to regular cocoa consumption (Table 8). For example, different glucuronide and sulfate conjugates of (-)-epicatechin, *O*-methyl-epicatechin, dihydroxyphenyl valerolactones, and methoxy hydroxyphenyl valerolactones increased their levels in plasma after regular consumption of cocoa powder in a non-healthy group of volunteers. Hydroxyphenylacetic acids, AMMU isomers, 5-(3′,4′-dihydroxyphenyl)-γ-valerolactone, and its glucuronides and sulfates have been proposed as biomarkers of regular consumption of cocoa [55,57]. Further on, one study reported that methylglutaryl carnitine, derived from the acylcarnitine pathway, decreased after a long-term cocoa intake [55].

More recent studies have supported this information as 5-(3′,4′-dihidroxyphenyl)-γ-valerolactone has also been proposed as a specific biomarker of flavan-3-ols consumption, deriving only from (epi) catechin-based mono and oligomeric cocoa flavanols [83].

There is also evidence that polyphenol consumption from cocoa products might change the gut microbiota, exerting prebiotic effects, and which could be related to the activation of anti-inflammatory pathways with benefits in the host and alter the obtained profile of metabolites. One study observed an increase of *Lactobacillus* spp. and *Bifidobacterium* spp. in contrast with a decrease of *Clostridium* spp. populations after daily consumption during one month of a flavanol-rich cocoa beverage, compared to a low flavanol cocoa beverage [33,97,98].

4.3. Other Metabolites

Metabolites different than methylxanthines and polyphenols metabolism have also been detected after cocoa intake. For example, trigonelline (*N*-methylnicotinic acid) and hydroxynicotinic acid may be a result of nicotinic acid metabolism. They were detected in urine only within 6 h after cocoa powder consumption. Also, diketopiperazines such as cyclo(ser-tyr) and cyclo(pro-pro) were reported for the first time associated to cocoa powder intake and are considered flavor and taste metabolites [86].

N-phenylpropenoyl-L-amino acids have also been metabolites of interest as there is some preliminary evidence of their biological activity. Thirteen of them have been reported in urine after cocoa intake [20].

Additionally, significant changes in plasma metabolites after 6 h of chocolate intake have been reported for β-hydroxybutyrate, acetone, acetoacetate, and aspartate. (Table 5) [84].

One study observed that daily dark chocolate intake for 2 weeks modulated energy and hormone metabolism of humans with low and high anxiety traits, and also modulated their microbial gut metabolism. Urine analysis showed a reduction of stress hormone metabolites, and other stress-related metabolites presented a trend to change towards the low anxiety profile. Some of these metabolites are glycine, citrate, trans-aconitate, proline, DOPA, β-alanine, hippurate, and *p*-cresol sulfate [99]. This can eventually help understand the impact of cocoa intake on health.

5. Conclusions

A diversity of metabolites has been identified in urine and plasma after consumption of cocoa or cocoa derived products. Theobromine and caffeine have been some of the main methylxanthines identified together with their metabolites. Polyphenolic phase I and phase II metabolism leads mainly to (epi)catechin sulphates, glucuronides, and sulfoglucuronides. Non absorbed polyphenols are transformed by colonic microorganisms yielding a diversity of metabolites that range from phenolic acids to different valerolactones, and more complex metabolites.

Biomarkers of consumption of cocoa have been proposed: AMMU, 3-methyluric acid, 7-methylxanthine, 3-methylxanthine, theobromine, 3,7-dimethyluric acid, hydroxyphenylacetic acid, and 5-(3′,4′-dihydroxyphenyl)-γ-valerolactone; the latter with its correspondant glucuronides and sulfates.

Evidence shows that absorption, metabolism, and excretion of cocoa metabolites depend on the food matrix, the dose, age, gender, overall health status and other factors such as the polymerization degree (e.g., procyanidins), and stereochemistry (e.g., flavanols).

The development of clinical studies is fundamental to understand the metabolic pathways of different metabolites present in cocoa and further on the effects on health. Special attention must be given to properly design the experiments, and instrumental methodologies for the extraction and quantification of metabolites in biological samples.

Author Contributions: Writing—original draft preparation: A.L.M.-G.; review and editing: P.E.

Funding: This research was done within the project 735-B5-050 "Valorización de matrices alimentarias mediante la evaluación del impacto del procesamiento sobre el perfil metabolómico", funded by the University of Costa Rica, Costa Rica.

Conflicts of Interest: The authors declare no conflict of interest.

References

1. Dand, R. 2-Agronomics of international cocoa production. In *The International Cocoa Trade*, 3rd ed.; Woodhead Publishing Limited: Cambridge, UK, 2011; pp. 23–64.

2. Integrated Taxonomic Information System (ITIS) Report: *Theobroma cacao* L. Available online: https://www.itis.gov/servlet/SingleRpt/SingleRpt?search_topic=TSN&search_value=505487#null (accessed on 30 December 2017).

3. Afoakwa, E.O. *Cocoa Production and Processing Technology*; CRC Press, Taylor & Francis: Boca Raton, FL, USA, 2014; pp. 173–178. [CrossRef]

4. Nair, K.P.P. 5–Cocoa (*Theobroma cacao* L.). In *The Agronomy and Economy of Important Tree Crops of the Developing World*; Elsevier: Burlington, MA, USA, 2010; pp. 131–180.

5. Biehl, B.; Ziegleder, G. Cococa: Chemistry of processing. In *Encyclopedia of Food Sciences and Nutrition*, 2nd ed.; Caballero, B., Trugo, L., Finglas, P.M., Eds.; Elsevier: Amsterdam, The Netherlands, 2003; pp. 1436–1448.

6. Biehl, B.; Ziegleder, G. Cocoa: Production, products and use. In *Encyclopedia of Food Sciences and Nutrition*, 2nd ed.; Caballero, B., Trugo, L., Finglas, P.M., Eds.; Elsevier: Amsterdam, The Netherlands, 2003; pp. 1448–1463.

7. Hernandez, B.; Castellote, A.I.; Permanyer, J.J. Triglyceride analysis of cocoa beans from different geographical origins. *Food Chem.* **1991**, *41*, 269–276. [CrossRef]

8. Parsons, J.G.; Keeney, P.G.; Patton, S. Identification and Quantitative Analysis of Phospholipids in Cocoa Beans. *J. Food Sci.* **1969**, *34*, 497–499. [CrossRef]

9. Aremu, C.Y.; Agiang, M.A.; Ayatse, J.O.I. Nutrient and antinutrient profiles of raw and fermented cocoa beans. *Plant Foods Hum. Nutr.* **1995**, *48*, 217–223. [CrossRef]

10. Afoakwa, E.O.; Quao, J.; Takrama, J.; Budu, A.S.; Saalia, F.K. Chemical composition and physical quality characteristics of Ghanaian cocoa beans as affected by pulp pre-conditioning and fermentation. *J. Food Sci. Technol.* **2013**, *50*, 1097–1105. [CrossRef] [PubMed]

11. Adeyeye, E.I.; Akinyeye, R.O.; Ogunlade, I.; Olaofe, O.; Boluwade, J.O. Effect of farm and industrial processing on the amino acid profile of cocoa beans. *Food Chem.* **2010**, *118*, 357–363. [CrossRef]

12. Jerkovic, V.; Bröhan, M.; Monnart, E.; Nguyen, F.; Nizet, S.; Collin, S. Stilbenic profile of cocoa liquors from different origins determined by RP-HPLC-APCI(+)-MS/MS: Detection of a new resveratrol hexoside. *J. Agric. Food Chem.* **2010**, *58*, 7067–7074. [CrossRef] [PubMed]

13. Cádiz-Gurrea, M.L.; Lozano-Sanchez, J.; Contreras-Gámez, M.; Legeai-Mallet, L.; Fernández-Arroyo, S.; Segura-Carretero, A. Isolation, comprehensive characterization and antioxidant activities of *Theobroma cacao* extract. *J. Funct. Foods* **2014**, *10*, 485–498. [CrossRef]

14. Cooper, K.A.; Campos-Giménez, E.; Alvarez, D.J.; Nagy, K.; Donovan, J.L.; Williamson, G. Rapid reversed phase ultra-performance liquid chromatography analysis of the major cocoa polyphenols and inter-relationships of their concentrations in chocolate. *J. Agric. Food Chem.* **2007**, *55*, 2841–2847. [CrossRef] [PubMed]

15. Hurst, W.J.; Glinski, J.A.; Miller, K.B.; Apgar, J.; Davey, M.H.; Stuart, D.A. Survey of the trans-resveratrol and trans-piceid content of cocoa-containing and chocolate products. *J. Agric. Food Chem.* **2008**, *56*, 8374–8378. [CrossRef]

16. Marseglia, A.; Palla, G.; Caligiani, A. Presence and variation of γ-aminobutyric acid and other free amino acids in cocoa beans from different geographical origins. *Food Res. Lnt.* **2014**, *63*, 360–366. [CrossRef]

17. Ortega, N.; Romero, M.P.; Macià, A.; Reguant, J.; Anglès, N.; Morelló, J.R.; Motilva, M.J. Comparative study of UPLC–MS/MS and HPLC–MS/MS to determine procyanidins and alkaloids in cocoa samples. *J. Food Compost. Anal.* **2010**, *23*, 298–305. [CrossRef]

18. Patras, M.A.; Milev, B.P.; Vrancken, G.; Kuhnert, N. Identification of novel cocoa flavonoids from raw fermented cocoa beans by HPLC–MSn. *Food Res. Lnt.* **2014**, *63*, 353–359. [CrossRef]

19. Sánchez-Rabaneda, F.; Jáuregui, O.; Casals, I.; Andrés-Lacueva, C.; Izquierdo-Pulido, M.; Lamuela-Raventós, R.M. Liquid chromatographic/electrospray ionization tandem mass spectrometric study of the phenolic composition of cocoa (*Theobroma cacao*). *J. Mass. Spectrom.* **2003**, *38*, 35–42. [CrossRef] [PubMed]

20. Stark, T.; Lang, R.; Keller, D.; Hensel, A.; Hofmann, T. Absorption of N-phenylpropenoyl-L-amino acids in healthy humans by oral administration of cocoa (*Theobroma cacao*). *Mol. Nutr. Food Res.* **2008**, *52*, 1201–1214. [CrossRef]

21. Kim, J.; Lee, K.W.; Lee, H.J. Cocoa (*Theobroma cacao*) seeds and phytochemicals in human health. In *Nuts and Seeds in Health and Disease Prevention*; Preedy, R.V., Watson, R.R., Patel, B.V., Eds.; Academic Press, Elsevier: Burlington, MA, USA, 2011; pp. 351–360.

22. Wollgast, J.; Anklam, E. Review on polyphenols in *Theobroma cacao*: Changes in composition during the manufacture of chocolate and methodology for identification and quantification. *J. Food Res. Int.* **2000**, *33*, 423–447. [CrossRef]

23. Niemenak, N.; Rohsius, C.; Elwers, S.; Omokolo-Ndoumou, D.; Lieberei, R. Comparative study of different cocoa (*Theobroma cacao* L.) clones in terms of their phenolics and anthocyanins contents. *J. Food Compost. Anal.* **2006**, *19*, 612–619. [CrossRef]

24. Ortega, N.; Romero, M.-P.P.; Macià, A.; Reguant, J.; Anglès, N.; Morelló, J.-R.R.; Motilva, M.-J.J. Obtention and characterization of phenolic extracts from different cocoa sources. *J. Agri. Food Chem.* **2008**, *56*, 9621–9627. [CrossRef]

25. Ardhana, M.; Fleet, G. The microbial ecology of cocoa bean fermentations in Indonesia. *Int. J. Food Microbiol.* **2003**, *86*, 87–99. [CrossRef]

26. Mozzi, F.; Ortiz, M.E.; Bleckwedel, J.; De Vuyst, L.; Pescuma, M. Metabolomics as a tool for the comprehensive understanding of fermented and functional foods with lactic acid bacteria. *Food Res. Int.* **2003**, *54*, 1152–1161. [CrossRef]

27. Nazaruddin, R.; Seng, L.K.; Hassan, O.; Said, M. Effect of pulp preconditioning on the content of polyphenols in cocoa beans (*Theobroma cacao*) during fermentation. *Ind. Crops Prod.* **2006**, *24*, 87–94. [CrossRef]

28. Hii, C.L.; Law, C.; Suzannah, S.; Cloke, M. Polyphenols in cocoa (*Theobroma cacao* L.). *As. J. Food Ag-Ind.* **2010**, *2*, 702–722.

29. Gutiérrez, T.J. State-of-the-art chocolate manufacture: A review. *Compr. Rev. Food Sci. Food Saf.* **2017**, *16*, 1313–1344. [CrossRef]

30. Ioannone, F.; Di Mattia, C.D.; De Gregorio, M.; Sergi, M.; Serafini, M.; Sacchetti, G. Flavanols, proanthocyanidins and antioxidant activity changes during cocoa (*Theobroma cacao* L.) roasting as affected by temperature and time of processing. *Food Chem.* **2015**, *174*, 256–262. [CrossRef]

31. Kothe, L.; Zimmermann, B.F.; Galensa, R. Temperature influences epimerization and composition of flavanol monomers, dimers and trimers during cocoa bean roasting. *Food Chem.* **2013**, *141*, 3656–3663. [CrossRef]

32. Steinberg, F.M.; Bearden, M.M.; Keen, C.L. Cocoa and chocolate flavonoids: Implications for cardiovascular health. *J. Am. Diet Assoc.* **2003**, *103*, 215–223. [CrossRef]

33. Magrone, T.; Russo, M.A.; Jirillo, E. Cocoa and dark chocolate polyphenols: From biology to clinical applications. *Front Immunol.* **2017**, *8*, 677. [CrossRef]

34. Aprotosoaie, A.C.; Miron, A.; Trifan, A.; Simon, L.V.; Costache, I.J. The cardiovascular effects of cocoa polyphenols–an overview. *Diseases* **2016**, *4*, 39. [CrossRef]

35. Hooper, L.; Kay, C.; Abdelhamid, A.; Kroon, P.A.; Cohn, J.S.; Rimm, E.B.; Cassidy, A. Effects of chocolate, cocoa, and flavan-3-ols on cardiovascular health: A systematic review and meta-analysis of randomized trials. *Am. J. Clin. Nutr.* **2012**, *95*, 740–751. [CrossRef]

36. Corti, R.; Flammer, A.J.; Hollenberg, N.K.; Lüscher, T.F. Cocoa and cardiovascular health. *Circulation* **2009**, *119*, 1433–1441. [CrossRef]

37. Latif, R. Chocolate/cocoa and human health: A review. *The Neth. J. Med.* **2013**, *71*, 63–68.

38. Martin, M.A.; Goya, L.; Ramos, S. Potential for preventive effects of cocoa and cocoa polyphenols in cancer. *Food Chem. Toxicol.* **2013**, *56*, 336–351. [CrossRef] [PubMed]

39. Decroix, L.; Soares, D.D.; Meeusen, R.; Heyman, E.; Tonoli, C. Cocoa flavanol supplementation and exercise: A systematic review. *Sports Med.* **2018**, *48*, 867–892. [CrossRef]

40. EFSA. Scientific Opinion on the modification of the authorisation of a health claim related to cocoa flavanols and maintenance of normal endothelium-dependent vasodilation pursuant to Article 13 (5) of Regulation (EC) No 1924/2006 following a request in accordance with article 19 of Regulation (EC) No 1924/2006. *EFSA J.* **2014**, *12*, 3654. [CrossRef]

41. Dicks, L.; Kirch, N.; Gronwald, D.; Wernken, K.; Zimmermann, B.; Helfrich, H.-P.; Ellinger, S. Regular intake of a usual serving size of flavanol-rich cocoa powder does not affect cardiometabolic parameters in stably treated patients with type 2 diabetes and hypertension—a double-blinded, randomized, placebo-controlled trial. *Nutrients* **2018**, *10*, 1435. [CrossRef]

42. Vlachojannis, J.; Erne, P.; Zimmermann, B.; Chrubasik-Hausmann, S. The impact of cocoa flavanols on cardiovascular health. *Phytother. Res.* **2016**, *30*, 1641–1657. [CrossRef] [PubMed]

43. Ellinger, S.; Stehle, P. Impact of cocoa consumption on inflammation processes—a critical review of randomized controlled trials. *Nutrients* **2016**, *8*, 321. [CrossRef]

44. Hollman, P.C.; Cassidy, A.; Comte, B.; Heinone, M.; Richelle, M.; Richling, E.; Serafini, M.; Scalbert, A.; Sies, H.; Vidry, S. The biological relevance of direct antioxidant effects of polyphenols for cardiovascular health in humans is not established. *J. Nutr.* **2011**, *141*, 989S–1009S. [CrossRef] [PubMed]

45. Morze, J.; Schwedhelm, C.; Bencic, A.; Hoffmann, G.; Boeing, H.; Przybylowics, K.; Schwingshackl, L. Chocolate and risk of chronic disease: A systematic review and dose-response meta analysis. *Eur. J. Nutr.* **2019**. [CrossRef]

46. González-Sarrías, A.; Combet, E.; Pinto, P.; Mena, P.; Dall'Asta, M.; García-Aloy, M.; Rodríguez-Mateos, A.; Gibney, E.R.; Dumont, J.; Massaro, M.; et al. A systematic review and meta-analysis of the effects of flavanol-containing tea, cocoa and apple products on body composition and blood lipids: Exploring the factors responsible for variability in their efficacy. *Nutrients* **2017**, *9*, 746. [CrossRef]

47. Tomás-Barberán, F.A.; Espín, J.C. Effect of food structure and processing on (poly)phenol-gut microbiota interactions and the effects on human health. *Annu. Rev. Food Sci. Technol.* **2019**, *10*. [CrossRef]

48. Fraga, C.G.; Croft, K.D.; Kennedy, D.O.; Tomás-Barberán, F.A. The effects of polyphenols and other bioactives on human health. *Food Funct.* **2019**, *10*, 514–528. [CrossRef]

49. Stellingwerff, T.; Godin, J.-P.; Chou, C.J.; Grathwohl, D.; Ross, A.B.; Cooper, K.A.; Williamson, G.; Actis-Goretta, L. The effect of acute dark chocolate consumption on carbohydrate metabolism and performance during rest and exercise. *Appl. Physiol. Nutr. Metab.* **2014**, *39*, 173–182. [CrossRef]

50. Schramm, D.D.; Karim, M.; Schrader, H.R.; Holt, R.R.; Kirkpatrick, N.J.; Polagruto, J.A.; Ensunsa, J.L.; Schmitz, H.H.; Keen, C.L. Food effects on the absorption and pharmacokinetics of cocoa flavanols. *Life Sci.* **2003**, *73*, 857–869. [CrossRef]

51. Machin, D.; Campbell, M.J.; Walters, S.J. *Medical Statistics: A Textbook for the Health Sciences*, 4th ed.; Wiley: West Sussex, UK, 2007; pp. 217–239.

52. Grimes, D.A.; Schulz, K.F. An overview of clinical research: The lay of the land. *Lancet* **2002**, *359*, 57–61. [CrossRef]

53. Friedman, L.M.; Furberg, C.D.; DeMets, D.L.; Reboussin, D.M.; Granger, C.B. *Fundamentals of clinical trials*, 5th ed.; Springer International Publishing: Cham, Switzerland, 2015.

54. Ferrara, M.; Sébédio, J.-L. 1- Challenges in nutritional metabolomics. In *Metabolomics as a Tool in Nutrition Research*; Elsevier and Woodhead Publishing: Cambridge, UK, 2015; pp. 3–16. [CrossRef]

55. Garcia-Aloy, M.; Llorach, R.; Urpi-Sarda, M.; Jáuregui, O.; Corella, D.; Ruiz-Canela, M.; Salas-Salvadó, J.; Fitó, M.; Estruch, R.; Andres-Lacueva, C. A metabolomics-driven approach to predict cocoa product consumption by designing a multimetabolite biomarker model in free-living subjects from the PREDIMED study. *Mol. Nutr. Food Res.* **2015**, *59*, 212–220. [CrossRef]

56. Rodriguez, A.; Costa-Bauza, A.; Saez-Torres, C.; Rodrigo, D.; Grases, F. HPLC method for urinary theobromine determination: Effect of consumption of cocoa products on theobromine urinary excretion in children. *Clin. Biochem.* **2015**, *48*, 1138–1143. [CrossRef]

57. Urpi-Sarda, M.; Monagas, M.; Khan, N.; Llorach, R.; Lamuela-Raventós, R.M.; Jáuregui, O.; Estruch, R.; Izquierdo-Pulido, M.; Andrés-Lacueva, C. Targeted metabolic profiling of phenolics in urine and plasma after regular consumption of cocoa by liquid chromatography-tandem mass spectrometry. *J. Chromatogr.* **2009**, *1216*, 7258–7267. [CrossRef]

58. Davinelli, S.; Corbi, G.; Zarrelli, A.; Arisi, M.; Calzavara-Pinton, P.; Grassi, D.; De Vivo, I.; Scapagnini, G. Short-term supplementation with flavanol-rich cocoa improves lipid profile, antioxidant status and positively influences the AA/EPA ratio in healthy subjects. *J. Nutr. Biochem.* **2018**, *61*, 33–39. [CrossRef]

59. Ptolemy, A.S.; Tzioumis, E.; Thomke, A.; Rifai, S.; Kellogg, M. Quantification of theobromine and caffeine in saliva, plasma and urine via liquid chromatography-tandem mass spectrometry: A single analytical protocol applicable to cocoa intervention studies. *J. Chromatogr. B Analyt. Thecnol. Biomed. Life Sci.* **2010**, *878*, 409–416. [CrossRef]

60. Wiese, S.; Esatbeyoglu, T.; Winterhalter, P.; Kruse, H.-P.; Winkler, S.; Bub, A.; Kulling, S.E. Comparative biokinetics and metabolism of pure monomeric, dimeric, and polymeric flavan-3-ols: A randomized cross-over study in humans. *Mol. Nutr. Food Res.* **2015**, *59*, 610–621. [CrossRef]

61. Heiss, C.; Kleinbongard, P.; Dejam, A.; Perré, S.; Schroeter, H.; Sies, H.; Kelm, M. Acute consumption of flavanol-rich cocoa and the reversal of endothelial dysfunction in smokers. *J. Am. Coll. Cardiol.* **2005**, *46*, 1276–1283. [CrossRef]

62. Richelle, M.; Tavazzi, I.; Enslen, M.; Offord, E. Plasma kinetics in man of epicatechin from black chocolate. *Eu. J. Clin. Nutr.* **1999**, *53*, 22–26. [CrossRef]

63. Wang, J.F.; Schramm, D.D.; Holt, R.R.; Ensunsa, J.L.; Fraga, C.G.; Schmitz, H.H.; Keen, C.L. A dose-response effect from chocolate consumption on plasma epicatechin and oxidative damage. *J. Nutr.* **2000**, *130*, 2115S–2119S. [CrossRef]

64. Baba, S.; Osakabe, N.; Yasuda, A.; Natsume, M.; Takizawa, T.; Nakamura, T.; Terao, J. Bioavailability of (-)-epicatechin upon intake of chocolate and cocoa in human volunteers. *Free Radic. Res.* **2000**, *33*, 635–641. [CrossRef]

65. Rios, L.Y.; Gonthier, M.-P.; Remesy, C.; Mila, I.; Lapierre, C.; Lazarus, S.A.; Williamson, G.; Scalbert, A. Chocolate intake increases urinary excretion of polyphenol-derived phenolic acids in healthy human subjects. *Am. J. Clin. Nutr.* **2003**, *77*, 912–918. [CrossRef]

66. Actis-Goretta, L.; Lévèques, A.; Giuffrida, F.; Romanov-Michailidis, F.; Viton, F.; Barron, D.; Duenas-Paton, M.; González-Manzano, S.; Santos-Buelga, C.; Williamson, G.; et al. Elucidation of (-)-epicatechin metabolites after ingestion of chocolate by healthy humans. *Free Radic. Biol. Med.* **2012**, *53*, 787–795. [CrossRef]

67. Costa-Bauza, A.; Grases, F.; Calvó, P.; Rodríguez, A.; Prieto, R.M. Effect of consumption of cocoa-derived products on uric acid crystallization in urine of healthy volunteers. *Nutrients* **2018**, *10*, 1516. [CrossRef]

68. Holt, R.R.; Lazarus, S.A.; Sullards, M.C.; Zhu, Q.Y.; Schramm, D.D.; Hammerstone, J.F.; Fraga, C.G.; Schmitz, H.H.; Keen, C.L. Procyanidin dimer B2 [epicatechin-(4β -8)-epicatechin] in human plasma after the consumption of a flavanol-rich cocoa. *Am. J. Clin. Nutr.* **2002**, *76*, 798–804. [CrossRef]

69. Ito, H.; Gonthier, M.-P.; Manach, C.; Morand, C.; Mennen, L.; Rémésy, C.; Scalbert, A. Polyphenol leveles in human urine after intake of six different polyphenol-rich beverages. *Br. J. Nutr.* **2005**, *94*, 500–509. [CrossRef]

70. Roura, E.; Andrés-Lacueva, C.; Jáuregui, O.; Badia, E.; Estruch, R.; Izquierdo-Pulido, M.; Lamuela-Raventós, R.M. Rapid liquid chromatography tandem mass spectrometry assay to quantify plasma (-)-epicatechin metabolites after ingestion of a standard portion of cocoa beverage in humans. *J. Agric. Food Chem.* **2005**, *53*, 6190–6194. [CrossRef]

71. Schroeter, H.; Heiss, C.; Balzer, J.; Kleinbongard, P.; Keen, C.L.; Hollenberg, N.K.; Sies, H.; Kwik-Uribe, C.; Schmitz, H.H.; Kelm, M. (-)-Epicatechin mediates beneficial effects of flavanol-rich cocoa on vascular function in humans. *Proc. Natl. Acad. Sci. USA* **2006**, *103*, 1024–1029. [CrossRef]

72. Roura, E.; Almajano, M.P.; Bilbao, M.L.M.; Andrés-Lacueva, C.; Estruch, R.; Lamuela-Raventós, R.M. Human urine: Epicatechin metabolites and antioxidant activity after cocoa beverage intake. *Free Radical. Res.* **2007**, *41*, 943–949. [CrossRef]

73. Roura, E.; Andrés-Lacueva, C.; Estruch, R.; Mata-Bilbao, M.L.; Izquierdo-Pulido, M.; Waterhouse, A.L.; Lamuela-Raventós, R.M. Milk does not affect the bioavailability of cocoa powder flavonoid in healthy human. *Ann. Nutr. Metab.* **2007**, *51*, 493–498. [CrossRef]

74. Keogh, J.B.; McInerney, J.; Clifton, P.M. The effect of milk protein on the bioavailability of cocoa polyphenols. *J. Food Sci.* **2007**, *72*, S230–S233. [CrossRef]

75. Roura, E.; Andrés-Lacueva, C.; Estruch, R.; Lourdes Mata Bilbao, M.; Izquierdo-Pulido, M.; Lamuela-Raventós, R.M. The effects of milk as a food matrix for polyphenols on the excretion profile of cocoa (-)-epicatechin metabolites in healthy human subjects. *Br. J. Nutr.* **2008**, *100*, 846–851. [CrossRef]

76. Urpi-Sarda, M.; Monagas, M.; Khan, N.; Lamuela-Raventos, R.M.; Santos-Buelga, C.; Sacanella, E.; Castell, M.; Permanyer, J.; Andres-Lacueva, C. Epicatechin, procyanidins, and phenolic microbial metabolites after cocoa intake in humans and rats. *Anal. Bioanal. Chem.* **2009**, *394*, 1545–1556. [CrossRef]

77. Mullen, W.; Borges, G.; Donovan, J.L.; Edwards, C.A.; Serafini, M.; Lean, M.E.J.; Crozier, A. Milk decreases urinary excretion but not plasma pharmacokinetics of cocoa flavan-3-ol metabolites in humans. *Am. J. Clin. Nutr.* **2009**, *89*, 1784–1791. [CrossRef]

78. Urpi-Sarda, M.; Llorach, R.; Khan, N.; Monagas, M.; Rotches-Ribalta, M.; Lamuela-Raventos, R.; Estruch, R.; Tinahones, F.J.; Andres-Lacueva, C. Effect of milk on the urinary excretion of microbial phenolic acids after cocoa powder consumption in humans. *J. Agric. Food Chem.* **2010**, *58*, 4706–4711. [CrossRef]

79. Ottaviani, J.I.; Momma, T.Y.; Heiss, C.; Kwik-Uribe, C.; Schroeter, H.; Keen, C.L. The stereochemical configuration of flavanols influences the level and metabolism of flavanols in humans and their biological activity in vivo. *Free Radic. Biol. Med.* **2011**, *50*, 237–244. [CrossRef]

80. Ottaviani, J.I.; Kwik-Uribe, C.; Keen, C.L.; Schroeter, H. Intake of dietary procyanidins does not contribute to the pool of circulating flavanols in humans. *Am. J. Clin. Nutr.* **2012**, *95*, 851–858. [CrossRef]

81. Martínez-López, S.; Sarriá, B.; Gómez-Juaristi, M.; Goya, L.; Mateos, R.; Bravo-Clemente, L. Theobromine, caffeine, and theophylline metabolites in human plasma and urine after consumption of soluble cocoa products with different methylxanthine contents. *J. Food Res. Int.* **2014**, *63*, 446–455. [CrossRef]

82. Rodriguez-Mateos, A.; Cifuentes-Gomez, T.; Gonzalez-Salvador, I.; Ottaviani, J.I.; Schroeter, H.; Kelm, M.; Heiss, C.; Spencer, J.P.E. Influence of age on the absorption, metabolism, and excretion of cocoa flavanols in healthy subjects. *Mol. Nutr. Food Res.* **2015**, *59*, 1504–1512. [CrossRef]

83. Ottaviani, J.I.; Fong, R.; Kimball, J.; Ensunsa, J.L.; Britten, A.; Lucarelli, D.; Luben, R.; Grace, P.B.; Mawson, D.H.; Tym, A.; et al. Evaluation at scale of microbiome-derived metabolites as biomarker of flavan-3-ol intake in epidemiological studies. *Sci. Rep.* **2018**, *8*, 9859. [CrossRef]

84. Ostertag, L.M.; Philo, M.; Colquhoun, I.J.; Tapp, H.S.; Saha, S.; Duthie, G.G.; Kemsley, E.K.; De Roos, B.; Kroon, P.A.; Le Gall, G. Acute consumption of flavan-3-ol-enriched dark chocolate affects human endogenous metabolism. *J. Proteome Res.* **2017**, *16*, 2516–2526. [CrossRef]

85. Ostertag, L.M.; Kroon, P.A.; Wood, S.; Horgan, G.W.; Cienfuegos-Jovellanos, E.; Saha, S.; Duthie, G.G.; De Roos, B. Flavan-3-ol-enriched dark chocolate and white chocolate improve acute measures of platelet function in a gender-specific way-a randomized-controlled human intervention trial. *Mol. Nutr. Food Res.* **2013**, *57*, 191–202. [CrossRef]

86. Llorach, R.; Urpi-Sarda, M.; Jauregui, O.; Monagas, M.; Andres-Lacueva, C. An LC-MS-based metabolomics approach for exploring urinary metabolome modifications after cocoa consumption. *J. Proteome Res.* **2009**, *8*, 5060–5068. [CrossRef]

87. Llorach-Asunción, R.; Jauregui, O.; Urpi-Sarda, M.; Andres-Lacueva, C. Methodological aspects for metabolome visualization and characterization: A metabolomic evaluation of the 24 h evolution of human urine after cocoa powder consumption. *J. Pharm. Biomed. Anal.* **2010**, *51*, 373–381. [CrossRef]
88. Smart, R.C.; Hodgson, E. *Molecular and Biochemical Toxicology*, 5th ed.; Wiley: Raleigh, NC, USA, 2018.
89. Caffeine metabolism—Reference pathway. Available online: http://www.genome.jp/kegg-bin/show_pathway?org_name=rn&mapno=00232&mapscale=&show_description=hide (accessed on 11 January 2017).
90. Kohlmeier, M. *Nutrient Metabolism*; Academic Press: London, UK, 2003.
91. Urpi-sarda, M.; Llorach, R.; Monagas, M.; Khan, N.; Rotches-Ribalta, M.; Roura, E.; Lamuela-Raventós, R.; Estruch, R.; Andres-Lacueva, C. 5. Effect of cocoa powder in the prevention of cardiovascular disease: Biological, consumption and inflammatory biomarkers. A metabolomic approach. In *Recent Advances in Pharmacutical Sciences*; Muñoz-Torrero, D., Ed.; Transworld Research Network: Kerata, India, 2011; pp. 121–132.
92. Scalbert, A.; Morand, C.; Manach, C.; Rémésy, C. Absorption and metabolism of polyphenols in the gut and impact on health. *Biomed. Pharmacother.* **2002**, *56*, 276–282. [CrossRef]
93. Cifuentes-Gomez, T.; Rodriguez-Mateos, A.; Gonzalez-Salvador, I.; Alañon, M.E.; Spencer, J.P.E. Factors affecting the absorption, metabolism, and excretion of cocoa flavanols in humans. *J. Agric. Food Chem.* **2015**, *63*, 7615–7623. [CrossRef]
94. Serafini, M.; Testa, M.F.; Villaño, D.; Pecorari, M.; Van Wieren, K.; Azzini, E.; Brambilla, A.; Maiani, G. Antioxidant activity of blueberry fruit is impaired by association with milk. *Free Radic. Biol. Med.* **2009**, *46*, 769–774. [CrossRef]
95. Patton, S.; Keenan, T.W. The Milk fat globule membrane. *Biochim. Biophys. Acta Biomembranes.* **1975**, *415*, 273–309. [CrossRef]
96. Borges, G.; Ottaviani, J.I.; Van der Hooft, J.J.J.; Schroeter, H.; Crozier, A. Absorption, metabolism, distribution and excretion of (−)-epicatechin: A review of recent findings. *Mol. Aspects Med.* **2018**, *16*, 18–30. [CrossRef]
97. Tzounis, X.; Vulevic, J.; Kuhnle, G.G.C.; George, T.; Leonczak, J.; Gibson, G.R.; Kwik-Uribe, C.; Spencer, J.P.E. Flavanol monomer-induced changes to the human faecal microflora. *Br. J. Nutr.* **2008**, *99*, 782–792. [CrossRef] [PubMed]
98. Tzounis, X.; Rodriguez-Mateos, A.; Vulevic, J.; Gibson, G.R.; Kwik-Uribe, C.; Spencer, J.P.E. Prebiotic evaluation of cocoa-derived flavanols in healthy humans by using a randomized, controlled, double-blind, crossover intervention study. *Am. J. Clin. Nutr.* **2011**, *93*, 62–72. [CrossRef]
99. Martin, F.-P.J.; Rezzi, S.; Peré-Trepat, E.; Kamlage, B.; Collino, S.; Leibold, E.; Kastler, J.; Rein, D.; Fay, L.B.; Kochhar, S. Metabolic effects of dark chocolate consumption on energy, gut microbiota, and stress-related metabolism in free-living subjects. *J. Proteome. Res.* **2009**, *8*, 5568–5579. [CrossRef] [PubMed]

 *nutrients*

Article

Flavanol Bioavailability in Two Cocoa Products with Different Phenolic Content. A Comparative Study in Humans

Miren Gómez-Juaristi, Beatriz Sarria *, Sara Martínez-López, Laura Bravo Clemente * and Raquel Mateos *

Department of Metabolism and Nutrition, Institute of Food Science, Technology and Nutrition (ICTAN-CSIC), Spanish National Research Council (CSIC), José Antonio Novais 10, 28040 Madrid, Spain
* Correspondence: beasarria@ictan.csic.es (B.S.); lbravo@ictan.csic.es (L.B.C.); raquel.mateos@ictan.csic.es (R.M.); Tel.: +34-915-492-300 (B.S. & L.B.C. & R.M.)

Received: 18 May 2019; Accepted: 19 June 2019; Published: 26 June 2019

Abstract: Cocoa has beneficial health effects partly due to its high flavanol content. This study was aimed at assessing the absorption and metabolism of polyphenols in two soluble cocoa products: a conventional (CC) and a flavanol-rich product (CC-PP). A crossover, randomized, blind study was performed in 13 healthy men and women. On two different days, after an overnight fast, volunteers consumed one serving of CC (15 g) or CC-PP (25 g) in 200 mL of semi-skimmed milk containing 19.80 mg and 68.25 mg of flavanols, respectively. Blood and urine samples were taken, before and after CC and CC-PP consumption, and analyzed by high-performance liquid chromatography coupled to electrospray ionisation and quadrupole time-of-flight mass spectrometry (HPLC-ESI-QToF-MS). Up to 10 and 30 metabolites were identified in plasma and urine, respectively. Phase II derivatives of epicatechin were identified with kinetics compatible with small intestine absorption, although the most abundant groups of metabolites were phase II derivatives of phenyl-γ-valerolactone and phenylvaleric acid, formed at colonic level. 5-(4′-Hydroxyphenyl)-γ-valerolactone-sulfate could be a sensitive biomarker of cocoa flavanol intake. CC and CC-PP flavanols showed a dose-dependent absorption with a recovery of 35%. In conclusion, cocoa flavanols are moderately bioavailable and extensively metabolized, mainly by the colonic microbiota.

Keywords: flavanols; soluble cocoa products; bioavailability; human; plasma nutrikinetics; liquid chromatography coupled to electrospray ionisation and quadrupole time-of-flight mass spectrometry (LC-ESI-QToF-MS); colonic bacteria

---

1. Introduction

Cocoa remains a popular foodstuff worldwide. Soluble cocoa products have been popular in Spain, among other countries, being commonly consumed twice a day, at breakfast and as part of an afternoon break known as "merienda". The potential health-promoting effects of cocoa products have gained extensive attention in the last few years. Most of these effects are attributed to the polyphenolic fraction of cocoa, mainly flavanols epicatechin and catechin, and low molecular weight procyanidins such as procyanidins B1 and B2 [1,2]. In fact, cocoa has been defined as a functional food due to its high flavanol content [1]. Moreover, soluble cocoa products are also a source of methylxanthines (mainly theobromine), magnesium and dietary fiber; all are biologically active substances that may also affect human health positively [2].

The biological activity of phenolic compounds depends on their bioavailability and metabolic fate, as well as on their digestive accessibility, which is determined by the release from the food matrix and efficiency in trans-epithelial passage. Recently Mena et al. [3] reviewed the rate and extent of absorption of cocoa polyphenols in humans, as well as the metabolic pathways involved. These polyphenols are partially absorbed in the upper gastrointestinal tract, being conjugated by phase II enzymes into methoxy, sulfated and/or glucuronidated metabolites, with maximum plasma concentrations around 2 h after cocoa intake in the nM range [4]. Controversy exists regarding whether procyanidins can be broken down in the stomach yielding monomers that may be absorbed or, conversely, whether intact procyanidins can be absorbed from the gastrointestinal tract. In this sense, Ottaviani et al. [5] observed that dietary procyanidins do not contribute to the systemic pool of flavanols in humans when test drinks that contained only flavanols, flavanols and procyanidins, or only procyanidins were consumed by humans.

Polyphenols not absorbed in the small intestine reach the colon where they are metabolized by the intestinal microbiota mainly to phenyl-γ-valerolactone and phenylvaleric acid metabolites [3]. These ring fission products, as well as phenylvaleric acid conjugates, were excreted primarily 5–10 h after ingestion of green tea [6,7]. In a recent report on the absorption, distribution, metabolism, and excretion in humans of radiolabeled and stereochemically pure [2-^{14}C](-)-epicatechin ([^{14}C]epicatechin) [8], 20 different metabolites were identified and quantified: phase II derivatives of epicatechin, hydroxyphenyl-γ-valerolactones and phenylvaleric acids. Moreover, it was confirmed that the gut microbiome is a key driver of epicatechin metabolism.

Nowadays, innovation in the elaboration of cocoa products includes using new delivery forms in an attempt to increase polyphenols' bioavailability, like microencapsulation of cocoa phenols in cocoa-nut creams [9]. New soluble cocoa products enriched with bioactive components such as dietary fiber, methylxanthines or polyphenols are also being introduced into the food market. This is the case of two novel soluble cocoa products, produced and commercialized by the same manufacturer as the products used in the present study. One is rich in dietary fiber and the other rich in cocoa, containing 1.16 and 3.02 mg/g of flavanols, respectively [10]—amounts similar to the flavanol content of the soluble cocoas used in the present study (see below). When consumed in realistic doses by healthy and subjects at cardiovascular risk (hypercholesterolemic), results showed that both soluble cocoa products had a positive effect on serum lipid profile, increasing HDL-cholesterol without inducing anthropometric changes. These effects could be associated in part to the flavanol content in the two commercial cocoa products, although in that study the circulating levels of phenolic metabolites was not quantified. However, other studies have shown a correlation between the effect of dark chocolate (50 g, providing 7.5 gallic acid equivalents (GAE) of polyphenols) improving platelet function and the increased plasma concentrations of structurally-related (epi)catechin metabolites (SREM). These data confirm that the potential health benefits of cocoa consumption may be mediated by flavan-3-ol circulating metabolites [11], in spite of the limited bioavailability of cocoa polyphenols.

Indeed, many studies have focused on determining the bioavailability and metabolism of cocoa phenolic compounds, although most bioavailability studies have been carried out with unrealistic doses (for some reviews see [3,12–14]). Therefore, the aim of the present work was to evaluate the bioavailability of flavanols in healthy humans after consuming a realistic amount of two soluble cocoa products: a conventional soluble cocoa (CC) and a flavanol-rich soluble cocoa (CC-PP). In addition, an important effort has been made to identify microbiota-derived metabolites, showing the importance of gut bacteria on polyphenol absorption and metabolism.

2. Materials and Methods

2.1. Chemical Reagents and Materials

The commercialized soluble cocoa products used in the study were provided by Idilia Foods (previous company name: Nutrexpa S.L.), one being a conventional soluble cocoa product, labeled as CC, and the other a flavanol-rich soluble cocoa product, labeled as CC-PP. All solvents and reagents were of analytical grade unless otherwise stated. Ascorbic acid, epicatechin, catechin, procyanidin B1, 3-(3,4-dihydroxyphenyl)propionic acid, 3-(4-hydroxyphenyl)propionic acid, 3-(4-

hydroxy-3-methoxyphenyl)propionic acid, 3,4-dihydroxyphenylacetic acid, 4-hydroxyphenylacetic acid, 4-hydroxy-3-methoxyphenylacetic acid, protocatechuic acid, 4-hydroxybenzoic acid and ferulic acid were from Sigma-Aldrich (Madrid, Spain). Procyanidin B2 and epicatechin-3-gallate were acquired from Extrasynthese Genay Cedex (France). Methanol, formic acid, and acetonitrile (high-performance liquid chromatography (HPLC) grade) were acquired from Panreac (Madrid, Spain).

2.2. Quantification of Total Polyphenols of the Soluble Cocoa Products. Characterization and Quantification by high-performance liquid chromatography-mass spectrometry (HPLC-MS) and high-performance liquid chromatography-diode array (HPLC-DAD).

Total polyphenols were measured spectrophotometrically using the Folin–Ciocalteau reagent and gallic acid as standard. Results were expressed as µg equiv gallic acid/g product.

As described by Bravo et al. [15], cocoa extracts were obtained by washing 1 g of defatted cocoa with 40 mL of 50% aqueous methanol (HPLC grade) containing 0.8% of 2 mol/L hydrochloric acid for 1 h at room temperature with constant shaking. Afterwards, samples were centrifuged (10 min, 3000× g) and supernatants were collected. The residues obtained were successively extracted with 40 mL of 70% acetone (v/v) in water (1 h, constant shaking). The samples were then centrifuged (10 min, 3000× g) and supernatants were collected. Finally, supernatants obtained after each extraction step were combined and made up to 100 mL. Polyphenolic composition was analyzed using an Agilent 1200 series liquid chromatographic system equipped with an autosampler, quaternary pump, diode array detector (DAD) and simple quadrupole (sQ) mass spectrometer (Agilent Technologies, Waldbronn, Germany). Samples (20 µL) were injected into a Superspher RP18 column (4.6 mm × 250 mm i.d., 4 µm; Agilent Technologies) preceded by an ODS RP18 guard column kept in a thermostatic oven at 37 °C.

Elution was performed at a flow rate of 0.6 mL/min using a binary system consisting of 1% formic acid in deionized water (solvent A) and 1% formic acid in acetonitrile (solvent B). The solvent gradient changed from 6% to 10% solvent B over 20 min, 10% to 13% solvent B over 5 min, 13% to 15% solvent B over 5 min, 15% to 10% in 10 min, 10% to 6% in 5 min and then maintained isocratically for 5 min. Chromatograms were recorded at 280 nm. The mass spectrometer was fitted to an atmospheric pressure electrospray ionization (ESI) source, which operated in negative ion mode. Capillary voltage was set to 3500 V, with nebulizing gas flow rate of 12 h/L, drying gas temperature of 350 °C and nebulizer pressure of 45 psi. Mass spectrometry data were acquired in scan mode (mass range m/z 100–1000). Data acquisition and analysis were carried out in an Agilent ChemStation.

An Agilent 1200 series liquid chromatographic system (Agilent Technologies) equipped with a quaternary pump, column oven, autosampler and DAD was used to quantify the identified polyphenols in soluble cocoa products by high-performance liquid chromatography (HPLC–DAD). The chromatographic conditions (column, guard column, binary gradient, injection volume, etc.) were as described above. For quantitative analysis the external standard method was used. Samples were prepared and analyzed in triplicate and the results were expressed as the mean value.

2.3. Subjects and Study Design

The study protocol was conducted in accordance with the ethical recommendations of the Declaration of Helsinki and approved by the Ethics Committee of Hospital Universitario Puerta de Hierro in Majadahonda (Madrid, Spain) (ACT ID 256, 28th of June 2010; Project Identification Code AGL2015-69986-R). Recruitment of volunteers was carried out through placing advertisements at the Institute of Food Science, Technology and Nutrition (ICTAN).

The study was carried out in thirteen healthy subjects (3 men and 10 women); the mens' average age and body mass index were 26.67 ± 3.21 year and 22.47 ± 2.97 kg/m^2, respectively, and womens' were 32.60 ± 9.85 year and 23.36 ± 3.73 kg/m^2, respectively. They were non-smoker, non-vegetarian, non-pregnant women, who were not taking any medication or nutritional supplements, and were not suffering from any chronic pathology or gastrointestinal disorder. The sample size was estimated attending to similar previous bioavailability studies [4,16]. The volunteers gave their informed consent prior to participation.

The present randomized, single-blind study was carried out at the Human Nutrition Unit (HNU) of the Institute of Food Science, Technology and Nutrition (ICTAN). Volunteers attended the HNU on two days, separated by two weeks. Three days previous to each visit, participants were instructed not to consume cocoa products (chocolate, soluble cocoa, etc.), juices, tea, wine, grape must, oranges, tangerines, apples, grapes, strawberries or other berries, beets, onions, soybeans and soy derivatives. In addition, volunteers were asked to complete a 24 h food intake recall the day before they attended the HNU in order to monitor any possible food restriction incompliance.

On each intervention day, volunteers arrived at the HNU after an overnight fast. Prior to the intake of the soluble cocoa product, a nurse inserted a cannula in the cubital vein of the non-prevailing arm of the volunteers and blood samples were collected into EDTA-coated tubes at baseline ($t = 0$) and 0.5, 1, 2, 3, 4, 6, and 8 h after consuming the corresponding cocoa product. The soluble cocoa products, either 15 g of CC or 25 g CC-PP, were dissolved in 200 mL of semi-skimmed milk following manufacturer's preparation instructions. Plasma was separated by centrifugation (10 min, 3000 rpm, 4 °C) and stored at −80 °C until further analysis. Urine samples were collected at different time intervals (t = −2–0, 0–4, 4–8, 8–12, and 12–24 h) in urine collection flasks that contained 0.5 g of ascorbic acid as preservative and were aliquoted and frozen at −20 °C until analysis. A polyphenol-free breakfast, lunch and afternoon snack were provided 2 h, 4 h, and 8 h after consumption of cocoa products, and water and isotonic beverages were available *ad libitum*.

2.4. Extraction of Phenolic Metabolites from Biological Samples

A liquid–liquid extraction and protein precipitation with acetonitrile was used to isolate metabolites from plasma. A 1 mL defrosted plasma sample was mixed with 50 μL of ascorbic acid (0.2 g/mL). After vortexing the aqueous mixture, it was added drop wise to 750 μL of cold acetonitrile and vortexed for 2 min before centrifuging at 12,000 rpm for 10 min at 4 °C. The supernatant was separated, and the pellet was re-extracted twice more following the same procedure. Supernatants were combined and reduced to dryness under a stream of nitrogen at 30 °C. The dried samples were resuspended in 150 μL of aqueous formic acid (0.1%) containing 10% acetonitrile acidified with 0.1% formic acid and centrifuged at 4 °C for 20 min at 14,000 rpm. The final supernatant was collected, filtered (0.45 μm pore-size, cellulose-acetate membrane filters, Albet, Dassel (Germany)) and 30 μL were analyzed by high-performance liquid chromatography coupled to electrospray ionisation and quadrupole time-of-flight mass spectrometry (HPLC-ESI-QToF-MS). Recoveries of the standards used to quantify metabolites ranged from 95 to 99%.

Urine samples were diluted with an equivalent volume of Milli-Q water (50%) and centrifuged at 14000 rpm (10 min, 4 °C). Supernatants were filtered (0.45 μm pore-size cellulose-acetate membrane filters) and a 5 μL aliquot was directly injected into the LC-ESI-QToF-MS equipment.

2.5. Metabolite Identification by HPLC-ESI-QToF-MS Analysis

Analyses were performed on an Agilent 1200 series LC system coupled to an Agilent 6530A Accurate-Mass Quadrupole Time-Of-Flight (Q-ToF) with ESI-Jet Stream Technology (Agilent Technologies). Compounds were separated on a reverse-phase Ascentis Express C18 (15 cm × 3 mm, 2.7 m) column (Sigma-Aldrich Química, Madrid) preceded by a Supelco 55215-U guard column at 30 °C. The test samples, either 30 μL of the plasma extract or 5 μL of diluted urine, were injected and separated using a mobile phase consisting of Milli-Q water (phase A) and acetonitrile (phase B), both containing 0.1% formic acid, at a flow rate of 0.3 mL/min. The mobile phase was initially programmed with 90% of solvent A and 10% of B. The elution program increased to 30% of solvent B in 10 min. Then, the initial conditions (10% solvent B) were recovered in 5 min and maintained for 5 min. The Q-ToF acquisition conditions were as follows: drying gas flow (nitrogen, purity > 99.9%) and temperature were 10 L/min and 325 °C, respectively; sheath gas flow and temperature were 6 L/min and 250 °C, respectively; nebulizer pressure was 25 psi; cap voltage was 3500 V and nozzle voltage was 500 V. Mass range selected was from 100 up to 970 *m/z* in negative mode and fragmentor voltage of 150 V. Data were processed in a Mass Hunter Workstation Software.

Due to the lack of standards for certain phase II metabolites, they were tentatively quantified using the calibration curves of their corresponding phenolic precursors. Thus, epicatechin was used to quantify epicatechin and 5-(3′,4′-dihydroxyphenyl)-γ-valerolactone (DHPVL) derivatives and 3,4-dihydroxyphenylpropionic acid to quantify phenylvaleric acid derivatives. The rest of microbial metabolites identified, derivatives of hydroxyphenylpropionic, hydroxyphenylacetic, hydroxybenzoic and hippuric acids, were quantified using their respective commercially available standards. Urine concentration of the excreted metabolites was normalized by the volume excreted in each studied interval. A linear response was obtained for all the standard curves (from 1 to 1,000 nM), as checked by linear regression analysis. Calibration curves were freshly prepared in a pool of both plasma and urine due to matrix effects. Limits of detection and quantification in plasma ranged from 1 to 5 nM and from 2 to 8 nM, respectively, while limits of detection and quantification in urine ranged from 2.5 to 30 nM and from 50 to 90 nM, respectively. The inter- and intra-day precision of the assay (as the coefficient of variation, ranging from 2.5 to 9.5%) were considered acceptable and allowed the quantification of phenolic compounds and their metabolites (quantified as equivalents of the respective parent molecules). The recovery ranged between 96% and 103% in plasma and between 92% and 97% in urine samples.

2.6. Nutrikinetic and Statistical Analysis

Statistical analyses were carried out using the program SPSS (version 23.0, SPSS, Inc., IBM Company, New York, NY, USA). Significant differences between metabolites excreted in urine after consumption of CC and CC-PP cocoa products were evaluated based on non-parametric Wilcoxon test ($p < 0.05$). To determine the absorption and elimination of epicatechin metabolites after consumption of the soluble cocoa products, metabolite nutrikinetics were studied using the pharmacokinetic functions of Microsoft Excel, calculating the maximum concentration (C_{max}), area under curve (AUC) and time to reach maximum concentration (T_{max}). Data are expressed as mean ± standard deviation.

3. Results

3.1. Phenolic Content of Soluble Cocoa Products

The total polyphenolic content of both cocoa products, according to the Folin–Ciocalteau assay, was 21.70 and 25.63 μg gallic acid equivalents (GAE)/g in the conventional (CC) and flavanol-rich (CC-PP) soluble cocoa products, respectively.

In addition, the phenolic composition of both cocoa products was analyzed by HPLC-DAD. The 200 mL serving prepared from 15 and 25 g of CC and CC-PP products, respectively, provided 68.2 and 235.1 μmoles (19.80 and 68.25 mg) of flavan-3-ols, respectively (Table 1; Table S1; Figure S1). As expected, the amount of flavanols was higher in the phenol-enriched CC-PP product than in the conventional cocoa (CC), contrary to the results obtained by the Folin-Ciocalteau assay. Epicatechin was the most abundant monomer in both products (43.2% and 42.1% of the total polyphenols in CC and CC-PP, respectively) compared to catechin (24.2% and 19.4% of the total polyphenols in CC and CC-PP, respectively). Regarding dimeric procyanidins, procyanidin B2 (PB2) was present in higher concentrations than catechin, with lower amounts of procyanidin B1 (PB1) (3.0–8.5% of the total polyphenols in CC and CC-PP, respectively).

Table 1. Phenolic composition of cocoa products (conventional soluble cocoa -CC- and flavanol-rich soluble cocoa -CC-PP-) determined by high-performance liquid chromatography-diode array (HPLC-DAD).

Flavanols	CC mg/g d.m. (%)	CC-PP mg/g d.m. (%)
Epicatechin	0.57 ± 0.07 (43.2%)	1.15 ± 0.06 (42.1%)
Catechin	0.32 ± 0.03 (24.2%)	0.53 ± 0.04 (19.4%)
Procyanidin B1	0.04 ± 0.02 (3.0%)	0.23 ± 0.02 (8.5%)
Procyanidin B2	0.39 ± 0.05 (29.6%)	0.82 ± 0.06 (30.0%)
Total Flavanols	1.32 ± 0.17 (100%)	2.73 ± 0.18 (100%)

Expressed in mg per gram of dry matter (d.m. 6.5% CC and 6.7% CC-PP moisture). Values in parenthesis represent the percentage of total flavanols quantified by high-performance liquid chromatography (HPLC). Mean ± standard deviation (*n* = 3).

3.2. LC-ESI-QToF-MS Identification of Flavanols and Metabolites in Plasma and Urine

Table 2 shows the retention time (RT), molecular formula, accurate mass of the quasimolecular ion $[M-H]^-$ after negative ionization, MS^2 fragments and location (U: urine or P: plasma) of the main compounds identified in plasma and urine samples by LC-ESI-QToF-MS. Characterization of the identified compounds was supported by commercial standards and/or previously published results.

Table 2. High-performance liquid chromatography coupled to electrospray ionisation and quadrupole time-of-flight mass spectrometry (HPLC-ESI-QToF-MS) identification of flavanol metabolites detected in plasma (P) and urine (U) samples obtained after the ingestion of soluble cocoa products.

Identified Compound	RT (min)	Molecular Formula	$[M-H]^-$	MS^2 Fragment	Location
Flavanols					
Epicatechin-3′-glucuronide	7.8	$C_{21}H_{22}O_{12}$	465.1038	289	P, U
Epicatechin-3′-methoxy-glucuronide	8.0	$C_{22}H_{24}O_{12}$	479.1195	303	U
Epicatechin-3′-sulfate	9.8	$C_{15}H_{14}O_9S$	369.0286	289	P, U
Epicatechin-methoxy-sulfate (isomer 1)	11.0	$C_{16}H_{16}O_9S$	383.0442	303	P, U
Epicatechin-methoxy-sulfate (isomer 2)	11.7	$C_{16}H_{16}O_9S$	383.0442	303	P, U
Epicatechin-methoxy-sulfate (isomer 3)	12.3	$C_{16}H_{16}O_9S$	383.0442	303	U
Phenyl-γ-Valerolactone (PVL) derivatives					
5-(3′,4′-Dihydroxyphenyl)-γ-valerolactone (DHPVL)	10.8	$C_{11}H_{12}O_4$	207.0663	163	P, U
5-(3′-Hydroxyphenyl)-γ-valerolactone-4′-glucuronide (HPVL-4′-glucuronide)	7.4	$C_{17}H_{20}O_{10}$	383.0984	207;163	U
5-(4′-Hydroxyphenyl)-γ-valerolactone-3′-glucuronide (HPVL-3′-glucuronide)	8.4	$C_{17}H_{20}O_{10}$	383.0984	207;163	P, U
5-(Hydroxyphenyl)-γ-valerolactone-sulfate (HPVL-sulfate)	12.0	$C_{11}H_{12}O_7S$	287.0231	207;163	P, U
5-Phenyl-γ-valerolactone-methoxy-glucuronide (PVL-methoxy-glucuronide)	8.6	$C_{18}H_{22}O_{10}$	397.1140	221	P, U
5-Phenyl-γ-valerolactone-methoxy-sulfate (PVL-methoxy-sulfate)	12.0	$C_{12}H_{14}O_7S$	301.0387	221	U
5-(3′-Hydroxyphenyl)-γ-valerolactone (HPVL)	11.6	$C_{11}H_{12}O_3$	191.0714	147	U
5-Phenyl-γ-valerolactone-3′-glucuronide (PVL-3′-glucuronide)	9.4	$C_{17}H_{20}O_9$	367.1035	191	U
5-Phenyl-γ-valerolactone-3′-sulfate (PVL-3′-sulfate)	11.7	$C_{11}H_{12}O_6S$	271.0282	191	P, U
Phenylvaleric acid derivatives					
4-Hydroxy-5-(3′,4′-dihydroxyphenyl)valeric acid (HDHPVA)	5.5	$C_{11}H_{14}O_5$	225.0768	179	U
4-Hydroxy-5-(hydroxyphenyl)valeric acid-glucuronide (HHPVA-glucuronide)	5.1	$C_{17}H_{22}O_{11}$	401.1089	225	U
4-Hydroxy-5-(hydroxyphenyl)valeric acid-sulfate (HHPVA-sulfate)	7.3	$C_{11}H_{14}O_8S$	305.0337	225	P, U

Table 2. *Cont.*

Identified Compound	RT (min)	Molecular Formula	[M-H]$^-$	MS^2 Fragment	Location
Other microbial metabolites					
3,4-Dihydroxyphenylpropionic acid	8.5	$C_9H_{10}O_4$	181.0506	137;122	P, U
3-Methoxy-4-hydroxyphenylpropionic acid	10.8	$C_{10}H_{12}O_4$	195.0663	137	P, U
3-Hydroxyphenylpropionic acid	11.1	$C_9H_{10}O_3$	165.0557	121	P, U
3,4-Dihydroxyphenylacetic acid	5.6	$C_8H_8O_4$	167.0350	123	P, U
3-Methoxy-4-hydroxyphenylacetic acid	6.5	$C_9H_{10}O_4$	181.0506	137	U
3-Hydroxyphenylacetic acid	7.4	$C_8H_8O_3$	151.0401	107	U
Ferulic acid	12.3	$C_{10}H_{10}O_4$	193.0506	134	P, U
*Iso*ferulic acid	15.4	$C_{10}H_{10}O_4$	193.0506	134	U
3,4-Dihydroxybenzoic acid	3.8	$C_7H_6O_4$	153.0193	109	P, U
4-Hydroxyhippuric acid	10.7	$C_9H_9O_4N$	194.0459	100	P, U
3-Hydroxyhippuric acid	14.1	$C_9H_9O_4N$	194.0459	150	P, U
Hydroxybenzoic acid	6.3	$C_7H_6O_3$	137.0244	93	P, U

No un-metabolized compounds originally present in both cocoa products (CC and CC-PP) were detected in plasma and urine samples. Phase II derivatives of epicatechin were detected in biological fluids after consuming both types of soluble cocoa products. In particular, glucuronidated epicatechin was assigned to the chromatographic peak that eluted at 7.8 min, with a quasimolecular ion at m/z 465.1038 and fragment ion at m/z 289 corresponding to epicatechin, present in both plasma and urine. Based on previous studies [4,16], this peak was assigned as epicatechin-3′-glucuronide.

Related to glucuronide derivatives, the presence of the epicatechin-methoxy-glucuronide derivative ([M-H]$^-$ at m/z 479.1195 and fragment ion at m/z 303 corresponding to methoxy derivative of epicatechin) was confirmed in urine. This compound showed a higher RT (8.0 min) than that described for the glucuronidated derivative (RT at 7.8 min), consistent with the lipophilicity that the methyl group provides to the molecule. This metabolite was tentatively assigned as epicatechin-3′-methoxy-glucuronide based on the study carried out by Actis-Goretta et al. [16]. Subsequently, the chromatographic peak present in both plasma and urine at 9.8 min was assigned to epicatechin-3′-sulfate, thanks to its MS spectrum ([M-H]$^-$ at m/z 369.0286 and fragment ion at m/z 289 corresponding to epicatechin). This metabolite has already been described after consumption of dark chocolate [16] and cocoa [4]. Three methoxy-sulfated isomers at 11.0, 11.7 and 12.3 min ([M-H]$^-$ at m/z 383.0442 and fragment ion at m/z 303) were tentatively identified as epicatechin-methoxy-sulfate (isomer 1, isomer 2, and isomer 3), based on previous studies [4,16]. The two metabolites that eluted earlier (at 11.0 and 11.7 min) have been detected in both plasma and urine, while the third metabolite at 12.3 min was only detected in urine.

After the intake of soluble cocoa products, derivatives of phenyl-γ-valerolactone are important compounds formed as a result of the microbial metabolism of flavanols [17]. Thus, 5-(3′,4′-dihydroxyphenyl)-γ-valerolactone (DHPVL) was identified in both plasma and urine at RT 10.8 min a quasimolecular ion at m/z 207.0663 and fragment ion at m/z 163, in agreement with the pattern fragmentation already described by other authors [17,18]. In addition, phase II derivatives of DHPVL were detected. Two glucuronidated (RT at 7.4 and 8.4 min) and one sulfated (RT 12.0 min) derivatives of DHPVL were detected in both plasma and urine, except the glucuronidated metabolite at RT 7.4 min, not found in plasma. Quasimolecular ions at m/z 383.0984 and 287.0231 were compatible with glucuronide and sulfate derivatives, respectively, in addition to common fragment ions at m/z 207 and 163 corresponding to DHPVL. Finally, chromatographic peaks at 7.4, 8.4, and 12.0 min were assigned as 5-(3′-hydroxyphenyl)-γ-valerolactone-4′-glucuronide (HPVL-4′-glucuronide), 5-(4′-hydroxyphenyl)-γ-valerolactone-3′-glucuronide (HPVL-3′-glucuronide), and 5-(4′-hydroxyphenyl)-γ-valerolactone-sulfate (HPVL-sulfate), respectively, based on Actis-Goretta et al. [16], who used similar chromatographic conditions and observed that the glucuronidated and sulfated isomers of epicatechin at the C4′ position elute earlier than those at the C3′ position, contrary to what occurs with methoxy derivatives [16]. Likewise, results reported by Ottaviani et al. [8] confirmed the identity of these metabolites.

Methoxy derivatives of DHPVL were also identified, in particular, 5-phenyl-γ-valerolactone-methoxy-glucuronide (PVL-methoxy-glucuronide) was assigned to the chromatographic peak that eluted at 8.6 min, with a quasimolecular ion at *m/z* 397.1140 and fragment ion at *m/z* 221 corresponding to 5-phenyl-γ-valerolactone-methoxy, found in plasma and urine. The chromatographic peak present in urine at 12.0 min was assigned to 5-phenyl-γ-valerolactone-methoxy-sulfate (PVL-methoxy-sulfate) thanks to its MS spectrum, ([M-H]⁻ at *m/z* 301.0387 and fragment ion at *m/z* 221).

Based on the preferred dehydroxylation route in the C4′ of ring B described for flavanols [19], the presence of 5-(3′-hydroxyphenyl)-γ-valerolactone (HPVL) and its phase II derivatives was postulated. The search for the [M-H]⁻ ion at *m/z* 191.0714 yielded a peak at RT 11.6 min supported by the fragment ion at *m/z* 147, which allowed the identification of HPVL. Likewise, the peaks at RT 9.4 and 11.7 min showed a consistent fragmentation pattern with a glucuronidated derivative ([M-H]⁻ at *m/z* 367.1035 and fragment at *m/z* 191) and a sulfated derivative ([M-H]⁻ at *m/z* 271.0282 and fragment at *m/z* 191), respectively. These compounds have been tentatively identified as 5-(phenyl)-γ-valerolactone-3′-glucuronide (PVL-3′-glucuronide) and 5-(phenyl)-γ-valerolactone-3′-sulfate (PVL-3′-sulfate), respectively.

Phenyl-γ-valerolactones evolve to phenylvaleric acid derivatives. Thus, 4-hydroxy-5-(3′,4′-dihydroxyphenyl)valeric acid (HDHPVA) was assigned to the chromatographic peak at RT 5.5 min detected in urine, due to its quasimolecular ion [M-H]⁻ at *m/z* 225.0768 and MS/MS fragmentation pattern coinciding with that described by Stoupi et al. [20], who identified this compound after in vitro fermentation of epicatechin and PB2 (fragment ion at *m/z* 179). Likewise, its glucuronidated derivative ([M-H]⁻ at *m/z* 401.1089 and fragment ion at *m/z* 225, corresponding to its precursor HDHPVA) and sulfated derivative ([M-H]⁻ at *m/z* 305.0337 and fragment ion at *m/z* 225) were identified, eluting at 5.1 and 7.3 min, respectively. While the sulfated metabolite was detected in both plasma and urine, the glucuronidated derivative was only detected in urine.

Finally, derivatives of hydroxyphenylpropionic, hydroxyphenylacetic, hydroxybenzoic, and hydroxyhippuric acids were detected in plasma and urine samples (Table 2).

3.3. Quantification of Plasma Metabolites and Nutrikinetic Parameters

Out of the 30 metabolites identified after consumption of the two soluble cocoa products, 8 and 10 were detected in plasma after CC and CC-PP consumption, respectively, although only 7 and 10 metabolites, respectively, showed levels above the limit of quantification. The kinetics of plasma appearance and clearance of these metabolites up to 8 h post-intake are represented in Figure 1. Nutrikinetic parameters are summarized in Table 3.

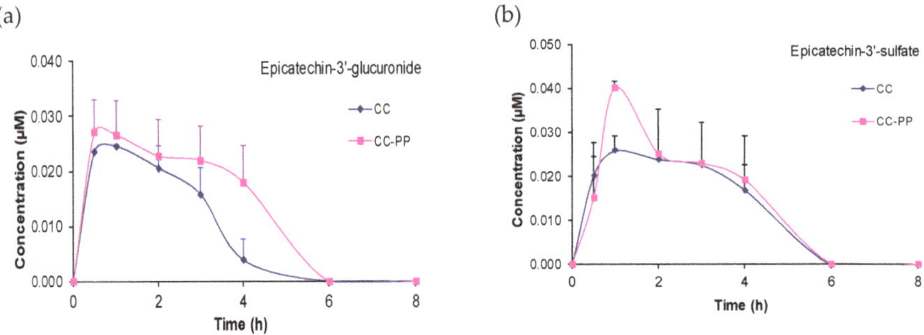

Figure 1. *Cont.*

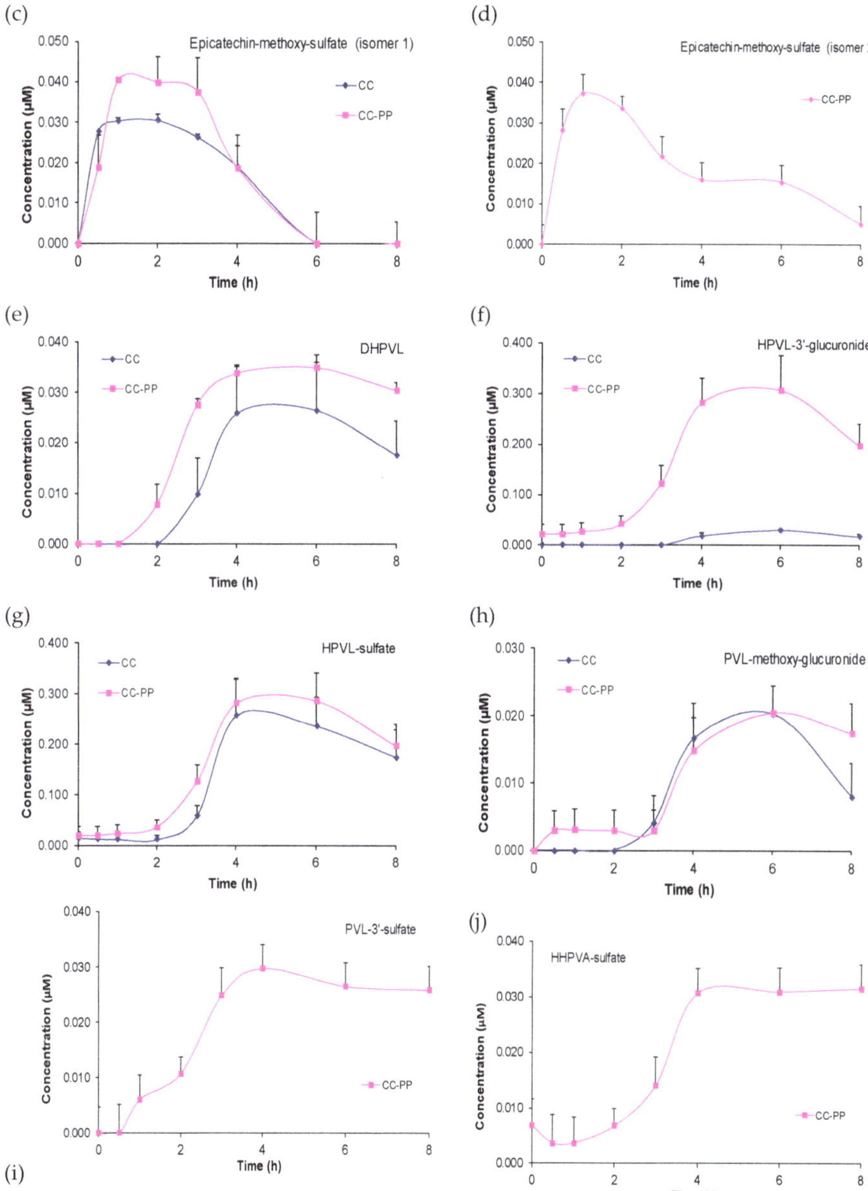

Figure 1. Plasma concentrations of the identified metabolites after consuming a conventional soluble cocoa product (CC) and a flavanol-rich soluble cocoa product (CC-PP) containing 19.80 and 68.25 mg of flavanols, respectively. (**a**) Epicatechin-3′-glucuronide; (**b**) epicatechin-3′-sulfate; (**c**) epicatechin- methoxy-sulfate (isomer 1); (**d**) epicatechin-methoxy-sulfate (isomer 2); (**e**) 5-(3′,4′-dihydroxyphenyl)- γ-valerolactone (DHPVL); (**f**) 5-(4′-hydroxyphenyl)-γ-valerolactone-3′-glucuronide (HPVL-3′-glucuronide); (**g**) 5-(hydroxyphenyl)-γ-valerolactone-sulfate (HPVL-sulfate); (**h**) 5-phenyl-γ- valerolactone-methoxy-glucuronide (PVL-methoxy-glucuronide); (**i**) 5-phenyl-γ-valerolactone-3′- sulfate (PVL-3′-sulfate) and (**j**) 4-hydroxy-5-(hydroxyphenyl)valeric acid-sulfate (HHPVA-sulfate). Results represent concentration (μM) as mean ± standard deviation (*n* = 13). The lower part of the error bars is not displayed for the sake of clarity.

Un-metabolized flavanols were not detected in plasma after the consumption of both products (CC and CC-PP). Phase II derivatives of flavanols, epicatechin-3′-glucuronide, epicatechin-3′-sulfate and epicatechin-methoxy-sulfate (isomer 1) after CC and CC-PP intake, and epicatechin-methoxy-sulfate (isomer 2) after CC-PP intake, were detected in plasma. Concentrations of these phase II derivatives showed a rapid increase between 1 and 1.5 h after the consumption soluble cocoa products, whilst their clearance was slow, maintaining or even showing a second maxima between 3 and 6 h, with subsequent clearance at 8 h post-intake (Figure 1).

Table 3. Nutrikinetic parameters of metabolites detected in human plasma after consuming a conventional soluble cocoa product (CC) and a flavanol-rich soluble cocoa product (CC-PP) containing 19.80 and 68.25 mg of flavanols, respectively. Values represent mean $\pm$ standard deviation (n = 13).

Metabolite	CC			CC-PP			p Value (CC vs CC-PP)		
	C_{max} (μM)	T_{max} (h)	AUC (μM min^{-1})	C_{max} (μM)	T_{max} (h)	AUC (μM min^{-1})	C_{max}	T_{max}	AUC
Intestinal absorption									
Epicatechin-3′-glucuronide	0.025 ± 0.001	1.5 ± 0.5	0.072 ± 0.028	0.037 ± 0.001	1.4 ± 0.8	0.110 ± 0.066	<0.05	N.S.	N.S.
Epicatechin-3′-sulfate	0.026 ± 0.003	1.0 ± 0.1	0.101 ± 0.023	0.042 ± 0.004	1.2 ± 0.5	0.109 ± 0.076	N.S.	N.S.	N.S.
Epicatechin-methoxy-sulfate (isomer 1)	0.031 ± 0.003	1.3 ± 0.6	0.122 ± 0.025	0.041 ± 0.002	1.1 ± 0.6	0.173 ± 0.107	<0.05	N.S.	N.S.
Epicatechin-methoxy-sulfate (isomer 2)	N.D.			0.041 ± 0.004	1.3 ± 0.5	0.145 ± 0.107	<0.05	<0.05	<0.05
Microbial metabolites									
DHPVL	0.035 ± 0.008	6.0 ± 1.6	0.124 ± 0.073	0.037 ± 0.007	5.5 ± 1.4	0.192 ± 0.029	N.S.	N.S.	N.S.
HPVL-3′-glucuronide	0.031 ± 0.004	5.6 ± 0.8	0.106 ± 0.021	0.357 ± 0.200	5.0 ± 1.2	1.433 ± 0.727	<0.05	N.S.	<0.05
HPVL-sulfate	0.336 ± 0.240	5.3 ± 2.4	1.150 ± 0.742	0.332 ± 0.161	4.9 ± 1.2	1.384 ± 0.648	N.S.	N.S.	N.S.
PVL-methoxy-glucuronide	0.022 ± 0.001	5.3 ± 1.0	0.078 ± 0.032	0.027 ± 0.003	5.6 ± 1.3	0.091 ± 0.059	<0.05	N.S.	N.S.
PVL-3′-sulfate [b]	Traces	(4–6) [a]		0.039 ± 0.014	5.2 ± 2.2	0.161 ± 0.105	<0.05	<0.05	<0.05
HHPVA-sulfate	N.D.			0.037 ± 0.008	6.3 ± 1.8	0.167 ± 0.068	<0.05	<0.05	<0.05

AUC: area under the curve; DHPVL: 5-(3′,4′-dihydroxyphenyl)-γ-valerolactone; HPVL: 5-(3′-hydroxyphenyl)-γ-valerolactone; PVL: 5-phenyl-γ-valerolactone; HDHPVA: 4-hydroxy-5-(3′,4′-dihydroxyphenyl)valeric acid; N.D.: Not detected; N.S.: Non-significant differences; p values were assessed using the general linear model of variance for repeated measures. [a] Range where the metabolite showed the highest value. [b] No pharmacokinetic parameters of these metabolites were determined because they were present at trace levels.

5-(3′,4′-dihydroxyphenyl)-γ-valerolactone (DHPVL) and its phase II derivatives; HPVL-3′-glucuronide, HPVL-3′-sulfate, PVL-methoxy-glucuronide, and PVL-3′-sulfate, formed the main group of metabolites detected in plasma (Figure 1). The plasmatic profile of these metabolites showed maxima concentrations between 4.9 and 6.3 h (T_{max}) post-intake, except PVL-3′-sulfate after CC intake which appeared at traces level (Table 3). HPVL-3′-sulfate was the predominant metabolite after CC and CC-PP intake, showing C_{max} value of 1.150 and 1.384 μM, respectively, followed by HPVL-3′-glucuronide with C_{max} ranging from 0.106 to 1.433 μM, respectively (Table 3).

A sulfated derivative of phenylvaleric acid, HHPVA-sulfate, was also detected in plasma but only after CC-PP intake with similar kinetic behavior than derivatives of phenyl-γ-valerolactones, with maxima concentration at 6.3 h post-intake (Figure 1).

In general, metabolites' C_{max} after CC consumption were significantly lower than after CC-PP intake (p < 0.05), consistent with the higher content of flavanols ingested with the polyphenol-rich cocoa.

3.4. Quantification of Urinary Metabolites

Up to 29 and 30 metabolites were quantified in 24 h urine samples after CC and CC-PP consumption, respectively (Table 4; Table 5). Neither epicatechin nor PB2 dimer was detected in urine in any of the interventions.

Phase II derivatives of epicatechin were preferentially excreted in the first two sampling intervals (0–4 and 4–8 h) after the ingestion of both soluble products. In the range of 0–4 h, the excretion of these metabolites was 58.3% and 62.0% (CC and CC-PP, respectively) of the total phase II derivatives of epicatechin excreted in urine; these percentages decreased to 36.0% and 34.8% in the second interval (4–8 h). Sulfated and/or methoxysulfated derivatives contributed 93% in both interventions with CC and CC-PP, showing that sulfation was the preferential route of biotransformation according to what was previously observed in the plasma. This group of metabolites represented 35.9% and 31.0% of the total urinary metabolites after CC and CC-PP intake, respectively.

DHPVL and its phase II derivatives, HPVL-4′-glucuronide, HPVL-3′-glucuronide and HPVL-sulfate, along with PVL-methoxy-glucuronide, PVL-methoxy-sulfate, PVL-3′-glucuronide and PVL-3′-sulfate conformed the most important group of metabolites quantified in urine after the ingestion of both soluble cocoa products. This group was largely excreted between 4 and 8 h post-intake and represented the 54.4% and 56.1% of the total urinary metabolites after CC and CC-PP intake, respectively. Excretion of HPVL-sulfate added up to 9.9 and 31.32 μmoles in 24 h followed by PVL-3′-sulfate (2.0 μmoles and 7.8 μmoles) after CC and CC-PP intake, respectively, evidencing that the sulfation was the preferential biotransformation pathway.

Regarding phenylvaleric acid derivatives, HHPVA-sulfate and HHPVA-glucuronide were also excreted preferentially between 4 and 8 h post-intake, accounting for 9.7% and 12.9% of the total urinary metabolites after CC and CC-PP intake, respectively.

Finally, derivatives of hydroxyphenylpropionic, hydroxyphenylacetic, and hydrophenylbenzoic acids, along with hydroxyhippuric acid, were present in basal urine before cocoa product intake but their levels increased after CC and CC-PP consumption (0–24 h), peaking between 4 and 8 h compared to baseline values (Table 4). However, there was little difference in the total content excreted for this group of metabolites (13.1 μmoles and 15.8 μmoles after CC and CC-PP consumption, respectively) despite the large difference of polyphenol intake between both soluble cocoa products (68.2 μmoles of CC and 235 μmoles of CC-PP). For this reason, this group of metabolites was not taken into account to determine flavanols recovery.

Attending to these results, it may be summarized that the total amount of metabolites in 24 h urine after the intake of a single serving of CC and CC-PP beverages added up to 24.1 μmol and 81.3 μmoles (Tables 4 and 5), respectively, corresponding to 35.3% and 34.6% of the 68.2 and 235.1 μmoles of polyphenols consumed, respectively.

Table 4. Metabolites excreted in urine (from 0 to 24 h) by healthy volunteers after consumption of the conventional soluble cocoa product (CC) containing 19.80 mg of flavanols.

	Metabolite	Basal (µmol)	0–4 h (µmol)	4–8 h (µmol)	8–12 h (µmol)	12–24 h (µmol)	Total (µmol)
Intestinal Absorption	Epicatechin-3′-glucuronide	N.D.	0.355 ± 0.071	0.267 ± 0.059	<L.Q.	<L.Q.	0.622 ± 0.130 *
	Epicatechin-3′-methoxy-glucuronide	N.D.	<L.Q.	<L.Q.	N.D.	N.D.	<L.Q. *
	Epicatechin-3′-sulfate	N.D.	1.323 ± 0.236	1.018 ± 0.178	0.076 ± 0.025	<L.Q.	2.417 ± 0.439 *
	Epicatechin-methoxy-sulfate (isomer 1)	N.D.	2.717 ± 0.457	1.300 ± 0.197	0.154 ± 0.031	0.257 ± 0.114	4.428 ± 0.800 *
	Epicatechin-methoxy-sulfate (isomer 2)	N.D.	0.181 ± 0.032	0.247 ± 0.030	<L.Q.	<L.Q.	0.428 ± 0.062 *
	Epicatechin-methoxy-sulfate (isomer 3)	N.D.	0.473 ± 0.089	0.285 ± 0.036	<L.Q.	N.D.	0.758 ± 0.125 *
	Total—intestinal absorption	N.D.	5.049 ± 0.885	3.117 ± 0.500	0.230 ± 0.056	0.257 ± 0.114	8.653 ± 1.556 *
Colonic Absorption	DHPVL	<L.Q.	<L.Q.	0.146 ± 0.032	<L.Q.	<L.Q.	0.146 ± 0.032 *
	HPVL	N.D.	N.D.	<L.Q.	<L.Q.	<L.Q.	<L.Q. *
	HDHPVA	N.D.	N.D.	<L.Q.	<L.Q.	<L.Q.	<L.Q.
	HPVL-4′-glucuronide	<L.Q.	<L.Q.	0.111 ± 0.049	<L.Q.	<L.Q.	0.111 ± 0.049 *
	HPVL-3′-glucuronide	<L.Q.	<L.Q.	0.497 ± 0.204	0.103 ± 0.028	<L.Q.	0.600 ± 0.232 *
	HPVL-sulfate	0.143 ± 0.064	0.456 ± 0.193	5.638 ± 1.922	1.945 ± 0.647	1.744 ± 0.374	9.926 ± 3.200 *
	PVL-methoxy-glucuronide	<L.Q.	<L.Q.	0.111 ± 0.039	<L.Q.	<L.Q.	0.111 ± 0.039 *
	PVL-methoxy-sulfate	<L.Q.	N.D.	0.085 ± 0.031	<L.Q.	<L.Q.	0.085 ± 0.031 *
	PVL-3′glucuronide	N.D.	<L.Q.	0.233 ± 0.155	0.126 ± 0.078	0.079 ± 0.022	0.438 ± 0.255
	PVL-3′-sulfate	<L.Q.	<L.Q.	0.860 ± 0.385	0.456 ± 0.284	0.375 ± 0.133	1.691 ± 0.802 *
	HHPVA-glucuronide	N.D.	N.D.	N.D.	N.D.	N.D.	N.D. *
	HHPVA-sulfate	0.089 ± 0.055	0.076 ± 0.055	1.112 ± 0.397	0.565 ± 0.171	0.503 ± 0.152	2.345 ± 0.830 *
	Total—colonic absorption	0.232 ± 0.119	0.532 ± 0.248	8.793 ± 3.214	3.195 ± 1.208	2.701 ± 0.681	15.453 ± 5.470 *
Other microbial metabolites	3,4-Dihydroxyphenylpropionic acid	0.117 ± 0.024	0.217 ± 0.051	0.243 ± 0.039	0.074 ± 0.017	0.215 ± 0.016	0.866 ± 0.148
	3-Methoxy-4-hydroxyphenylpropionic acid	0.104 ± 0.027	0.144 ± 0.044	<L.Q.	<L.Q.	<L.Q.	0.248 ± 0.071
	Hydroxyphenylpropionic acid	0.118 ± 0.023	0.226 ± 0.046	0.259 ± 0.043	0.077 ± 0.018	0.203 ± 0.025	0.883 ± 0.154
	3,4-Dihydroxyphenylacetic acid	0.094 ± 0.032	0.648 ± 0.505	0.556 ± 0.379	0.094 ± 0.033	0.133 ± 0.020	1.525 ± 0.974
	3-Methoxy-4-hydroxyphenylacetic acid	0.103 ± 0.014	0.099 ± 0.054	0.093 ± 0.028	<L.Q.	0.103 ± 0.022	0.398 ± 0.118
	Hydroxyphenylacetic acid	0.369 ± 0.077	0.408 ± 0.084	0.605 ± 0.092	0.225 ± 0.052	0.600 ± 0.100	2.208 ± 0.328
	Ferulic acid	<L.Q.	<L.Q.	0.094 ± 0.009	0.099 ± 0.059	N.D.	0.193 ± 0.068
	Isoferulic acid	0.170 ± 0.137	0.086 ± 0.028	0.096 ± 0.040	0.088 ± 0.069	0.082 ± 0.016	0.522 ± 0.290
	Protocatechuic acid	0.161 ± 0.071	0.227 ± 0.142	0.373 ± 0.237	0.094 ± 0.047	0.104 ± 0.044	0.959 ± 0.217 *
	Hydroxyhippuric acid	0.821 ± 0.251	0.864 ± 0.154	1.171 ± 0.387	0.242 ± 0.067	0.995 ± 0.316	4.093 ± 1.175
	Hydroxyhippuric acid	0.075 ± 0.031	0.085 ± 0.043	0.082 ± 0.055	0.075 ± 0.031	0.077 ± 0.046	0.394 ± 0.206
	Hidroxybenzoic acid	0.188 ± 0.041	0.145 ± 0.027	0.200 ± 0.051	0.070 ± 0.018	0.195 ± 0.025	0.798 ± 0.162
	Total other microbial metabolites	2.320 ± 0.728	3.149 ± 1.178	3.772 ± 1.360	1.138 ± 0.411	2.707 ± 0.630	13.086 ± 3.911
	INTESTINAL + COLONIC METABOLITES	0.232 ± 0.119	5.581 ± 1.133	11.910 ± 3.714	3.425 ± 1.264	2.958 ± 0.795	24.106 ± 7.026

HDHPVA: 4-hydroxy-5-(3′,4′-dihydroxyphenyl)valeric acid; DHPVL: 5-(3′,4′-dihydroxyphenyl)-γ-valerolactone; HPVL: 5-(3′-hydroxyphenyl)-γ-valerolactone; PVL: 5-phenyl-γ-valerolactone. * Significant differences respect to CC-PP intervention was observed based on non-parametric Wilcoxon test at $p < 0.05$. Values represent mean ± standard deviation ($n = 13$). N.D.: not detected; <L.Q. lower than quantification limit.

Table 5. Metabolites excreted in urine (from 0 to 24 h) by healthy volunteers after consumption of the flavanol-rich soluble cocoa product (CC-PP) containing 68.25 mg of flavanols.

	Metabolite	Basal (μmol)	0–4 h (μmol)	4–8 h (μmol)	8–12 h (μmol)	12–24 h (μmol)	Total (μmol)
Intestinal Absorption	Epicatechin-3′-glucuronide	N.D.	0.958 ± 0.272	0.695 ± 0.190	<L.Q.	<L.Q.	1.653 ± 0.462 *
	Epicatechin-3′-methoxy-glucuronide	N.D.	0.121 ± 0.036	<L.Q.	<L.Q.	N.D.	0.121 ± 0.036 *
	Epicatechin-3′-sulfate	N.D.	3.627 ± 0.797	3.572 ± 0.662	0.158 ± 0.038	<L.Q.	7.357 ± 1.523 *
	Epicatechin-methoxy-sulfate (isomer 1)	N.D.	8.642 ± 1.434	3.277 ± 0.644	0.330 ± 0.095	0.232 ± 0.070	12.481 ± 2.243 *
	Epicatechin-methoxy-sulfate (isomer 2)	N.D.	0.566 ± 0.152	0.557 ± 0.120	0.077 ± 0.021	<L.Q.	1.200 ± 0.293 *
	Epicatechin-methoxy-sulfate (isomer 3)	N.D.	1.702 ± 0.243	0.652 ± 0.148	<L.Q.	<L.Q.	2.354 ± 0.402 *
	Total—intestinal absorption	N.D.	15.616 ± 2.934	8.753 ±1.764	0.565 ± 0.154	0.232 ± 0.070	25.166 ± 4.959 *
Colonic Absorption	DHPVL	<L.Q.	<L.Q.	0.628 ± 0.087	0.174 ± 0.029	0.282 ± 0.082	1.084 ± 0.199 *
	HPVL	N.D.	<L.Q.	0.104 ± 0.041	<L.Q.	<L.Q.	0.104 ± 0.041 *
	HDHPVA	N.D.	<L.Q.	<L.Q.	<L.Q.	<L.Q.	<L.Q.
	HPVL-4′-glucuronide	<L.Q.	<L.Q.	0.330 ± 0.111	0.550 ± 0.015	0.081 ± 0.025	0.961 ± 0.151 *
	HPVL-3′-glucuronide	<L.Q.	<L.Q.	1.802 ± 0.567	0.274 ± 0.060	0.282 ± 0.130	2.358 ± 0.766 *
	HPVL-sulfate	0.332 ± 0.164	1.965 ± 0.636	15.979 ± 2.963	5.386 ± 0.934	7.656 ± 2.125	31.318 ± 6.822 *
	PVL-methoxy-glucuronide	<L.Q.	<L.Q.	0.340 ± 0.074	0.084 ± 0.014	0.131 ± 0.046	0.555 ± 0.134 *
	PVL-methoxy-sulfate	N.D.	<L.Q.	0.177 ± 0.036	<L.Q.	<L.Q.	0.177 ± 0.047 *
	PVL-3′glucuronide	<L.Q.	0.108 ± 0.045	0.521 ± 0.191	0.179 ± 0.063	0.508 ± 0.349	1.316 ± 0.648
	PVL-3′-sulfate	<L.Q.	0.651 ± 0.389	4.188 ± 1.535	1.223 ± 0.373	1.690 ± 0.572	7.752 ± 2.868 *
	HHPVA-glucuronide	N.D.	<L.Q.	0.714 ± 0.088	0.087 ± 0.019	0.251 ± 0.075	1.052 ± 0.182 *
	HHPVA-sulfate	0.089 ± 0.032	0.584 ± 0.223	5.018 ± 1.090	1.798 ± 0.376	1.987 ± 0.566	9.476 ± 2.287 *
	Total—colonic absorption	0.421 ± 0.196	3.308 ± 1.293	29.801 ± 6.783	9.755 ± 1.883	12.868 ± 3.970	56.153 ± 14.145 *
Other microbial metabolites	3,4-Dihydroxyphenylpropionic acid	0.213 ± 0.031	0.256 ± 0.027	0.238 ± 0.028	0.109 ± 0.016	0.301 ± 0.029	1.117 ± 0.131
	3-Methoxy-4-hydroxyphenylpropionic acid	0.077 ± 0.032	0.175 ± 0.054	<L.Q.	<L.Q.	0.199 ± 0.052	0.461 ± 0.138
	Hydroxyphenylpropionic acid	0.074 ± 0.013	0.350 ± 0.038	0.356 ± 0.076	0.140 ± 0.041	0.284 ± 0.075	1.204 ± 0.243
	3,4-Dihydroxyphenylacetic acid	0.109 ± 0.015	0.751 ± 0.057	0.556 ± 0.096	0.198 ± 0.037	0.539 ± 0.196	2.153 ± 0.401
	3-Methoxy-4-hydroxyphenylacetic acid	0.198 ± 0.071	0.188 ± 0.026	0.124 ± 0.050	<L.Q.	0.138 ± 0.016	0.648 ± 0.178
	Hydroxyphenylacetic acid	0.585 ± 0.070	0.631 ± 0.140	0.678 ± 0.231	0.434 ± 0.070	0.534 ± 0.143	2.862 ± 0.654
	Ferulic acid	<L.Q.	0.137 ± 0.092	0.150 ± 0.094	0.151 ± 0.094	0.312 ± 0.290	0.750 ± 0.570
	*Iso*ferulic acid	<L.Q.	0.164 ± 0.044	0.160 ± 0.066	0.152 ± 0.094	0.210 ± 0.152	0.686 ± 0.356
	Protocatechuic acid	0.088 ± 0.014	0.105 ± 0.016	0.113 ± 0.020	0.095 ± 0.040	0.127 ± 0.015	0.528 ± 0.105 *
	Hydroxyhippuric acid	0.871 ± 0.264	0.789 ± 0.100	0.579 ± 0.096	0.328 ± 0.063	0.987 ± 0.193	3.554 ± 0.716
	Hydroxyhippuric acid	0.086 ± 0.013	0.090 ± 0.009	0.082 ± 0.009	0.075 ± 0.063	0.129 ± 0.017	0.462 ± 0.111
	Hidroxybenzoic acid	0.212 ± 0.037	0.396 ± 0.060	0.298 ± 0.036	0.172 ± 0.063	0.302 ± 0.169	1.380 ± 0.365
	Total other microbial metabolites	2.513 ± 0.560	4.032 ± 0.663	3.334 ± 0.802	1.854 ± 0.581	4.062 ± 1.347	15.805 ± 3.968
	INTESTINAL + COLONIC METABOLITES	0.421 ± 0.196	18.924 ± 4.227	38.554 ± 8.547	10.320 ± 2.037	13.100 ± 4.040	81.319 ± 19.104

HDHPVA: 4-hydroxy-5-(3′,4′-dihydroxyphenyl)valeric acid; DHPVL: 5-(3′,4′-dihydroxyphenyl)-γ-valerolactone; HPVL: 5-(3′-hydroxyphenyl)-γ-valerolactone; PVL: 5-phenyl-γ-valerolactone. * Significant differences respect to CC-PP intervention was observed based on non-parametric Wilcoxon test at $p < 0.05$. Values represent mean ± standard deviation ($n = 13$). N.D.: not detected; <L.Q. lower than quantification limit.

4. Discussion

Relevant studies on the absorption and metabolism of flavanols in different cocoa products have been recently published [3,13,14,21]. However, the particularity of the present work lies in the realistic dose administered to the participants, following the recommendations of the cocoa product manufacturer. Furthermore, the present study compares the bioavailability of flavanols in two cocoa products: one conventional, naturally rich in cocoa (CC) and the other enriched in flavanols (CC-PP). This enabled us to check if the higher phenol doses administered with CC-PP would alter the kinetics of appearance, clearance and biotransformation. Furthermore, special attention has been paid to identify novel microbial metabolites, considering that the identification could contribute to shed light on the biotransformation pathway of flavanols in cocoa in general terms, and soluble cocoa products in particular, therefore allowing to further understand the bioactivity of these products.

The results show that flavanols present in soluble cocoa products are partially absorbed and extensively metabolized, so that most of the metabolites are produced by the intestinal microbiota. Thus, phenyl-γ-valerolactones and phenylvaleric acid derivatives, mainly as phase II conjugated metabolites, formed after absorption in the colon, were the predominant metabolites in plasma and urine, underlying the importance of the microbiota in the metabolism of flavanols (Figure 2).

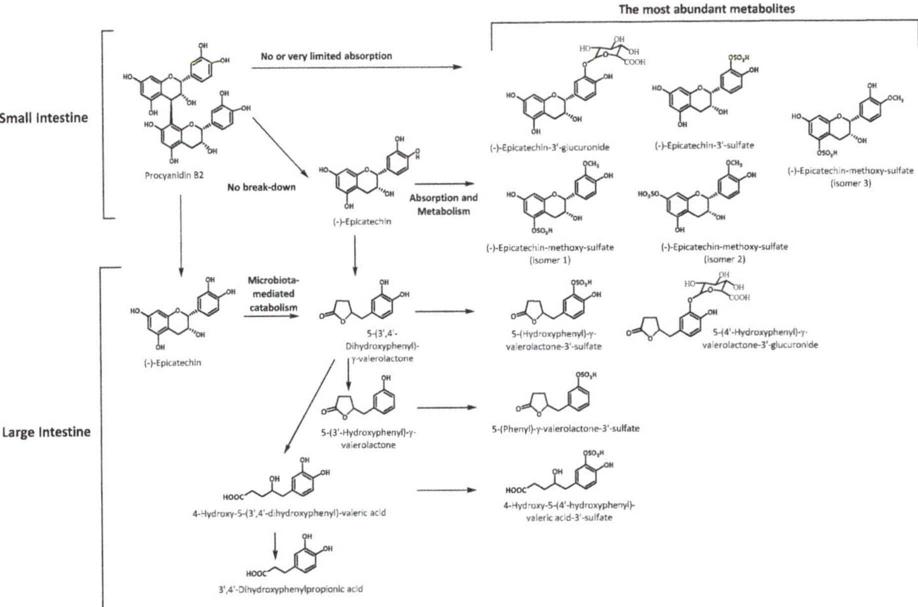

Figure 2. Biotransformation pathways in humans of the main flavanols contained in a conventional soluble cocoa product (CC) and a flavanol-rich soluble cocoa product (CC-PP).

Neither un-metabolized epicatechin and catechin nor procyanidins B1 and B2 were detected in the collected biological fluids in agreement with previous studies [4,11,16,22–25]. Only Schroeter et al. [26] described the presence of epicatechin and catechin in plasma after the ingestion of powdered cocoa drinks with a high flavanol content (917 mg). Flavanols followed two different pathways: a minor part of metabolites was subsequently metabolized by phase II enzymes into sulfated, glucuronidated and methoxy derivatives in the intestinal epithelium, after entering the bloodstream, whereas most flavanols reached the colon and were transformed by microbial enzymes prior to absorption and conjugation into phase II metabolites.

Regarding the compounds identified in the systemic circulation, the present results are in agreement with those described by Actis-Goretta et al. [16], Ottaviani et al. [4], Borges et al. [21], and Montagnana et al. [11], who identified glucuronidated, sulfated and methoxy-sulfated derivatives of epicatechin in plasma. These metabolites were quantified in nM concentrations with C_{max} values from 25 to 31 nM and 37 to 42 nM after CC and CC-PP intake, respectively. As can be seen, there were no remarkable differences in the plasma concentrations of the major phase II epicatechin derivatives, in spite of the different intake of polyphenols with both cocoa products (CC, providing 19.80 mg, and CC-PP, with a total intake of 68.2 mg). In contrast, plasma concentrations as high as ~300 nM were reported by Actis-Goretta et al. [16] after the ingestion of 154 mg of flavanols in 100 g of dark chocolate, or up to 590 nM after the intake of 1,100 mg of flavanols in a soluble cocoa product [4]. Mullen et al. [25] demonstrated the interference of milk proteins in the absorption of flavanols, which might explain the lower plasmatic level of metabolites described in this study compared to Actis-Goretta et al. [16] or Ottaviani et al. [4], among others. However, it is noteworthy that these high intakes correspond to non-realistic doses that are difficult to maintain on a daily basis within a balanced diet. The time that these metabolites took to reach the maximum concentration in plasma (T_{max}), ranged from 1.0 h and 1.5 h with both soluble cocoa products, pointing to the absorption in the proximal gastrointestinal tract, in agreement with other studies [4,8,24,25,27,28].

On reaching the colon, flavanols undergo microbiota-mediated conversion yielding the 5C-ring fission metabolites, 5-(hydroxypheynyl)-γ-valerolactones and 5-(hydroxyphenyl)-γ-hydroxyvaleric acids that appear in plasma preferably as phase II metabolites, being HPVL-sulfate and HPVL-3′-glucuronide the most abundant metabolites. C_{max} of the total amount of these metabolites was 424 nM and 829 nM after CC and CC-PP intake, respectively, and their T_{max} ranged from 5.0 to 6.0 h, distinctive of colon-derived products. These results are in line with those obtained by Ottaviani et al. [8], who evaluated the absorption, metabolism, distribution, and excretion of radiolabeled and stereochemically pure [2-^{14}C](-)-epicatechin ([^{14}C]epicatechin) in 8 male volunteers that consumed a drink containing 207 μmol (60 mg) of flavanols and reported in plasma a total concentration of 588 nM of this group of metabolites. Recently, Montagnana et al. [11] also has detected phenyl-γ-valerolactone metabolites (glucuronidated, sulfated and methoxy derivatives) in plasma 4 h after the ingestion of 50 g of 90% cocoa chocolate (7.5 gallic acid equivalents of polyphenols).

There were substantially higher levels of metabolites in urine than plasma. The main urinary metabolites were epicatechin-methoxy-sulfate (isomer 1) and epicatechin-3′-sulfate along with epicatechin-methoxy-sulfate (isomer 3), epicatechin-3′-glucuronide and epicatechin-methoxy-sulfate (isomer 2) followed by epicatechin-3′-methoxy-glucuronide. Sulfated and/or methylsulfated derivatives of epicatechin represented 92% of this group of metabolites after consumption of both soluble cocoa products, confirming sulfation as the preferential biotransformation pathway. These results are in line with previous studies [4,8,16,25] although the proportion of metabolites was dependent on the amount of epicatechin ingested, so that a greater proportion of sulfated derivatives in all their forms (sulfated and methoxy-sulfated) was observed at lower doses of epicatechin, as in Mullen et al. [25], after the intake of 13 mg of flavanols, whereas a higher proportion of glucuronidated derivatives was observed at higher doses of epicatechin as in Ottaviani et al. [4], who administered 1100 mg of flavanols. Later Ottaviani et al. [8] detected a balanced amount of glucuronidated and sulfated derivatives of epicatechin after the intake of 60 mg of epicatechin. Phase II derivatives of epicatechin were excreted mainly in the initial 0–4 h urine collection period, followed by the interval 4–8 h and then rapidly decreased in the following intervals, in keeping with the plasma pharmacokinetic profiles (Figure 1). Phase II derivatives of epicatechin accounted for 35.9% and 31.0% of total urinary metabolites after CC and CC-PP intake, respectively, which suggests a limited bioavailability at the intestinal level.

Regarding microbial metabolites derived from epicatechin, phase II derivatives of phenyl-γ-valerolactones and phenylvaleric acid were the most important group of metabolites quantified in urine after the ingestion of both soluble cocoa products (64.1% and 69% of the total urinary metabolites) and were largely excreted between 4–8 h post-intake. These compounds have been less described in the literature than phase II derivatives of epicatechin. Urpi-Sarda et al. [17] identified in 24 h urine mainly DHPVL and its phase II derivatives (glucuronides, sulfates, methoxy-glucuronides and methoxy-sulfates), with some being also found in plasma after the intake of 46.4 mg of flavanols in cocoa powder. Llorach et al. [29] revealed the presence of 4-hydroxy-5-(3′,4′-dihydroxyphenyl) valeric acid in 24 h urine along with some phase II derivatives of DHPVL after the consumption of a single dose of cocoa powder. Vitaglione et al. [9] also detected DHPVL in urine after the intake of different products prepared with cocoa cream. Recently, Ottaviani et al. [8] completed the characterization of phenyl-γ-valerolactones and phenylvaleric acid derivatives after consumption of 60 mg of [^{14}C]epicatechin identifying sulfated and glucuronidated forms of DHPVL, HPVL and HDHPVA.

It is worth noting that excretion of HPVL-sulfate added up to 9.9 and 31.32 μmoles in 24 h after CC and CC-PP intake, respectively, evidencing that sulfation was the preferential biotransformation pathway followed by methylation and glucuronidation, in agreement with the profile described in plasma. Consequently, HPVL-sulfate as the main metabolite in both plasma and urine, could be a very sensitive biomarker of flavanol intake, considering that the excreted amount of this metabolite reached 41.1% and 38.5% of the total metabolites excreted after the consumption of CC and CC-PP, respectively.

Finally, derivatives of hydroxyphenylpropionic, hydroxyphenylacetic, hydroxybenzoic, hydroxycinnamic, and hydroxyhippuric acids were also characterized. These compounds are not exclusively formed during the biotransformation of flavanols, since most were present in basal urine before soluble cocoa product intake (Tables 4 and 5) and, therefore, they were not taken into account to determine flavanol recovery.

The total urinary excretion of metabolites derived from flavanols in CC and CC-PP presented a recovery of 35.3% and 34.6% of the phenols ingested, respectively, pointing to a moderate bioavailability of flavanols. It is worth noting that, although urinary recovery of metabolites, both from intestinal and colonic absorption, showed a dose-dependent increase after CC and CC-PP intake, the total amount of excreted metabolites was similar in both cases, around 35% of the ingested polyphenols, pointed out the limited bioavailability of cocoa flavanols. There is a marginal difference between the data reported in the literature regarding the recovery of flavanols. For instance, Baba et al. [27] showed an epicatechin excretion of 29.8% and 25.3% after the ingestion of chocolate and cocoa, respectively, which provided 220 mg of flavanols. Ito et al. [30] reported an excretion of 1.9% after the ingestion of 289 mg of flavanols in a soluble cocoa administered with water. Mullen et al. [25] described excretions of 18% and 10% after the ingestion of a beverage containing 13 mg of cocoa flavanols dissolved in water or milk, respectively. Afterwards, Actis-Goretta et al. [16] reported excretion values of 21.7% following the ingestion of 100 g of dark chocolate that provided 154 mg of flavanols. Recently, jejunal absorption of (−)-epicatechin in humans assessed by an intestinal perfusion technique revealed an average of ~46% (−)-epicatechin absorption based on recovery in the perfusion fluid, with high inter-individual variability among the eight volunteers participating in the study, ranging from 31% to 90% [31]. More recently, Ottaviani et al. [8] determined that the mean total recovery of radioactivity in urine from 8 volunteers in a 0–48 h period after ingestion of [^{14}C]EC was 82.5 ± 4.7% of intake, with individual values ranging from 49.9% to 90.2%. In most volunteers, only a relatively small amount of radioactivity was excreted after 48 h (8; 21). It is remarkable that in all these studies there are important differences in the designs of the interventions, treatments applied to the samples before the analysis, quantification methods and even in the form of administering the flavanols (dark chocolate, milk chocolate, cocoa powder dissolved in water or milk), so that there are variables that can significantly affect the results, making comparisons difficult. Of all the factors mentioned, the matrix effect of the food has been the subject of numerous studies and, although they have generated controversial results, there are more studies that suggest an interference of milk proteins in the absorption of flavanols, as described in Mullen et al. [25], among others. High inter-individual differences in the production of metabolites have been detected in this study, as in most previous studies here cited, since different factors such as sex, age, dietary habits, and gut microbiota may significantly influence the absorption and metabolism of phenolic compounds. Among the mentioned factors, the colonic microbiota is arguably the most important factor affecting the inter-individual variability, considering that metabolites formed at colonic level constitute the predominant circulating metabolites. Nevertheless, all studies, including the present work, suggest modest recovery of flavanols in urine considering only epicatechin-derived phase II metabolites.

A limitation of this study was the reduced urinary collection time. Most of the microbial metabolites show relevant amounts in the 12–24 h interval, not returning to basal levels, and therefore it would have been interesting to extend the collection time to at least 48 h. Therefore, it is likely that the amount of urinary metabolites has been underestimated and thus a higher bioavailability of cocoa polyphenols cannot be ruled out. Another limitation was the lack of certain metabolite standards, mainly phase II derivatives, forcing to express the results as equivalents of the corresponding precursor compound. Therefore, the results here indicated did not accurately measure the concentrations of the metabolites described in the biological samples. Nevertheless, the results are in line with other studies on the bioavailability of cocoa flavanols [4,16,25].

In summary, polyphenols contained in two commercial, soluble cocoa products were partially absorbed and extensively metabolized. Phase II derivatives of epicatechin were identified and their pharmacokinetic profiles were compatible with epicatechin absorption at small intestine level. However, the predominant group of metabolites identified corresponded to those formed by the microbiota, hydroxyphenyl-γ-valerolatones and phenylvaleric acid, which were absorbed and metabolized into phase II derivatives. Among these metabolites, 5-(hydroxyphenyl)-γ-valerolactone 3′-sulfate showed a high amount in urine, which could be used as biomarker of intake of flavanol-rich food. In all, although flavanol bioavailability, determined by urinary recovery, showed a dose-dependent absorption after consuming both cocoa products, this was partial and limited to 35% of the ingested polyphenols, irrespective of the initial dose.

In conclusion, the bioavailability of flavanols in soluble cocoa products is moderate, these compounds are extensively metabolized, mainly by the microbiota, and remain in the body of cocoa consumers for a long time, which favors the possible bioactivity of these compounds.

Supplementary Materials: The following are available online at http://www.mdpi.com/2072-6643/11/7/1441/s1, Figure S1: Representative high-performance liquid chromatography-diode array (HPLC-DAD) chromatograms of phenolic compounds of a conventional soluble cocoa (CC) (upper panel, **a**) and a flavanol-rich soluble cocoa (CC-PP) (lower panel, **b**) registered at 280 nm; Table S1: Retention times and λmax of the compounds identified by high-performance liquid chromatography-diode array (HPLC-DAD) and the masses of each of the flavonoid identified determined by LC-ESI-QToF-MS.

Author Contributions: Conceptualization, R.M. and L.B.C.; methodology, M.G.-J., S.M.-L. and R.M.; software, M.G.-J. and B.S.; validation, M.G.-J., S.M.-L. and R.M.; formal analysis, M.G.-J. and S.M.-L.; investigation, R.M. and L.B.C.; resources, R.M. and L.B.C.; data curation, R.M.; writing—original draft preparation, R.M.; writing—review and editing, R.M., B.S. and L.B.C.; visualization, R.M.; supervision, R.M. and L.B.C.; project administration, R.M. and L.B.C.; funding acquisition, R.M. and L.B.C.

Funding: This research was funded by the Spanish Ministry of Science, Innovation and Universities (MINECO-FEDER); gran numbers AGL2010-18269 and AGL2015-69986-R. S.M.-L. was a predoctoral fellow of the JAE Program (JAEPre097) co-funded by CSIC and the European Social Fund. M. G.-J. was a predoctoral FPI fellow BES2008-007138 of the Spanish Ministry of Science and Innovation.

Acknowledgments: We are grateful to the volunteers who participated in the study. Authors thank I. Alvarez and M.A. Martinez at the Analytical Techniques Service (USTA) facilities of ICTAN for their technical assistance with the mass spectrometry analyses. All authors revised and approved the final version of the manuscript.

Conflicts of Interest: Samples for this study were provided by Idilia Foods (formerly Nutrexpa S.L.), although this company did not take part in the study, in any form. Authors declare no conflict of interest.

References

1. Selmi, C.; Cocchi, C.A.; Lanfredini, M.; Keen, C.L.; Gershwin, M.E. Chocolate at heart: The anti-inflammatory impact of cocoa flavonoids. *Mol. Nutr. Food Res.* **2008**, *52*, 1340–1348. [PubMed]
2. Ellam, S.; Williamson, G. Cocoa and human health. *Annu. Rev. Nutr.* **2013**, *33*, 105–128. [PubMed]

3. Mena, P.; Bresciani, L.; Brindani, N.; Ludwig, I.A.; Pereira-Caro, G.; Angelino, D.; Llorach, R.; Calani, L.; Brighenti, F.; Clifford, M.N.; et al. Phenyl-γ-valerolactones and phenylvaleric acids, the main colonic metabolites of flavan-3-ols: Synthesis, analysis, bioavailability, and bioactivity. *Nat. Prod. Rep.* **2019**, *36*, 714–752. [CrossRef] [PubMed]

4. Ottaviani, J.I.; Momma, T.Y.; Kuhnle, G.K.; Keen, C.L.; Schroeter, H. Structurally related (−)-epicatechin metabolites in humans: Assessment using de novo chemically synthesized authentic metabolites. *Free Radic. Biol. Med.* **2012**, *52*, 1403–1412. [PubMed]

5. Ottaviani, J.I.; Kwik-Uribe, C.; Keen, C.L.; Schroeter, H. Intake of dietary procyanidins does not contribute to the pool of circulating flavanols in humans. *Am. J. Clin. Nutr.* **2012**, *95*, 851–858. [CrossRef] [PubMed]

6. Calani, L.; Del Rio, D.; Luisa Callegari, M.; Morelli, L.; Brighenti, F. Updated bioavailability and 48 h excretion profile of flavan-3-ols from green tea in humans. *Int. J. Food Sci. Nutr.* **2012**, *63*, 513–521. [PubMed]

7. van der Hooft, J.J.; de Vos, R.C.; Mihaleva, V.; Bino, R.J.; Ridder, L.; de Roo, N.; Jacobs, D.M.; van Duynhoven, J.P.; Vervoot, J. Structural elucidation and quantification of phenolic conjugates present in human urine after tea intake. *Anal. Chem.* **2012**, *84*, 7263–7271. [CrossRef]

8. Ottaviani, J.I.; Borges, G.; Momma, T.Y.; Spencer, J.P.; Keen, C.L.; Crozier, A.; Schroeter, H. The metabolome of [2-(^{14}C)(-)-epicatechin in humans: Implications for the assessment of efficacy, safety, and mechanisms of action of polyphenolic bioactives. *Sci. Rep.* **2016**, *1*, 6:29034.

9. Vitaglione, P.; Lumaga, R.B.; Ferracane, R.; Sellitto, S.; Morello, J.R.; Miranda, J.R.; Shimoni, E.; Fogliano, V. Human bioavailability of flavanols and phenolic acids from cocoa-nut creams enriched with free or microencapsulated cocoa polyphenols. *Br. J. Nutr.* **2013**, *109*, 1832–1843.

10. Sarriá, B.; Martínez-López, S.; Sierra-Cinos, J.L.; García-Diz, L.; Goya, L.; Mateos, R.; Bravo, L. Effects of bioactive constituents in functional cocoa products on cardiovascular health in humans. *Food Chem.* **2015**, *174*, 214–218. [CrossRef]

11. Montagnana, M.; Danese, E.; Angelino, D.; Mena, P.; Rosi, A.; Benati, M.; Gelati, M.; Salvagno, G.L.; Favaloro, E.J.; Del Rio, D.; et al. Dark chocolate modulates platelet function with a mechanism mediated by flavan-3-ol metabolites. *Medicine* **2018**, *97*, e13432. [PubMed]

12. Cifuentes-Gomez, T.; Rodríguez-Mateos, A.; Gonzalez-Salvador, I.; Alañon, M.E.; Spencer, J.P.E. Factors affecting the absorption, metabolism and excretion of cocoa flavanols in humans. *J. Agric. Food Chem.* **2015**, *63*, 7615–7623. [PubMed]

13. Rodríguez-Mateos, A.; Vauzour, D.; Krueger, C.G.; Shanmuganayagam, D.; Reed, J.; Calani, L.; Mena, P.; Del Río, D.; Crozier, A. Bioavailability, bioactivity and impact on health of dietary flavonoids and related compounds: An update. *Arch. Toxicol.* **2014**, *88*, 1803–1853. [PubMed]

14. Oracz, J.; Nebesny, E.; Zyzelewicz, D.; Budryn, G.; Luzak, B. Bioavailability and metabolism of selected cocoa bioactive compounds: A comprehensive review. *Crit. Rev. Food Scin. Nutr.* **2019**, 1–39. [CrossRef] [PubMed]

15. Bravo, L.; Abia, R.; Saura-Calixto, F. Polyphenols as dietary fiber associated compounds. Comparative study on in vivo and in vitro properties. *J. Agric. Food Chem.* **1994**, *42*, 1481–1487. [CrossRef]

16. Actis-Goretta, L.; Leveques, A.; Giuffrida, F.; Romanov-Michailidis, F.; Viton, F.; Barron, D.; Dueñas-Paton, M.; González-Manzano, S.; Santos-Buelga, C.; Williamson, G.; et al. Elucidation of (-)-epicatechin metabolites after ingestion of chocolate by healthy humans. *Free Radic. Biol. Med.* **2012**, *53*, 787–795. [CrossRef] [PubMed]

17. Urpí-Sarda, M.; Monagas, M.; Khan, N.; Llorach, R.; Lamuela-Raventós, R.M.; Jáuregui, O.; Estruch, R.; Izquierdo-Pulido, M.; Andrés-Lacueva, C. Targeted metabolic profiling of phenolics in urine and plasma after regular consumption of cocoa by liquid chromatography-tandem mass spectrometry. *J. Chromatogr. A* **2009**, *1216*, 7258–7267. [PubMed]

18. Sánchez-Patán, F.; Chioua, M.; Garrido, I.; Cueva, C.; Samadi, A.; Marco-Contelles, J.; Moreno-Arribas, M.V.; Bartolomé, B.; Monagas, M. Synthesis, analytical features, and biological relevance of 5-(3′,4′-dihydroxyphenyl)-γ-valerolactone, a microbial metabolite derived from the catabolism of dietary flavan-3-ols. *J. Agric. Food Chem.* **2011**, *59*, 7083–7091. [CrossRef] [PubMed]

19. Monagas, M.; Urpí-Sarda, M.; Sánchez-Patán, F.; Llorach, R.; Garrido, I.; Gómez-Cordovés, C.; Andrés-Lacueva, C.; Bartolomé, B. Insights into the metabolism and microbial biotransformation of dietary flavan-3-ols and the bioactivity of their metabolites. *Food Funct.* **2010**, *1*, 233–253. [CrossRef] [PubMed]

20. Stoupi, S.; Williamson, G.; Drynan, J.W.; Barron, D.; Clifford, M.N. Procyanidin B2 catabolism by human fecal microflora: Partial characterization of "dimeric" intermediates. *Arch. Biochem. Biophys.* **2010**, *501*, 73–78. [CrossRef]

21.	Borges, G.; Ottaviani, J.I.; van der Hooft, J.J.J.; Schroeter, H.; Crozier, A. Absorption, metabolism, distribution and excretion of (-)-epicatechin: A review of recent findings. *Mol. Asp. Med.* **2018**, *61*, 18–30. [CrossRef] [PubMed]

22.	Richelle, M.; Tavazzi, I.; Enslen, M.; Offord, E.A. Plasma kinetics in man of epicatechin from black chocolate. *Eur. J. Clin. Nutr.* **1999**, *53*, 22–26. [CrossRef] [PubMed]

23.	Roura, E.; Andrés-Lacueva, C.; Jauregui, O.; Badia, E.; Estruch, R.; Izquierdo-Pulido, M.; Lamuela-Raventós, R.M. Rapid liquid chromatography tandem mass spectrometry assay to quantify plasma (-)-epicatechin metabolites after ingestion of a standard portion of cocoa beverage in humans. *J. Agric. Food Chem.* **2005**, *53*, 6190–6194. [PubMed]

24.	Roura, E.; Andrés-Lacueva, C.; Estuch, R.; Mata-Bilbao, M.L.; Izquierdo-Pulido, M.; Waterhouse, A.L.; Lamuela-Raventós, R.M. Milk does not affect the bioavailability of cocoa powder flavonoid in healthy human. *Ann. Nutr. Metab.* **2007**, *51*, 493–498. [PubMed]

25.	Mullen, W.; Borges, G.; Donovan, J.L.; Edwards, C.A.; Serafini, M.; Lean, M.E.J. Milk decreases urinary excretion but not plasma pharmacokinetics of cocoa flavan-3-ol metabolites in humans. *Am. J. Clin. Nutr.* **2009**, *89*, 1784–1791. [CrossRef] [PubMed]

26.	Schroeter, H.; Heiss, C.; Balzer, J.; Kleinbongard, P.; Keen, C.L.; Hollenberg, N.K.; Sies, H.; Kwik–Uribe, C.; Schmitz, H.H.; Kelm, M. (-)-Epicatechin mediates beneficial effects of flavanol-rich cocoa on vascular function in humans. *Proc. Natl. Acad. Sci. USA* **2006**, *103*, 1024–1029. [CrossRef] [PubMed]

27.	Baba, S.; Osakabe, N.; Yasuda, A.; Natsume, M.; Takizawa, T.; Nakamura, T.; Terao, J. Bioavailability of (-)epicatechin upon intake of chocolate and cocoa in human volunteers. *Free Radic. Res.* **2000**, *33*, 635–641.

28.	Tomás-Barberán, F.; Cienfuegos-Jovellanos, E.; Marín, A.; Muguerza, B.; Gil-Izquierdo, A.; Cerdá, B.; Espín, J.C. A new process to develop a cocoa powder with higher flavonoid monomer content and enhanced bioavailability in healthy humans. *J. Agric. Food Chem.* **2007**, *55*, 3926–3935.

29.	Llorach-Asunción, R.; Jauregui, O.; Urpí-Sarda, M.; Andrés-Lacueva, C. Methodological aspects for metabolome visualization and characterization: A metabolomic evaluation of the 24 h evolution of human urine after cocoa powder consumption. *J. Pharm. Biomed. Anal.* **2010**, *51*, 373–381.

30.	Ito, H.; Gonthier, M.; Manach, C.; Morand, C.; Mennen, L.; Remesy, C.; Scalbert, A. Polyphenol levels in human urine after intake of six different polyphenol-rich beverages. *Brit. J. Nutr.* **2005**, *94*, 500–509. [CrossRef]

31.	Actis-Goretta, L.; Leveques, A.; Rein, M.T.; Teml, A.; Schafer, C.; Hofmann, U.; Li, H.; Schwab, M.; Eichelbaum, M.; Williamson, G. Intestinal absorption, metabolism, and excretion of (-)-epicatechin in healthy humans assessed by using an intestinal perfusion technique. *Am. J. Clin. Nutr.* **2013**, *98*, 924–933. [CrossRef] [PubMed]

 *nutrients*

Article

Low Plasma Appearance of (+)-Catechin and (−)-Catechin Compared with Epicatechin after Consumption of Beverages Prepared from Nonalkalized or Alkalized Cocoa—A Randomized, Double-Blind Trial

Sabine Ellinger [1,2,*], **Andreas Reusch** [2], **Lisa Henckes** [3], **Christina Ritter** [3], **Benno F. Zimmermann** [3], **Jörg Ellinger** [4], **Rudolf Galensa** [3], **Peter Stehle** [2] and **Hans-Peter Helfrich** [5]

1   Faculty of Food, Nutrition and Hospitality Sciences, Hochschule Niederrhein, University of Applied Sciences, Rheydter Str. 277, 41065 Mönchengladbach, Germany
2   Institute of Nutritional and Food Sciences, Nutritional Physiology, University of Bonn, Nußallee 9, 53115 Bonn, Germany; areusch@uni-bonn.de (A.R.); p.stehle@uni-bonn.de (P.S.)
3   Institute of Nutritional and Food Sciences, Food Sciences, University of Bonn, Endenicher Allee 11-13, 53115 Bonn, Germany; lisa.henckes@gmx.de (L.H.); ritter.c@gmx.de (C.R.); benno.zimmermann@uni-bonn.de (B.F.Z.); galensa@uni-bonn.de (R.G.)
4   Department of Urology and Children Urology, University Hospital Bonn, Sigmund-Freud-Str. 25, 53127 Bonn, Germany; joerg.ellinger@ukbonn.de
5   Institute of Numerical Simulation, University of Bonn, Endenicher Allee 60, 53115 Bonn, Germany; helfrich@uni-bonn.de
*   Correspondence: Sabine.Ellinger@hs-niederrhein.de; Tel.: +49-2161-186-5406

Received: 17 November 2019; Accepted: 13 January 2020; Published: 16 January 2020

Abstract: Flavan-3-ols are claimed to be responsible for the cardioprotective effects of cocoa. Alkalized cocoa powder (ALC), commonly used for many non-confectionary products, including beverages, provides less (+)-catechin, (−)-epicatechin, and procyanidins and more (−)-catechin than nonalkalized cocoa powder (NALC). This may affect the plasma appearance of monomeric flavan-3-ol stereoisomers after consumption of NALC vs. ALC. Within a randomized, crossover trial, 12 healthy nonsmokers ingested a milk-based cocoa beverage providing either NALC or ALC. Blood was collected before and within 6 h postconsumption. (+)-Catechin, (−)-catechin, and epicatechin were analyzed in plasma by HPLC as sum of free and glucuronidated metabolites. Pharmacokinetic parameters were obtained by a one-compartment model with nonlinear regression methods. For epicatechin in plasma, total area under the curve within 6 h postconsumption (AUC_{0-6h}) and incremental AUC_{0-6h} were additionally calculated by using the linear trapezoidal method. After consumption of NALC and ALC, (+)-catechin and (−)-catechin were mostly not detectable in plasma, in contrast to epicatechin. For epicatechin, total AUC_{0-6h} was different between both treatments, but not incremental AUC_{0-6h}. Most kinetic parameters were similar for both treatments, but they varied strongly between individuals. Thus, epicatechin is the main monomeric flavan-3-ol in plasma after cocoa consumption. Whether NALC should be preferred against ALC due to its higher (−)-epicatechin content remains unclear with regard to the results on incremental AUC_{0-6h}. Future studies should investigate epicatechin metabolites in plasma for a period up to 24 h in a larger sample size, taking into account genetic polymorphisms in epicatechin metabolism and should consider all metabolites to understand inter-individual differences after cocoa intake.

Keywords: cocoa; flavan-3-ol stereoisomers; (−)-epicatechin; (+)-catechin; (−)-catechin; plasma appearance; chiral separation; pharmacokinetics; one-compartment model

---

1. Introduction

Epidemiological studies have shown that cocoa consumption lowers cardiometabolic risk [1]. This may be explained by a reduction in blood pressure [2–6] and serum lipids [2,7,8] and by an increase in insulin sensitivity [2,7], which could be observed in several meta-analyses of randomized controlled trials (RCTs). The beneficial effects of cocoa consumption on cardiometabolic biomarkers have been ascribed to flavan-3-ols [9,10]. In this respect, monomeric flavan-3-ols (catechin, epicatechin) are of particular interest due to their relatively high bioavailability from cocoa compared with oligomeric flavan-3-ols [11,12], which occurred in plasma only in traces (procyanidin B2) [11,12] or were even not detectable (procyanidin B5) [13].

Cocoa powder which is used for non-confectionary products, e.g., beverages, is usually alkalized to improve solubility and sensory properties. However, by alkalization, flavan-3-ols are largely destroyed, thereby reducing total flavan-3-ol content up to 80%, depending on the extent of alkalization [14–16]. Moreover, alkalization of cocoa powder induces an epimerization of (−)-epicatechin to (−)-catechin, an atypical stereoisomer that is also generated by roasting of cocoa beans. Consequently, alkalized cocoa powder (ALC) provides less flavan-3-ols and a lower ratio of (−)-epicatechin to (−)-catechin than nonalkalized cocoa powder (NALC) [17]. (+)-Epicatechin, also an atypical stereoisomer in processed cocoa, derives from epimerization of (+)-catechin. However, in contrast to (−)-catechin, which accounted on average for 89% of total catechin in chocolate, the amount of (+)-epicatechin was comparably low (5% of total epicatechin content) [18,19]. (+)-Epicatechin was detectable in a single alkalized cocoa powder by Ritter et al. [20], but not in the powders analyzed by Kofink et al. [17].

In rats, in situ perfusion of a solution of either pure (+)-catechin or pure (−)-catechin revealed a 3-fold lower intestinal absorption as well as 7-fold lower plasma concentrations of (−)-catechin compared with (+)-catechin metabolites [18]. In a self-experiment of a healthy volunteer of our group, the intake of a mixture of 103.5 mg (+)-catechin and 121.5 mg (−)-catechin (ratio 0.85) in milk revealed a plasma ratio of (+)-catechin to (−)-catechin metabolites to a value of 4.05. Furthermore, after consumption of a commercially available cocoa powder (46 g, dissolved in milk) providing 3.6 mg (+)-catechin and 16.6 mg (−)-catechin (ratio 0.22), the plasma ratio of (+)-catechin to (−)-catechin reached 0.8. These observations suggest a 5-fold lower bioavailability of (−)-catechin compared with (+)-catechin [20].

It is known from experimental studies that the presence of other cocoa flavan-3-ols like (−)-epicatechin or procyanidins can modulate the plasma appearance of individual flavan-3-ol stereoisomers, probably due to different affinity to metabolic enzymes or due to interactions with transport mechanisms [21,22]. Since the flavan-3-ol composition varies strongly between ALC and NALC, carefully guided human intervention studies are mandatory to monitor the plasma appearance of biologically active flavan-3-ols after consumption of cocoa products and, thus, to evaluate their potential to support cardiovascular health.

The aim of our intervention study was, thus, to follow the plasma appearance of (−)-catechin (primary aim) and of further stereoisomers, that is, (+)-catechin and (−)-epicatechin (secondary aim), after consumption of cocoa-rich beverages prepared with NALC and ALC and to compare plasma appearance of these compounds by means of pharmacokinetic parameters.

2. Materials and Methods

2.1. Study Design and Intervention

This pilot study was conducted as double-blind RCT with crossover design at the Institute of Nutritional and Food Sciences, Nutritional Physiology, University of Bonn. The study was performed according to the guidelines of the Declaration of Helsinki 2004, and the protocol was approved by the ethics committee of the University of Bonn (project-ID: 309/08). The study was registered in the German Clinical Trials Register (trial-ID: DRKS00017550) and was part of an interdisciplinary PhD program. Written informed consent was obtained from all participants.

The participants were recruited between 09.01.2010 and 08.02.2010 and were randomly assigned to groups A and B by permuted block randomization (block size of four, ratio 2:2; sequence determined by drawing lots by an uninvolved person). Both groups ingested cocoa beverages, prepared from either NALC or ALC, after at least 12 h overnight fast on two different occasions in different order. Both treatments were separated by at least one week washout. The cocoa beverages were prepared by an uninvolved person and were provided in a covered cup with a drinking straw to ensure blinding of researcher and participants. If the subjects noted that their cups were empty, a researcher removed the lids, filled the cups with a little water, and asked the participants to drink the remaining volume to ensure complete cocoa consumption. An allocation list revealing the order of both treatments (NALC, ALC) to groups A and B was sealed in an envelope, which was opened after finishing the statistical analysis.

On each study day, venous blood samples were collected before and 0.5, 1.0, 1.5, 2.0, 3.0, and 6.0 h after consumption of the cocoa beverages. After the 3 h blood sampling, the participants received a standardized meal (two rolls, each topped with 12 g butter and 40 g Gouda cheese, 48% fat in dry matter). Water was allowed to drink ad libitum.

The participants were instructed to abstain from flavonoid-rich foods (e.g., cocoa-containing foods and beverages, green tea, black tea, fruits, and juices; handout) 24 h before each treatment.

2.2. Participants

Twelve healthy subjects were included in the present study. They were eligible if they were between 18 and 50 years old and nonsmoker. Exclusion criteria comprised any known diseases (milk allergy, lactose intolerance, hepatic, renal and gastrointestinal diseases, metabolic and eating disorders), pregnancy, lactation, drug abuse, regular supplementation of vitamin C, vitamin E, and phytochemicals (e.g., tea or red wine extract), and participation in another trial 30 days before or throughout this study. Inclusion and exclusion criteria were checked by questionnaire.

2.3. Cocoa Beverages

NALC (pH 5.4–6.0) and ALC (pH 7.4–7.8) were produced following industrial standards by Schwartauer Werke, Kakao Verarbeitung Berlin, Berlin, Germany. The composition of NALC and ALC is shown in Table 1. According to manufacturer, both cocoa powders provided equal amounts of protein, fat, and carbohydrates per 100 g, and thus an equal amount of energy. As expected, the flavan-3-ol contents of ALC and NALC differed both in quantities and quality, as analyzed in our lab [20,23]. NALC contained higher amounts of flavan-3-ols compared with ALC, that is, 1.33 times more total catechin (sum of (+)-catechin and (−)-catechin), 5 times more epicatechin, and about 7 and 14 times higher amounts of procyanidin B2 and procyanidin C1, respectively. NALC provided a lower ratio of (−)-catechin to (−)-epicatechin (0.32) compared with ALC (1.26). This was expected as the ratio of (−)-catechin to (−)-epicatechin is usually <1 in nonalkalized and >1 in alkalized cocoa [24]. (+)-Epicatechin was not detectable in both cocoa powders. Hence, (−)-epicatechin was quantitatively the dominating flavan-3-ol monomer in NALC and (−)-catechin in ALC, respectively. Moreover, own analysis on methylxanthines, which was conducted according to the official methods of the German Federal Health Office [25], revealed that NALC provided more theobromine than ALC and less caffeine.

The dose of cocoa powder used for cocoa treatment was adjusted to individual body weight (BW). On the basis of our earlier pilot bioavailability studies [20], we decided to provide for both treatments equal amounts of total catechin (i.e., sum of (+)-catechin and (−)-catechin; 0.240 mg/kg BW) to allow reliable evaluation of their absorption kinetics; consequently, we used 0.6 g ALC and 0.8 g NALC per kg BW, respectively. Both cocoa powders were dissolved in heated, skimmed milk (6 mL/kg BW). All cocoa beverages were sweetened with sucrose (80 g/L) to improve taste. The total intake of energy, macronutrients, flavan-3-ols, and methylxanthines from the cocoa beverages for a subject weighing 67.7 kg is shown in Table 2.

Table 1. Composition of the cocoa powders.

	Per 100 g Nonalkalized Cocoa Powder	Per 100 g Alkalized Cocoa Powder
Energy (kJ) [1]	1045	1045
Macronutrients [1]		
Protein (g)	23.5	23.5
Fat (g)	10.0	10.0
Carbohydrates (g)	14.0	14.0
Flavan-3-ols [2]		
Catechin		
(+)-Catechin (mg)	3	1
(−)-Catechin (mg)	37	29
Epicatechin		
(+)-Epicatechin (mg)	n.d.	n.d.
(−)-Epicatechin (mg)	115	23
Oligomers (procyanidins)		
Procyanidin B2 (mg)	54	7
Procyanidin C1 (mg)	29	2
Methylxanthines [2]		
Theobromine (mg)	3125	2326
Caffeine (mg)	97	111

[1] According to manufacturer. [2] According to own analyses. Own data represent mean values, which were based on measurements in duplicate. n.d.: not detectable.

Table 2. Mean intake of energy, macronutrients, flavan-3-ols, and methylxanthines from beverages prepared with nonalkalized or alkalized cocoa powder in milk [1].

	Mean Intake from the Beverage Prepared with Nonalkalized Cocoa Powder	Mean Intake from the Beverage Prepared with Alkalized Cocoa Powder
Energy (kJ) [2]	1398	1418
Macronutrients [2]		
Protein (g)	22	25
Fat (g)	12	14
Carbohydrates (g)	58	60
Flavan-3-ols [3]		
Catechin		
(+)-Catechin (mg)	1.2	0.5
(−)-Catechin (mg)	15.0	15.7
Epicatechin		
(+)-Epicatechin (mg)	n.d.	n.d.
(−)-Epicatechin (mg)	46.7	12.5
Procyanidins		
Procyanidin B2 (mg)	21.9	3.8
Procyanidin C1 (mg)	11.8	1.1
Methylxanthines [3]		
Theobromine (mg)	1269	1261
Caffeine (mg)	39	111

[1] Considering a mean body weight (BW) of 67.7 kg. The intake of the milk-based beverages was adapted to individual BW (6 mL/kg BW) and the use of 0.6 g nonalkalized cocoa and 0.8 g alkalized cocoa per kg BW, respectively. [2] Considering the intake from milk, sugar (EBISpro), and cocoa (manufacturer). [3] According to own analyses. Own data represent mean values, which were based on measurements in duplicate. n.d.: not detectable.

2.4. Blood Sampling and Treatment of Plasma Samples

Venous blood samples were collected into tubes coated with EDTA (S-Monovette, Sarstedt, Nümbrecht, Germany) and were placed on ice immediately. Thereafter, the blood samples were centrifuged (3000× g, 20 min, 4 °C), and the plasma was stabilized with a solution of ascorbic acid and EDTA, as described previously [26]. All samples were stored at −80 °C for two days until (+)-catechin, (−)-catechin, and epicatechin were analyzed.

2.5. (+)-Catechin, (−)-Catechin, and Epicatechin in Plasma

When analyzing the concentration of monomeric flavan-3-ols in plasma, chiral separation was performed to differentiate between (+)-catechin and (−)-catechin. Since both cocoa powders provided only (−)-epicatechin and no detectable amounts of (+)-epicatechin (Table 1), only (−)-epicatechin was expected to occur in plasma after cocoa treatment. Therefore, we decided to abstain from chiral separation in case of epicatechin.

After thawing the plasma samples, each sample (500 µL) was incubated with 6250 units of β-glucuronidase type VII-A, EC 232-606-8 from *Escherichia coli* (Sigma-Aldrich, Steinheim, Germany) at 37 °C for 45 min. Thereafter, a solid-phase extraction was performed according to the procedure described previously [20]. Chiral HPLC with coulometric electrode array detection was used to determine the concentration of (+)-catechin and (−)-catechin according to the protocol of Ritter et al. [20], and HPLC with coulometric electrode array detection was also used to quantify the concentration of epicatechin according to the method of Zimmermann et al. [27]. The concentration of each flavan-3-ol in plasma was determined as sum of free, unmetabolized flavan-3-ol and glucuronidated metabolites. The limit of detection (LOD) of (+)-catechin and (−)-catechin was 23.43 nmol/L and 20.33 nmol/L, respectively. The coefficient of variation (CV) was 0.97% for (+)-catechin and 1.54% for (−)-catechin, respectively [28]. For epicatechin, the LOD was 8.27 nmol/L [27] and the CV 5.1%. For the calculation of pharmacokinetic data, concentrations below the LOD were set at the half value of corresponding detection limit to reduce bias. In addition to pharmacokinetic data which were calculated on the basis of a one-compartment model described below, the area under the curve *(AUC)* was determined for each subject as total *AUC* ($tAUC_{0-6h}$) and incremental *AUC* ($iAUC_{0-6h}$) by using the linear trapezoidal method based on measurements taken over 6 h.

2.6. Pharmacokinetic Data

The dependency of the concentration of the flavan-3-ol on time in plasma was fitted for each subject by using a one-compartment model with first-order absorption and elimination. For this, the following equation was applied [29]:

$$c(t) = \frac{D70 \times k1 \times k2x^2}{Cl \times (k1 - k2)} \exp(-k2 \times (t + tsh) - \exp(-k1 \times (t + tsh))),$$

where $c(t)$ is the flavan-3-ol plasma concentration (µmol/L) at time t, D_{70} the dose of the flavan-3-ol that would be ingested by a person weighing 70 kg with NALC ((+)-catechin: 4 µmol ≈ 1.3 mg; (−)-catechin: 54 µmol ≈ 15.5 mg; epicatechin: 166 µmol ≈ 48.3 mg) and ALC ((+)-catechin: 2 µmol ≈ 0.6 mg; (−)-catechin: 56 µmol ≈ 16.2 mg; epicatechin: 44 µmol ≈ 12.88 mg), k_1 the absorption rate (per hour), k_2 the elimination rate (per hour), and t the time (h) recorded after complete ingestion of the cocoa drink. We included a time shift (t_{sh}, h) in our model to consider the individual period necessary for complete ingestion of the cocoa beverages. The clearance (Cl) is defined as the product $V_{hyp}/F_{abs} \times k_2$, where V_{hyp}/F_{abs} denotes the hypothetical distribution volume over absolute bioavailability.

The fitting by nonlinear regression was performed with IBM SPSS for Windows, version 19.0 (IBM Corp., Armonk, NY, USA). We used the bootstrap option with 1000 bootstrap samples in order to calculate unbiased estimates of the standard deviations [30] of the parameters on plasma kinetics. We used the logarithms of k_1, k_2, and Cl as parameters to avoid negative estimates of k_1, k_2, and Cl. Further kinetic parameters were computed based on k_1, k_2, and Cl: maximum plasma concentration C_{max} (µmol/L), the time to reach C_{max} after cocoa intake (t_{max}, h), absorption half-life time $t_{1/2a}$ (h), elimination half-life time $t_{1/2e}$ (h), $tAUC$ (µmol/L × h), V_{hyp}/F_{abs} (l), and the clearance (l/h).

In order to account for the heterogeneity of the variances within the bootstrap samples for each parameter and the variances between the subjects, weighted means and variances of the parameters were determined. The weights for each parameter were chosen as $\tau_i = (\sigma_i^2 + \Delta^2)^{-1}$, where the within variances σ_i^2 were obtained by the bootstrap estimates, and the between variance Δ^2 by the DerSimonian and Laird procedure [31], commonly used in meta-analyses.

To characterize the plasma kinetics of the flavan-3-ol stereoisomers for NALC and ALC, logarithmized parameters k_1, k_2, V_{hyp}/F_{abs}, $t_{1/2a}$, $t_{1/2e}$, $tAUC$, C_{max}, and Cl were used, leading to geometric means. For t_{max}, arithmetic means were calculated. The variation between individuals was indicated by 95% reference intervals (95% RI). To compare the plasma kinetics of the flavan-3-ols for NALC and ALC, the ratios (k_1, k_2, V_{hyp}/F_{abs}, $t_{1/2a}$, $t_{1/2e}$, $tAUC$, C_{max}) and the difference (t_{max}) of the parameters were calculated.

2.7. Anthropometric Data and Dietary Intake

On each study day, body height and weight were determined in fasting state by using a calibrated digital weighing and measuring station (seca 764, Hamburg, Germany; accuracy 0.1 kg for weight and 0.1 cm for height, respectively). On the day before each treatment, food consumption was documented by the participants in a standardized 24 h dietary record. The intake of energy, macronutrients, and dietary fiber was calculated by using the software EBISpro for Windows, version 8.0 (Erhardt, University of Hohenheim, Stuttgart, Germany), and the intake of flavonoids by using the USDA Flavonoid Database, release 3 [32].

2.8. Statistics

Statistical evaluation was performed by using IBM SPSS for Windows, version 19.0 (IBM Corp., Armonk, NY, USA). For details on pharmacokinetic data, see Section 2.6. Data on BW, nutritional intake before each treatment, $tAUC_{0-6h}$, $iAUC_{0-6h}$, and the time needed for the ingestion of each cocoa beverage were checked for normal distribution by Kolmogorov–Smirnov test. If normality could be assumed, data were compared by paired t-tests. Otherwise, data were log-transformed and checked again for normal distribution. If normal distribution could not be achieved, data were compared by Wilcoxon signed-rank test. Correlations between metric variables without normal distribution were investigated by Spearman's rank correlation test. Metric data are presented as means ± standard error of the mean (SEM) if not indicated otherwise. Nominal data are given as absolute or relative frequencies.

3. Results

This trial was performed with 12 obviously healthy participants (6 women and 6 men; age 26.9 ± 4.1 years, mean ± SD) with a body mass index (BMI) between 18.6 and 26.4 kg/m^2 (22.7 ± 2.6 kg/m^2; mean ± SD). All of them completed the study. On average, both treatments were performed in intervals of 15 days. Considering a BW between 48.8 and 86.6 kg (67.5 ± 11.3 kg; mean ± SD) and a beverage consumption of 6 mL/kg BW, the volume of the ingested beverages ranged from 293 to 520 mL (405 ± 68 mL; mean ± SD). The time that was needed for complete consumption of the beverages prepared with either NALC (median 27 min, interquartile range 19–33 min) or ALC (median 34 min, interquartile range 29–42 min) was not different between both treatments ($p = 0.168$) and did not correlate with the corresponding amount of cocoa powder ingested (both p-values > 0.05).

Before ingestion of NALC and ALC, BW (67.7 ± 3.3 kg vs. 67.5 ± 3.3 kg) was not different, and the intake of energy (2154 ± 189 kcal vs. 2453 ± 189 kcal), protein (90 ± 7 g vs. 103 ± 11 g), fat (101 ± 11 g vs. 122 ± 16 g), carbohydrates (217 ± 27 g vs. 235 ± 28 g), dietary fiber (10 ± 2 g vs. 12 ± 2 g), and flavonoids (1.1 ± 0.6 g vs. 0.6 ± 0.2 g) on the previous day was comparable (all p-values > 0.05).

In fasting plasma obtained before cocoa treatment, (+)-catechin and (−)-catechin were not detectable in any sample (data not shown) and the concentration of epicatechin was below or near to the LOD in most subjects. After cocoa treatment, (+)-catechin was only detectable in 5 out of 144 plasma samples. Three of them were obtained after consumption of NALC, and two samples after ingestion of

ALC. (−)-Catechin was found in 13 out of 144 plasma samples, mostly after ingestion of ALC ($n = 10$) and only in a few samples after ingestion of NALC ($n = 3$) (data not shown). Thus, pharmacokinetic parameters were not calculated for (+)-catechin and (−)-catechin.

In contrast to catechin enantiomers, the concentration of epicatechin was quantifiable in all plasma samples after intervention and increased in all subjects after ingestion of both NALC (Figure 1a–m) and ALC (Figure 2a–m). Each subfigure (a–m) presents the measured concentrations and the predicted concentration time curves for a single participant. For most subjects, the individual concentration time curves of epicatechin in plasma which were obtained by the model indicate a good prediction for epicatechin concentration in plasma after ingestion of NALC and ALC, respectively. In single cases, the prediction was not good, indicated by the high sum of squares of residuals (NALC9, ALC4) and the fact that the model predicts a plasma rise of only 75% (NALC1), 50% (ALC3) and <50% (ALC8, ALC12) of the maximum measured concentration. For NALC5 and ALC12, the measured concentrations do not provide a clear curve shape due to the relatively high baseline values. The increase, determined by k_1, as well as the decline of the concentration–time curves, determined by k_2, differed strongly between individuals. Large individual variations were also observed for $t_{1/2a}$, which depends on k_1, for $t_{1/2e}$, which depends on k_2, and for t_{max}, which is determined by k_1 and k_2. Strong inter-individual variations also occurred for C_{max} and for Cl due to the differences in V_{hyp}/F_{abs} and k_2 (Table 3). The model predicts that epicatechin is almost completely eliminated from plasma by 10 h after cocoa consumption in all subjects except for one (ALC5) (Figures 1 and 2).

Table 3. Plasma kinetics of epicatechin in individual subjects after ingestion of a beverage prepared with nonalkalized and alkalized cocoa.

	k_1 (h^{-1})	k_2 (h^{-1})	C_{max} (µmol/L)	t_{max} (h)	$t_{1/2a}$ (h)	$t_{1/2e}$ (h)	$tAUC$ (µmol/L × h)	V_{hyp}/F_{abs} (l)	Cl (l/h)
NALC1	1.85	0.40	0.71	1.07	0.37	1.74	2.71	154	61
NALC2	0.67	0.68	0.90	1.50	1.03	1.03	3.61	68	46
NALC3	0.64	0.64	0.14	1.60	1.08	1.08	0.59	438	280
NALC4	0.72	0.72	0.18	1.39	0.96	0.96	0.66	348	250
NALC5	0.42	0.42	0.07	1.28	1.64	1.64	0.45	880	371
NALC6	0.60	0.60	0.54	1.73	1.16	1.16	2.45	113	68
NALC7	0.81	0.49	0.17	1.44	0.86	1.41	0.74	454	224
NALC8	1.58	0.24	1.37	1.08	0.44	2.85	7.93	86	21
NALC9	1.04	1.04	2.79	1.00	0.67	0.67	7.29	22	23
NALC10	0.84	0.59	0.26	1.38	0.83	1.18	1.02	279	164
NALC11	1.40	0.38	0.82	1.29	0.50	1.82	3.52	124	47
NALC12	1.13	1.12	0.78	0.89	0.61	0.62	1.88	79	89
ALC1	0.54	0.54	0.06	1.66	1.29	1.29	0.29	283	152
ALC2	1.60	0.38	0.04	1.15	0.43	1.81	0.17	700	268
ALC3	0.52	0.52	0.25	2.04	1.32	1.32	1.27	67	35
ALC4	1.26	1.26	3.28	0.82	0.55	0.55	7.05	5	6
ALC5	1.78	0.06	0.06	1.90	0.39	11.99	1.14	673	39
ALC6	45.62	0.31	0.14	0.10	0.02	2.27	0.49	297	91
ALC7	0.90	0.90	0.11	0.93	0.77	0.77	0.34	147	132
ALC8	1.19	1.20	0.08	0.78	0.58	0.58	0.19	194	232
ALC9	0.33	0.33	0.85	1.47	2.10	2.10	6.97	19	6
ALC10	6.62	0.12	0.29	0.61	0.10	5.64	2.53	143	18
ALC11	0.83	0.34	0.03	1.75	0.84	2.03	0.15	881	300
ALC12	0.48	0.8	0.04	1.83	1.43	1.43	0.24	383	185

Data present maximum-likelihood estimates which were gained by nonlinear regression from our pharmacokinetic model. ALC: alkalized cocoa; $tAUC$: total area under the curve; C_{max}: maximum plasma concentration; Cl: clearance; k_1: absorption rate; k_2: elimination rate; NALC: nonalkalized cocoa; $t_{1/2a}$: absorption half-life time; $t_{1/2e}$: elimination half-life time; t_{max}: time to reach maximum plasma concentration; V_{hyp}/F_{abs}: hypothetical distribution volume over absolute bioavailability.

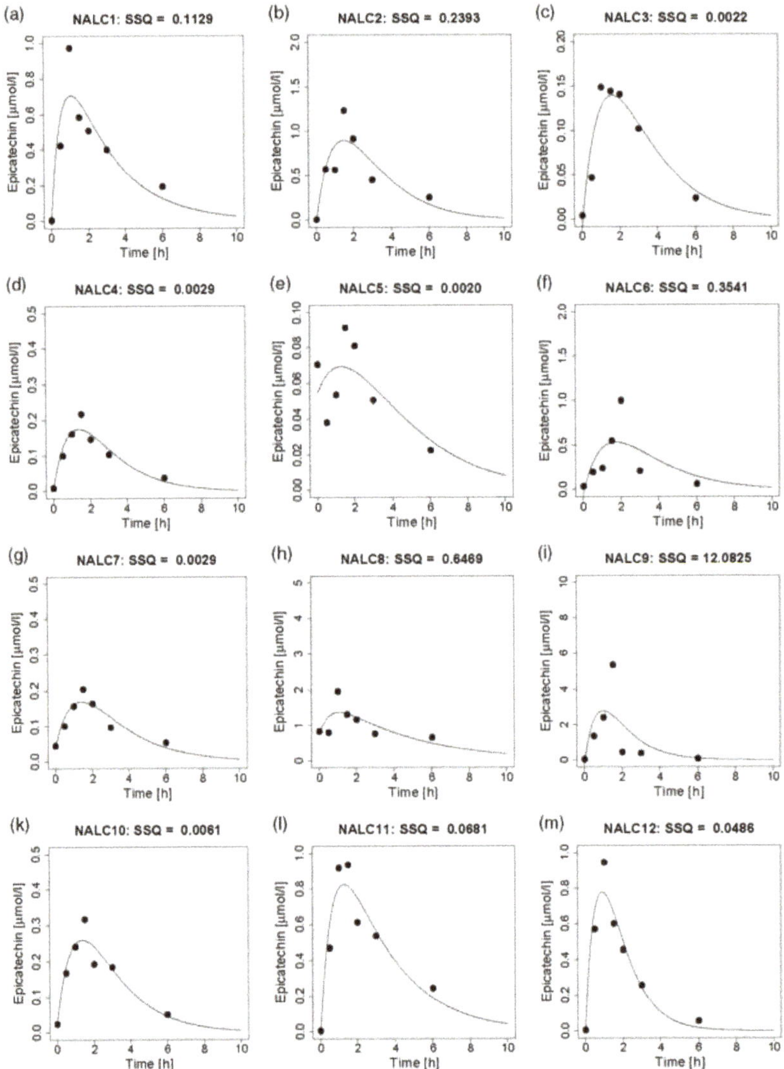

Figure 1. Individual epicatechin concentration time curves obtained after consumption of a beverage prepared with nonalkalized cocoa. Each panel (**a–m**) represents one participant. The dots represent the measured values. The curves reflect the predicted concentrations in plasma within 10 h after ingestion of nonalkalized cocoa, based on the one-compartment model. NALC, nonalkalized cocoa; SSQ, sum of squares of residuals. Note the different scaling of the y-axes.

For the comparison of both treatments (beverages prepared from NALC vs. ALC), ALC6 was excluded from our analysis as the implausible values for k_1 (45.6/h; reflected by the rapid increase of the curve in Figure 2f), t_{max} (0.10 h), and $t_{1/2a}$ (0.02 h) (Table 3) would considerably confound the mean values for ALC. As shown in Table 4, mean maximum plasma epicatechin concentration after ingestion of NALC was 0.39 (95% RI: 0.24, 0.65) µmol/L (C_{max}) and reached after 1.13 (95% RI: 0.74, 1.53) h (t_{max}). After ingestion of ALC, mean C_{max} was only 0.11 (95% RI: 0.05, 0.27) µmol/L and obtained after 0.95 (95% RI: 0.36, 1.54) h (t_{max}). As expected, the mean values of C_{max} and $tAUC$ were 3.51 (95% RI: 0.95, 12.98) and 2.82 (95% RI: 0.59, 13.55) times higher for NALC than for ALC, respectively. K_1, k_2,

$t_{1/2a}$, V_{hyp}/F_{abs}, and Cl were comparable between NALC and ALC, except for $t_{1/2e}$ (Table 4). For t_{max}, arithmetic means were also comparable, considering a difference between NALC and ALC values of 0.64 and the 95% RI (−0.67, 1.96).

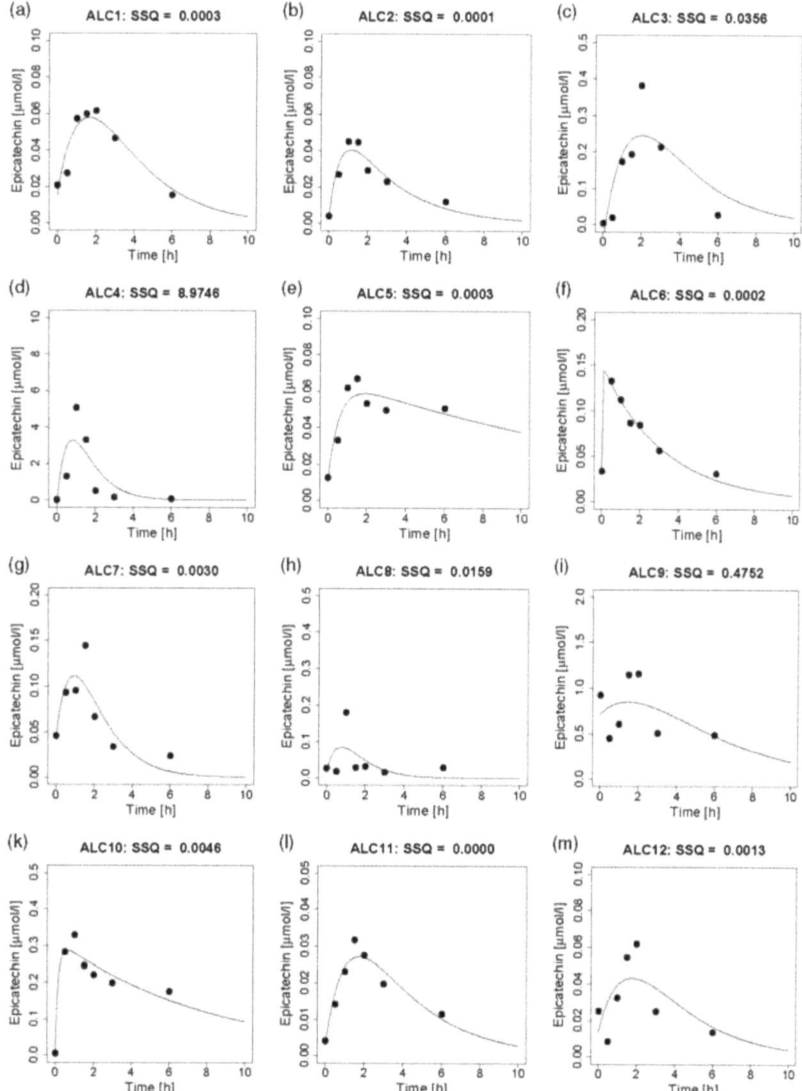

Figure 2. Individual epicatechin concentration time curves obtained after consumption of a beverage prepared with alkalized cocoa. Each panel (**a–m**) represents one participant. The dots represent the measured values. The curves reflect the predicted concentrations in plasma within 10 h after ingestion of alkalized cocoa, based on the one-compartment model. ALC, alkalized cocoa; SSQ, sum of squares of residuals. Note the different scaling of the y-axes.

Table 4. Plasma kinetics of epicatechin after consumption of a beverage prepared with nonalkalized and alkalized cocoa—results from the pharmacokinetic model.

	NALC ($n = 12$)	**ALC ($n = 11$)**	**NALC/ALC ($n = 11$)**
k_1 (h^{-1})	0.86 (0.66, 1.13)	0.95 (0.54, 1.68)	0.91 (0.42, 1.97)
k_2 (h^{-1})	0.63 (0.49, 0.80)	0.34 (0.22, 0.52)	1.85 (1.04, 3.38)
$t_{1/2a}$ (h)	0.81 (0.61, 1.05)	0.73 (0.41, 1.28)	0.76 (0.35, 1.65)
$t_{1/2e}$ (h)	1.10 (0.87, 1.41)	2.04 (1.33, 3.15)	0.37 (0.21, 0.67)
t_{max} (h)	1.13 (0.74, 1.53)	0.95 (0.36, 1.54)	n.d.
C_{max} (µmol/L)	0.39 (0.24, 0.65)	0.11 (0.05, 0.27)	3.51 (0.95, 12.98)
$tAUC$ (µmol/L × h)	1.61 (1.00, 2.58)	0.57 (0.19, 1.70)	2.82 (0.59, 13.55)
V_{hyp}/F_{abs} (l)	192 (120, 308)	211 (101, 442)	0.91 (0.32, 2.56)
Cl (l/h)	103 (64, 166)	78 (26, 231)	1.33 (0.28, 6.38)

Data are weighted geometric means, except for t_{max}, which is given as weighted arithmetic means and 95% reference intervals in parentheses. ALC: alkalized cocoa; $tAUC$: total area under the curve; C_{max}: maximum plasma concentration; Cl: clearance; k_1: absorption rate; k_2: elimination rate; NALC: nonalkalized cocoa; NALC/ALC: ratio after treatment with NALC compared with treatment with ALC; n.d.: not determined; $t_{1/2a}$: absorption half-life time; $t_{1/2e}$: elimination half-life time; t_{max}: time to reach maximum concentration; V_{hyp}/F_{abs}: hypothetical distribution volume over absolute bioavailability.

The $tAUC_{0-6h}$ for epicatechin in plasma obtained by the trapezoidal rule was higher after consumption of NALC compared with ALC (2.23 ± 0.54 vs. 1.14 ± 0.52 µmol/L × h; $p = 0.045$, based on the comparison of logarithmized values), whereas $iAUC_{0-6h}$ was not different after treatment with NALC vs. ALC (1.75 ± 0.46 vs. 0.81 ± 0.45 µmol/L × h; $p = 0.096$, based on the comparison of logarithmized values).

4. Discussion

To the best of our knowledge, this is the first comparative human study evaluating the plasma appearance of catechin and epicatechin stereoisomers after consumption of NALC and ALC as milk drink ingredients representing common food items. Repeated plasma analyses enabled a reliable calculation of pharmacokinetic parameters based on a one-compartment model. The data confirmed our working hypothesis that (−)-catechin from cocoa powders appears only in relatively low concentrations in plasma compared with other flavan-3-ol stereoisomers if considering also the amounts ingested.

Before cocoa consumption, (+)-catechin and (−)-catechin were not detectable in fasting plasma and the concentration of epicatechin was mostly below or near to the LOD. This corresponds to our expectations with regard to the low flavonoid intake on the pre-study day and demonstrates good compliance of our participants with dietary restrictions. As BW remained stable throughout the study, changes in nutrition status which might affect pharmacokinetic parameters can be excluded.

After cocoa consumption, (+)-catechin was only detectable in 3% and (−)-catechin in 7% out of 144 plasma samples, respectively. At the first glance, this was surprising as another RCT observed an increase in catechin metabolites in plasma (no chiral separation) after ingestion of 3.05 mg catechin from cocoa up to concentrations of 0.49 µmol/L, as determined by HPLC with CoulArray detection [11]. However, in contrast to Steinberg et al. [11], sulfated metabolites were not assessed in our study, which might have underestimated the plasma appearance of both (+)-catechin and (−)-catechin.

In the meantime, the first signs on stereochemical differences in the absorption and plasma appearance of (+)-catechin and (−)-catechin, obtained from a study with rats [18] and a self-experiment of Ritter et al. [28], were confirmed by an RCT with healthy volunteers. In this study, the concentration of (−)-catechin metabolites in plasma 2 and 4 h postconsumption as well as their excretion in 24 h urine was lower after ingestion of pure (−)-catechin (1.5 mg/kg BW) compared with those metabolites obtained in plasma and urine after ingestion of equal amounts of either pure (+)-catechin or pure (−)-epicatechin, respectively. Moreover, a stereoisomeric interconversion of flavan-3-ols in vivo was not observed [33]. Thus, the lack of (−)-catechin in 90% of our plasma samples after cocoa consumption might be explained by the relatively low bioavailability of (−)-catechin, whereas the lack of (+)-catechin metabolites in more than 98% of our plasma samples may result from the comparably low intake of (+)-catechin from both NALC and ALC (Table 2).

(−)-Catechin was found in 10 plasma samples after ingestion ALC, but only in 5 samples after ingestion of NALC, despite the intake of similar amounts of (−)-catechin from NALC (0.222 mg/kg BW) and ALC (0.232 mg/kg BW; Table 2). Combined ingestion of (+)-catechin with equal doses of (−)-epicatechin (about 50 mg/kg BW) reduced bioavailability of (+)-catechin in rats [21]. If the higher intake of (−)-epicatechin relative to (−)-catechin from NALC vs. ALC (ratio 3.1 vs. 0.8) could have reduced the absorption of (−)-catechin, is questionable. For low doses, as those used in our study, absorption by a carrier-mediated transport protein as previously suggested for high doses of (+)-catechin and (−)-epicatechin in pig brush border [34] is rather unlikely. Moreover, transferability of the results from rats and *in vitro* studies to humans as well as from (+)-catechin to (−)-catechin remains questionable. If the higher intake of procyanidins from NALC might affect the absorption of catechin in our study, as suggested from a study with Caco-2 cells [22], is not clear as the procyanidins' content was not the only difference between NALC and ALC.

Our results on plasma appearance of epicatechin after cocoa consumption are quite different to those of (+)-catechin and (−)-catechin. In contrast to catechin enantiomers, the concentration of epicatechin in all plasma samples obtained after ingestion of NALC and ALC was above the LOD, although lower amounts of (−)-epicatechin were ingested from ALC compared with (−)-catechin (Table 2). It has to be kept in mind that the LOD of epicatechin was lower (8.27 nmol/L [27]) compared with (+)-catechin (23.43 nmol/L) and (−)-catechin (20.33 nmol/L) [28]. In the study of Steinberg et al. which provided 12.2 mg monomeric flavan-3-ols from cocoa with a ratio of epicatechin to catechin of 3:1, the concentration of epicatechin metabolites in plasma was 10 times higher than those of catechin (4.11 vs. 0.4 μmol/L) [11]. Thus, our results clearly suggest that (−)-epicatechin is the major flavan-3-ol stereoisomer in plasma after cocoa treatment due to higher bioavailability compared with (−)-catechin and due to its higher abundance in cocoa than (+)-catechin.

By using a one-compartment model, absorption and excretion rates of (−)-epicatechin as well as V_{hyp}/F_{abs} could be estimated, which allows to calculate the plasma concentration time course of (−)-epicatechin and to characterize the plasma kinetics by a range of pharmacokinetic parameters. Interestingly, bioavailability of (−)-epicatechin as suggested by C_{max} and $tAUC$ obtained by the model (Table 4) roughly corresponds to the ratio of epicatechin (3.75) ingested by NALC vs. ALC. However, with regard to the $AUCs$ determined by using the trapezoidal rule, the results on bioavailability are less clear as significantly higher values were only detectable for $tAUC_{0–6h}$, but not for $iAUC_{0–6h}$. In single cases, postprandial concentrations of (−)-epicatechin in plasma were not higher than in fasting state, which may explain the lack of significant differences in $iAUC_{0–6h}$. However, inter-individual variability in epicatechin response was high, and the sample size relatively small, which makes it difficult to detect significant differences in $iAUC_{0–6h}$ between both treatments. On the basis of our results, 94 subjects per group (188 in total) would be needed according to the formula of Ott [35] to detect a statistically significant difference in $iAUC_{0–6h}$ of ≥0.930 ± 2.563 μmol/L × h (mean ± SD) between both treatments, presuming a power of 80%, an α of 0.05, and a β of 0.20. Hence, our study was underpowered with regard to $iAUC_{0–6h}$. Shorter intervals for determining $iAUC_{0–6h}$ by the trapezoidal rule would have reduced the error. When planning our study, sample size calculation was not possible as expectations on the difference (as mean ± SD) between both treatments were not available for any flavan-3-ol monomer. Nevertheless, in contrast to rats, where bioavailability of (−)-epicatechin was lower when given together with an equal dose (17.2 mmol/kg BW) of (+)-catechin [21], a lower bioavailability of (−)-epicatechin from ALC compared with NALC due to competitive effects between the absorption of catechin enantiomers and (−)-epicatechin can be nearly ruled out due to their low plasma appearance. Moreover, it is rather unlikely that the different amounts of procyanidins ingested by NALC compared with ALC (Table 2) increased the absorption of (−)-epicatechin, as suggested from a study with Caco-2 cells, as the ratio for C_{max} (3.5) and $tAUC$ (2.8) between the treatment with NALC and ALC in our model roughly corresponds to the higher intake of (−)-epicatechin intake from NALC than from ALC

(Table 4). In the meantime, Caco-2 cells are no longer considered as a good model to study epicatechin absorption as the profile of (−)-epicatechin metabolites from Caco-2 cells and human enterocytes in vivo is different and determines the efflux of (−)-epicatechin metabolites back into the apical side of Caco-2 cells and to the gut lumen, respectively [36]. An RCT with healthy volunteers has recently shown that methylxanthines (1.48 mg theobromine and 0.15 mg caffeine per kg BW) enhance the absorption of (−)-epicatechin [37]. As our subjects ingested similar amounts of methylxanthines from NALC (18.55 mg theobromine, 0.58 mg caffeine) and ALC (18.42 mg theobromine, 0.88 mg caffeine) per kg BW, an impact of methylxanthines on the plasma appearance of (−)-epicatechin in our study is unlikely.

The results on V_{hyp}/F_{abs} were in the same range for NALC and ALC (Table 4). A mean value about 200 L for V_{hyp}/F_{abs} exceeds fluid compartments such as total plasma volume and total body water. After an oral ingestion of 60 mg pure $[2\text{-}^{14}C]\text{-}(-)$-epicatechin in healthy men, only 20% of total radioactivity detected in 24 h urine was due to structurally related epicatechin metabolites (glucuronidated, sulfated, and methylated metabolites), and about half of this radioactivity refers to glucuronidated (−)-epicatechin [38]. Hence, even when sulfated and methylated metabolites were not considered, which leads to an overestimation of V_{hyp}/F_{abs}, the percentage of ingested (−)-epicatechin that reaches systemic circulation seems to be rather low.

In contrast to our hypothesis, k_1, k_2, $t_{1/2a}$, t_{max}, V_{hyp}/F_{abs}, and Cl were comparable after consumption of the beverages which had been prepared with NALC or ALC, except for $t_{1/2e}$ (Table 4), suggesting that the kind of cocoa beverage which differs largely in the flavan-3-ol composition (Table 2) does not affect the velocity of absorption and metabolism of (−)-epicatechin. This is reasonable as C_{max} was comparable after treatment with NALC and ALC, if we consider the 3.7 times larger amount of (−)-epicatechin ingested. It is important to mention that the flavan-3-ol composition was not the only factor which differed between both treatments; the intake of macronutrients was also different by using different amounts of cocoa (Table 2). In an RCT with crossover design, co-ingestion of flavanol-rich cocoa together with sugar (0.75 and 17.5 kJ/kg BW; referring to amounts about 35 and 70 g) or carbohydrate-rich foods (0.75 and 17.5 kJ/kg BW; leading to a carbohydrate intake up to 70 g) increased the AUC of epicatechin in plasma of healthy subjects. This effect was dependent on carbohydrate intake and could not be observed by co-ingestion of foods rich in protein or fat [39]. However, in our study, confounding effects on plasma kinetics of (−)-epicatechin by the use of different amounts of NALC and ALC are unlikely as the difference in the intake of carbohydrates between both cocoa beverages was quite low (58 g vs. 60 g) (Table 2).

The distinct kinetic curve shapes (Figures 1 and 2) after both treatments reflect strong inter-individual differences in the plasma appearance of epicatechin (Table 3). Such variations were also reported to occur after ingestion of pure (−)-epicatechin [33]. Genetic polymorphisms are known to exist for sulfotransferases, but also for UDP-glucuronosyltransferases [40]. Thus, the individual variability in the extent and velocity of conjugation of (−)-epicatechin might contribute to the strong variations in absorption, metabolism, and excretion between our subjects.

A strength of our study is the chiral separation between catechin enantiomers and the use of a nonlinear model which, however, was not applicable to describe the plasma appearance of (+)-catechin and (−)-catechin due to the lack of detectable amounts of catechin metabolites in most samples. However, this model provides a good prediction on plasma kinetics of (−)-epicatechin in most subjects. Moreover, weighing individual values by considering both between and within variances has shown to be a valuable tool to compare the plasma kinetics of (−)-epicatechin after consumption of different cocoa beverages. With regard to the strong inter-individual variations and the fact that most kinetic parameters depend on each other, weighted geometric means, as calculated in our study, are more suitable than unweighted arithmetic means, which were provided in other studies investigating the plasma kinetics of (−)-epicatechin metabolites from cocoa [41–43].

As we did not consider sulfated flavan-3-ol metabolites, the plasma appearance of flavan-3-ol stereoisomers was underestimated, which is a clear limitation of our study. Moreover, the use of different amounts of cocoa powder to ensure an equal intake of total catechin (i.e., sum of (+)-catechin and (−)-catechin) from both treatments might introduce another confounding factor by the intake of different amounts of macronutrients. However, as stated above, this was rather unlikely in our study as the differences between both treatments were negligible (Table 2). The time required for complete ingestion of the cocoa beverages was not different between both treatments, maybe due to the strong inter-individual variability. However, the time required was considered by implication of t_{sh} in our pharmacokinetic model. The strong inter-individual variability cannot be explained by the ingested amount of cocoa powder as it did not correlate with the time needed for complete consumption of each cocoa drink. Inter-individual differences in appetite sensations (e.g., hunger, satiety, fullness) might have affected time for ingestion of both beverages, but, of course, this remains speculative as appetite sensations were not assessed. Even if the model mostly provided good predictions, we observed for single values a discrepancy between predicted and measured concentrations in plasma (e.g., NALC1, NALC9, ALC 3, ALC8). As our model was based on seven plasma samples that were obtained within 6 h, single values with atypically high measured concentrations did not strongly affect the shape of the curve and therefore, these measured values are distant from the predicted values. The collection of blood samples was restricted to 6 h as a return to baseline value was observed in previous studies within 6 h after consumption of cocoa-containing beverages [42,44]. As the measured plasma concentrations did not return to baseline value in all participants within 6 h, we extended the predicted concentration time curves for epicatechin in plasma up to 10 h. If possible, the plasma concentrations should be determined in shorter intervals within the first 6 h and thereafter in larger intervals up to 24 h to completely assess all structurally related epicatechin metabolites [38,45].

5. Conclusions

In conclusion, (−)-epicatechin is the main monomeric flavan-3-ol stereoisomer in plasma after bolus ingestion of a beverage prepared from either NALC or ALC, although ALC provided lower amounts of (−)-epicatechin than of (−)-catechin. Plasma appearance of both (+)-catechin and (−)-catechin is much lower than that of epicatechin, which suggests low bioavailability of both catechin enantiomers from both NALC and ALC. Whether NALC should be preferred against ALC due to its higher (−)-epicatechin content remains unclear considering the results on $iAUC_{0-6h}$. Future studies addressing similar questions should provide equal amounts of cocoa in an equal time frame to a larger sample size. They should investigate all epicatechin metabolites (i.e., glucuronidated and sulfated ones, ideally in nonmethylated and methylated form) as well as genetic polymorphisms of (−)-epicatechin metabolizing enzymes. Better predictions may be achieved by collecting more blood samples within 24 h after cocoa consumption and by considering the individual dose of epicatechin ingested in the model. The role of single factors on plasma appearance of flavan-3-ol monomers, such as the ratio of catechin to epicatechin and the impact of procyandins from cocoa, should be elucidated by matching the treatments.

Author Contributions: The authors' contributions were as follows: S.E., P.S., A.R., B.F.Z., and R.G. contributed to the conception and the design of the study. A.R. recruited the subjects and was responsible for data acquisition. L.H. determined the flavan-3-ol composition of the cocoa powders, except for the chiral analysis, which was done by C.R., B.F.Z. determined the content of caffeine and theobromine in cocoa. C.R. determined (+)-catechin and (−)-catechin in plasma and determined also epicatechin in plasma together with B.F.Z., J.E. was the principal clinical investigator who also collected the blood samples. S.E. calculated the pharmacokinetic parameters with support of H.-P.H. and prepared the manuscript. H.-P.H. was responsible for the application of the one-compartment model and the nonlinear regression. He proposed the bootstrap methods for getting the variances of the parameters in connection with the one-compartment model and the DerSimonian–Laird procedure. S.E., A.R., B.F.Z., P.S. and H.-P.H. interpreted the data and drafted the manuscript. All authors read and approved the final manuscript.

Funding: The study was supported by a research funding by Institute Danone Nutrition for Health, Munich, Germany (S.E., grant No. 2009/19).

Acknowledgments: We thank the Central College of the German Confectionery Industry, Solingen, Germany, for kind provision of NALC and ALC.

Conflicts of Interest: The authors declare no conflict of interest.

References

1. Buitrago-Lopez, A.; Sanderson, J.; Johnson, L.; Warnakula, S.; Wood, A.; Di Angelantonio, E.; Franco, O.H. Chocolate consumption and cardiometabolic disorders: Systematic review and meta-analysis. *BMJ* **2011**, *343*, d4488. [CrossRef]
2. Hooper, L.; Kay, C.; Abdelhamid, A.; Kroon, P.A.; Cohn, J.S.; Rimm, E.B.; Cassidy, A. Effects of chocolate, cocoa, and flavan-3-ols on cardiovascular health: A systematic review and meta-analysis of randomized trials. *Am. J. Clin. Nutr.* **2012**, *95*, 740–751. [CrossRef] [PubMed]
3. Ried, K.; Sullivan, T.R.; Fakler, P.; Frank, O.R.; Stocks, N.P. Effect of cocoa on blood pressure. *Cochrane Database Syst. Rev.* **2012**, *8*, CD008893. [CrossRef]
4. Ellinger, S.; Reusch, A.; Stehle, P.; Helfrich, H.P. Epicatechin ingested via cocoa products reduces blood pressure in humans: A nonlinear regression model with a bayesian approach. *Am. J. Clin. Nutr.* **2012**, *95*, 1365–1377. [CrossRef] [PubMed]
5. Desch, S.; Schmidt, J.; Kobler, D.; Sonnabend, M.; Eitel, I.; Sareban, M.; Rahimi, K.; Schuler, G.; Thiele, H. Effect of cocoa products on blood pressure: Systematic review and meta-analysis. *Am. J. Hypertens.* **2010**, *23*, 97–103. [CrossRef] [PubMed]
6. Hooper, L.; Kroon, P.A.; Rimm, E.B.; Cohn, J.S.; Harvey, I.; Le Cornu, K.A.; Ryder, J.J.; Hall, W.L.; Cassidy, A. Flavonoids, flavonoid-rich foods, and cardiovascular risk: A meta-analysis of randomized controlled trials. *Am. J. Clin. Nutr.* **2008**, *88*, 38–50. [CrossRef] [PubMed]
7. Shrime, M.G.; Bauer, S.R.; McDonald, A.C.; Chowdhury, N.H.; Coltart, C.E.; Ding, E.L. Flavonoid-rich cocoa consumption affects multiple cardiovascular risk factors in a meta-analysis of short-term studies. *J. Nutr.* **2011**, *141*, 1982–1988. [CrossRef]
8. Tokede, O.A.; Gaziano, J.M.; Djousse, L. Effects of cocoa products/dark chocolate on serum lipids: A meta-analysis. *Eur. J. Clin. Nutr.* **2011**, *65*, 879–886. [CrossRef]
9. Heiss, C.; Finis, D.; Kleinbongard, P.; Hoffmann, A.; Rassaf, T.; Kelm, M.; Sies, H. Sustained increase in flow-mediated dilation after daily intake of high-flavanol cocoa drink over 1 week. *J. Cardiovasc. Pharmacol.* **2007**, *49*, 74–80. [CrossRef]
10. Schroeter, H.; Heiss, C.; Balzer, J.; Kleinbongard, P.; Keen, C.L.; Hollenberg, N.K.; Sies, H.; Kwik-Uribe, C.; Schmitz, H.H.; Kelm, M. (-)-epicatechin mediates beneficial effects of flavanol-rich cocoa on vascular function in humans. *Proc. Natl. Acad. Sci. USA* **2006**, *103*, 1024–1029. [CrossRef]
11. Steinberg, F.M.; Holt, R.R.; Schmitz, H.H.; Keen, C.L. Cocoa procyanidin chain length does not determine ability to protect ldl from oxidation when monomer units are controlled. *J. Nutr. Biochem.* **2002**, *13*, 645–652. [CrossRef]
12. Holt, R.R.; Lazarus, S.A.; Sullards, M.C.; Zhu, Q.Y.; Schramm, D.D.; Hammerstone, J.F.; Fraga, C.G.; Schmitz, H.H.; Keen, C.L. Procyanidin dimer b2 [epicatechin-(4beta-8)-epicatechin] in human plasma after the consumption of a flavanol-rich cocoa. *Am. J. Clin. Nutr.* **2002**, *76*, 798–804. [CrossRef] [PubMed]
13. Ottaviani, J.I.; Kwik-Uribe, C.; Keen, C.L.; Schroeter, H. Intake of dietary procyanidins does not contribute to the pool of circulating flavanols in humans. *Am. J. Clin. Nutr.* **2012**, *95*, 851–858. [CrossRef] [PubMed]
14. Miller, K.B.; Hurst, W.J.; Payne, M.J.; Stuart, D.A.; Apgar, J.; Sweigart, D.S.; Ou, B. Impact of alkalization on the antioxidant and flavanol content of commercial cocoa powders. *J. Agric. Food Chem.* **2008**, *56*, 8527–8533. [CrossRef] [PubMed]
15. Andres-Lacueva, C.; Monagas, M.; Khan, N.; Izquierdo-Pulido, M.; Urpi-Sarda, M.; Permanyer, J.; Lamuela-Raventós, R.M. Flavanol and flavonol contents of cocoa powder products: Influence of the manufacturing process. *J. Agric. Food Chem.* **2008**, *56*, 3111–3117. [CrossRef]
16. Gu, L.; House, S.E.; Wu, X.; Ou, B.; Prior, R.L. Procyanidin and catechin contents and antioxidant capacity of cocoa and chocolate products. *J. Agric. Food Chem.* **2006**, *54*, 4057–4061. [CrossRef]
17. Kofink, M.; Papagiannopoulos, M.; Galensa, R. (-)-catechin in cocoa and chocolate: Occurrence and analysis of an atypical flavan-3-ol enantiomer. *Molecules* **2007**, *12*, 1274–1288. [CrossRef]
18. Donovan, J.L.; Crespy, V.; Oliveira, M.; Cooper, K.A.; Gibson, B.B.; Williamson, G. (+)-catechin is more bioavailable than (-)-catechin: Relevance to the bioavailability of catechin from cocoa. *Free Radic. Res.* **2006**, *40*, 1029–1034. [CrossRef]

19. Cooper, K.A.; Campos-Giménez, E.; Jiménez Alvarez, D.; Nagy, K.; Donovan, J.L.; Williamson, G. Rapid reversed phase ultra-performance liquid chromatography analysis of the major cocoa polyphenols and inter-relationships of their concentrations in chocolate. *J. Agric. Food Chem.* **2007**, *55*, 2841–2847. [CrossRef]

20. Ritter, C.; Zimmermann, B.F.; Galensa, R. Chiral separation of (+)/(-)-catechin from sulfated and glucuronidated metabolites in human plasma after cocoa consumption. *Anal. Bioanal. Chem.* **2010**, *397*, 723–730. [CrossRef]

21. Baba, S.; Osakabe, N.; Natsume, M.; Muto, Y.; Takizawa, T.; Terao, J. In vivo comparison of the bioavailability of (+)-catechin, (-)-epicatechin and their mixture in orally administered rats. *J. Nutr.* **2001**, *131*, 2885–2891. [CrossRef] [PubMed]

22. Kosinska, A.; Andlauer, W. Cocoa polyphenols are absorbed in caco-2 cell model of intestinal epithelium. *Food Chem.* **2012**, *135*, 999–1005. [CrossRef] [PubMed]

23. Kothe, L.; Zimmermann, B.F.; Galensa, R. Temperature influences epimerization and composition of flavanol monomers, dimers and trimers during cocoa bean roasting. *Food Chem.* **2013**, *141*, 3656–3663. [CrossRef] [PubMed]

24. Hurst, W.J.; Krake, S.H.; Bergmeier, S.C.; Payne, M.J.; Miller, K.B.; Stuart, D.A. Impact of fermentation, drying, roasting and dutch processing on flavan-3-ol stereochemistry in cacao beans and cocoa ingredients. *Chem. Cent. J.* **2011**, *5*, 53. [CrossRef] [PubMed]

25. Investigation of foods: Determination of theobromine and caffeine in pastries. In *Federal Health Office Collection of Official Methods under Article 35 of German Federal Foods Act*; Methods of Sampling and Analysis of Foods, Tobacco Products, Cosmetics and Commodity Goods; Beuth: Berlin, Germany, 1998. (In German)

26. Kirch, N.; Berk, L.; Liegl, Y.; Adelsbach, M.; Zimmermann, B.F.; Stehle, P.; Stoffel-Wagner, B.; Ludwig, N.; Schieber, A.; Helfrich, H.P.; et al. A nutritive dose of pure (-)-epicatechin does not beneficially affect increased cardiometabolic risk factors in overweight-to-obese adults-A randomized, placebo-controlled, double-blind crossover study. *Am. J. Clin. Nutr.* **2018**, *107*, 948–956. [CrossRef]

27. Zimmermann, B.F.; Papagiannopoulos, M.; Brachmann, S.; Lorenz, M.; Stangl, V.; Galensa, R. A shortcut from plasma to chromatographic analysis: Straightforward and fast sample preparation for analysis of green tea catechins in human plasma. *J. Chromatogr. B Anal. Technol. Biomed Life Sci.* **2009**, *877*, 823–826. [CrossRef]

28. Ritter, C. Method Development for Enantioselective Separation of Polyphenols in Different Matrices. Ph.D. Thesis, University of Bonn, Faculty of Mathematics and Natural Sciences, Bonn, Germany, 2011. (In German).

29. Gabrielsson, J.; Weimer, D. *Pharmacokinetic and Pharmacodynamic Data Analysis: Concepts and Applications*; Swedish Pharmaceutical Press: Stockholm, Sweden, 2000; pp. 27–69.

30. Efron, B. Nonparametric estimates of standard error: The jackknife, the bootstrap and other methods. *Biometrika* **1981**, *63*, 589–599. [CrossRef]

31. DerSimonian, R.; Laird, N. Meta-analysis in clinical trials. *Control. Clin. Trials* **1986**, *7*, 177–188. [CrossRef]

32. Bhagwat, S.; Haytowitz, D.B.; Holden, J.M. *Usda Database for the Flavonoid Content of Selected Foods-Release 3*; US Department of Agriculture, Agricultural Research Service: Baltimore, MD, USA, 2011.

33. Ottaviani, J.I.; Momma, T.Y.; Heiss, C.; Kwik-Uribe, C.; Schroeter, H.; Keen, C.L. The stereochemical configuration of flavanols influences the level and metabolism of flavanols in humans and their biological activity in vivo. *Free Radic. Biol. Med.* **2011**, *50*, 237–244. [CrossRef]

34. Starp, C.; Alteheld, B.; Stehle, P. Characteristics of (+)-catechin and (-)-epicatechin transport across pig intestinal brush border membranes. *Ann. Nutr. Metab.* **2006**, *50*, 59–65. [CrossRef]

35. Ott, L. *Introduction to Statistical Methods and Data Analysis*; Duxberg Press: Belmont, CA, USA, 1977.

36. Actis-Goretta, L.; Leveques, A.; Rein, M.; Teml, A.; Schafer, C.; Hofmann, U.; Li, H.; Schwab, M.; Eichelbaum, M.; Williamson, G. Intestinal absorption, metabolism, and excretion of (-)-epicatechin in healthy humans assessed by using an intestinal perfusion technique. *Am. J. Clin. Nutr.* **2013**, *98*, 924–933. [CrossRef] [PubMed]

37. Sansone, R.; Ottaviani, J.I.; Rodriguez-Mateos, A.; Heinen, Y.; Noske, D.; Spencer, J.P.; Crozier, A.; Merx, M.W.; Kelm, M.; Schroeter, H.; et al. Methylxanthines enhance the effects of cocoa flavanols on cardiovascular function: Randomized, double-masked controlled studies. *Am. J. Clin. Nutr.* **2017**, *105*, 352–360. [CrossRef] [PubMed]

38. Ottaviani, J.I.; Borges, G.; Momma, T.Y.; Spencer, J.P.; Keen, C.L.; Crozier, A.; Schroeter, H. The metabolome of [2-^{14}c] (-)-epicatechin in humans: Implications for the assessment of efficacy, safety, and mechanisms of action of polyphenolic bioactives. *Sci. Rep.* **2016**, *6*, 29034. [CrossRef] [PubMed]

39. Schramm, D.D.; Karim, M.; Schrader, H.R.; Holt, R.R.; Kirkpatrick, N.J.; Polagruto, J.A.; Ensunsa, J.L.; Schmitz, H.H.; Keen, C.L. Food effects on the absorption and pharmacokinetics of cocoa flavanols. *Life Sci.* **2003**, *73*, 857–869. [CrossRef]

40. Burchell, B.; Coughtrie, M.W. Genetic and environmental factors associated with variation of human xenobiotic glucuronidation and sulfation. *Environ. Health Perspect.* **1997**, *105* (Suppl. 4), 739–747. [CrossRef]

41. Richelle, M.; Tavazzi, I.; Enslen, M.; Offord, E.A. Plasma kinetics in man of epicatechin from black chocolate. *Eur. J. Clin. Nutr.* **1999**, *53*, 22–26. [CrossRef]

42. Neilson, A.P.; George, J.C.; Janle, E.M.; Mattes, R.D.; Rudolph, R.; Matusheski, N.V.; Ferruzzi, M.G. Influence of chocolate matrix composition on cocoa flavan-3-ol bioaccessibility in vitro and bioavailability in humans. *J. Agric. Food Chem.* **2009**, *57*, 9418–9426. [CrossRef]

43. Mullen, W.; Borges, G.; Donovan, J.L.; Edwards, C.A.; Serafini, M.; Lean, M.E.; Crozier, A. Milk decreases urinary excretion but not plasma pharmacokinetics of cocoa flavan-3-ol metabolites in humans. *Am. J. Clin. Nutr.* **2009**, *89*, 1784–1791. [CrossRef]

44. Roura, E.; Andres-Lacueva, C.; Estruch, R.; Mata-Bilbao, M.L.; Izquierdo-Pulido, M.; Waterhouse, A.L.; Lamuela-Raventos, R.M. Milk does not affect the bioavailability of cocoa powder flavonoid in healthy human. *Ann. Nutr. Metab.* **2007**, *51*, 493–498. [CrossRef]

45. Actis-Goretta, L.; Leveques, A.; Giuffrida, F.; Romanov-Michailidis, F.; Viton, F.; Barron, D.; Duenas-Paton, M.; Gonzalez-Manzano, S.; Santos-Buelga, C.; Williamson, G.; et al. Elucidation of (-)-epicatechin metabolites after ingestion of chocolate by healthy humans. *Free Radic. Biol. Med.* **2012**, *53*, 787–795. [CrossRef]

Review

Effect of Cocoa Products and Its Polyphenolic Constituents on Exercise Performance and Exercise-Induced Muscle Damage and Inflammation: A Review of Clinical Trials

Marika Massaro [1,*], Egeria Scoditti [1], Maria Annunziata Carluccio [1], Antonia Kaltsatou [2] and Antonio Cicchella [3]

[1] National Research Council—Institute of Clinical Physiology, Laboratory of Nutrigenomic and Vascular Biology, 73100 Lecce, Italy
[2] FAME Laboratory, Department of Physical Education and Sport Science, University of Thessaly, 42100 Trikala, Greece
[3] Department for Quality of Life Studies, University of Bologna, 40126 Bologna, Italy
* Correspondence: marika.massaro@ifc.cnr.it; Tel.: +39-0832-298-860

Received: 6 May 2019; Accepted: 25 June 2019; Published: 28 June 2019

Abstract: In recent years, the consumption of chocolate and, in particular, dark chocolate has been "rehabilitated" due to its high content of cocoa antioxidant polyphenols. Although it is recognized that regular exercise improves energy metabolism and muscle performance, excessive or unaccustomed exercise may induce cell damage and impair muscle function by triggering oxidative stress and tissue inflammation. The aim of this review was to revise the available data from literature on the effects of cocoa polyphenols on exercise-associated tissue damage and impairment of exercise performance. To this aim, PubMed and Web of Science databases were searched with the following keywords: "intervention studies", "cocoa polyphenols", "exercise training", "inflammation", "oxidative stress", and "exercise performance". We selected thirteen randomized clinical trials on cocoa ingestion that involved a total of 200 well-trained athletes. The retrieved data indicate that acute, sub-chronic, and chronic cocoa polyphenol intake may reduce exercise-induced oxidative stress but not inflammation, while mixed results are observed in terms of exercise performance and recovery. The interpretation of available results on the anti-oxidative and anti-inflammatory activities of cocoa polyphenols remains questionable, likely due to the variety of physiological networks involved. Further experimental studies are mandatory to clarify the role of cocoa polyphenol supplementation in exercise-mediated inflammation.

Keywords: athlete; cocoa; chocolate; exercise performance; oxidative stress; performance; physical exercise; polyphenol; skeletal muscle; inflammation

---

1. Introduction

Skeletal muscle exerts a dominant role in postural control, the protection of internal organs, locomotion, and other physiological functions requiring energy-mediated mechanical activity based on muscle fiber contraction [1]. Since skeletal muscle is the largest organ in the body [2] and acts as a thermal machine, skeletal muscle energy consumption plays a key role in the regulation of whole-body energy homeostasis [1]. However, recently, skeletal muscle has been proposed as a potential "endocrine organ" that is able to orchestrate the release of an array of muscle-derived signaling molecules or myokines, including interleukin(IL)-6, IL-7, IL-8, IL-10, IL-15, and IL-1 receptor antagonists, irisin, and myostatin [3]. The tightly regulated release of such myokines is involved in many exercise-induced metabolic adaptations, such as those related to glycemic control and lipid homeostasis [4], as well as in the regulation of muscle fiber composition and contractility [5]. Furthermore, the increases in muscle mass and vascularization appear to be regulated by myokines [5]

From a pathogenic point of view, it is well known that physical inactivity predisposes individuals to developing a variety of chronic diseases [6], while the practice of regular physical exercise exerts a range of beneficial effects on body health [7]. Regular exercise is indeed recognized to improve whole-body energy metabolism and to make muscles stronger and more resistant to fatigue [8]. However, this relationship can be described by an inverted U-shaped dose–response curve (exercise hormesis theory [9]), as follows: when the stressor (the exercise bout) is absent, no adaptation occurs. Positive, healthful adaptation starts when the dose of exercise is within a specific intensity and duration range and is followed by a rest period [10]. However, if exercise bouts are too heavy or extended, as in athletes involved in prolonged and intensive exercise activities [11], or are not followed by rest periods, muscle fiber damage and inflammation may occur, requiring exogenous intervention to re-establish normal muscle functionality [10] or to delay the onset of fatigue through substances acting on the central nervous system (CNS) [12]. Mechanical and metabolic stress accompanying excessive exercise may result in damage to the integrity of myofibers and may cause non-permanent reductions in contractile function [13]. At the same time, muscle injury triggers a complex tissue inflammatory reaction featured by the release of pro-inflammatory myokines and stress hormones with the final aim of repairing damaged muscle and orchestrating regenerative and adaptive processes [14]. Under these conditions, the activation of resident endothelial cells and the recruitment of inflammatory leukocytes, including neutrophils and macrophages, occurs, leading to cumulative production of pro-inflammatory and anti-inflammatory cytokines that, together with myokines, actively contribute to the removal of necrotic tissue and cellular debris [10]. Once necrotic tissue is removed, satellite cells proliferate to regenerate muscle tissue [15]. By this multi-step process, exercise-induced inflammation supports cellular remodeling, promotes hypertrophic adaptations, and restores tissue homeostasis and contractile function [16].

Besides muscle glycogen breakdown and increased levels of calcium, catecholamines, growth hormone, and cortisol [17], cytokine expression and release during exercise is mediated by the overproduction of reactive oxygen species (ROS) [18]. Exercise may indeed increase oxygen consumption (VO_2) up to 20 times above resting values [18]. In the mitochondria of muscle cells, this translates to 200-fold greater oxygen utilization and the subsequent production of a large amount of ROS [18]. ROS generated through this route may lead to oxidative damage to the mitochondria and muscle contractile proteins, with subsequent direct induction of muscle damage and fatigue after exercise [10]. Furthermore, ROS also orchestrate the activation of redox-sensitive signal pathways that control cytokine production and muscle adaptation, such as those involved in the activation of nuclear factor(NF)-κB, Nuclear Factor of Activated T-cells (NFAT), Nuclear factor erythroid 2-related factor(Nrf2) and heat shock proteins (HSPs), and Peroxisome Proliferator-Activated Receptor-gamma Coactivator (PGC)-1α [17]. In particular, the expression of most immune-related genes requires activation of the pleiotropic, redox-sensitive NF-κB and its binding to specific consensus sequences to activate gene transcription [19]. Among these genes, there are many myokines including IL-1, IL-8, IL-6, chemotactic factors (such as the monocyte chemotactic protein (MCP)-1, many endothelial leukocytes, adhesion molecules (including the vascular cell adhesion molecule (VCAM)-1), and the pro-inflammatory prostaglandin producing enzyme cyclooxygenase (COX)-2 [20]. In this way, NF-κB promotes the release of pro-inflammatory mediators, including chemokines, cytokines, acute phase proteins, and adhesion molecules, to facilitate the regenerative response in damaged skeletal muscle [21]. Furthermore, several studies have clearly indicated that muscle ROS and NF-κB also activate other important cell signaling pathways, leading to skeletal muscle adaptations to exercise such as mitochondrial biogenesis [22,23] and endogenous antioxidant defense [24].

However, due to overwhelming ROS production and the overcoming of antioxidant tissue defense during periods of intensified physical training, chronically activated NF-κB and the following dysregulated production of inflammatory myokines may lead to skeletal muscle atrophy and soreness [20].

Prolonged, high-intensity, strenuous or unaccustomed bouts of exercise have been experimentally associated with increases in contractile-induced damage and inflammation reactions in skeletal muscle [25]. For example, dysregulation in the inflammatory system has been observed in athletes undergoing intense periods of physical training, as highlighted by excessive delayed-onset muscle soreness, muscle stiffness, a reduction in muscle strength, increased creatine kinase (CK) activity, and impaired immune function [25]. Under these conditions, the functional performance of skeletal muscle has been shown to be reduced for at least 24–96 h [26], with serious health and economic implications for professional athletes and related clubs and societies [27]. Therefore, while exercise-induced inflammation is necessary for muscle repair and adaptation, the uncontrolled proliferation of inflammatory cells and oxidants can exacerbate muscle damage and impair muscle function.

The growing evidence on exercise-induced oxidative damage and impairment of athlete performance has spurred intense research on the evaluation of muscle protection by antioxidant supplementation in exercising individuals [25]. Although many studies have shown potential therapeutic effects by antioxidant supplementation [28–30], results from several others remain inconsistent. For example, supplementation with vitamin C and N-acetylcysteine was found to increase oxidative stress and cell damage following eccentric exercise [31]. Similarly, in a recent study by Bailey et.al, a combination of vitamins C and E was shown to increase markers of oxidative stress and inflammation [32]. Since low amounts of free radicals may act as cellular signals to enhance antioxidant defenses, rather than being deleterious, as occurs at higher concentrations, it has been suggested that high antioxidant supplementation may attenuate some of the exercise-induced cellular signals that stimulate adaptations in skeletal muscle [18]. This has been proven to be true upon oral supplementation with vitamins C and E [22,33] after an endurance trial and after high intensity exercise.

Based on these findings, dietary recommendations in exercising individuals should emphasize the consumption of a well-balanced diet and/or natural antioxidant-rich foods such as cocoa and chocolate, rather than taking antioxidant supplements. This "nutraceutical strategy" has been increasingly proposed as a potential suitable tool for preventing or reducing oxidative stress and related inflammation during intensive physical training. In particular, besides being a high energy-dense foods, cocoa and cocoa products including chocolate are a rich source of antioxidant polyphenols that have proven to possess health-promoting effects through their antioxidant, anti-inflammatory and metabolic properties [34].

The overall goal of the current review was therefore to examine the effects of cocoa polyphenols and on different redox-related outcome parameters, including oxidative and inflammatory biomarkers, and muscle performance and recovery.

2. Polyphenol Antioxidant Profile in Cocoa and Chocolate

Chocolate and cocoa are foodstuffs derived from the beans of the Theobromacacao tree [35]. After harvesting, cacao pods are processed to form a liquid paste known as "cocoa liquor", the cocoa component is richer in healthful bioactives [35]. The healthiest types of chocolate are indeed those containing higher amounts of cocoa liquor, namely dark chocolate, and, to a lesser extent, milk chocolate [36]. Most of the health effects of cocoa-rich chocolate are due to the high content of nutritional polyphenols in cocoa liquor [37]. In the last thirty years, polyphenols have attracted much interest owing to their antioxidant capacity (free radical scavenging and metal chelating ability) and their possible beneficial implications in human health, such as in the treatment and prevention of cancer and cardiovascular disease [38]. Hence, cocoa has the health effects generally ascribed to polyphenol consumption [34].

Cocoa seeds contain many bioactive compounds including high levels of polyphenols (12–18% of dry weight) as well as fatty acids, vitamins, minerals, fiber, and several methylxanthine alkaloids

(4% of dry weight), which are psychoactive dopaminergic substances such as caffeine, theobromine, theophylline, phenylethylamine, and paraxanthine (Figure 1).

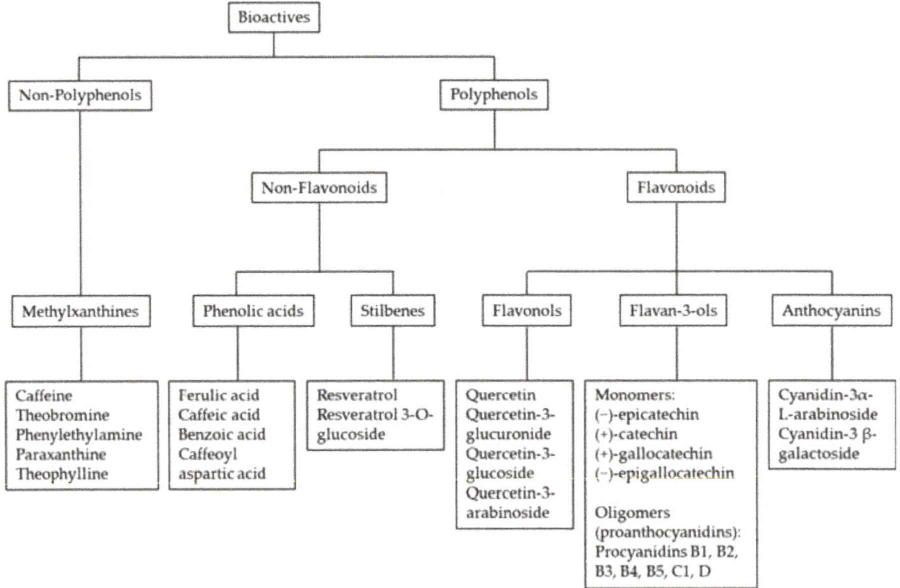

Figure 1. Main bioactive compounds of cocoa and chocolate.

From a bromatological point of view, the most abundant polyphenolic classes in cocoa and chocolate are:

(a) the monomeric flavan-3-ols or catechins (up to 29–38% of total cocoa polyphenols) including catechin, gallocatechin, and epigallocatechin (Figure 1). In particular, epicatechin may represent up to 35% of total polyphenols in both cocoa powder and chocolate [39];

(b) the proanthocyanidins (up to 58–65% of total cocoa polyphenols), which are polymers of epicatechin and catechin and include mainly dimeric and trimeric molecules [39];

(c) the anthocyanins (up to 4% of total cocoa polyphenols), which include cyanidin-3α-L-arabinoside and cyanidin-3 β-galactoside as the most represented compounds [39].

Finally, some other phenolic compounds can be found at very low concentrations including phenolic acids, caffeoyl-conjugates, and stilbenes [39]. Notably, the proportions of these compounds depend on the cultivar, origin, agricultural practices, and postharvest practices and processing. In particular, the phenolic content and profile of cocoa-derived products change considerably during the manufacturing process, especially during fermentation and alkalization [40].

The metabolic fate of phenolic compounds after ingestion is a critical aspect in determining the health effects of such compounds and the mechanisms through which they exert their biological activities. While oligomers larger than trimers are unlikely to be absorbed, dimers have been reported in human plasma after cocoa ingestion [41]. Catechin and epicatechin are well absorbed in the small intestine. They are transiently detectable in plasma as glucuronide-conjugated metabolites and sulfate groups with a maximum plasma concentration at around 2 h and a return to baseline values within 24 h (in general, in 6–8 h) of the consumption of a flavonoid-rich chocolate meal (such as dark chocolate, dose range 40–100 g) [42–44]. Recently, a fine metabolomic assessment on human plasma and urine samples collected at 2 and 6 h after the consumption of flavan-3-ol-enriched dark chocolate highlighted the presence of discriminant epicatechin derivatives in the urine, thus confirming the high bioavailability of cocoa flavanols [45]. Also, in athletes, the assessment of the plasma epicatechin concentration is often used as a biomarker of chocolate intake to correlate the effects of chocolate on metabolic endpoints to the effective polyphenol absorption [46]. However, to the best of our knowledge, no study has specifically addressed the metabolism and pharmacokinetics of chocolate polyphenols in athletes following physical exercise.

3. Literature Search Strategy

Two electronic databases were consulted: PubMed and Web of Science. Key terms that were included and combined were "intervention studies", "cocoa polyphenols", "exercise training", "inflammation", "oxidative stress", and "exercise performance". The final search was carried out on 24 April 2019. The search strategies were combined, and duplicates were removed by Endnote X7 (Clarivate Analytics, previously Thomson Reuters, Philadelphia, PA, US) and manually. Studies in this section needed to fulfill the following inclusion criteria: (i) research conducted with human participants and (ii) original data from randomized clinical trials (RCTs) on cocoa ingestion with an acute or long-term exercise intervention. The exclusion criteria were (i) studies written in languages other than English, (ii) animal or in vitro studies, (iii) congress or workshop publications, (iv) studies in which no exercise was performed, (v) studies in which no supplementation was given, and (vi) studies in which mixed supplements were given. No limits were used concerning the year of publication. The inclusion or exclusion of articles was determined by applying the above criteria on the title and abstract as a first screening and on full texts as a second screening. Case studies and reviews were excluded, although the respective references were consulted and integrated into this revision if responding to the above-mentioned criteria.

4. Effects of Cocoa Polyphenols on Exercise-Induced Oxidative Stress, Inflammation, and Recovery

Thirteen intervention studies with parallel or crossover design met all inclusion criteria and were included in the analysis. Overall, they involved a total of 200 participants, mostly well-trained athletes, and examined the effects of cocoa polyphenol intake on exercise-induced changes in plasma markers of inflammation, oxidative stress, and performance.

Eight studies administered polyphenols in the form of solid chocolate [46–52], while five studies examined the effects of cocoa polyphenols in a liquid formulation [53–57]. The number of polyphenols tested was rather varied, ranging from 36 to 1000 mg, administered in an acute (2 h before exercise), sub-chronic (for 2 weeks before exercise), and chronic (up to three months before exercise) fashion. No study involved female athletes, and tested sports and exercise disciplines were mostly cycling, football, and running. The results are compared and summarized in Table 1 for acute studies, and Table 2 for sub-chronic and chronic studies.

The selected studies examined the effects of cocoa polyphenol supplementation on the exercise-mediated deregulation of oxidative redox status by evaluating markers of oxidative damage (including malondialdehyde [MDA], 4-hydroxynonenal [4HNE], F2-isoprostane, and thiobarbituric acid reactive substances [TBARS], and protein carbonylation) and antioxidant protection (glutathione and total antioxidant status—TAS) [46–48,50,52–55,57,58]. Three studies analyzed the effects of the acute administration of cocoa polyphenols on exercise-induced oxidative stress (Table 1). In particular, Davison et al. observed that the consumption of a single low amount (248 mg) of cocoa flavanols as solid dark chocolate 2 h before exercise was able to enhance pre-exercise antioxidant status and, correspondently, tended to reduce the post-exercise plasma F2-isoprostane content as compared with a control condition, thus suggesting an improvement in antioxidant defense for athletes following the consumption of dark chocolate enriched in polyphenols [58]. Similarly, Wiswedel et al. observed that a single low intake (186 mg) of cocoa flavanols did not increase the total antioxidant capacity but was able to prevent increases in F2-isoprostane and MDA induced by 30 min of cycling [57]. On the other hand, Decroix at al. evaluated the consumption of a single but higher dose (900 mg) of cocoa flavanols in the form of a beverage given 30 min before exercise. Again, the total antioxidant capacity was significantly increased by cocoa but without a clear effect in terms of MDA [54]. Furthermore, Davidson et al. [58] and Decroix et al. [54] also evaluated the effect of polyphenol ingestion on a classical marker of inflammation, IL-6, and specifically evaluated whether cocoa supplementation affects exercise performance and recovery. Both authors independently observed that cocoa flavanols had no effects on IL-6 plasma levels or on exercise performance and recovery. In agreement with Davidson [58] and Decroix [54], Stellingwerff et al. [51] and Peschek et al. [56] did not find any improvement in physical performance following the acute administration of low or medium concentrations of cocoa polyphenols following downhill running [56] or cycling [51] (Table 1).

Seven studies instead examined the effects of sub-chronic administration (from 5 days up to 2 weeks) of cocoa flavanols before or during exercise (Table 2). Allgrove et al. [47], Patel et al. [49], and Singh et al. [50] evaluated the effects of low amounts of cocoa polyphenols (about 200 mg/day) on cycling-mediated oxidative stress and inflammation. After two weeks of supplementation plus a double dose of cocoa polyphenols on the testing day, 2 h before cycling exercise for 1.5 h, Allgrove et al. [47] found a significant increase in the total antioxidant activity capacity but no effects on exercise performance or on the plasma levels of several pro-inflammatory myokines, including IL-6, IL-1RA, and IL-10, were evident. However, under similar experimental conditions (in terms of polyphenol concentration and loading time), Patel et al. observed an improvement in muscle physical performance as indicated by a lower oxygen demand during analogous moderate intensity exercise [49]. Finally, Singh evaluated the effects of a polyphenol loading time of 7 days on trained and untrained cyclists [50]. However, he failed to show improvements in the total antioxidant status in both sub-types of athletes [50]. Slightly different were the effects shown for sub-chronic cocoa polyphenol supplementations in well-trained football players. Under these conditions, both Fraga et al. [48] and Gonzalez-Garrido et al. [55] showed significant improvements in antioxidant defense and, interestingly, in exercise performance following low and high polyphenol administration, respectively [48,55]. However, these positive results are not supported by the recent findings of de Carvalho et al. in rugby players [53]. Here, the sub-chronic administration of cocoa polyphenols neither improved oxidative stress markers and plasma CK levels, nor enhanced athletic performance [53]. Only two studies have examined the effects of the very long chronic administration (three months) of cocoa polyphenols on oxidative stress and inflammation markers in football players [46] and cyclists [52] (Table 2). Both studies concordantly observed improvements in different oxidative stress markers including the hydrogen peroxide breakdown activity level, reduced glutathione, and protein carbonyls [46,52]. However, discordant effects were observed for exercise performance outcomes

Table 1. Acute effects of cocoa polyphenols on exercise-mediated oxidative stress, inflammation, and exercise performance.

First Author, Year, (Reference)	Study Design	Number of Participants, Sex, Age, Weight	Exercise Protocol	Interventions (mg Polyphenols Daily)	Measurements	Circulating Leucocyte	IL-6	CK; LDH	Oxidative Stress Markers	EP and R Improvement (Yes/No)
Davidson et al. [58]	RCT, crossover design	14, male, 22 ± 1 years, 71.6 ± 1.6 kg	2.5 h cycling	100 g DCHO (39.1 mg catechin, 96.8 mg epicatechin); cocoa-liquor-free control bar (0 mg); 2 h before exercise	Rest, pre-exercise, post-exercise, after 1 h recovery	↔ by DCHO	↔ in DCHO	NE	TAS: ↑ in DCHO; F2-isoprostane: ↓ by DCHO	No
Decroix et al. [54]	RCT, crossover design	12, male, 30.5 ± 3 years, 72.8 ± 7.8 kg	2 cycling sessions of 30 min;	HFCD (900 mg) or LFCD (15 mg) before first exercise session	Baseline, after 100 and 230 min from supplementation; after training sessions	NE	↔ by CF	NE	TEAC: ↑ by HFCD; MDA: ↔ by HFCD	No
Wiswedel et al. [57]	RCT, double-blind, cross-over	10, male, 20–40 years	Cycling, 29 min	HFCD (187 mg flavanols) or LFCD (14 mg flavanols) 2 h pre-exercise	At baseline and after 2, 4 and 6 h from ingestion	NE	NE	NE	F2-isoprostanes: ↔ by HFCD; ↑ by LFCD	NE
Peschek et al. [56]	RCT, single-blind, cross-over	8, male, 24.6 ± 5.6 years, 73.4 ± 7.0 kg	30 min downhill running; after 48 h, 5 km running	HFCD (2 × 350 mg) or placebo (0 mg) after downhill running	At baseline, after 24 h, after 48 h, after 5 km running	NE	NE	CK and LDH: ↔ by HFCD	NE	No
Stellingwerff et al. [51]	Randomized, single blind, crossover	9, male, 30.0 ± 6.1 years, 72.8 ± 6.0 kg	2.5 h cycling, plus a time trial of 15 min	HFCHO (262 mg) or LFCHO (<0.05 mg) acutely (2 h pre-exercise)	After time trial	NE	NE	NE	NE	No

Abbreviations: CK, creatine kinase; DCHO, dark chocolate; EP, exercise performance; HFCD, high-flavanol cocoa drink; HFCHO, high-flavanol dark chocolate; IL-6, interleukin 6; LDH, lactate dehydrogenase; LFCD, low-flavanol cocoa drink; LFCHO, low-flavanol chocolate; MDA, malondialdehyde; NE, not evaluated; R, recovery; RCT, randomized controlled trial; TAS, Total Antioxidant Status; TEAC, Trolox Equivalent Antioxidant Capacity; ↓, decrease; ↑ increase; ↔ no change.

Table 2. Sub-chronic and chronic effects of cocoa polyphenols on exercise-mediated oxidative stress, inflammation, and exercise performance.

First Author, Year, (Reference)	Study Design	Number of Participants, Sex, Age, Weight	Exercise Protocol	Interventions, (Polyphenols Daily mg)	Measurements	Circulating Leucocyte	IL-6	CK; LDH	Oxidative Stress Markers	EP and R Improvement (Yes/No)
Allgrove et al. [47]	RCT, crossover design	20, male, 22 ± 4 years, 74.6 ± 8 kg	Cycling for 90 min followed by 25 min time trial	80 g of DCHO (197.4 mg) or 56.8 g of cocoa liquor-free chocolate control (0 mg) bar for 2 weeks before the trial and on the trial day	Before exercise, post-exercise bout, post- exhaustion, and after 1 h of resting recovery	↔ by DCHO	↔ by DCHO	NE	TAS: ↔ by DCHO; F2-isoprostanes: ↔ in DCHO; ↑ in control	No
Singh et al. [50]	RCT, double-blind, cross-over	16, male, 23 ± 5 years, 79 ± 11 kg	Cycling, 60 min	Cocoa polyphenol supplement (240 mg) or placebo (0 mg) for 7 days	At baseline and after 8 days	NE	NE	NE	TAS: ↔ by cocoa polyphenol supplement	NE
Patel et al. [49]	Randomized, single blind, cross-over	9, male, 21 ± 1 years, 76 ± 9.3 kg	Cycling, 20 min plus a 2 min time trial	40 g DCHO (259 mg) or 40 g white chocolate, 2 weeks	At baseline, after 14 days	NE	NE	NE	NE	Yes
Fraga et al. [48]	RCT, crossover design	28, male, 18 ± 1 years, 74.0 ± 0.2 kg	Football, 2 times a week training	FCMCHO (168 mg) or cocoa butter white chocolate for 2 weeks	At baseline and after 2 weeks	NE	NE	CK:↔ by FCMCHO; LDH: ↓ FCMCHO	MDA: ↓ by FCMCHO; OXOd and TRAP: ↔ by FCMCHO	Yes
Gonzalez-Garrido et al. [55]	Intervention study with pre/post-design	15, male, 17.0 ± 1.11 years, 66.98 ± 6.52 kg	Football training five days/week, and 90 min match/week	HFCD (1050 mg) for 5 days	At baseline and after 6 days	NE	NE	CK and LDH: ↓ by HFCD	TBARS: ↔ by HFCD; MDA: ↓ by HFCD; 4-HNE: ↓ by HFCD; Carbonyl groups: ↓ by HFCD; TAS: ↑ by HFCD	Yes
de Carvalho et al. [53]	RCT, double-blind	13, male, 20.69 ± 1.49 years, 87.02 ± 8.03 kg	Daily rugby match for 5 days	HFCD (2 × 308 mg) or LFCD for 5 days	At baseline and after 6 days	NE	NE	CK: ↔ by HFCD	F2-isoprostanes: ↔ by HFCD	No
Cavarretta et al. [46]	RCT, double-blind	20, male, 17.8 ± 0.9 years (control); 17.4 ± 0.5 years (intervention)	120 min football training, 6 times/week, and 90 min match/week	Normal diet plus 40 g DCHO in tablet (36 mg) or normal diet for 30 days	At baseline and after 60 days	NE	NE	CK and LDH: ↓ by DCHO	HBA: ↓ by DCHO	No
Taub et al. [52]	RCT, double-blind, cross-over	17, male, 49.5 ± 1.6 years, 79 ± 11 kg	Cycling	HFCHO (175 mg) or LFCHO (1.2 mg) for 3 months	At baseline and after 3 months	NE	NE	NE	GSH/GSSG: ↑ by HFCHO; Protein carbonyl: ↓ by HFCHO	Yes

Abbreviations: 4-HNE, 4-hydroxynonenal; CK, creatine kinase; DCHO, dark chocolate; EP, exercise performance; FCMCHO, flavanol-containing milk chocolate; GSH/GSSG, reduced glutathione(GSH) to oxidized glutathione ratio; HBA, hydrogen peroxide (H_2O_2) breakdown activity; HFCD, high-flavanol cocoa drink; HFCHO, high-flavanol dark chocolate; IL-6, interleukin-6; LDH, lactate dehydrogenase; LFCD, low-flavanol cocoa drink; LFCHO, low-flavanol chocolate; MDA, malondialdehyde; NE, not evaluated; OXOdg, 8-oxo-20-deoxyguanosine; R, recovery; TAS, Total Antioxidant Status; TBARS, thiobarbituric acid reactive substance; TRAP, total relative antioxidant potency; ↓, decrease; ↑ increase; ↔ no change.

5. Discussion

In this review, we examined the effects of cocoa polyphenol intake on exercise-mediated oxidative stress and inflammation, performance, and recovery.

Examining thirteen studies, we observed that acute, sub-chronic, and chronic cocoa polyphenol intakes reduce exercise-induced oxidative stress but not inflammatory markers, while the effects on exercise performance and recovery remain controversial.

Both acute and long-term exercise are now widely recognized as potential pro-oxidative and pro-inflammatory promoters [25]. Although the generation of ROS represents a physiological process in most human metabolic reactions, when ROS production and related endogenous antioxidant abilities are imbalanced, a maladaptive biological response may occur, leading to both oxidative stress and inflammation [59]. In muscle cells, aerobic energy production generates a significant amount of ROS, which can increase by up to 10–20-fold during physical exercise [60]. Although moderate levels of ROS may serve as signaling molecules that mediate muscle repair and adaptation, protein turn-over, mitochondrial biogenesis, and the upregulation of antioxidant enzyme levels [9], increasing evidence suggests that high and unbalanced ROS levels are able to deregulate the redox state and induce a high rate of pro-inflammatory myokine release in muscles, which may lead to contractile muscle dysfunction, accelerated muscle fatigue, longer recovery time, and reduced exercise performance [61].

Therefore, although inflammation provoked by physical activity is directly related to the intensity of the effort, if the exercise bout is excessive in terms of intensity and/or duration, the human body may need some kind of external "aid" to activate the recovery process. A list of forbidden substances is issued and continuously updated by the anti-doping agency [62,63], leaving many possibilities open for the introduction and use of non-doping supplements. A number of substances of vegetal origin have been proposed as potential tools to delay fatigue onset during physical activity and/or to promote the recovery process.

Chocolate and cocoa polyphenols are among these substances. Different from other proposed food supplements, chocolate, and in particular dark chocolate, has been postulated to possess a dual function: it might act as an ergogenic support due to its characteristic richness in saturated fat and sugar, but being also rich in methylxanthines (including theobromine, and caffeine), it might directly act as a stimulant on the CNS [38]. Furthermore, due to its richness in antioxidant polyphenols, it might impact exercise-mediated redox deregulation and inflammation. Evidence on the effect of chocolate on brain functions has been reported, especially in terms of delayed perception of fatigue [12]. Some positive evidence exists on the influence of the consumption of a chocolate milk beverage between three bouts of intense exercise in cycling [64] and after endurance sport in general [65]. Subsequent research failed to demonstrate the effectiveness of chocolate milk in comparison with raw milk plus honey in reducing delayed muscle soreness [66], making evidence of the utility of chocolate in exercise recovery weak. In these studies, however, the chocolate polyphenol content was very low or absent and/or not specifically addressed, thus introducing a bias in the interpretation of results.

The data reviewed here are clearly indicative of a dual effect by cocoa polyphenols. On the one hand, in a rather homogeneous fashion, they collectively highlight the ability of cocoa polyphenols to exert antioxidant effects, as indicated by reduced accumulation of lipid and protein oxidation products and by the strengthening of the athlete antioxidant capacities. On the other hand, the data do not show a clear improvement in post-exercise pro-inflammatory status, and the evidence is rather faint regarding the effect of cocoa polyphenols on exercise performance and post-exercise recovery.

Therefore, although the protective effects of polyphenol supplementation have been largely documented in humans as well as in animal and in vitro studies [67], the available data reported here show contradictory results and, in agreement with other revisions [68–70], do not sustain a clear benefit of cocoa polyphenol supplementation in ameliorating the inflammatory changes induced by exercise.

Notably, most studies are not directly comparable because they used low-flavanol cocoa or milk chocolate as controls, with different contents of not only flavanols but also other potential bioactives including methylxanthines. Moreover, potential interactions between methylxanthines and cocoa polyphenols have been found, resulting in synergistic/additive effects and increased bioavailability of flavanols when co-ingested with methylxanthines [71]. It seems, therefore, appropriate to match the test and control treatments for methylxanthine content to separate the effects of cocoa polyphenols. This aspect may in part contribute to some inconsistencies in studies' results. However, among the selected papers, Stellingwerff et al., even observing an increase in theobromine and epicatechin plasma level after dark chocolate consumption, did not find improvement in exercise performance [51], this uggesting the necessity of further investigations.

Many other factors may contribute to explaining these disappointing results. First of all, previous positive results mainly come from studies investigating polyphenol activity in pathological conditions (in particular cardiovascular disease and cancer) where the redox status is highly unbalanced and the addition of dietary antioxidants may effectively favor the restoration of the right equilibrium [72,73]. On the other hand, finely tuned ROS production during exercise is essential to promote the expression of several proteins that are crucial for exercise-induced adaptation, and the use of antioxidants in supra-physiological doses may be detrimental and/or alter the oxidative stress response in terms of pro-inflammatory gene induction. Thus, antioxidant supplementation might produce adverse consequences by decreasing the ROS concentration beyond the required homeostatic level. Finally, the additional effects of exogenous supplemental antioxidants on different types of exercise are difficult to predict, because exercise itself is a positive stimulus that generally drives the antioxidant capacity. Exercise itself is a strong masking agent that is able to obscure any possible effects of a single substance [9].

It is also important to consider different biological responses related to the type of exercise and, also, that antioxidants might be effective during specific periods of training and that the requirements may vary according to different seasonal needs [74]. For all of these reasons, a personalized plan that considers all of the specific requirements of athletes during the different phases of training would represent the best option to improve global performance, since the training process is highly variable and dependent on a wide range of factors [74]. All of these management aspects should be taken into consideration when planning future experimental studies to ascertain the protective role of cocoa polyphenols in professional and amateur athletes.

Finally, between-studies comparisons are sometimes difficult because of the often-limited sample size, as well as the different populations, types of training and physical activity levels, and background diets of the participants that may influence the effects of chocolate supplementation and exercise. In general, selected studies recruited no more than 10–15 subjects, with the exception of only three studies [46–48] that recruited 20 or more subjects. A small sample size reduces the power of the study and increases the margin of error, rendering the study meaningless. Relatively more homogeneous conditions were observed under acute cocoa administration. Here, the most recurrent kind of physical exercise was cycling but practiced for different times ranging from 2.5 h [51,58] to 30 min [54,57].

Levels of serum polyphenol concentrations, as an index of adherence to the protocol, were evaluated in all acute studies [51,54,57,58], except in the study of Peschek [56]. Data on cocoa polyphenols availability suggest a more frequent increase in epicatechin levels [51,54,58] with respect to catechin [57]. On the other hand, under sub-chronic and chronic cocoa administration, the assessed physical activities were more diversified including football [46,48,55] and rugby [53] besides cycling [47,49,50,52]. Moreover, levels of serum polyphenol concentrations were evaluated in only two studies with discordant results since, under the same sport activity, i.e., the football, Fraga et al. did not find any increase in the plasma levels of epicatechin and catechin [48], while Caravetta et al. did find a clear increase in the epicatechin levels [46]. However, independent of the increase in the plasma levels of cocoa polyphenols, both studies observed a clear improvement in the oxidative stress status.

It is conceivable that the effects of cocoa flavonoids, as well as of other food supplements, may be more evident as the severity and temporal extension of inflammation increases, which mostly results from exhausting low-intensity aerobic exercise. So, the standardization of physical activity intervention, sample size, and administered supplement should be carefully addressed in further studies concerning cocoa flavonoid usage in professional and amateur sport.

Few studies have tried to deepen the understanding of the mechanisms of polyphenol activity by examining the activation of pro-inflammatory transcriptional factors through muscle biopsies before and after antioxidant supplementation and training [75,76]. Nieman et al. obtained disappointing results regarding muscle NF-κB activation, which was shown to be unaffected by exercise and following quercetin supplementation [11]. Such mechanistic exploration should be reproduced and confirmed. Finally, only four studies [47,54,55,58] evaluated the plasma levels of pro-inflammatory markers, including IL-6, IL-1, TNF-α, IL-1 RA, and CRP. It is possible that the tested inflammatory biomarkers are not sensitive or specific enough to be modulated through the cocoa-mediated nutraceutical approaches. The array of myokines tested should be broadened in the search for potential more sensitive and specific biomarkers.

In conclusion, the evidence supporting the effects of the consumption of cacao or dark chocolate on exercise performance and/or exercise-mediated inflammation remains weak.

At present, there is no evidence supporting the use of cacao or dark chocolate as an ergogenic aid. Evidence on the antioxidative and anti-inflammatory effects of cocoa polyphenols in athletes remains weak due to the variety of physiological networks involved. Further experimental studies are necessary in order to clarify the interaction of exercise training and cocoa antioxidant supplementation and the beneficial effects of cocoa polyphenols in exercise-mediated inflammation.

Author Contributions: Conceptualization, M.M.; methodology, M.M. and A.K; investigation, M.M. and E.S.; writing—original draft preparation, M.M. and E.S.; writing—review and editing, M.M., A.C., A.K. and M.A.C.; supervision, M.M., E.S., A.C.

Funding: This research received no external funding.

Conflicts of Interest: The authors declare no conflict of interest.

References

1. Ivanenko, Y.; Gurfinkel, V.S. Human Postural Control. *Front. Neurosci.* **2018**, *12*, 171. [CrossRef] [PubMed]
2. Pedersen, B.K. Muscle as a secretory organ. *Compr. Physiol.* **2013**, *3*, 1337–1362. [PubMed]
3. Hoffmann, C.; Weigert, C. Skeletal Muscle as an Endocrine Organ: The Role of Myokines in Exercise Adaptations. *Cold Spring Harb. Perspect. Med.* **2017**, *7*, a029793. [CrossRef] [PubMed]
4. Pedersen, B.K. Anti-inflammatory effects of exercise: Role in diabetes and cardiovascular disease. *Eur. J. Clin. Investig.* **2017**, *47*, 600–611. [CrossRef] [PubMed]
5. Serrano, A.L.; Baeza-Raja, B.; Perdiguero, E.; Jardi, M.; Munoz-Canoves, P. Interleukin-6 is an essential regulator of satellite cell-mediated skeletal muscle hypertrophy. *Cell Metab.* **2008**, *7*, 33–44. [CrossRef] [PubMed]
6. González, K.; Fuentes, J.; Márquez, J.L. Physical Inactivity, Sedentary Behavior and Chronic Diseases. *Korean J. Fam. Med.* **2017**, *38*, 111–115. [CrossRef] [PubMed]
7. Lin, X.; Zhang, X.; Guo, J.; Roberts, C.K.; McKenzie, S.; Wu, W.C.; Liu, S.; Song, Y. Effects of Exercise Training on Cardiorespiratory Fitness and Biomarkers of Cardiometabolic Health: A Systematic Review and Meta-Analysis of Randomized Controlled Trials. *J. Am. Heart Assoc.* **2015**, *4*. [CrossRef] [PubMed]
8. Mujika, I.; Ronnestad, B.R.; Martin, D.T. Effects of Increased Muscle Strength and Muscle Mass on Endurance-Cycling Performance. *Int. J. Sports Physiol. Perform.* **2016**, *11*, 283–289. [CrossRef] [PubMed]
9. Ji, L.L.; Kang, C.; Zhang, Y. Exercise-induced hormesis and skeletal muscle health. *Free Radic. Biol. Med.* **2016**, *98*, 113–122. [CrossRef] [PubMed]
10. Peake, J.M.; Neubauer, O.; Della Gatta, P.A.; Nosaka, K. Muscle damage and inflammation during recovery from exercise. *J. Appl. Physiol.* **2017**, *122*, 559–570. [CrossRef] [PubMed]
11. Nieman, D.C. Marathon training and immune function. *Sports Med.* **2007**, *37*, 412–415. [CrossRef] [PubMed]
12. Meeusen, R. Exercise, nutrition and the brain. *Sports Med.* **2014**, *44* (Suppl. 1), S47–S56. [CrossRef] [PubMed]
13. Proske, U.; Morgan, D.L. Muscle damage from eccentric exercise: Mechanism, mechanical signs, adaptation and clinical applications. *J. Physiol.* **2001**, *537*, 333–345. [CrossRef] [PubMed]

14. Pedersen, B.K.; Hoffman-Goetz, L. Exercise and the immune system: Regulation, integration, and adaptation. *Physiol. Rev.* **2000**, *80*, 1055–1081. [CrossRef] [PubMed]

15. Dumont, N.A.; Bentzinger, C.F.; Sincennes, M.C.; Rudnicki, M.A. Satellite Cells and Skeletal Muscle Regeneration. *Compr. Physiol.* **2015**, *5*, 1027–1059. [PubMed]

16. Stupka, N.; Tarnopolsky, M.A.; Yardley, N.J.; Phillips, S.M. Cellular adaptation to repeated eccentric exercise-induced muscle damage. *J. Appl. Physiol. (1985)* **2001**, *91*, 1669–1678. [CrossRef]

17. Peake, J.M.; Suzuki, K.; Coombes, J.S. The influence of antioxidant supplementation on markers of inflammation and the relationship to oxidative stress after exercise. *J. Nutr. Biochem.* **2007**, *18*, 357–371. [CrossRef]

18. Peternelj, T.T.; Coombes, J.S. Antioxidant Supplementation during Exercise Training Beneficial or Detrimental? *Sports Med.* **2011**, *41*, 1043–1069. [CrossRef]

19. Peterson, J.M.; Bakkar, N.; Guttridge, D.C. NF-kappaB signaling in skeletal muscle health and disease. *Curr. Top. Dev. Biol.* **2011**, *96*, 85–119.

20. Kramer, H.F.; Goodyear, L.J. Exercise, MAPK, and NF-kappaB signaling in skeletal muscle. *J. Appl. Physiol. (1985)* **2007**, *103*, 388–395. [CrossRef]

21. Aoi, W.; Naito, Y.; Takanami, Y.; Kawai, Y.; Sakuma, K.; Ichikawa, H.; Yoshida, N.; Yoshikawa, T. Oxidative stress and delayed-onset muscle damage after exercise. *Free Radic. Biol. Med.* **2004**, *37*, 480–487. [CrossRef] [PubMed]

22. Gomez-Cabrera, M.C.; Domenech, E.; Romagnoli, M.; Arduini, A.; Borras, C.; Pallardo, F.V.; Sastre, J.; Vina, J. Oral administration of vitamin C decreases muscle mitochondrial biogenesis and hampers training-induced adaptations in endurance performance. *Am. J. Clin. Nutr.* **2008**, *87*, 142–149. [CrossRef] [PubMed]

23. Paulsen, G.; Cumming, K.T.; Holden, G.; Hallen, J.; Ronnestad, B.R.; Sveen, O.; Skaug, A.; Paur, I.; Bastani, N.E.; Ostgaard, H.N.; et al. Vitamin C and E supplementation hampers cellular adaptation to endurance training in humans: A double-blind, randomised, controlled trial. *J. Physiol.* **2014**, *592*, 1887–1901. [CrossRef] [PubMed]

24. Gomez-Cabrera, M.C.; Domenech, E.; Vina, J. Moderate exercise is an antioxidant: Upregulation of antioxidant genes by training. *Free Radic. Biol. Med.* **2008**, *44*, 126–131. [CrossRef] [PubMed]

25. Pyne, D.B.; Smith, J.A.; Baker, M.S.; Telford, R.D.; Weidemann, M.J. Neutrophil oxidative activity is differentially affected by exercise intensity and type. *J. Sci. Med. Sport* **2000**, *3*, 44–54. [CrossRef]

26. Ascensao, A.; Rebelo, A.; Oliveira, E.; Marques, F.; Pereira, L.; Magalhaes, J. Biochemical impact of a soccer match - analysis of oxidative stress and muscle damage markers throughout recovery. *Clin. Biochem.* **2008**, *41*, 841–851. [CrossRef] [PubMed]

27. Hagglund, M.; Walden, M.; Magnusson, H.; Kristenson, K.; Bengtsson, H.; Ekstrand, J. Injuries affect team performance negatively in professional football: An 11-year follow-up of the UEFA Champions League injury study. *Br. J. Sports Med.* **2013**, *47*, 738–742. [CrossRef] [PubMed]

28. Davis, J.M.; Carlstedt, C.J.; Chen, S.; Carmichael, M.D.; Murphy, E.A. The dietary flavonoid quercetin increases VO(2max) and endurance capacity. *Int. J. Sport Nutr. Exerc. Metab.* **2010**, *20*, 56–62. [CrossRef] [PubMed]

29. McAnulty, L.S.; Nieman, D.C.; Dumke, C.L.; Shooter, L.A.; Henson, D.A.; Utter, A.C.; Milne, G.; McAnulty, S.R. Effect of blueberry ingestion on natural killer cell counts, oxidative stress, and inflammation prior to and after 2.5 h of running. *Appl. Physiol. Nutr. Metab.* **2011**, *36*, 976–984. [CrossRef]

30. Nieman, D.C.; Williams, A.S.; Shanely, R.A.; Jin, F.; McAnulty, S.R.; Triplett, N.T.; Austin, M.D.; Henson, D.A. Quercetin's influence on exercise performance and muscle mitochondrial biogenesis. *Med. Sci. Sports Exerc.* **2010**, *42*, 338–345. [CrossRef]

31. Childs, A.; Jacobs, C.; Kaminski, T.; Halliwell, B.; Leeuwenburgh, C. Supplementation with vitamin C and N-acetyl-cysteine increases oxidative stress in humans after an acute muscle injury induced by eccentric exercise. *Free Radic. Biol. Med.* **2001**, *31*, 745–753. [CrossRef]

32. Bailey, D.M.; Williams, C.; Betts, J.A.; Thompson, D.; Hurst, T.L. Oxidative stress, inflammation and recovery of muscle function after damaging exercise: Effect of 6-week mixed antioxidant supplementation. *Eur. J. Appl. Physiol.* **2011**, *111*, 925–936. [CrossRef]

33. Morrison, D.; Hughes, J.; Della Gatta, P.A.; Mason, S.; Lamon, S.; Russell, A.P.; Wadley, G.D. Vitamin C and E supplementation prevents some of the cellular adaptations to endurance-training in humans. *Free Radic. Biol. Med.* **2015**, *89*, 852–862. [CrossRef] [PubMed]

34. Magrone, T.; Russo, M.A.; Jirillo, E. Cocoa and Dark Chocolate Polyphenols: From Biology to Clinical Applications. *Front. Immunol.* **2017**, *8*, 677. [CrossRef]

35. McShea, A.; Leissle, K.; Smith, M.A. The essence of chocolate: A rich, dark, and well-kept secret. *Nutrition* **2009**, *25*, 1104–1105. [CrossRef] [PubMed]

36. Katz, D.L.; Doughty, K.; Ali, A. Cocoa and chocolate in human health and disease. *Antioxid. Redox Signal.* **2011**, *15*, 2779–2811. [CrossRef] [PubMed]

37. Sies, H.; Schewe, T.; Heiss, C.; Kelm, M. Cocoa polyphenols and inflammatory mediators. *Am. J. Clin. Nutr.* **2005**, *81*, 304S–312S. [CrossRef]

38. Gammone, M.A.; Efthymakis, K.; Pluchinotta, F.R.; Bergante, S.; Tettamanti, G.; Riccioni, G.; D'Orazio, N. Impact of chocolate on the cardiovascular health. *Front. Biosci. (Landmark Ed.)* **2018**, *23*, 852–864. [CrossRef]

39. Aprotosoaie, A.C.; Luca, S.V.; Miron, A. Flavor Chemistry of Cocoa and Cocoa Products—An Overview. *Compr. Rev. Food Sci. Food Saf.* **2016**, *15*, 73–91. [CrossRef]

40. Andres-Lacueva, C.; Monagas, M.; Khan, N.; Izquierdo-Pulido, M.; Urpi-Sarda, M.; Permanyer, J.; Lamuela-Raventós, R.M. Flavanol and Flavonol Contents of Cocoa Powder Products: Influence of the Manufacturing Process. *J. Agric. Food Chem.* **2008**, *56*, 3111–3117. [CrossRef]

41. Holt, R.R.; Lazarus, S.A.; Sullards, M.C.; Zhu, Q.Y.; Schramm, D.D.; Hammerstone, J.F.; Fraga, C.G.; Schmitz, H.H.; Keen, C.L. Procyanidin dimer B2 [epicatechin-(4beta-8)-epicatechin] in human plasma after the consumption of a flavanol-rich cocoa. *Am. J. Clin. Nutr.* **2002**, *76*, 798–804. [CrossRef] [PubMed]

42. Flammer, A.J.; Sudano, I.; Wolfrum, M.; Thomas, R.; Enseleit, F.; Périat, D.; Kaiser, P.; Hirt, A.; Hermann, M.; Serafini, M.; et al. Cardiovascular effects of flavanol-rich chocolate in patients with heart failure. *Eur. Heart J.* **2011**, *33*, 2172–2180. [CrossRef] [PubMed]

43. Rusconi, M.; Conti, A. Theobroma cacao L., the Food of the Gods: A scientific approach beyond myths and claims. *Pharmacol. Res.* **2010**, *61*, 5–13. [CrossRef] [PubMed]

44. Williamson, G. Bioavailability and health effects of cocoa polyphenols. *Inflammopharmacology* **2009**, *17*, 111. [CrossRef] [PubMed]

45. Ostertag, L.M.; Philo, M.; Colquhoun, I.J.; Tapp, H.S.; Saha, S.; Duthie, G.G.; Kemsley, E.K.; de Roos, B.; Kroon, P.A.; Le Gall, G. Acute Consumption of Flavan-3-ol-Enriched Dark Chocolate Affects Human Endogenous Metabolism. *J. Proteome Res.* **2017**, *16*, 2516–2526. [CrossRef] [PubMed]

46. Cavarretta, E.; Peruzzi, M.; Del Vescovo, R.; Di Pilla, F.; Gobbi, G.; Serdoz, A.; Ferrara, R.; Schirone, L.; Sciarretta, S.; Nocella, C.; et al. Dark Chocolate Intake Positively Modulates Redox Status and Markers of Muscular Damage in Elite Football Athletes: A Randomized Controlled Study. *Oxid. Med. Cell Longev.* **2018**, *2018*, 4061901. [CrossRef] [PubMed]

47. Allgrove, J.; Farrell, E.; Gleeson, M.; Williamson, G.; Cooper, K. Regular dark chocolate consumption's reduction of oxidative stress and increase of free-fatty-acid mobilization in response to prolonged cycling. *Int. J. Sport Nutr. Exerc. Metab.* **2011**, *21*, 113–123. [CrossRef]

48. Fraga, C.G.; Actis-Goretta, L.; Ottaviani, J.I.; Carrasquedo, F.; Lotito, S.B.; Lazarus, S.; Schmitz, H.H.; Keen, C.L. Regular consumption of a flavanol-rich chocolate can improve oxidant stress in young soccer players. *Clin. Dev. Immunol.* **2005**, *12*, 11–17. [CrossRef]

49. Patel, R.K.; Brouner, J.; Spendiff, O. Dark chocolate supplementation reduces the oxygen cost of moderate intensity cycling. *J. Int. Soc. Sports Nutr.* **2015**, *12*, 47. [CrossRef]

50. Singh, I.; Quinn, H.; Mok, M.; Southgate, R.J.; Turner, A.H.; Li, D.; Sinclair, A.J.; Hawley, J.A. The effect of exercise and training status on platelet activation: Do cocoa polyphenols play a role? *Platelets* **2006**, *17*, 361–367. [CrossRef]

51. Stellingwerff, T.; Godin, J.P.; Chou, C.J.; Grathwohl, D.; Ross, A.B.; Cooper, K.A.; Williamson, G.; Actis-Goretta, L. The effect of acute dark chocolate consumption on carbohydrate metabolism and performance during rest and exercise. *Appl. Physiol. Nutr. Metab.* **2014**, *39*, 173–182. [CrossRef] [PubMed]

52. Taub, P.R.; Ramirez-Sanchez, I.; Patel, M.; Higginbotham, E.; Moreno-Ulloa, A.; Roman-Pintos, L.M.; Phillips, P.; Perkins, G.; Ceballos, G.; Villarreal, F. Beneficial effects of dark chocolate on exercise capacity in sedentary subjects: Underlying mechanisms. A double blind, randomized, placebo controlled trial. *Food Funct.* **2016**, *7*, 3686–3693. [CrossRef] [PubMed]

53. de Carvalho, F.G.; Fisher, M.G.; Thornley, T.T.; Roemer, K.; Pritchett, R.; Freitas, E.C.; Pritchett, K. Cocoa flavanol effects on markers of oxidative stress and recovery after muscle damage protocol in elite rugby players. *Nutrition* **2019**, *62*, 47–51. [CrossRef] [PubMed]

54. Decroix, L.; Tonoli, C.; Soares, D.D.; Descat, A.; Drittij-Reijnders, M.J.; Weseler, A.R.; Bast, A.; Stahl, W.; Heyman, E.; Meeusen, R. Acute cocoa Flavanols intake has minimal effects on exercise-induced oxidative stress and nitric oxide production in healthy cyclists: A randomized controlled trial. *J. Int. Soc. Sports Nutr.* **2017**, *14*, 28. [CrossRef] [PubMed]

55. Gonzalez-Garrido, J.A.; Garcia-Sanchez, J.R.; Garrido-Llanos, S.; Olivares-Corichi, I.M. An association of cocoa consumption with improved physical fitness and decreased muscle damage and oxidative stress in athletes. *J. Sports Med. Phys. Fitness* **2017**, *57*, 441–447. [PubMed]

56. Peschek, K.; Pritchett, R.; Bergman, E.; Pritchett, K. The effects of acute post exercise consumption of two cocoa-based beverages with varying flavanol content on indices of muscle recovery following downhill treadmill running. *Nutrients* **2013**, *6*, 50–62. [CrossRef]

57. Wiswedel, I.; Hirsch, D.; Kropf, S.; Gruening, M.; Pfister, E.; Schewe, T.; Sies, H. Flavanol-rich cocoa drink lowers plasma F(2)-isoprostane concentrations in humans. *Free Radic. Biol. Med.* **2004**, *37*, 411–421. [CrossRef]

58. Davison, G.; Callister, R.; Williamson, G.; Cooper, K.A.; Gleeson, M. The effect of acute pre-exercise dark chocolate consumption on plasma antioxidant status, oxidative stress and immunoendocrine responses to prolonged exercise. *Eur. J. Nutr.* **2012**, *51*, 69–79. [CrossRef]

59. Sies, H. Oxidative stress: A concept in redox biology and medicine. *Redox Biol.* **2015**, *4*, 180–183. [CrossRef]

60. Sjodin, B.; Hellsten Westing, Y.; Apple, F.S. Biochemical mechanisms for oxygen free radical formation during exercise. *Sports Med.* **1990**, *10*, 236–254. [CrossRef]

61. Becatti, M.; Mannucci, A.; Barygina, V.; Mascherini, G.; Emmi, G.; Silvestri, E.; Wright, D.; Taddei, N.; Galanti, G.; Fiorillo, C. Redox status alterations during the competitive season in elite soccer players: Focus on peripheral leukocyte-derived ROS. *Intern. Emerg. Med.* **2017**, *12*, 777–788. [CrossRef] [PubMed]

62. Mazzoni, I.; Barroso, O.; Rabin, O. The list of prohibited substances and methods in sport: Structure and review process by the world anti-doping agency. *J. Anal. Toxicol.* **2011**, *35*, 608–612. [CrossRef] [PubMed]

63. World anti-doping Agency. The World Anti-Doping Code International Standard: Substances & Methods Prohibited at All Times. Available online: https://www.wada-ama.org/sites/default/files/wada_2019_english_prohibited_list.pdf (accessed on 1 January 2019).

64. Karp, J.R.; Johnston, J.D.; Tecklenburg, S.; Mickleborough, T.D.; Fly, A.D.; Stager, J.M. Chocolate milk as a post-exercise recovery aid. *Int. J. Sport Nutr. Exerc. Metab.* **2006**, *16*, 78–91. [CrossRef]

65. Pritchett, K.; Pritchett, R. Chocolate milk: A post-exercise recovery beverage for endurance sports. *Med. Sport Sci.* **2012**, *59*, 127–134. [PubMed]

66. Hatchett, A.; Berry, C.; Oliva, C.; Wiley, D.; St Hilaire, J.; LaRochelle, A. A Comparison between Chocolate Milk and a Raw Milk Honey Solution's Influence on Delayed Onset of Muscle Soreness. *Sports (Basel, Switz.)* **2016**, *4*, 18. [CrossRef]

67. Perez-Vizcaino, F.; Fraga, C.G. Research trends in flavonoids and health. *Arch. Biochem. Biophys.* **2018**, *646*, 107–112. [CrossRef] [PubMed]

68. Decroix, L.; Soares, D.D.; Meeusen, R.; Heyman, E.; Tonoli, C. Cocoa Flavanol Supplementation and Exercise: A Systematic Review. *Sports Med.* **2018**, *48*, 867–892. [CrossRef] [PubMed]

69. Malaguti, M.; Angeloni, C.; Hrelia, S. Polyphenols in exercise performance and prevention of exercise-induced muscle damage. *Oxid. Med. Cell Longev.* **2013**, *2013*, 825928. [CrossRef]

70. Pingitore, A.; Lima, G.P.; Mastorci, F.; Quinones, A.; Iervasi, G.; Vassalle, C. Exercise and oxidative stress: Potential effects of antioxidant dietary strategies in sports. *Nutrition* **2015**, *31*, 916–922. [CrossRef]

71. Sansone, R.; Ottaviani, J.I.; Rodriguez-Mateos, A.; Heinen, Y.; Noske, D.; Spencer, J.P.; Crozier, A.; Merx, M.W.; Kelm, M.; Schroeter, H.; et al. Methylxanthines enhance the effects of cocoa flavanols on cardiovascular function: Randomized, double-masked controlled studies. *Am. J. Clin. Nutr.* **2017**, *105*, 352–360. [CrossRef]

72. Vauzour, D.; Rodriguez-Mateos, A.; Corona, G.; Oruna-Concha, M.J.; Spencer, J.P. Polyphenols and human health: Prevention of disease and mechanisms of action. *Nutrients* **2010**, *2*, 1106–1131. [CrossRef]

73. Zhang, P.Y. Polyphenols in Health and Disease. *Cell Biochem. Biophys.* **2015**, *73*, 649–664. [CrossRef] [PubMed]

74. Potgieter, S. Sport nutrition: A review of the latest guidelines for exercise and sport nutrition from the American College of Sport Nutrition, the International Olympic Committee and the International Society for Sports Nutrition. *S. Afr. J. Clin. Nutr.* **2013**, *26*, 6–16. [CrossRef]

75. Nieman, D.C.; Henson, D.A.; Davis, J.M.; Angela Murphy, E.; Jenkins, D.P.; Gross, S.J.; Carmichael, M.D.; Quindry, J.C.; Dumke, C.L.; Utter, A.C.; et al. Quercetin's influence on exercise-induced changes in plasma cytokines and muscle and leukocyte cytokine mRNA. *J. Appl. Physiol. (1985)* **2007**, *103*, 1728–1735. [CrossRef]

76. Ristow, M.; Zarse, K.; Oberbach, A.; Kloting, N.; Birringer, M.; Kiehntopf, M.; Stumvoll, M.; Kahn, C.R.; Bluher, M. Antioxidants prevent health-promoting effects of physical exercise in humans. *Proc. Natl. Acad. Sci. USA* **2009**, *106*, 8665–8670. [CrossRef] [PubMed]

 *nutrients*

Article

Impact of a Usual Serving Size of Flavanol-Rich Cocoa Powder Ingested with a Diabetic-Suitable Meal on Postprandial Cardiometabolic Parameters in Type 2 Diabetics—A Randomized, Placebo-Controlled, Double-Blind Crossover Study

Janina Rynarzewski [1], **Lisa Dicks** [1], **Benno F. Zimmermann** [2], **Birgit Stoffel-Wagner** [3], **Norbert Ludwig** [1,4], **Hans-Peter Helfrich** [5] **and Sabine Ellinger** [1,*]

[1] Faculty of Food, Nutrition and Hospitality Sciences, Hochschule Niederrhein, University of Applied Sciences, Rheydter Str. 277, 41065 Mönchengladbach, Germany; ja.rynarzewski@gmail.com (J.R.); Lisa.Dicks@hs-niederrhein.de (L.D.); Norbert.Ludwig@hs-niederrhein.de (N.L.)

[2] Department of Nutrition and Food Sciences, University of Bonn, Endenicher Allee 11-13, 53115 Bonn, Germany; benno.zimmermann@uni-bonn.de

[3] Institute of Clinical Chemistry and Clinical Pharmacology, University Hospital Bonn, Sigmund-Freud-Str. 25, 53127 Bonn, Germany; birgit.stoffel-wagner@ukbonn.de

[4] Praxis Anrath, Pastoratstr. 9-11, 47877 Willich, Germany

[5] Institute of Numerical Simulation, University of Bonn, Endenicher Allee 60, 53115 Bonn, Germany; helfrich@uni-bonn.de

* Correspondence: Sabine.Ellinger@hs-niederrhein.de; Tel.: +49-2161-186-5406; Fax: +49-2161-186-5313

Received: 12 January 2019; Accepted: 7 February 2019; Published: 16 February 2019

Abstract: Randomized controlled trials indicate that flavanol-rich cocoa intake may improve postprandial glucose and lipid metabolism in patients with type 2 diabetes (T2D), based on studies with meals that impose a strong metabolic load. Hence, the aim of the present study was to investigate whether flavanol-rich cocoa powder ingested as part of a diabetic-suitable meal may beneficially affect glucose, lipid metabolism, and blood pressure (BP) in patients with T2D. Twelve adults with T2D, overweight/obesity, and hypertension ingested capsules with 2.5 g of flavanol-rich cocoa or microcrystalline cellulose with a diabetic-suitable breakfast in a randomized, placebo-controlled, double-blind crossover study. BP was measured and blood samples were taken before, 2 and 4 h after breakfast and capsule intake. Cocoa treatment did not affect glucose, insulin, homeostasis model assessment for insulin resistance (HOMA-IR), triglycerides, total cholesterol, low density lipoprotein-cholesterol, high density lipoprotein-cholesterol, and BP. For glucose, insulin and HOMA-IR, only effects by time were observed after both treatments. Thus, 2.5 g of flavanol-rich cocoa powder ingested as part of a diabetic-suitable meal does not seem to affect postprandial glucose and lipid metabolism and BP in stably-treated diabetics. Nevertheless, future studies with close-meshed investigations are desirable, providing realistic amounts of cocoa together with realistic meals rich in carbohydrates to subjects with T2D or metabolic syndrome, which do not afford pharmacological treatment.

Keywords: type 2 diabetes; flavanol-rich cocoa; postprandial; meal; glucose metabolism; lipids; blood pressure

---

1. Introduction

The prevalence of type 2 diabetes (T2D) is increasing globally [1]. In patients with T2D, postprandial hyperglycemia has shown to raise the incidence of cardiovascular disease (CVD)

and all-cause mortality [2,3]. Postprandial hyperglycemia is often accompanied by postprandial hypertriglyceridemia, which acts as a further risk factor for CVD. Dietary modifications to lower postprandial glucose and triglyceride values are recommended [4]. Thus, functional food and food ingredients which may improve metabolic and vascular biomarkers could be favorable for patients with T2D [5].

Regular consumption of flavanol-rich cocoa may be beneficial for patients with T2D. Cocoa flavanols have shown to enhance insulin secretion, improve insulin sensitivity in peripheral tissues, lower lipids [6] and increase nitric oxide availability [7]. A decrease in insulin resistance [8], triglycerides [8], low density lipoprotein-cholesterol (LDL-cholesterol) [9], and blood pressure (BP) [8], as well as an increase in insulin sensitivity and high density lipoprotein-cholesterol (HDL-cholesterol) [8] could be observed in meta-analyses of randomized controlled trials (RCTs) after regular cocoa treatment. These effects were even more substantial in morbid subjects than in healthy ones without any comorbidities [8].

In patients with T2D, the postprandial effect of flavanol-rich cocoa on cardiometabolic parameters has only been investigated in a small number of RCTs to date. Most participants received oral hypoglycemic drugs [10,11] as well as lipid- and BP-lowering drugs [11] to ensure stable metabolic and BP control. HDL-cholesterol and endothelial function were improved after 4 h when the fast-food breakfast was ingested together with a flavanol-rich cocoa drink and not with a flavanol-poor placebo drink [11]. Acute hyperglycemia-induced endothelial dysfunction was reduced in individuals with T2D after consumption of flavanol-rich chocolate compared to chocolate low in flavanols [10]. However, a 75-g pure glucose load as provided by Mellor et al. [10] is given as an oral glucose tolerance test and used for diagnostic purposes [12,13]. The fast-food breakfast used by Basu et al. [11] provided 766 kcal, with fat as 59% of total energy. However, meals rich in isolated carbohydrates, saturated fat, and cholesterol and low in dietary fiber are not recommended in T2D [14]. Moreover, 20 g of cocoa powder (about eight tablespoons) for the preparation of a cocoa-rich drink is an unrealistic amount and such a drink is not a typical component of a high-fat-fast-food meal.

Thus, the aim of this study was to investigate whether a usual serving size of flavanol-rich cocoa powder, ingested together with a diabetic-suitable meal, may improve postprandial changes in glucose and lipid metabolism as well as BP.

2. Materials and Methods

2.1. Study Design and Intervention

This randomized, placebo-controlled, double-blind crossover study was performed between July and October 2017 in a medical practice (Praxis Anrath, Willich, Germany) according to the Declaration of Helsinki. The study was approved by the ethics committees of the University of Bonn (project identification code: 051/17; date of approval: 22 February 2017) and the Medical Association of North Rhine (project identification code: 2017138; date of approval: 30 May 2017). The trial was registered at German Clinical Trial Register (DRKS-ID: DRKS00012561) on 6 June 2017. Written informed consent was obtained from all participants for inclusion before they participated in the study.

Participants were consecutively recruited and allocated to two different groups by permuted block randomization (block size of four, sequence generated by drawing lots by an uninvolved person). After an overnight fast, both groups ingested five A- and B-capsules, respectively, in different order together with a diabetic-suitable breakfast. Each capsule was filled with 0.5 g ACTICOA™ cocoa powder (Barry Callebaut, Zurich, Switzerland, lot no. 100-F017906-AC-796) or pure microcrystalline cellulose (J. Rettenmaier and Söhne, Rosenberg, Germany) by KP Productions (Koblenz, Germany; certified to ISO 9001/2008, FSSC 22000, and HACCP). Hence, five cocoa-containing capsules provided 2.5 g of cocoa in total, corresponding to one tablespoon and one serving size as recommended by the manufacturer. Both interventions were separated by at least two weeks washout (14–17 days in 75% of all participants; min-max 14–52 days). For both treatments, opaque green capsules of hydroxypropyl methylcellulose were used as they disintegrate and dissolve quickly and completely in the upper gastrointestinal tract [15]. A- and B-capsules were opened after statistical analysis had been finished. Thus, participants as well as researchers were blinded to treatment. The breakfast with A-capsules was ingested at 8.48 ± 0.03 a.m. and the breakfast with B-capsules at 8.47 ± 0.03 a.m. (means ± SEMs), respectively. The breakfast was standardized and consisted of one rye bread roll, 4 g margarine and a homemade avocado spread prepared from 60 g avocado, 1 g lemon juice, and 7 g honey. In case of beneficial effects due to the intake of the cocoa powder, it could be used as valuable ingredient for the avocado spread. The energy and nutrient content of this breakfast is listed in Table 1.

Table 1. Energy and nutrient content of the breakfast suitable for diabetes.

Ingredients	Energy (kcal)	Protein (g)	Fat (g)	Carbo-Hydrates (g)	Dietary Fiber (g)
Rye bread roll, 60 g [1]	167	6.3	1.0	31.0	3.4
Margarine, 4 g [2]	28	0.0	3.0	0.0	–
Avocado, 60 g	83	0.9	8.0	2.0	2.5
Lemon juice, 1 g	0	0.0	0.0	0.0	–
Honey, 7 g [3]	21	0.0	–	5.0	–
Vanilla flavor, one drop	–	–	–	–	–
Σ	299	7.2	12.0	38.0	5.9

Data were calculated by using the nutrition software Prodi (Nutri-Science, Freiburg, Germany). [1] Bakery Stinges, Brüggen, Germany; [2] Bellasan®, Walter Rau Food Factories, Hilter, Germany; [3] Goldland®, Dr. Krieger's, Magdeburg, Germany.

On both study days, body weight and height, waist and hip circumference as well as body fat mass (FM) were determined in fasting state. Venous blood samples were collected and BP was measured before as well as 2 and 4 h after completing the breakfast and capsules' intake. All measurements were done by a single trained investigator. The subjects were instructed to maintain their diet on the day before each study day, but to abstain from cocoa products, red wine, green/black tea and fruit/vegetables including juices/nectars. Furthermore, they were advised to take their drugs on both experimental days at the same time as usual to exclude potential confounding effects.

2.2. Participants

Twelve patients, aged at least 18 years, suffering from T2D according to the criteria of the World Health Organization (WHO)/International Diabetes Federation (IDF) [16] for at least one year, with dietary or pharmacological treatment, good glucose control (i.e., HbA$_{1c}$ 6.5–7.5%, taking into account individual therapy goals), overweight/obesity (body mass index (BMI) ≥ 25.0 kg/m^2 [17]) and hypertension (based on the criteria of the European Society of Hypertension/the European Society of Cardiology (ESH/ESC) [18]) were included in the study provided that metabolic control, as ensured by pharmacological treatment, was stable. Insulin therapy, gastrointestinal diseases associated with malabsorption, history of cardiovascular events, pregnancy/lactation, smoking in the last three months, excessive consumption of chocolate (> 100 g/day), red wine (> one glass/day), cocoa drink, green or black tea (> one cup/day), regular use of vitamin preparations or flavanol-rich food supplements (e.g., red wine extract) and drug consumption were exclusion criteria. Eligibility was checked by questionnaire.

2.3. Cocoa Powder

ACTICOA™ cocoa powder was used for cocoa treatment, a functional food in which 80% of the natural flavanol content of raw cocoa is preserved due to gentle cocoa processing. According to the product specification, 2.5 g ACTICOA™ cocoa provided 0.6 g protein, 0.4 g fat, and 0.6 g carbohydrates, thus providing 9 kcal (38 kJ) of energy, 52.5 mg theobromine, and 5.0 mg caffeine. The average flavanol content (sum of mono- to decamers) was 8.3% according to the manufacturer (Barry Callebaut). Further details on mono- and oligomeric flavanols were obtained from our own analysis using ultra-high performance liquid chromatography according to the method of Damm et al. [19]. Results on flavanol composition are shown in Table 2.

Table 2. Flavanol composition of the cocoa.

Flavanols	Content per 2.5 g Cocoa Powder
Flavanols, degree of polymerization (per NP-HPLC)	
Monomers (mg)	49.7
Dimers (mg)	13.9
Trimers (mg)	5.5
Tetramers (mg)	4.7
Pentamers (mg)	3.1
Individual flavanols (per RP-HPLC)	
Epicatechin (mg)	40.4
Catechin (mg)	13.6
A-Dimers (mg)	4.3
Procyanidin B2 (mg)	12.3
Procyanidin B5 (mg)	1.3
Procyanidin C1 (mg)	3.1
Trimers [1] (mg)	4.1
Trimers [1] (mg)	4.4
Procyanidin D (mg)	4.1

[1] Trimeric procyanidins not further specified. NP-HPLC: normal-phase high-pressure liquid chromatography; RP-HPLC: reversed-phase high-pressure liquid chromatography.

2.4. Blood Pressure Investigation

BP was measured after a 5 min rest in a seated position in a quiet room as recommended by the ESH/ESC guidelines [18]. For each investigation, three measurements were performed by using a fully automatic BP monitor (OMRON M 500, OMRON Healthcare Europe, Mannheim, Germany) at 1–2 min intervals. The first reading was always discarded as it has been shown to be higher than subsequent readings [20]. A mean value of the second and third reading was calculated and used for statistical analysis.

2.5. Blood Sampling

Venous blood was collected into Vacutainer® tubes (Becton Dickinson, Heidelberg, Germany) with sodium fluoride and sodium heparin (for plasma glucose analysis) and in tubes without anticoagulant (for analysis of insulin, triglycerides, total-, LDL- and HDL-cholesterol in serum). All samples were kept in an insulated polystyrene box at room temperature. After collection of the 4-h blood sample, all samples (the fasting one and those obtained 2 and 4 h after the meal) were transported to Niederrhein, University of Applied Sciences, Mönchengladbach, Germany. The whole blood was centrifuged (4 °C, 3000× *g*, 20 min) within 0.5–4.75 h after blood sampling. Glucose, insulin and lipids have shown to remain stable in whole blood up to 24 h at room temperature in Vacutainer® tubes that were also used in our study [21]. Plasma and serum samples were aliquoted and frozen at –80 °C. After completion of the study, they were transported on dry ice to the Institute of Clinical Chemistry and Clinical Pharmacology at the University Hospital Bonn, Germany, for analysis.

2.6. Laboratory Investigations

Glucose, triglycerides, as well as total-, LDL-, and HDL cholesterol were determined by cobas c 701/702 and insulin by cobas e 801 (both Roche/Hitachi, Mannheim, Germany) by means of test kits as described previously [22]. Glucose and insulin were used to calculate homeostasis model assessment for insulin resistance (HOMA-IR) [23].

2.7. Anthropometric Investigations

Body weight and height were determined on a calibrated digital column scale (seca 910, Hamburg, Germany; accuracy 0.1 kg and 0.1 cm, respectively) to calculate BMI. Waist circumference was measured at the belly button and hip circumference at the maximal circumference (accuracy 0.1 cm) twice with a non-extensible tape. The mean values of both were used to calculate the waist-to-hip ratio (WHR) to determine the body fat distribution according to WHO [24]. FM was determined according to the guidelines of the European Society for Clinical Nutrition and Metabolism (ESPEN) [25]. Resistance and reactance were measured at 800 μA and 50 kHz with the BIA 2000-1 device (Data Input, Pöcking, Germany) as described by Kirch et al. [26]. FM was calculated by using the equation of Kyle et al. [27].

2.8. Food Intake

Before each study day, food intake, including beverages, was documented in a standardized one-day dietary record. The intake of energy and selected nutrients was determined by Prodi 6.4.0.1 software (Nutri-Science, Freiburg, Germany).

2.9. Statistical Analysis

Statistical analysis was performed by using IBM-SPSS Statistics, version 23 (IBM Crop., Armonk, NY, USA). Metric data were investigated for normal distribution with the Kolmogorov-Smirnov test. Variables without normal distribution were logarithmized. If normal distribution could be assumed, data which were only investigated in fasting state on each study day were compared with each other by using the *t*-test for paired samples. The influence of treatment and time on glucose and lipid metabolism as well as on BP was tested by using repeated-measures ANOVA. In case of significant effects by time, post-hoc tests were carried out. If the variances were homogeneous, Tukey's test was used. Otherwise, Dunnett's T3 test was performed. If repeated-measures ANOVA was not applicable due to the lack of normality, changes before and after each treatment were investigated by using Friedman's test. In case of significant changes, a post-hoc test (two-factorial analysis of variances according to ranks) was done, followed by Wilcoxon signed rank test to compare data of the same points of time between both treatments. p-values < 0.05 were considered statistically significant.

Baseline characteristics of the participants are presented as means ± SDs (metric data) and as frequencies (nominal/ordinal data). Pre- and postintervention data are shown as means ± SEMs and medians (interquartile ranges), respectively.

3. Results

All 12 participants (nine men, three women) completed the study and were included in the statistical analysis. On average, the patients were 68.0 ± 9.0 years old and had been suffering from diabetes for 8.9 ± 5.2 years. Almost two-third of the participants were treated with oral antidiabetic agents, one third with lipid-lowering drugs and all of them with antihypertensiva, mostly providing different active pharmaceutical ingredients (Table 3). Pharmacological treatment did not change during the study. Nutrition status (body weight, BMI, waist circumference, WHR, and FM) was not significantly different between the two study days (Table 4). Furthermore, no significant differences in the intake of energy, nutrients (protein, fat, carbohydrates, dietary fiber, cholesterol, iron, vitamin C) were found on the days before each investigation (Table 4).

Effects by treatment on glucose metabolism (glucose, insulin, HOMA-IR), lipid metabolism (triglycerides, total cholesterol, LDL-cholesterol, HDL-cholesterol) and on BP could not be observed (Table 5). No interactions between treatment and time occurred. Temporal changes were found for glucose after 2 and 4 h compared to fasting values, irrespective whether the breakfast was ingested together with the capsules providing cocoa (p = 0.007) or not (p = 0.003). Effects by time could be detected for insulin and HOMA-IR ($p \leq$ 0.001). Insulin and HOMA-IR increased 2 h after both treatments compared to pre-consumption values (insulin: p = 0.032 for cocoa treatment; p = 0.031 for placebo treatment according to Dunnett's T3 test; HOMA-IR: $p < 0.001$ for cocoa treatment; $p = 0.002$ for placebo treatment according to Tukey's test). Afterwards, HOMA-IR decreased significantly (cocoa treatment: $p = 0.002$; placebo treatment: $p = 0.001$; Tukey's test). Repeated-measures ANOVA revealed effects by time on LDL-cholesterol ($p = 0.048$), LDL/HDL-cholesterol ratio ($p = 0.004$) and on systolic BP ($p = 0.002$), whereas no differences between time points could be detected with Tukey's test.

Unintended effects by the ingestion of cocoa-containing or cocoa-free capsules together with the meals were not reported.

Table 3. Demographic and clinical data.

	Participants ($n = 12$)
Demographic data	
Men/women (n/n)	3/9
Age (years)	68.0 ± 8.7
Diabetes duration (years)	8.9 ± 5.2
Comorbidities (n)	
Hypertension	12
Hyperlipidemia	7
Coronary heart disease	3
Vascular diseases	1
Antihyperglycemic drugs (n)	7
Glimepiride (sulfonylurea)	4
Gliptine (dipeptidyl peptidase-4 inhibitor)	4
Metformin (biguanide)	7
Repaglinide (meglitinide analogue)	1
Antihypertensive drugs (n)	12
Lisinopril (ACE inhibitor)	1
Ramipril (ACE inhibitor)	5
Candesartan (AT1 receptor antagonist)	2
Olmesartan (AT1 receptor antagonist)	2
Valsartan (AT1 receptor antagonist)	1
Amlodipine (calcium channel blocker)	6
Lercanidipine (calcium channel blocker)	1
Bisoprolol (beta-receptor blocker)	3
Carvedilol (beta-receptor blocker)	1
Metoprolol (beta-receptor blocker)	1
Nebivolol (beta-receptor blocker)	2
Hydrochlorothiazide (diuretic)	6
Moxonidine (imidazoline-receptor agonist)	2
Lipid-lowering drugs (n)	4
Atorvastatin (statin)	1
Simvastatin (statin)	1
Ezetimibe (NPC1L1 inhibitor)	3

Data are means $\pm$ SDs, unless otherwise specified. Results on medication according to individual medication list. The sum of active pharmacological ingredients from antiglycemic, antihypertensive and lipid-lowering drugs exceeded the number of participants as most patients received a combination of different pharmacological ingredients. ACE inhibitor, angiotensin converting enzyme inhibitor; AT1 receptor antagonist, angiotensin II type 1 receptor antagonist; NPC1L1 inhibitor, Niemann-Pick C1-like 1 protein inhibitor.

Table 4. Nutrition status in fasting state and nutritional intake on the previous day before each treatment.

	Before Cocoa Treatment (*n* = 12)	Before Placebo Treatment (*n* = 12)	*p*
Nutrition status			
Body weight (kg)	97.4 ± 3.3	98.1 ± 3.0	0.169
BMI (kg/m^2)	33.5 ± 0.9	33.7 ± 0.9	0.174
BMI classification [1] (*n*)			
Overweight	2	2	–
Obesity	10	10	–
Fat mass (% body weight)	36.3 ± 1.6	36.5 ± 1.6	0.571
Waist circumference (cm)	112.1 ± 2.0	113.0 ± 2.0	0.089
Waist-to-hip ratio	1.0 ± 0.0	1.0 ± 0.0	0.555
Nutritional intake			
Energy (kcal)	1479 ± 149	1675 ± 208	0.354
Protein (g) [2]	71.8 ± 9.6	80.5 ± 15.2	0.633
Fat (g)	79.5 ± 9.4	81.4 ± 11.7	0.865
Carbohydrates (g)	100.6 ± 11.7	142.7 ± 17.2	0.058
Dietary fiber (g)	10.4 ± 1.4	12.2 ± 1.0	0.365
Cholesterol (mg) [2]	390.9 ± 87.5	406.2 ± 109.6	0.922
Iron (mg)	7.7 ± 1.0	8.6 ± 1.3	0.582
Vitamin C (mg)	43.4 ± 10.5	44.1 ± 5.1	0.934

p-values according to *t*-test for paired samples. [1] According to the body mass index classification of the World Health Organization (WHO) (2004). [2] Logarithmized values were used for paired *t*-test. Data are means ± SEMs unless indicated otherwise. BMI, body mass index.

Table 5. Results on glucose and lipid metabolism and on blood pressure.

	Cocoa Treatment (n = 12)			Placebo Treatment (n = 12)			Repeated-Measures ANOVA			Friedman Test
	0 h	2 h	4 h	0 h	2 h	4 h	TR	Time	Time x TR	
Glucose metabolism										
Glucose (mmol/L)	7.3 (6.1, 8.3) [ab]	8.7 (6.9, 9.8) [a]	6.2 (5.0, 6.6) [b]	7.5 (6.8, 8.4) [ab]	9.1 (7.6, 10.6) [a]	6.5 (5.5, 7.4) [b]	—	—	—	0.009 [1]
Insulin (mU/L)	14.2 ± 1.8 [a]	42.1 ± 7.2 [b]	22.7 ± 4.3 [ab]	16.5 ± 1.8 [a]	40.5 ± 6.1 [b]	19.8 ± 3.4 [ab]	0.891	<0.001 [2]	0.644	—
HOMA-IR (mmol/L) [3]	4.6 ± 0.6 [a]	16.7 ± 3.6 [b]	6.5 ± 1.5 [a]	5.4 ± 0.6 [a]	16.8 ± 3.1 [b]	5.9 ± 1.2 [a]	0.762	<0.001 [4]	0.463	—
Lipid status										
Triglycerides (mmol/L)	1.93 ± 0.22	2.03 ± 0.19	1.93 ± 0.19	1.75 ± 0.17	1.96 ± 0.21	2.03 ± 0.23	0.846	0.184	0.280	—
Total cholesterol (mmol/L)	5.01 ± 0.25	5.00 ± 0.25	5.02 ± 0.25	5.05 ± 0.23	4.99 ± 0.25	5.11 ± 0.25	0.904	0.182	0.346	—
LDL-cholesterol (mmol/L)	2.98 ± 0.20 [a]	2.97 ± 0.20 [a]	3.01 ± 0.20 [a]	3.07 ± 0.20 [a]	3.00 ± 0.20 [a]	3.08 ± 0.21 [a]	0.825	0.048 [4]	0.372	—
HDL-cholesterol (mmol/L)	1.20 ± 0.07	1.19 ± 0.07	1.19 ± 0.07	1.22 ± 0.08	1.18 ± 0.08	1.20 ± 0.08	0.924	0.157	0.473	—
LDL-chol/HDL chol ratio	2.55 ± 0.19 [a]	2.56 ± 0.19 [a]	2.59 ± 0.20 [a]	2.59 ± 0.18 [a]	2.60 ± 0.18 [a]	2.64 ± 0.19 [a]	0.858	0.004 [4]	0.891	—
Blood pressure										
Systolic (mmHg)	145.2 ± 5.4 [a]	139.4 ± 5.4 [a]	138.3 ± 4.7 [a]	153.0 ± 3.8 [a]	139.8 ± 4.7 [a]	139.8 ± 4.3 [a]	0.591	0.002 [4]	0.333	—
Diastolic (mmHg)	78.0 ± 2.7	75.2 ± 3.4	77.8 ± 2.8	82.0 ± 2.1	77.9 ± 3.9	78.8 ± 2.5	0.503	0.089	0.600	—

[1] In case of significant changes, a post-hoc test (two-factorial analysis of variances according to ranks) was done. [2] In case of significant effects by time and missing variance homogeneity, Dunnett's T3 test was performed. [3] Logarithmized values were used for repeated-measures ANOVA. [4] If effects by time were significant and variance homogeneity was given, Tukey test was performed. Values within the same treatment with different superscript letters differ significantly ($p \leq 0.05$). Plasma glucose at 2 h and at 4 h was not significantly different between cocoa and placebo treatment (Wilcoxon signed rank test). Data are means ± SEMs and medians (interquartile ranges), respectively. Chol, cholesterol; HDL-cholesterol, high density lipoprotein-cholesterol; HOMA-IR, homeostasis model assessment for insulin resistance; LDL-cholesterol, low density lipoprotein-cholesterol; TR, treatment.

4. Discussion

To the best of our knowledge, this is the first study investigating the postprandial effect of a usual serving-size of a flavanol-rich cocoa powder, provided in addition to a diabetic-suitable meal, on glucose and lipid metabolism and on BP in hypertensive type 2 diabetics with stable metabolism. Contrary to our hypothesis, glucose and lipid metabolism and BP were not influenced by the additional intake of 2.5 g cocoa (Table 5). Only a significant decrease in plasma glucose was observed 4 h vs. 2 h after both treatments. Since the values at 2 h and at 4 h were not significantly different between cocoa and placebo treatment, these changes in plasma glucose might simply reflect the glycemic response induced by the breakfast.

Our results on plasma glucose are similar to those of Mellor et al. [10] and Basu et al. [11]. Both provided a flavanol-rich chocolate 60 min before a glucose challenge (75 g pure glucose) [10] or a flavanol-rich cocoa drink together with a high-fat fast food meal (providing 50 g of carbohydrates, predominantly as starch) [11], and who did not observe any effects either. They also investigated patients who suffered from T2D as well as from overweight or obesity (n = 18 [11], n = 10 [10]). The patients were similar regarding age and pharmacological treatment (some using oral antidiabetic agents [10,11] as well as lipid- and BP-lowering drugs [11]) to the participants in our study. For cocoa treatment, 13.5 g ACTICOA™ chocolate [10] and a cocoa drink prepared from 20 g cocoa powder [11] were used, respectively. Both studies provided 40 mg epicatechin, which is claimed to be responsible for the vasoprotective effects of cocoa [28], and which in our study was also ingested by 2.5 g cocoa (Table 2). Since the amount of epicatechin in cocoa correlates strongly with the sum of catechin, epicatechin, procyanidins B2, B5, C1, and D (R^2 = 0.993) [29], the intake of cocoa flavanols in our study and in those of Basu et al. and Mellor et al. were probably comparable. Cocoa flavanols have shown to inhibit digestive enzymes (e.g., α-amylase, α-glucosidase), sodium/glucose cotransporter 1 (SGLT1), dipeptidyl peptidase-IV, and to stimulate incretin secretion (glucagon-like peptide 1, GLP-1; glucose-dependent insulinotropic polypeptide, GIP), which may reduce glycemic response. However, these mechanisms were only investigated in vitro and in animal studies, always using much higher concentrations of flavanols and larger amounts of cocoa than in our study [30].

Regarding insulin and HOMA-IR, we observed only changes by time, but not by treatment. Our result on insulin agrees with the findings of Mellor et al. [10], whereas Basu et al. [11] detected a significant effect of cocoa treatment, leading to higher insulin concentrations and HOMA-IR values at 4 h after cocoa compared to the placebo drink. Cocoa flavanols have been shown to enhance insulin signaling (IRS-1, IRS-2, PI3K-Akt-signaling pathway, AMP kinase) and the translocation of GLUT-4 in vitro and in animals [30]. However, as the dose of epicatechin (40 mg) ingested from cocoa in our study and in those of Basu et al. [11] and Mellor et al. [10] was comparable, ingredients of cocoa other than flavanols may be relevant for changes in insulin secretion and resistance. Brand-Miller et al. [31] observed a higher postprandial insulin secretion in healthy subjects after ingestion of chocolate, cakes, breakfast cereals, ice cream, flavored milk and pudding if these were enriched with cocoa powder. The results of Brand-Miller et al. and Basu et al. may be explained by specific insulinogenic amino acids [11,31]. Their intake might have been much higher in the study of Basu et al. [11] than in our study and in the study of Mellor et al. [10], considering that higher amounts of protein from cocoa (Basu et al.: 2.7 g [11]) were ingested than in our study (0.5 g) (no data available from Mellor et al. [10]). Stearic acid, a fatty acid in cocoa butter, is discussed to stimulate specifically insulin secretion [31]. We provided less cocoa powder (2.5 g) than Basu et al. (20 g) with a lower fat content (16% of dry mass, corresponding to 0.4 g fat compared to Basu et al. [11] (23.5% fat of dry mass, corresponding to 4.7 g; no data on fat content in chocolate available from Mellor et al. [10]). This may further explain the lack of changes in insulin concentration and HOMA-IR by cocoa treatment in our study.

Cocoa treatment affected neither triglycerides nor total, LDL- and HDL-cholesterol in our study. At first glance, this is astonishing as cocoa flavanols may activate catabolic pathways (e.g., β-oxidation) and inhibit anabolic ones (e.g., acetyl CoA carboxylase, HMG-CoA-reductase) in concentrations that can be reached via diet by stimulating 5′-AMP-activated protein kinase [32]. However, the

5′-AMP-activated protein kinase is diminished by insulin through the Akt signaling pathway [33]. Consequently, the insulin response induced by our breakfast might have overwhelmed the effect of cocoa flavanols on 5′-AMP-activated protein kinase. This possibly explains the lack of changes in serum lipids in our study.

Our findings on triglycerides, total and LDL-cholesterol agree with findings of Basu et al. [11], except for HDL-cholesterol which increased after 4 h vs. 1 h if the cocoa drink was ingested with the fast food meal. Since high doses of pure theobromine (850 mg/d) increased HDL-cholesterol in healthy subjects [34], theobromine may be responsible for the increase in HDL-cholesterol, as observed by Basu et al. [11] after cocoa consumption provided 220 mg of theobromine. In our study, the theobromine intake from cocoa (52.5 mg) was possibly too low to induce any beneficial changes in HDL-cholesterol, which may explain the lack of a treatment effect on HDL-cholesterol in our study.

BP and flow-mediated dilation (FMD) could be improved after regular cocoa consumption according to a meta-analysis of RCTs [35]. An increase in FMD was even observed after acute cocoa intake [35]. Hence, a decrease in BP by a cocoa-enriched meal was expected due to an increased arterial elasticity, but was not detectable in our study. However, the increase in FMD after acute cocoa intake was not accompanied by changes in BP in healthy subjects [36] and in subjects with T2D [10,11]. Consequently, FMD seems to be more a sensitive vascular parameter with a stronger response to cocoa intake compared to BP.

Most of our patients were pharmacologically treated for diabetes, lipometabolic disorders, and hypertension (Table 3) as were those of Basu et al. [11] and Mellor et al. [10]. As the cellular and molecular mode of action of cocoa flavanols is partly similar to those of pharmaceuticals, such as metformin and ACE inhibitors (e.g., improving insulin release, secretion and sensitivity in fat, liver tissue, and in muscles [32,37] by increasing the translocation of the GLUT-4 transporter [32], inhibition of the angiotensin converting enzyme [38]), almost no significant effects by cocoa treatment could be achieved. While similar studies on postprandial effects of cocoa with pharmacologically-untreated subjects with T2D would be interesting, these have not become available to date. Gutiérrez-Salmeán et al. have shown that pure (–)-epicatechin (1 mg/kg body weight) ingested within a ready-to-drink oral nutritional supplement rich in carbohydrates (63% of total energy) lowers postprandial glycemic and lipemic response in healthy subjects after 2 and 4 h compared to epicatechin-free treatment, in overweight/obese subjects (*n* = 8) even more so than in normal-weight subjects (*n* = 12) [39]. In another trial, also performed with overweight/obese subjects, cocoa extract (1.4 g, providing 153 mg of epicatechin), in addition to a meal rich in fat and low in carbohydrates (53% and 38% of total energy, respectively) did not change postprandial glucose and lipid metabolism or blood pressure compared to a meal without cocoa extract [40]. In both studies, none of the participants suffered from diabetes and hypertension and did not use glucose, lipid, and BP lowering drugs [39,40]. Thus, in overweight/obese subjects with metabolic disturbances and without pharmacological treatments, postprandial metabolism may be improved by flavanol-rich cocoa as part of a meal rich in low-molecular carbohydrates.

We investigated postprandial changes after 2 and 4 h for the following reasons. First, epicatechin usually achieves maximum concentration in plasma 2 h after cocoa consumption [28]. Second, cardiometabolic effects were observed 2 h [10,11] and 4 h [11] after cocoa ingestion together with a metabolic challenge in type 2 diabetics. Third, beneficial effects in glucose and lipid metabolism were found in obese/overweight subjects 2 and 4 h after a meal if this meal was ingested with pure (–)-epicatechin [39]. However, taking also into account the different plasma kinetics of glucose and insulin in subjects with and without T2D after an oral glucose challenge [41], postprandial changes in glucose metabolism should be investigated in a close-meshed procedure, i.e., every 30–60 min for 2–3 h, to precisely assess the glycemic response in patients with T2D.

The strengths of our RCT include the double-blind, placebo-controlled crossover study design. Moreover, nutritional status and dietary intake were investigated before each treatment. Therefore, confounding effects of lifestyle changes that might have affected our outcome variables are unlikely.

The collection of only two blood samples after a meal makes it difficult to assess the postprandial changes in glucose metabolism as precisely as necessary to detect possible differences. This is the major limitation of our study. However, it remains speculative whether this is the reason why we did not detect significant effects by cocoa. As sample size estimation was not possible due to the lack of data, our study might have been underpowered. A larger sample size would allow an adjustment to factors that might be relevant for the response to cocoa. Nevertheless, the results of the present study may be used to calculate the sample size for future studies.

In conclusion, administration of 2.5 g of flavanol-rich cocoa powder together with a diabetic-suitable breakfast does not seem to modulate glucose and lipid metabolism or BP in patients suffering from T2D and hypertension. Nevertheless, future studies are desirable. These should include close-meshed investigations to exactly assess the glycemic and lipemic response in subjects with T2D or metabolic syndrome, but without pharmacological treatment. Moreover, realistic amounts of cocoa should be provided with realistic meals rich in carbohydrates. Additionally, further vascular parameters, such as FMD, should be considered.

Author Contributions: J.R., L.D., and S.E. designed the study, J.R. and N.L. recruited the participants, and N.L. enrolled the subjects. J.R. performed the investigations of anthropometric parameters and of blood pressure, analyzed the food records and performed statistical analysis with support from H.P.H. N.L. was the medical advisor. B.F.Z. investigated the flavanol composition of the cocoa. B.S.W. was responsible for the analysis of glucose, insulin, and lipids. J.R., L.D., and S.E. drafted the manuscript. S.E. had primary responsibility for the final content of the manuscript. All authors read and approved the final version of the manuscript.

Funding: The study was funded by Hochschule Niederrhein, University of Applied Sciences. Barry Callebaut, Zurich, Switzerland, kindly provided the cocoa as unrestricted gift.

Acknowledgments: We thank all patients who participated in this study and thank Ines Diezemann, Praxis Anrath, Willich, for collecting the blood samples.

Conflicts of Interest: The authors declare no conflict of interest. Barry Callebaut, Zurich, Switzerland, provided the cocoa as an unrestricted gift, but had no role in the design of the study, in the collection, analyses, or interpretation of data, in the writing of the manuscript, or in the decision to publish the results.

References

1. International Diabetes Federation. IDF Diabetes Atlas. Available online: http://diabetesatlas.org (accessed on 22 December 2018).
2. Cavalot, F.; Pagliarino, A.; Valle, M.; Di Martino, L.; Bonomo, K.; Massucco, P.; Anfossi, G.; Trovati, M. Postprandial blood glucose predicts cardiovascular events and all-cause mortality in type 2 diabetes in a 14-year follow-up: Lessons from the San Luigi Gonzaga Diabetes Study. *Diabetes Care* **2011**, *34*, 2237–2243. [CrossRef] [PubMed]
3. Takao, T.; Suka, M.; Yanagisawa, H.; Iwamoto, Y. Impact of postprandial hyperglycemia at clinic visits on the incidence of cardiovascular events and all-cause mortality in patients with type 2 diabetes. *J. Diabetes Investig.* **2017**, *8*, 600–608. [CrossRef]
4. Pappas, C.; Kandaraki, E.A.; Tsirona, S.; Kountouras, D.; Kassi, G.; Diamanti-Kandarakis, E. Postprandial dysmetabolism: Too early or too late? *Hormones (Athens)* **2016**, *15*, 321–344. [CrossRef] [PubMed]
5. Alkhatib, A.; Tsang, C.; Tiss, A.; Bahorun, T.; Arefanian, H.; Barake, R.; Khadir, A.; Tuomilehto, J. Functional foods and lifestyle approaches for diabetes prevention and management. *Nutrients* **2017**, *9*, 1310. [CrossRef] [PubMed]
6. Ramos, S.; Martín, M.A.; Goya, L. Effects of cocoa antioxidants in type 2 diabetes mellitus. *Antioxidants (Basel)* **2017**, *6*, 84. [CrossRef] [PubMed]
7. Ludovici, V.; Barthelmes, J.; Nagele, M.P.; Enseleit, F.; Ferri, C.; Flammer, A.J.; Ruschitzka, F.; Sudano, I. Cocoa, blood pressure, and vascular function. *Front. Nutr.* **2017**, *4*, 36. [CrossRef] [PubMed]
8. Lin, X.; Zhang, I.; Li, A.; Manson, J.E.; Sesso, H.D.; Wang, L.; Liu, S. Cocoa flavanol intake and biomarkers for cardiometabolic health: A systematic review and meta-analysis of randomized controlled trials. *J. Nutr.* **2016**, *146*, 2325–2333. [CrossRef]
9. Tokede, O.A.; Gaziano, J.M.; Djousse, L. Effects of cocoa products/dark chocolate on serum lipids: A meta-analysis. *Eur. J. Clin. Nutr.* **2011**, *65*, 879–886. [CrossRef]

10. Mellor, D.D.; Madden, L.A.; Smith, K.A.; Kilpatrick, E.S.; Atkin, S.L. High-polyphenol chocolate reduces endothelial dysfunction and oxidative stress during acute transient hyperglycaemia in type 2 diabetes: A pilot randomized controlled trial. *Diabet. Med.* **2013**, *30*, 478–483. [CrossRef]

11. Basu, A.; Betts, N.M.; Leyva, M.J.; Fu, D.; Aston, C.E.; Lyons, T.J. Acute cocoa supplementation increases postprandial HDL cholesterol and insulin in obese adults with type 2 diabetes after consumption of a high-fat breakfast. *J. Nutr.* **2015**, *145*, 2325–2332. [CrossRef]

12. American Diabetes Association. 2. Classification and diagnosis of diabetes: Standards of medical care in diabetes-2019. *Diabetes Care* **2019**, *42*, S13–S18. [CrossRef] [PubMed]

13. Kerner, W.; Bruckel, J.; German Diabetes Association. Definition, classification and diagnosis of diabetes mellitus. *Exp. Clin. Endocrinol. Diabetes* **2014**, *122*, 384–386. [CrossRef] [PubMed]

14. Evert, A.B.; Boucher, J.L. New diabetes nutrition therapy recommendations: What you need to know. *Diabetes Spectr.* **2014**, *27*, 121–130. [CrossRef] [PubMed]

15. Al-Tabakha, M.M. HPMC capsules: Current status and future prospects. *J. Pharm. Pharm. Sci.* **2010**, *13*, 428–442. [CrossRef] [PubMed]

16. World Health Organization (WHO); International Diabetes Federation (IDF). *Definition and Diagnosis of Diabetes Mellitus and Intermediate Hyperglycaemia—Report of a WHO/IDF Consultation*; WHO: Geneva, Switzerland, 2006. Available online: http://whqlibdoc.who.int/publications/2006/9241594934_eng.pdf?ua=1 (accessed on 22 December 2018).

17. World Health Organization. *Obesity: Preventing and Managing the Global Epidemic—Report of a WHO Consultation (WHO Technical Report Series 894)*; WHO: Geneva, Switzerland, 2000; Available online: http://apps.who.int/iris/bitstream/10665/42330/1/WHO_TRS_894.pdf?ua=1&ua=1 (accessed on 22 December 2018).

18. Mancia, G.; Fagard, R.; Narkiewicz, K.; Redon, J.; Zanchetti, A.; Bohm, M.; Christiaens, T.; Cifkova, R.; De Backer, G.; Dominiczak, A.; et al. 2013 ESH/ESC guidelines for the management of arterial hypertension: The Task Force for the Management of Arterial Hypertension of the European Society of Hypertension (ESH) and of the European Society of Cardiology (ESC). *Eur. Heart J.* **2013**, *34*, 2159–2219. [CrossRef] [PubMed]

19. Damm, I.; Enger, E.; Chrubasik-Hausmann, S.; Schieber, A.; Zimmermann, B.F. Fast and comprehensive analysis of secondary metabolites in cocoa products using ultra high-performance liquid chromatography directly after pressurized liquid extraction. *J. Sep. Sci.* **2016**, *39*, 3113–3122. [CrossRef] [PubMed]

20. Pickering, T.G.; Hall, J.E.; Appel, L.J.; Falkner, B.E.; Graves, J.W.; Hill, M.N.; Jones, D.H.; Kurtz, T.; Sheps, S.G.; Roccella, E.J.; et al. Recommendations for blood pressure measurement in humans: An AHA scientific statement from the Council on High Blood Pressure Research Professional and Public Education Subcommittee. *J. Clin. Hypertens. (Greenwich)* **2005**, *7*, 102–109. [CrossRef]

21. Oddoze, C.; Lombard, E.; Portugal, H. Stability study of 81 analytes in human whole blood, in serum and in plasma. *Clin. Biochem.* **2012**, *45*, 464–469. [CrossRef]

22. Dicks, L.; Kirch, N.; Gronwald, D.; Wernken, K.; Zimmermann, B.F.; Helfrich, H.P.; Ellinger, S. Regular intake of a usual serving size of flavanol-rich cocoa powder does not affect cardiometabolic parameters in stably treated patients with type 2 diabetes and hypertension—A double-blinded, randomized, placebo-controlled trial. *Nutrients* **2018**, *10*, 1435. [CrossRef]

23. Matthews, D.R.; Hosker, J.P.; Rudenski, A.S.; Naylor, B.A.; Treacher, D.F.; Turner, R.C. Homeostasis model assessment: Insulin resistance and beta-cell function from fasting plasma glucose and insulin concentrations in man. *Diabetologia* **1985**, *28*, 412–419. [CrossRef]

24. World Health Organization. *Waist Circumference and Waist-Hip Ratio: Report of a WHO Expert Consultation*; WHO: Geneva, Switzerland, 2011; Available online: https://www.who.int/nutrition/publications/obesity/WHO_report_waistcircumference_and_waisthip_ratio/en/ (accessed on 22 December 2018).

25. Kyle, U.G.; Bosaeus, I.; De Lorenzo, A.D.; Deurenberg, P.; Elia, M.; Manuel Gomez, J.; Lilienthal Heitmann, B.; Kent-Smith, L.; Melchior, J.C.; Pirlich, M.; et al. Bioelectrical impedance analysis-part II: Utilization in clinical practice. *Clin. Nutr.* **2004**, *23*, 1430–1453. [CrossRef] [PubMed]

26. Kirch, N.; Berk, L.; Liegl, Y.; Adelsbach, M.; Zimmermann, B.F.; Stehle, P.; Stoffel-Wagner, B.; Ludwig, N.; Schieber, A.; Helfrich, H.P.; et al. A nutritive dose of pure (−)-epicatechin does not beneficially affect increased cardiometabolic risk factors in overweight-to-obese adults—A randomized, placebo-controlled, double-blind crossover study. *Am. J. Clin. Nutr.* **2018**, *107*, 948–956. [CrossRef] [PubMed]

27. Kyle, U.G.; Genton, L.; Karsegard, L.; Slosman, D.O.; Pichard, C. Single prediction equation for bioelectrical impedance analysis in adults aged 20–94 years. *Nutrition* **2001**, *17*, 248–253. [CrossRef]

28. Kirch, N.; Ellinger, S. Cocoa flavanols and cardioprotective effects. What flavanols may contribute to vascular health? *Ernahrungs Umschau* **2014**, *61*, M482–M489. [CrossRef]

29. Cooper, K.A.; Campos-Gimenez, E.; Jimenez Alvarez, D.; Nagy, K.; Donovan, J.L.; Williamson, G. Rapid reversed phase ultra-performance liquid chromatography analysis of the major cocoa polyphenols and inter-relationships of their concentrations in chocolate. *J. Agric. Food Chem.* **2007**, *55*, 2841–2847. [CrossRef] [PubMed]

30. Strat, K.M.; Rowley, T.J.T.; Smithson, A.T.; Tessem, J.S.; Hulver, M.W.; Liu, D.; Davy, B.M.; Davy, K.P.; Neilson, A.P. Mechanisms by which cocoa flavanols improve metabolic syndrome and related disorders. *J. Nutr. Biochem.* **2016**, *35*, 1–21. [CrossRef] [PubMed]

31. Brand-Miller, J.; Holt, S.H.; de Jong, V.; Petocz, P. Cocoa powder increases postprandial insulinemia in lean young adults. *J. Nutr.* **2003**, *133*, 3149–3152. [CrossRef]

32. Martin, M.A.; Goya, L.; Ramos, S. Antidiabetic actions of cocoa flavanols. *Mol. Nutr. Food Res.* **2016**, *60*, 1756–1769. [CrossRef]

33. Valentine, R.J.; Coughlan, K.A.; Ruderman, N.B.; Saha, A.K. Insulin inhibits AMPK activity and phosphorylates AMPK Ser(4)(8)(5)/(4)(9)(1) through Akt in hepatocytes, myotubes and incubated rat skeletal muscle. *Arch. Biochem. Biophys.* **2014**, *562*, 62–69. [CrossRef]

34. Neufingerl, N.; Zebregs, Y.E.; Schuring, E.A.; Trautwein, E.A. Effect of cocoa and theobromine consumption on serum HDL-cholesterol concentrations: A randomized controlled trial. *Am. J. Clin. Nutr.* **2013**, *97*, 1201–1209. [CrossRef]

35. Hooper, L.; Kay, C.; Abdelhamid, A.; Kroon, P.A.; Cohn, J.S.; Rimm, E.B.; Cassidy, A. Effects of chocolate, cocoa, and flavan-3-ols on cardiovascular health: A systematic review and meta-analysis of randomized trials. *Am. J. Clin. Nutr.* **2012**, *95*, 740–751. [CrossRef] [PubMed]

36. Westphal, S.; Luley, C. Flavanol-rich cocoa ameliorates lipemia-induced endothelial dysfunction. *Heart Vessels* **2011**, *26*, 511–515. [CrossRef] [PubMed]

37. Cordero-Herrera, I.; Martín, M.A.; Bravo, L.; Goya, L.; Ramos, S. Cocoa flavonoids improve insulin signalling and modulate glucose production via AKT and AMPK in HepG2 cells. *Mol. Nutr. Food Res.* **2013**, *57*, 974–985. [CrossRef] [PubMed]

38. Bernstein, K.E. Views of the renin-angiotensin system: Brilling, mimsy, and slithy tove. *Hypertension* **2006**, *47*, 509–514. [CrossRef] [PubMed]

39. Gutiérrez-Salmeán, G.; Ortiz-Vilchis, P.; Vacaseydel, C.M.; Rubio-Gayosso, I.; Meaney, E.; Villarreal, F.; Ramírez-Sánchez, I.; Ceballos, G. Acute effects of an oral supplement of (−)-epicatechin on postprandial fat and carbohydrate metabolism in normal and overweight subjects. *Food Funct.* **2014**, *5*, 521–527. [CrossRef] [PubMed]

40. Ibero-Baraibar, I.; Suárez, M.; Arola-Arnal, A.; Zulet, M.A.; Martinez, J.A. Cocoa extract intake for 4 weeks reduces postprandial systolic blood pressure response of obese subjects, even after following an energy-restricted diet. *Food Nutr. Res.* **2016**, *60*, 30449. [CrossRef] [PubMed]

41. Kramer, C.K.; Vuksan, V.; Choi, H.; Zinman, B.; Retnakaran, R. Emerging parameters of the insulin and glucose response on the oral glucose tolerance test: Reproducibility and implications for glucose homeostasis in individuals with and without diabetes. *Diabetes Res. Clin. Pract.* **2014**, *105*, 88–95. [CrossRef]

Article

Regular Intake of a Usual Serving Size of Flavanol-Rich Cocoa Powder Does Not Affect Cardiometabolic Parameters in Stably Treated Patients with Type 2 Diabetes and Hypertension—A Double-Blinded, Randomized, Placebo-Controlled Trial

Lisa Dicks [1,2], Natalie Kirch [1], Dorothea Gronwald [2], Kerstin Wernken [2], Benno F. Zimmermann [3], Hans-Peter Helfrich [4] and Sabine Ellinger [1,*]

[1] Faculty of Food, Nutrition and Hospitality Sciences, Hochschule Niederrhein, University of Applied Sciences, Rheydter Str. 277, 41065 Mönchengladbach, Germany; Lisa.Dicks@hs-niederrhein.de (L.D.); Natalie.Kirch@hs-niederrhein.de (N.K.)

[2] diabetesPRAXIS Rathausallee, Rathausallee 6-8, 47239 Duisburg, Germany; doro@dr-gronwald.com (D.G.); K.Wernken@gmx.de (K.W.)

[3] Department of Nutrition and Food Sciences, University of Bonn, Endenicher Allee 11-13, 53115 Bonn, Germany; benno.zimmermann@uni-bonn.de

[4] Institute of Numerical Simulation, University of Bonn, Endenicher Allee 60, 53115 Bonn, Germany; helfrich@uni-bonn.de

* Correspondence: Sabine.Ellinger@hs-niederrhein.de; Tel.: +49-2161-186-5406

Received: 6 September 2018; Accepted: 28 September 2018; Published: 5 October 2018

Abstract: Regular cocoa consumption has been shown to improve blood pressure (BP), insulin sensitivity, and lipid levels in patients with type 2 diabetes (T2D), using up to 100 g of chocolate or 54 g of cocoa. These effects, attributed to cocoa flavanols, would be beneficial for patients with T2D if they could be achieved by a usual serving size of flavanol-rich cocoa. Forty-two hypertensive patients with T2D (stable pharmacological treatment, with good adjustment for glucose metabolism, lipids, and BP) ingested capsules with 2.5 g/day of a flavanol-rich cocoa or cocoa-free capsules for 12 weeks in a double-blinded, randomized, placebo-controlled study with parallel group design. Participants had to maintain diet, lifestyle, and medication. Before and after intervention, fasting blood samples were collected; BP and nutritional status were investigated. Cocoa treatment did not affect BP, nor glucose metabolism (glucose, HbA_{1c}, insulin, HOMA-IR) and lipids (triglycerides, total cholesterol, low-density lipoprotein cholesterol, high-density lipoprotein cholesterol). Body weight, fat mass, and nutrient supply remained unchanged. Changes in the placebo group did not occur. Regular intake of a usual serving size of flavanol-rich cocoa does not improve cardiometabolic parameters in stably treated patients with T2D and hypertension. As the medication modulates partly the same targets as cocoa flavanols, future studies should focus on the preventive effect of cocoa against diabetes and other cardiometabolic diseases in individuals with preexisting abnormalities that do not require any pharmacological treatment.

Keywords: type 2 diabetes; flavanol-rich cocoa; blood pressure; glucose metabolism; lipid status

---

1. Introduction

Type 2 diabetes (T2D) is accompanied by an increased cardiovascular risk, partly due to comorbidities such as dyslipidemia and hypertension [1,2]. Incidence as well as the progression of these chronic diseases are strongly affected by diet. These findings explain the current interest in functional food and food ingredients which may improve cardiometabolic health [3].

Randomized controlled trials (RCTs) suggest that patients with T2D may benefit from regular cocoa consumption [4–9]. Vascular elasticity increased in patients with T2D after cocoa intake (54 g/day) [4] and in hypertensive subjects with impaired glucose tolerance (IGT) after ingestion of dark chocolate (100 g/day) [5]. In two RCTs, blood pressure (BP) was reduced after intake of 100 g [5] and 25 g [9] flavanol-rich chocolate, respectively. In another study providing cocoa (40 g/day), biomarkers of endothelial inflammation decreased [7]. Single parameters of glucose metabolism like HbA_{1c} [4,9], fasting blood glucose (FBG) [9], and insulin resistance and sensitivity [5] were improved after cocoa consumption. Favorable changes in serum lipids were also found: a decrease in low-density lipoprotein cholesterol (LDL-C) [4,5,8] and an increase in high-density lipoprotein cholesterol (HDL-C) [6–8] after regular intake of cocoa (20 g [8], 54 g [4]) and chocolate (45 g [6], 100 g [5]), respectively. However, these studies provided up to 100 g chocolate and 54 g cocoa daily, which cannot be recommended to patients with T2D due to the high energy content [10].

Cocoa products which provide $\geq$ 200 mg flavanols with a degree of polymerization (DP) of 1–10 per daily portion (2.5 g cocoa; 10 g chocolate) were approved by the European Food Safety Authority (EFSA) health claim that "cocoa flavanols (CF) help maintain the elasticity of blood vessels, which contributes to normal blood flow" [11]. Endothelial dysfunction promotes cardiovascular disorders such as atherosclerosis and hypertension in T2D [12]. Therefore, cocoa, which provides $\geq$ 200 mg flavanols per daily portion and little energy due to the lack of sugar and fat, may be advantageous for patients suffering from T2D.

Thus, the present study should investigate whether regular ingestion of 2.5 g/day of such an unsweetened, strongly defatted, and flavanol-rich cocoa powder might improve BP (the primary outcome measure) as well as glucose and lipid metabolism (the secondary outcome parameters) in stably treated subjects with T2D.

2. Materials and Methods

2.1. Study Design

This double-blinded, randomized, placebo-controlled trial with parallel group design was performed between September 2016 and April 2017 in a specialized medical office for diabetology (diabetes PRAXIS Rathausallee, Duisburg, Germany). The study was conducted according to the guidelines of the Declaration of Helsinki and the protocol was approved by the ethics committees of the University of Bonn (project identification code 037/16, date of approval 10 February 2016) and of the Medical Association of North Rhine (project identification code 2016143, date of approval 25 May 2016). The study was registered in the German Clinical Trials Register (DRKS-ID: DRKS00011007) on 24 August 2016. Written informed consent for inclusion was obtained from all subjects before enrollment.

Participants were consecutively recruited (August 2016–January 2017) and allocated to groups A and B (ratio 1:1) by permuted block randomization (block size of 4, sequence generated by drawing lots by an uninvolved person). They received five A- or B-capsules daily for 12 weeks. Each provided 0.5 g ACTICOA™ cocoa powder (Barry Callebaut, Zurich, Switzerland; lot no. 100-F017906-AC-796) or pure microcrystalline cellulose (J. Rettenmaier and Söhne, Rosenberg, Germany), respectively. For both treatments, nontransparent capsules of hydroxypropyl methylcellulose with identical appearance were chosen which disintegrate and dissolve quickly and completely in the upper gastrointestinal tract [13]. KPProductions (Koblenz, Germany) filled both types of capsules. The document revealing the allocation of A- and B-capsules to cocoa and placebo treatment was sealed in an envelope which was opened after statistical analysis had been finished. Thus, participants and researchers were blinded to treatment. Before and after treatment, participants were examined, and blood samples were drawn after $\geq$ 10 h overnight fasting. The capsules were noted in the subjects' medication plan and were recommended to be taken in the morning (three capsules) and in the evening (two capsules) along with the prescribed long-term medication (e.g., metformin) to ensure regular ingestion. Subjects were advised not to take the capsules with milk. Furthermore, they were instructed to maintain diet and lifestyle during intervention, but to abstain from any other cocoa products and to limit the intake of further flavanol-rich foods.

2.2. Participants

Forty-two patients with T2D (based on the criteria of the World Health Organization (WHO) and the International Diabetes Federation (IDF) [14], diabetes duration $\geq$ one year) and hypertension (according to the European Society of Hypertension/the European Society of Cardiology (ESH/ESC) [15]), under dietetic and/or pharmacological treatment and with good glycemic control (HbA$_{1c}$ 48–58 mmol/mol with consideration of individual therapeutic aims) were included in the study. Exclusion criteria comprised treatment with insulin, any changes in chronic medication in the previous three months, history of cardiovascular events, malabsorption disorders, smoking, pregnancy or lactation, present/former alcohol or drug abuse, supplementation of vitamins/antioxidants, and daily consumption of excessive amounts of other flavanol-rich foods (> one glass of red wine, one cup of green/black tea or 100 g chocolate). Eligibility was checked by questionnaire.

2.3. Cocoa Powder

According to manufacturer, 2.5 g ACTICOA™ cocoa provided 0.6 g protein, 0.4 g fat, and 0.6 g carbohydrates, thus delivering 38 kJ (9 kcal). The average flavanol content was 207.5 mg (8.3%) (DP 1–10). Own analysis using ultra high-performance liquid chromatography provided quantitative data on mono- and oligomeric flavanols according to their DP as well as data on individual flavanols [16] (Table 1).

Table 1. Nutritional and flavanol composition of the cocoa.

Ingredients	Content Per Daily Portion (2.5 g)
Manufacturer Analysis	
Energy (kJ/kcal)	38/9
Macronutrients	
Protein (g)	0.6
Fat (g)	0.4
Carbohydrates (g)	0.6
Micronutrients	
Sodium (mg)	0.5
Potassium (mg)	37.5
Calcium (mg)	11.4
Iron (mg)	1.1
Phosphorus (mg)	18.1
Magnesium (mg)	11.4
Methylxanthines	
Caffeine (mg)	5.0
Theobromine (mg)	52.5
Flavanols, degree of polymerization 1–10 (mg)	207.5
Laboratory Analysis [a]	
Flavanols, degree of polymerization (per NP-HPLC)	
Monomers (mg)	49.7
Dimers (mg)	13.9
Trimers (mg)	5.5
Tetramers (mg)	4.7
Pentamers (mg)	3.1

Table 1. Cont.

Ingredients	Content Per Daily Portion (2.5 g)
Individual flavanols (per RP-HPLC)	
Epicatechin (mg)	40.4
Catechin (mg)	13.6
A-Dimers (mg)	4.3
Procyanidin B2 (mg)	12.3
Procyanidin B5 (mg)	1.3
Procyanidin C1 (mg)	3.1
Trimers [b] (mg)	4.1
Trimers [b] (mg)	4.4
Procyanidin D (mg)	4.1

[a] Performed at the Department of Nutrition and Food Sciences, University of Bonn. [b] Trimeric procyanidins not further specified. NP-HPLC: normal phase HPLC; RP-HPLC: reversed phase HPLC.

2.4. Blood Pressure Investigation

BP was determined with a full-automatic BP monitor (OMRON Healthcare Europe, Mannheim, Germany) by a single trained investigator according to the ESH/ESC guidelines [15]. Two measurements were performed at 1- to 2-min intervals; a third measurement was done if the first two values in systolic blood pressure (SBP) differed by ≥ 5 mm Hg. The first measurement was always discarded. The second value was used for statistical analysis. If a third measurement was required, the mean value of the second and third measurement was considered.

2.5. Laboratory Investigations

Venous blood was collected between 7:30 and 9:00 into tubes with fluoride and citrate or tubes without anticoagulant. Fresh blood samples were analyzed by the Medical Care Center, Dr. Stein and Colleagues Laboratory Medicine, Mönchengladbach, Germany, which is accredited according to DIN EN ISO 15189. Glucose was determined in plasma. Total cholesterol (total-C), LDL-C, HDL-C, triglycerides, insulin, and creatinine were determined in serum. Glucose, lipids and creatinine were analyzed photometrically by Cobas® c 701/702 and insulin by ECLIA with Cobas® e 801 (both from Roche/Hitachi, Mannheim, Germany) using test kits for glucose (GLUC3; coefficient of variation (CV, < 1.3%), total-C (CHOL2; CV < 1.6%), LDL-C (LDLC3; CV < 2.0%), HDL-C (HDLC4; CV < 1.6%), triglycerides (TRIGL; CV < 2.0%), creatinine (CREJ2; CV < 2.2%) and insulin (Elecsys; CV < 2.0%). Glucose and insulin were used to calculate HOMA-IR [17]. HbA_{1c} (CV < 2.0%) was analyzed in capillary blood with the Alere Afinion® AS100 Analyzer (Cologne, Germany) immediately.

2.6. Anthropometric Investigations

Body weight (BW), height, and waist and hip circumference as well as fat mass (FM) were investigated by a single trained examiner. Weight and height were used to calculate body mass index (BMI). Waist-to-hip ratio was determined to characterize body fat distribution according to the WHO [18]. FM was examined by bioelectric impedance analysis corresponding to the European Society for Clinical Nutrition and Metabolism (ESPEN) guidelines [19] Resistance and reactance were determined with BIA 2000-1 (Data Input, Pöcking, Germany) as described by Kirch et al. [20]. FM was computed by using the equation of Kyle et al. [21].

2.7. Food Intake

Before each visit, participants documented their consumption of food and beverages in standardized 3-day estimated food records (2 weekdays, 1 weekend day). The intake of energy and selected nutrients was calculated by using Prodi® 6.4.0.1 (Nutri-Science, Freiburg, Germany) and the intake of epicatechin by using the United States Department of Agriculture (USDA) Database for the Flavonoid Content of Selected Foods (release 3.1 [22]).

2.8. Compliance

Subjects documented their capsules' intake in a diary and returned all remaining capsules. The compliance was calculated as ratio of ingested capsules to the number which should have been ingested. On average, an intake of ≥ 4 capsules per day qualified to be included in per-protocol analysis (compliance rate $\geq 80\%$).

2.9. Sample Size Calculation

The sample size calculation was based on the estimated changes in SBP by cocoa treatment. According to a former meta-regression analysis of our group, a mean decrease of 4.3 mm Hg was expected after daily ingestion of 40 mg epicatechin with 2.5 g cocoa compared to placebo. Based on this value, we considered a mean decrease of 2 mm Hg as statistically significant [23] and clinically relevant [24]. To detect a decrease in SBP of ≥ 2 mm Hg, 16 participants per group were needed presuming a power of 75%, an alpha of 0.05, and a standard deviation of 2.8 mm Hg. The latter was calculated by weighing the variances of four studies which account for 81% of total weight in meta-analysis [25]. Assuming a dropout rate of 25%, 21 subjects were included in each group.

2.10. Statistical Analysis

Metric data were investigated for normal distribution by using the Kolmogorov–Smirnov test and were logarithmized if necessary. When normal distribution could be assumed, data were compared with each other by applying the *t*-test for paired and unpaired samples, respectively. Otherwise, Mann–Whitney U- or Wilcoxon test were used. Nominal data were compared by using χ^2 test or Fisher's exact test. Differences indicated by *p*-values < 0.05 were considered to be statistically significant. Metric data are presented as means and standard error of the means (SEMs) and as medians and quartiles, respectively, and nominal data are given as frequencies. Statistical analysis was performed by using IBM-SPSS Statistics, version 23.0 (IBM Corp., Armonk, NY, USA).

3. Results

All 42 participants finished the study. As shown in Figure 1, one subject of each group was excluded due to changes in chronic medication (levothyroxine, cortisone) which might have affected our outcome markers, changes in BW of $\geq 5\%$ and due to a compliance < 80%. Moreover, one participant of the cocoa group was excluded due to not being in fasted state at the second visit. Thus, 35 subjects were included in per-protocol analysis. BP values of three subjects were excluded from statistical evaluation because they had taken their antihypertensives not equally before both investigations.

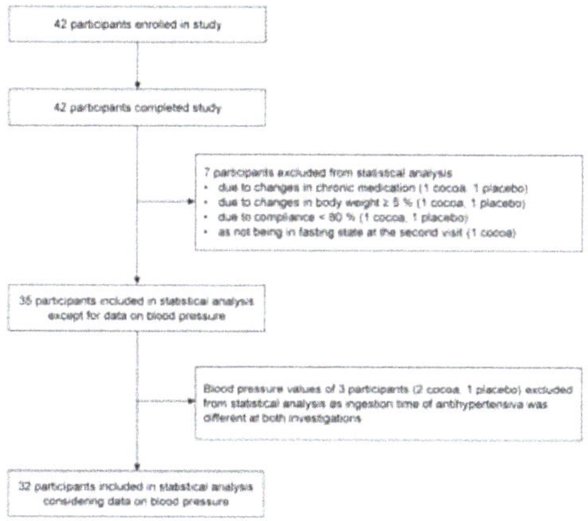

Figure 1. Flow of participants throughout the trial.

The participants (18 men, 17 women) were 64.2 ± 1.5 years old and had suffered from T2D for 6.9 ± 0.8 years. Details on demographic and clinical characteristics are shown in Table 2. There were no significant differences between the groups at baseline with regard to gender, age, body height and weight, BMI, and diabetes duration.

There were no changes in BW and FM. Waist circumference (103.6 ± 4.8 cm vs. 102.3 ± 4.6 cm, $p = 0.047$) and waist-to-hip ratio (0.97 ± 0.02 vs. 0.96 ± 0.02, $p = 0.011$) decreased significantly in the cocoa group (Table 3). The intake of energy, nutrients and epicatechin from food remained unchanged throughout intervention (Table 3). The median compliance with capsules intake (%) was 99.0 (97.6; 100.0) in the cocoa group and 100.0 (98.0; 100.0) in the placebo group. Unintended effects were not reported.

As shown in Table 4, no significant differences in BP, glucose metabolism (FBG, insulin, HbA_{1c}, HOMA-IR) and lipid status (total-C, LDL-C, HDL-C, triglycerides) could be detected between the groups at baseline. Changes in these parameters after both treatments did not occur (Table 4). Results on primary and secondary outcome markers of intention-to-treat analysis ($n = 42$) were not different from those of per-protocol analysis ($n = 35$).

Table 2. Demographic and clinical data.

	Cocoa group (n = 17)	Placebo group (n = 18)	p Baseline
Sex (n; %)			
Female	10 (58.8)	7 (38.9)	ns [a]
Male	7 (41.2)	11 (61.1)	ns [a]
Age (years)	65.6 ± 2.6	62.8 ± 1.6	ns [c]
Diabetes duration (years)	6.7 ± 1.4	7.2 ± 1.0	ns [c]
Antihyperglycemic drugs (n)			
Metformin	11	14	ns [b]
DPP4 inhibitors	5	3	ns [b]
SGLT2 inhibitors	2	2	ns [b]
Antihypertensive drugs (n)			
Beta-receptor blockers	7	7	ns [a]
AT1 receptor blockers	3	10	0.020 [a]
ACE inhibitors	8	9	ns [a]
Calcium-channel blockers	6	8	ns [a]
Diuretics	9	10	ns [a]
Lipid-lowering drugs (n)			
HMG-CoA reductase inhibitors	6	11	ns [a]
Fibrates	1	0	ns [b]

Data: means ± standard error of the means (SEMs) or frequencies; [a] χ^2 test, [b] Fisher's exact test, [c] t-test for unpaired samples. ACE: angiotensin-converting enzyme; AT: angiotensin; DPP4: dipeptidyl peptidase-4; HMG-CoA: 3-hydroxy-3-methylglutaryl-coenzyme A; ns: not significant; SGLT2: sodium-dependent glucose cotransporter 2.

Table 3. Data on nutrition status and on daily nutritional intake.

	Cocoa group (n = 17)			Placebo group (n = 18)			p Baseline
	Baseline	Week 12	p	Baseline	Week 12	p	
Nutrition status							
Body weight (kg)	89.9 ± 7.0	89.4 ± 7.0	ns [c]	91.3 ± 4.6	91.3 ± 4.6	ns [c]	ns [a]
BMI (kg/m^2) [#]	30.2 (26.5; 34.7)	29.8 (26.3; 34.8)	ns [c]	29.3 (26.0; 33.8)	29.5 (26.0; 33.4)	ns [c]	ns [a]
Waist circumference (cm)	103.6 ± 4.8	102.3 ± 4.6	0.047 [c]	103.4 ± 2.9	103.5 ± 2.9	ns [c]	ns [a]
Waist-to-hip ratio	0.97 ± 0.02	0.96 ± 0.02	0.011 [c]	0.99 ± 0.02	0.99 ± 0.02	ns [c]	ns [a]
Fat mass (kg)	34.7 ± 3.8	34.2 ± 3.7	ns [c]	33.5 ± 3.0	33.5 ± 3.1	ns [c]	ns [a]
Fat mass (% BW)	37.7 ± 1.8	37.4 ± 1.8	ns [c]	36.0 ± 1.9	36.0 ± 2.0	ns [c]	ns [a]
Nutritional intake [§]							
Energy (kcal)	2132 ± 227	2074 ± 153	ns [c]	1859 ± 128	2021 ± 149	ns [c]	ns [a]
Protein (g)	91 ± 11	85 ± 8	ns [c]	84 ± 6	89 ± 7	ns [c]	ns [a]
Protein (g/kg BW)	0.8 (0.7; 1.6)	0.9 (0.7; 1.2)	ns [d]	0.9 (0.8; 1.0)	1.1 (0.7; 1.3)	ns [d]	ns [b]
Fat (g)	98 ± 13	91 ± 8	ns [c]	81 ± 6	89 ± 9	ns [c]	ns [a]
SFAs (g) [#]	35 (22; 44)	36 (20; 45)	ns [c]	31 (22; 37)	30 (25; 45)	ns [c]	ns [a]
MUFAs (g)	33 ± 5	29 ± 3	ns [c]	28 ± 3	31 ± 4	ns [c]	ns [a]
PUFAs (g) [#]	14 (8; 24)	15 (11; 32)	ns [c]	14 (11; 21)	13 (9; 24)	ns [c]	ns [a]
Cholesterol (mg)	415 (300; 507)	401 (254; 588)	ns [d]	389 (197; 455)	343 (254; 485)	ns [d]	ns [b]
Carbohydrates (g) [#]	180 (122; 271)	198 (161; 239)	ns [c]	169 (123; 206)	179 (143; 218)	ns [c]	ns [a]
Dietary fiber (g)	22 ± 2	24 ± 2	ns [c]	20 ± 2	22 ± 2	ns [c]	ns [a]
Saccharose (g)	31 (22; 56)	34 (29; 52)	ns [d]	27 (19; 43)	33 (24; 28)	ns [d]	ns [b]
Alcohol (g)	0.5 (0.0; 8.2)	0.6 (0.0; 4.3)	ns [d]	0.2 (0.0; 7.7)	0.5 (0.1; 11.6)	ns [d]	ns [b]
Sodium (mg) [$]	3140 ± 489	2944 ± 342	ns [c]	3203 ± 328	3270 ± 363	ns [c]	ns [a]
Sodium chloride (g) [$]	7 ± 1	7 ± 1	ns [c]	7 ± 1	7 ± 1	ns [c]	ns [a]
Epicatechin (mg)	1.1 (0.3; 9.3)	4.8 (0.7; 9.4)	ns [d]	4.9 (0.8; 6.5)	4.5 (0.8; 6.4)	ns [d]	ns [b]

Data: Means ± standard error of the means (SEMs) or medians (25th-percentile; 75th-percentile); [a] *t*-test for unpaired samples, [b] Mann–Whitney U test, [c] *t*-test for paired samples, [d] Wilcoxon test; [#] logarithmized data used for statistical tests, [§] based on 3-day food records performed before each investigation [$] salt to taste not considered. BW: body weight; MUFA: monounsaturated fatty acids; ns: not significant; PUFA: polyunsaturated fatty acids; SFA: saturated fatty acids.

Table 4. Data on blood pressure and on laboratory investigation.

	Cocoa group (n = 17)			Placebo group (n = 18)			P Baseline
	Baseline	Week 12	p	Baseline	Week 12	p	
Blood pressure							
Systolic (mmHg) [$]	139.1 ± 3.2	138.5 ± 3.7	ns [c]	141.6 ± 4.2	140.4 ± 4.1	ns [c]	ns [a]
Diastolic (mmHg) [$]	78.1 ± 2.9	78.2 ± 2.4	ns [c]	79.1 ± 1.8	78.2 ± 2.6	ns [c]	ns [a]
Glucose metabolism							
Fasting blood glucose (mmol/l)	7.6 ± 0.3	7.5 ± 0.2	ns [c]	7.6 ± 0.3	7.8 ± 0.2	ns [c]	ns [a]
HbA$_{1c}$ (mmol/mol)	46.5 (43.2; 49.7)	46.5(41.0; 50.8)	ns [d]	47.5(44.3; 55.2)	48.6(43.2; 53.0)	ns [d]	ns [b]
Insulin (pmol/l)	99.6 ± 11.0	83.1 ± 9.0	ns [c]	89.6 ± 10.1	91.8 ± 7.7	ns [c]	ns [a]
HOMA-IR	4.7 ± 0.5	3.8 ± 0.4	ns [c]	4.4 ± 0.6	4.5 ± 0.4	ns [c]	ns [a]
Lipid status							
Total cholesterol (mmol/l)	5.0 ± 0.2	4.9 ± 0.2	ns [c]	4.7 ± 0.2	4.6 ± 0.2	ns [c]	ns [a]
LDL-cholesterol (mmol/l)	3.0 ± 0.2	2.9 ± 0.2	ns [c]	2.8 ± 0.2	2.9 ± 0.2	ns [c]	ns [a]
HDL-cholesterol (mmol/l) [#]	1.3(1.2; 1.5)	1.4(1.2; 1.8)	ns [c]	1.3(1.1; 1.4)	1.2(1.2; 1.4)	ns [c]	ns [a]
LDL/HDL cholesterol ratio	2.3 ± 0.2	2.1 ± 0.2	ns [c]	2.3 ± 0.2	2.3 ± 0.2	ns [c]	ns [a]
Triglycerides (mmol/l) [#]	1.3(0.9; 1.9)	1.4(0.9; 1.8)	ns [c]	1.8(1.3; 2.3)	1.5(1.1; 2.0)	ns [c]	ns [a]
Creatinine (μmol/l)	61.0 ± 3.8	61.0 ± 3.8	ns [c]	61.0 ± 3.1	61.0 ± 3.1	ns [c]	ns [a]

Data: Means ± standard error of the means (SEMs) or medians (25th-percentile; 75th-percentile); [a] *t*-test for unpaired samples, [b] Mann–Whitney U test, [c] *t*-test for paired samples, [d] Wilcoxon test; [#] logarithmized data used for statistical tests, [$] data refer to n = 15 (cocoa group) and n = 17 (placebo group). ns: not significant. HDL: high-density lipoprotein; LDL: low-density lipoprotein.

4. Discussion

To the best of our knowledge, this was the first RCT which investigated the cardiometabolic effects of a usual serving size (2.5 g/day, corresponding to one tablespoon) of a flavanol-rich, unsweetened and strongly defatted cocoa for 12 weeks in hypertensive patients with T2D and stable adjustment for BP, glucose and lipid metabolism. Contrary to our hypothesis, BP, glucose, and lipid metabolism were not affected by regular consumption of 2.5 g cocoa (Table 4). Diet and nutrition status remained unchanged except for a decrease in waist circumference in the cocoa group (Table 3). However, the mean change of 1.35 cm was within the inter-measurer error of 1.56 cm [26].

Contrary to our study, two studies with a similar panel of participants showed a reduction in BP after regular ingestion of flavanol-rich cocoa products [5,9]. However, studies which found a decrease in SBP and DBP [5] or an increase in flow-mediated dilatation (FMD) [4,5] after cocoa consumption provided about 3–4 times higher amounts of epicatechin daily (203 mg [4], 111 mg [5]) compared to 40 mg in our study (not specified in [9]). In RCTs without changes in SBP and DBP, the epicatechin intake by cocoa was in a similar range (46 mg [7], 17 mg [6]) as in our study (Table 1). This can also be assumed for flavanol intake as the amount of epicatechin in cocoa products correlates strongly with the sum of catechin, epicatechin, procyanidins B2, B5, C1 and D (R^2 = 0.993) [27]. Grassi et al. [28] have shown that CF dose-dependently improve BP and further vascular parameters like FMD and arterial stiffness even in amounts providing 17 mg epicatechin daily, but they investigated healthy subjects without cardiovascular risk factors and any medication. Potential confounders on BP such as changes in BW [9], body composition [5,9], or diet can be excluded in our study, but not in studies which found a reduction in BP [5,9]. Moreover, the studies of Grassi et al. [5] and Rostami et al. [9] were not double-blinded, in contrast to our study and those of Mellor et al. [6], and failed to demonstrate any effect on BP. Thus, an impact on BP due to different expectations of the participants with regard to treatment [29] can be ruled out in our study due to placebo-controlled study design.

In contrast to our study, two RCTs found an improvement in single parameters of glucose metabolism after regular cocoa consumption [5,9]. However, Grassi et al. [5] investigated subjects with IGT and observed changes in HOMA-IR and in markers of insulin sensitivity only after an oral glucose tolerance test, but not in the fasting state. In participants with T2D, changes in FBG [4,6,7], insulin [6,9], and HbA$_{1c}$ [4,6,9] were not significant except for the decrease in FBG observed by Rostami et al. [9].

We did not find any changes in serum lipids after cocoa consumption as observed in other RCTs. However, our subjects initially had a good lipid status (triglycerides 1.7 ± 0.1 mmol/l; total-C 4.8 ± 1.1 mmol/l; LDL-C 2.9 ± 0.6 mmol/l; HDL-C 1.3 ± 0.1 mmol/l; means ± SEM) compared to those of Parsaeyan et al. [9] (triglycerides 2.6 ± 0.1 mmol/l; total-C 6.3 ± 0.2 mmol/l; LDL-C 3.5 ± 0.1 mmol/l; HDL-C 0.9 ± 0.1 mmol/l) and Grassi et al. [5] (total-C 5.5 ± 0.7 mmol/l; LDL-C 3.4 ± 0.5 mmol/l) whose status was able to be improved by cocoa. In contrast to our subjects, their participants did not receive lipid-lowering drugs [5]. Trials which did not detect changes in triglycerides nor in total-C and LDL-C investigated subjects with an adequate lipid status who were partly pharmacologically treated [6,7,9]. Results in HDL-C were different between the studies [4–9], but HDL-C is affected by lifestyle (e.g., physical exercise, changes in weight, and diet [30]) which was not sufficiently controlled in most RCTs.

Our participants were adequately treated for diabetes, hypertension, and dyslipidemia, often by using pharmacological polytherapy (Table 2). The action of some of these pharmacological agents (e.g., metformin, angiotensin-converting enzyme inhibitors, statins) is partly based on the same mechanisms that are modulated by CFs [31,32]. CFs may lower carbohydrate absorption, protect β-cell function, enhance insulin secretion, and may improve insulin sensitivity through upregulation of glucose transporters and key elements of the insulin signaling pathway [10]. Further mechanisms include lowering cholesterol absorption and synthesis, increasing nitric oxide availability, reducing endothelin-1 and inhibiting angiotensin-converting enzyme (ACE) [33].

Studies which investigated only subjects with T2D without medical treatment [8] or subjects with IGT [5,8] were able to detect a decrease in BP, HOMA-IR [5], total-C, and LDL-C [5,8]. Thus, pharmacological treatment, as simultaneously used by all of our participants, might have maximally affected the insulin signaling cascade and nitric oxide availability. This could have exhausted the potential effects of CFs. Parsaeyan et al. [8] included patients suffering from T2D and increased lipids (total-C ≥ 6.22 mmol/l; triglycerides ≥ 2.26 mmol/l), but without dietary or pharmacological treatment of hyperlipidemia. Hence, the potential for metabolic improvement by flavanol-rich cocoa products seems to be higher in individuals with some degree of dysfunction such as IGT, in insufficiently adjusted patients with T2D and in the postprandial state. Since cardiometabolic parameters did not change after cocoa treatment in pharmacologically well-treated subjects, except for an increase in HDL-C [6,7], effects by CF may occur in participants without pharmacological therapy.

BP was the primary outcome marker of our study, and daily consumption of 2.5 g of a highly flavanol-rich cocoa for 12 weeks did not affect BP in stably-treated patients suffering from T2D and hypertension. For secondary outcome markers, which did not show any effects by cocoa treatment, the study might have been underpowered.

Strengths of our RCT are the double-blinded, placebo-controlled design and the excellent compliance with treatment, probably due to encapsulation of cocoa. Adherence to dietary restrictions was controlled by investigations on nutrition status and on nutritional intake. Therefore, confounding effects by changes in lifestyle, which might have affected our outcome variables, are unlikely even if physical activity was not assessed. A limitation of our study is the lack of biomarkers for flavanol exposure such as γ-valerolactones which have shown to be the predominating metabolites of monomeric [34,35] and oligomeric [34] CFs in human plasma and in urine in the fasting state. Vascular parameters addressing endothelial function would have been interesting, but investigation was not feasible.

5. Conclusions

In conclusion, daily intake of 2.5 g of flavanol-rich, unsweetened and strongly defatted cocoa powder does not affect BP, glucose and lipid metabolism in stably-treated patients with T2D and hypertension in a fasting state. This may be due to pharmaceutical polytherapy which partly modulates the same molecular targets as CFs. Future studies should focus on the preventive effect of such a cocoa against diabetes and further cardiometabolic diseases in individuals with preexisting abnormalities that do not require any pharmacological treatment. This may be interesting for the fasting state as well as for the postprandial state which is associated with metabolic stress.

Author Contributions: The authors' contributions were as follows: L.D., N.K. and S.E. designed the study. L.D., D.G., and K.W. recruited and enrolled the participants. L.D. performed anthropometric measurements, analyzed the food records, and performed statistical analysis with support of S.E. S.E. calculated the sample size together with H.-P.H. D.G. and K.W. were the medical advisors. B.F.Z. analyzed the flavanol composition of the cocoa. L.D. and S.E. drafted the manuscript. S.E. had primary responsibility for the final content of the manuscript. All authors approved the submitted version of the manuscript.

Funding: The study was funded by Hochschule Niederrhein, University of Applied Sciences. Barry Callebaut, Zurich, Switzerland, kindly provided the cocoa as unrestricted gift.

Acknowledgments: We thank our participants without whom this study would not have been possible.

Conflicts of Interest: The authors declare no conflict of interest. Barry Callebaut had no role in the design of the study; in the collection, analyses, or interpretation of data; in the writing of the manuscript, and in the decision to publish the results.

References

1. American Diabetes Association (ADA). Standards of medical care in diabetes-2017. *Diabetes Care* **2017**, *40*, 1–135.
2. Turner, R.C.; Millns, H.; Neil, H.A.W.; Stratton, I.M.; Manley, S.E.; Matthews, D.R.; Holman, R.R. Risk factors for coronary artery disease in non-insulin dependent diabetes mellitus: United Kingdom prospective diabetes study (UKPDS: 23). *BMJ* **1998**, *316*, 823–828. [CrossRef] [PubMed]
3. Alkhatib, A.; Tsang, C.; Tiss, A.; Bahorun, T.; Arefanian, H.; Barake, R.; Khadir, A.; Tuomilehto, J. Functional foods and lifestyle approaches for diabetes prevention and management. *Nutrients* **2017**, *9*, 1310. [CrossRef] [PubMed]
4. Balzer, J.; Rassaf, T.; Heiss, C.; Kleinbongard, P.; Lauer, T.; Merx, M.; Heussen, N.; Gross, H.B.; Keen, C.L.; Schroeter, H.; et al. Sustained benefits in vascular function through flavanol-containing cocoa in medicated diabetic patients a double-masked, randomized, controlled trial. *J. Am. Coll. Cardiol.* **2008**, *51*, 2141–2149. [CrossRef] [PubMed]
5. Grassi, D.; Desideri, G.; Necozione, S.; Lippi, C.; Casale, R.; Properzi, G.; Blumberg, J.B.; Ferri, C. Blood pressure is reduced and insulin sensitivity increased in glucose-intolerant, hypertensive subjects after 15 days of consuming high-polyphenol dark chocolate. *J. Nutr.* **2008**, *138*, 1671–1676. [CrossRef] [PubMed]

6.   Mellor, D.D.; Sathyapalan, T.; Kilpatrick, E.S.; Beckett, S.; Atkin, S.L. High-cocoa polyphenol-rich chocolate improves HDL cholesterol in type 2 diabetes patients. *Diabet. Med.* **2010**, *27*, 1318–1321. [CrossRef] [PubMed]

7.   Monagas, M.; Khan, N.; Andres-Lacueva, C.; Casas, R.; Urpí-Sardà, M.; Llorach, R.; Lamuela-Raventós, R.M.; Estruch, R. Effect of cocoa powder on the modulation of inflammatory biomarkers in patients at high risk of cardiovascular disease. *Am. J. Clin. Nutr.* **2009**, *90*, 1144–1150. [CrossRef] [PubMed]

8.   Parsaeyan, N.; Mozaffari-Khosravi, H.; Absalan, A.; Mozayan, M.R. Beneficial effects of cocoa on lipid peroxidation and inflammatory markers in type 2 diabetic patients and investigation of probable interactions of cocoa active ingredients with prostaglandin synthase-2 (PTGS-2/COX-2) using virtual analysis. *J. Diabetes Metab. Disord.* **2014**, *13*, 1–9. [CrossRef] [PubMed]

9.   Rostami, A.; Khalili, M.; Haghighat, N.; Eghtesadi, S.; Shidfar, F.; Heidari, I.; Ebrahimpour-Koujan, S.; Eghtesadi, M. High-cocoa polyphenol-rich chocolate improves blood pressure in patients with diabetes and hypertension. *ARYA Atheroscler.* **2015**, *11*, 21–29. [PubMed]

10. Ramos, S.; Martín, M.A.; Goya, L. Effects of cocoa antioxidants in type 2 diabetes mellitus. *Antioxidants* **2017**, *6*, 1–16. [CrossRef] [PubMed]

11. European Food Safety Authority (EFSA). Scientific opinion on the modification of the authorisation of a health claim related to cocoa flavanols and maintenance of normal endotheliumdependent vasodilation pursu-ant to Article 13(5) of Regulation (EC) No 1924/2006 following request in accordance with Article 19 of Regulation (EC) No 1924/2006. *EFSA J.* **2014**, *12*, 1–13.

12. Sena, C.M.; Pereira, A.M.; Seiça, R. Endothelial dysfunction-a major mediator of diabetic vascular disease. *Biochim. Biophys. Acta.* **2013**, *1832*, 2216–2231. [CrossRef] [PubMed]

13. Al-Tabakha, M.M. HPMC capsules: Current status and future prospects. *J. Pharm. Pharm. Sci.* **2010**, *13*, 428–442. [CrossRef] [PubMed]

14. World Health Organization (WHO); International Diabetes Federation (IDF). Definition and diagnosis of diabetes mellitus and intermediate hyperglycaemia: Report of a WHO/IDF consultation. In Proceedings of the WHO Document Production Services, Geneva, Switzerland, 2006.

15. European Society of Hypertension (ESH); European Society of Cardiology (ESC). 2013 ESH/ESC guidelines for the management of arterial hypertension: The task force for the management of Arterial Hypertension of the European Society of Hypertension (ESH) and of the European Society of Cardiology (ESC). *Eur. Heart J.* **2013**, *34*, 2159–2219. [CrossRef] [PubMed]

16. Damm, I.; Enger, E.; Chrubasik-Hausmann, S.; Schieber, A.; Zimmermann, B.F. Fast and comprehensive analysis of secondary metabolites in cocoa products using ultra high-performance liquid chromatography directly after pressurized liquid extraction. *J. Sep. Sci.* **2016**, *39*, 3113–3122. [CrossRef] [PubMed]

17. Matthews, D.R.; Hosker, J.P.; Rudenski, A.S.; Naylor, B.A.; Treacher, D.F.; Turner, R.C. Homeostasis model assessment: Insulin resistance and beta-cell function from fasting plasma glucose and insulin concentrations in man. *Diabetologia* **1985**, *28*, 412–419. [CrossRef] [PubMed]

18. World Health Organization (WHO). Waist Ccrcumference and waist-hip ratio: Report of a WHO expert consultation. 2011. Available online: http://apps.who.int/iris/bitstream/10665/44583/1/9789241501491_eng.pdf (accessed on 7 May 2017).

19. Kyle, U.G.; Bosaeus, I.; de Lorenzo, A.D.; Deurenberg, P.; Elia, M.; Manuel Gomez, J.; Lilienthal Heitmann, B.; Kent-Smith, L.; Melchior, J.-C.; Pirlich, M.; et al. Bioelectrical impedance analysis-part II: Utilization in clinical practice. *Clin. Nutr.* **2004**, *23*, 1430–1453. [CrossRef] [PubMed]

20. Kirch, N.; Berk, L.; Liegl, Y.; Adelsbach, M.; Zimmermann, B.F.; Stehle, P.; Stoffel-Wagner, B.; Ludwig, N.; Schieber, A.; Helfrich, H.-P.; et al. A nutritive dose of pure (-)-epicatechin does not beneficially affect increased cardiometabolic risk factors in overweight-to-obese adults-a randomized, placebo-controlled, double-blind crossover study. *Am. J. Clin. Nutr.* **2018**, *107*, 948–956. [CrossRef] [PubMed]

21. Kyle, U.G.; Genton, L.; Karsegard, L.; Slosman, D.O.; Pichard, C. Single prediction equation for bioelectrical impedance analysis in adults aged 20–94 years. *Nutrition* **2001**, *17*, 248–253. [CrossRef]

22. U.S. Department of Agriculture (USDA). Database for the Flavonoid Content of Selected Foods, Release 3.1 (December 2013). 2014. Available online: https://www.ars.usda.gov/northeast-area/beltsville-md/beltville-human-nutrition-research-center/nutrient-data-laboratory/docs/usda-database-for-the-flavonoid-content-of-selected-foods-release-31-december-2013/ (accessed on 5 September 2016).

23. Ott, L. *An Introduction to Statistical Methods and Data Analysis*; Duxbury Press: North Scituate, MA, USA, 1977.

24. Ellinger, S.; Reusch, A.; Stehle, P.; Helfrich, H.-P. Epicatechin ingested via cocoa products reduces blood pressure in humans: A nonlinear regression model with a Bayesian approach. *Am. J. Clin. Nutr.* **2012**, *95*, 1365–1377. [CrossRef] [PubMed]

25. Turnbull, F.; Blood Pressure Lowering Treatment Trialists' Collaboration. Effects of different blood-pressure-lowering regimens on major cardiovascular events: Results of prospectively-designed overviews of randomised trials. *Lancet* **2003**, *362*, 1527–1535. [PubMed]

26. Lohman, T.G.; Roche, A.F.; Martorell, R. *Anthropometric Standardization Reference Manual*; Human Kinetics Books: Champaign, IL, USA, 1988.

27. Cooper, K.A.; Campos-Gimenez, E.; Jimenez Alvarez, D.; Nagy, K.; Donovan, J.L.; Williamson, G. Rapid reversed phase ultra-performance liquid chromatography analysis of the major cocoa polyphenols and inter-relationships of their concentrations in chocolate. *J. Agric. Food Chem.* **2007**, *55*, 2841–2847. [CrossRef] [PubMed]

28. Grassi, D.; Desideri, G.; Necozione, S.; Di Giosia, P.; Barnabei, R.; Allegaert, L.; Bernaert, H.; Ferri, C. Cocoa consumption dose-dependently improves flow-mediated dilation and arterial stiffness decreasing blood pressure in healthy individuals. *J. Hypertens.* **2015**, *33*, 294–303. [CrossRef] [PubMed]

29. Meissner, K. Placebo responses on cardiovascular, gastrointestinal, and respiratory organ functions. *Handb. Exp. Pharmacol.* **2014**, *225*, 183–203. [PubMed]

30. Escolà-Gil, J.C.; Julve, J.; Griffin, B.A.; Freeman, D.; Blanco-Vaca, F. HDL and lifestyle interventions. *Handb. Exp. Pharmacol.* **2015**, *224*, 569–592. [PubMed]

31. Martin, M.Á.; Goya, L.; Ramos, S. Antidiabetic actions of cocoa flavanols. *Mol. Nutr. Food Res.* **2016**, *60*, 1727–1896. [CrossRef] [PubMed]

32. Strat, K.M.; Rowley, T.J.; Smithson, A.T.; Tessem, J.S.; Hulver, M.W.; Liu, D.; Davy, B.M.; Davy, K.P.; Neilson, A.P. Mechanisms by which cocoa flavanols improve metabolic syndrome and related disorders. *J. Nutr. Biochem.* **2016**, *35*, 1–21. [CrossRef] [PubMed]

33. Aprotosoaie, A.C.; Miron, A.; Trifan, A.; Luca, V.S.; Costache, I.-I. The cardiovascular effects of cocoa polyphenols -an overview. *Diseases* **2016**, *4*, 1–25. [CrossRef] [PubMed]

34. Ottaviani, J.I.; Kwik-Uribe, C.; Keen, C.L.; Schroeter, H. Intake of dietary procyanidins does not contribute to the pool of circulating flavanols in humans. *Am. J. Clin. Nutr.* **2012**, *95*, 851–858. [CrossRef] [PubMed]

35. Ottaviani, J.I.; Borges, G.; Momma, T.Y.; Spencer, J.P.E.; Keen, C.L.; Crozier, A.; Schroeter, H. The metabolome of (2-^{14}C) (-)-epicatechin in humans: Implications for the assessment of efficacy, safety, and mechanisms of action of polyphenolic bioactives. *Sci. Rep.* **2016**, *6*, 29034. [CrossRef] [PubMed]

Review

Cocoa Flavanols: Natural Agents with Attenuating Effects on Metabolic Syndrome Risk Factors

Maria Eugenia Jaramillo Flores *

Departamento de Ingeniería Bioquímica, Escuela Nacional de Ciencias Biológicas-Instituto Politecnico Nacional, Wilfrido Massieu s/n esq, Manuel Stampa, Unidad Profesional Adolfo López Mateos, Alcaldía G. A. Madero, Ciudad de México CP 07738, Mexico

Received: 21 February 2019; Accepted: 27 March 2019; Published: 30 March 2019

Abstract: The interest in cacao flavanols is still growing, as bioactive compounds with potential benefits in the prevention of chronic diseases associated with inflammation, oxidative stress and metabolic disorders. Several analytical methodologies support that the flavanols in cacao-derived products can be absorbed, have bioactive properties, and thus can be responsible for their beneficial effects on human health. However, it must be considered that their biological actions and underlying molecular mechanisms will depend on the concentrations achieved in their target tissues. Based on the antioxidant properties of cacao flavanols, this review focuses on recent advances in research regarding their potential to improve metabolic syndrome risk factors. Additionally, it has included other secondary plant metabolites that have been investigated for their protective effects against metabolic syndrome. Studies using laboratory animals or human subjects represent strong available evidence for biological effects of cacao flavanols. Nevertheless, in vitro studies are also included to provide an overview of these phytochemical mechanisms of action. Further studies are needed to determine if the main cacao flavanols or their metabolites are responsible for the observed health benefits and which are their precise molecular mechanisms.

Keywords: cocoa; bioactive compounds; flavanols bioavailability; anti-inflammatory properties; metabolic syndrome; oxidative stress

1. Introduction

The three main cultivars of cacao beans are: Criollo, Forastero and Trinitario. However, with the aim of increasing production and resistance to pests, new varieties have been created from the three original cultivars. Forastero is the most widespread (mainly in Africa) and used around the world, due to its high adaptability and resistance to pests. Criollo is native from Mexico, Central and South America. It is considered the most ancient cultivar and is appreciated for its high quality, flavor and aroma. However, its production represents only 5% of the world cacao beans production, due to its low resistance to pests. Trinitario is a hybrid between the Criollo and Forastero trees that combines good-quality flavor and aroma with pest and disease resistance [1,2].

The cacao bean is a fruit widely recognized as one of the main sources of phenolic compounds with the highest flavanols content of all foods on a per weight basis [3]. The content and profile of bioactive compounds of the cocoa depends on a number of factors such as type and quality of the crop, place of culture, type of process (fermentation, drying and roasting), such that in 6 samples of cocoa liquor around the world, the highest to least quantity of the following components: (-)-epicatechin is presented in samples from Ghana, followed by Mexico and Venezuela; (+)-catechin from Sao Tome and Ghana; caffeic acid derivatives from Venezuela, Ecuador and Ghana; (-)-gallocatechin (GC) from Ecuador, Madagascar and Mexico, without presence in the rest of the samples; (-) - epigallocatechin (EGC) from Madagascar, Ecuador and Sao Tome without the presence of these compounds in Mexico and Venezuela; caffeine from Ecuador, Venezuela and Mexico; and theobromine from Sao Tome, Ghana and Madagascar [4]. About 13 flavanols have been detected and quantified, mainly (-)–epicatechin (0.12–2.83 mg/g), (+)–catechin (0.040-0.090 mg/g), epigallocatechin, epigallocatechin-3-gallate and procyanidins B1(0.035 mg/g), B2 with three different isomers (0.13–0.97 mg/g), B3, B4 with 2 isomers, C1 and D; 7 flavones—luteolin orientin, isoorientin, apigenin, vitexin, ixovitexin; 4 flavanones—naringenin, prunin, hesperidin, eriodyctyol; 4 flavonols—quercetin (0.21–3.25 µg/g), quercetin 3-o-arabinoside (2.1–3.2 µg/g), isoquercitrin (4–4.3 µg/g), and hyperoside; 4 anthocyanidins—cyanidin, and 3 different glycosylated cyanidins; and finally, 8 fenolic acids—vanillic acid, syringic acid, chlorogenic acid, phlorectic acid, coumaric acid, caffeic acid, ferulic acid and phenilacetic acid [3].

Epidemiological and clinical studies show and confirm that regular intake of cocoa powder and/or dark chocolate (50%–70% cacao) is related to a decrease in systolic blood pressure (SBP) (–3.2 to –5.88 mmHg) and diastolic blood pressure (DBP) (–2.0 to –3.30 mmHg), as well as to an improvement of the vascular endothelial function (measured as a function of endothelium vasodilatation and an increase in the production of nitric oxide) in groups with some type of cardiovascular disease or with multiple risk factors [5–7]. Cocoa flavanols reduce the blood pressure by increasing the availability of nitric oxide (increasing the nitric oxide synthase activity and reducing the oxidative stress), consequently vasodilatation increases and finally blood pressure is reduced; or, by inhibition of angiotensin-converting enzyme, interrupting the chain of reactions of angiotensinogen that by the action of renin produces angiotensin I which in turn by action of angiotensin converting enzyme (ACE) is transformed into angiotensin II and finally increases blood pressure [8].

Metabolic syndrome (MS) is a heterogeneous group of correlated metabolic disorders that occur together and raise the risk for diabetes type II (DM2) and cardiovascular diseases with high rates of morbi-mortality [9–11]. To diagnose MS, the International Diabetes Federation (IDF) considers the presence of abdominal obesity (defined as waist circumference with ethnicity specific values) as the main risk factor, plus two additional symptoms: i) high blood pressure (SBP: ≥10 mmHg; DBP: ≥85 mmHg or a specific treatment for hypertension arterial (HTA)); ii) high plasmatic triglycerides (≥150 mg/dL or a treatment specific for this disorder); iii) low HDL-cholesterol (M: <40 mg/dL; W: <50 mg/dL); iv) impaired fasting glycemia (≥100 mg/dL) or diabetes mellitus type 2 (DM2) diagnosis [12].

Several clinical and epidemiological studies have demonstrated that the intake of flavonoids found in vegetables, fruits and oilseeds reduce the risk of developing several types of non-communicable chronic diseases derived from metabolic disorders [13,14].

Among the vast group of flavonoids found in nature, flavanols are a sub-group of particular interest since it has shown multiple protective effects against diseases associated with metabolic and oxidative stress [15].

2. Cacao Flavanols

More than 200 chemical compounds have been identified in cacao beans, and most of them are stored in the vacuoles of the so-called "polyphenolic cells" [7]. The polyphenol content makes up about 12%–18% of the whole bean's dry weight. Approximately 60% of the total polyphenols content in non-fermented cacao beans corresponds to monomeric (catechin and epicatechin) and oligomeric flavanols. The main monomeric flavanol is (−)-epicatechin (with up to 35% of polyphenol content), followed by (+)-catechin and procyanidin B2 (epicatechin-(4β-8)-epicatechin) [16]. Flavanols are secondary metabolites that belong to a sub-class of a larger group of plant compounds known as flavonoids. They share a general chemical structure that includes two rings (A and B) linked through three carbons that form an oxygenated heterocyclic ring (C). As a particular feature, flavanols have multiple hydroxyl groups on the A, B and C rings that have been associated with a decrease in oxidative stress markers. As with all bioactive substances, flavanols and procyanidins mechanisms are largely dependent on their bioavailability at their target tissue [17,18].

3. Flavanol Bioavailability

Upon ingestion, flavanols bioavailability depends on their absorption, metabolism at the gastrointestinal tract, tissue and cellular distribution, and tissue metabolism.

In vitro study showed that procyanidins are hydrolyzed into oligomers when passing through the gastric lumen, due to the high acidity of the medium (pH 2) [19]. Given the above, it was hypothesized that gastric de-polymerization of procyanidins favors their absorption by the small intestine. However, in vivo studies in both animals and humans have shown that both monomeric and oligomeric flavanols remain stable during gastric digestion [20–23]. While the gastric concentrations of epicatechin and procyanidins B2, B5 (dimeric) and C1 (trimeric) did not significantly change over the stomach transit period [19].

Depolymerization of procyanidins (composed mainly by epicatechin monomers) would have resulted in an increase in the epicatechin gastric concentrations and a change in the ratios of oligomers to epicathechin and catechin to epicatechin. However, neither hydrolysis of procyanidins nor change in the aforementioned ratios were observed [20].

Once monomeric and oligomeric flavanols reach the small intestine, they can undergo a series of biotransformations (mainly of phase II) that produce O-methylated, O-sulfated, and O-glucuronidated metabolites, which can be absorbed into the blood stream. Procyanidins with a high degree of polymerization cannot be absorbed in the small intestine, so they reach the colon to be used for microbial catabolism, which leads to the formation of smaller phenolic compounds capable of reaching the liver and then undergo phase II conjugation [24,25]. This has been demonstrated in previous in vivo absorption kinetic studies, in which the presence of these metabolites was observed in the plasma (in concentrations of macro to nanomoles) 30 to 60 minutes after the ingestion of cocoa-based beverages [16,20,21]. The monomers epicatechin and catechin showed the highest absorption rates (22%–55%), while dimeric and trimeric procyanidins were less absorbed (equal or less than 0.5%) [16]. Using liquid chromatography-mass spectrometry, total concentrations of (-)-epicatechin, (+)-catechin and procyanidin B2 were quantified after 30 and 120 minutes after the intake of 0.375 g of cocoa/kg of body weight in healthy subjects (average intake of 26.4 g of cocoa: 323 mg monomers; 256 mg dimers). At both times, it was observed that the plasma epicatechin levels were higher than those of procyanidin B2 (0.5 h: 2.61 ± 0.46 µmol/L as compared to 16 ± 5 nmol/L and 2 h: 5.92 ± 0.60 µmol/L as compared to 41 ± 4 nmol/L, respectively) [26].

Table 1 shows the results of different studies on the bioavailability of pure flavanol compounds (catechin, epicatechin and procyanidin B2). As observed, the maximum plasma concentration is reached about 1 h after administration and is dose-dependent. Regarding (-)-epicatechin, their methylated and non-methylated metabolites were produced almost in the same proportion. On the other hand, catechin produces three more times non-methylated than methylated metabolites. When catechin and epicatechin are given as a mix, their metabolite profile remains similar to the one observed when both compounds are administered alone. After the oral administration of (+)-catechin, (−)-epicatechin and a mixture of the two, it can be shown that (-)-epicatechin is the flavanol with the highest absorption, even though (+)-catechin is also a monomer and its plasma metabolites are similar to the ones produced by (-)-epicatechin (glucuronides, sulfates and sulfoglucuronides) [27]. Similar results have been reported in bioavailability studies of epicatechin and procyanidin B2, when administered individually. The absorption of epicatechin was evaluated after oral administration of different doses of cocoa powder or the pure compound. Results showed that bioavailability of (-)-epicatechin present in cocoa powder was absorbed as efficiently as (-)-epicatechin administered alone [28].

A study in healthy volunteers analyzed the postprandial profile of (-)-epicatechin plasma metabolite profiles after oral consumption of a cocoa beverage. Through the combination of different enzymatic hydrolysis (arylsulfatase and β-glucuronidase) and the use of de novo chemically synthesized reference standards, it was possible to identify and measure 8 circulating postprandial (-)-epicatechin metabolites. As reported in previous studies, (-)-epicatechin-3′-β-D-glucuronide was the most abundant metabolite, followed by (-)-epicatechin-3′-sulfate and 3′-O-methyl-(−)-epicatechin-5/7-sulfate. It was also shown that O-sulfonation is a key conjugating reaction in the metabolism of (−)-epicatechin in humans, since the group of (−)-epicatechin sulfates resulted in being the most diverse [29].

After the intake of 100 g of dark chocolate (70% cacao), have been identified (-)-epicatechin metabolites in human plasma and urine [30]. In this study were identified 10 (-)-epicatechin metabolites, of which (-)-epicatechin-3′-β-D-glucuronide, (-)-epicatechin-3′-sulfate and 3′-O-methyl-(-)-epicatechin 5-sulfate were the major metabolites. Finally, (-)-epicatechin metabolite profile could be divided into three groups (glucuronides, sulfates, and O-methyl sulfates) and their distribution might be modified depending on the amount of (-)-epicatechin ingested, due to enzymatic activity [30].

Table 1. Flavanols bioavailability.

Type of Study	Product/Compound	Dose	Plasma Metabolites	Plasma Cmax (μmol/L)	Plasma Tmax (H)	Area Under Curve AUC	Urinary Excretion	T ½	Reference
In vivo Sprague–Dawley male rats ($n = 30$)	(-)-epicatechin	1, 5 and 10 mg/kg	Total 3′-O-methylated forms (conjugated + no conjugated)	1 ± 0.02; 3.05 ± 0.15; 4.5 ± 0.22	1	-	Total (-)-epicatechin nonmethylated and 3′-O-methylated metabolites (nM/18 h): 397 ± 35 nM; 1870 ± 101 nM; 3003 ± 212nM	-	[27]
			Total nonmethylated forms	0.97 ± 0.14; 3.21 ± 0.29; 4.41 ± 0.50					
	cocoa poder	150, 750 and 1500 mg/kg	Total 3′-O-methylated forms (conjugated + no conjugated)	0.12 ± 0.04; 1.05 ± 0.05; 2.49 ± 0.16	1	-	Total (-)-epicatechin metabolites (non-methylated and 3′-O-methylated): 415±18 nM; 1523±120 nM; 3074±218 nM/18 h	-	
			Total nonmethylated forms	0.35 ± 0.04; 2.12 ± 0.05; 5.08 ± 0.43					
In vivo Sprague–Dawley male rats ($n = 20$)	(-)-epicatechin	172 μmol/kg	Total 3′-O-methylated forms	-	-	78.3 ± 4.9 μmol.h/L	9.45 ± 0.56 μmol/24 h	-	[28]
			Total non-methylated forms			88.3 ± 12.4 Mm ol.h/L	16.6 ± 2.3 μmol/24 h		
	(+)-catechin	172 μmol/kg	Total 3′-O-methylated forms	-	-	23 ± 1.1 μmol.h/L	3.60 ± 0.07 μmol/24 h	-	
			Total non- methylated forms			66.4 ± 2.8 μmol.h/L	8.85 ± 0.76 μmol/24 h		
	Mix	345 μmol/kg	Total epicatechin 3′-O-methylated forms	-	-	76.5 ± 6.8 μmol.h/L	4.51 ± 0.45 μmol/24 h	-	
			Total epicatechin nonmethylated forms			78.7 ± 4 μmol.h/L	9.43 ± 0.58 μmol/24 h		
			Total catechin 3′-O-methylated forms			18.9 ± 0.4 μmol.h/L	2.53 ± 0.34 μmol/24 h		
			Total catechin nonmethylated forms			56.5 ± 3.5 μmol.h/L	7.21 ± 0.51 μmol/24 h		

Table 1. *Cont.*

Type of Study	Product/Compound	Dose	Plasma Metabolites	Plasma Cmax (μmol/L)	Plasma Tmax (H)	Area Under Curve AUC	Urinary Excretion	T ½	Reference
In vivo Wistar albino male rats	[^{14}C] procyanidin B2	21 mg/kg IV	-	-	AUC$_{(0-24)}$: 149 ± 21μg.h/min	75.6 ± 5.4 % of total dose/24 h	6.67 ± 0.95		[23]
		21 mg/kg IG	2.60 ± 0.93 μg/Ml	6.11 ± 0.43	AUC$_{(0-24)}$: 17 ± 2.7μg.h/min	62.9 ± 5.48 % of dose	7.3 ± 2.07		
		10.5 mg/kg IG	1.38 ± 0.28 μg/mL	5.56 ± 0.98	AUC$_{(0-24)}$: 5.18 ± 1.35μg.h/min	62.2 ± 7.6 % of dose	4.57 ± 1.46		
		(–)-epicatechin-3′-sulfate; (–)-epicatechin-5-sulfate; (–)-epicatechin-7-sulfate	331 ± 26 nM; 37 ± 3 nM; 12 ± 1 nM assessed using authentic standards	2					
		Unmetabolized (–)-epicatechin	4 ± 1 nM	1					
Healthy volunteers (*n* = 5; 23.47 ± 3.3 years)	100g of Nestle' Noir 70% chocolate	Content: 79mg (-)-epicatechin 26mg (+)-catechin 49mg procyanidin B2	(-)-epicatechin-3′-β-D-glucuronide; (-)-epicatechin-4′-β-D-glucuronide; (-)-epicatechin-7-β-D-glucuronide	290 ± 49 nM; 44 ± 11 nM; 22 ± 6 nM	3.2 ± 0.2; 3.4 ± 0.3; 12.8 ± 4.8	1276 ± 182 nM/h; 164 ± 38 nM/h; 360 ± 50 nM/h	13.3 ± 3.85 μmol/24 h; 1.03 ± 0.06 μmol/24 h; 7.27 ± 1.35 μmol/24 h	3.8 ± 1.0; 1.8 ± 0.3; 5.6 ± 1.1		[30]
			(-)-epicatechin 3′-sulfate; (-)-epicatechin 4′-sulfate	233 ± 60 nM; 11 ± 3nM	3.2 ± 0.2; 3.5 ± 0.3	954 ± 207 nM/h; 66 ± 8 nM/h	8.53 ± 2.71 μmol/24 h; 0.56 ± 0.13μmol/24 h; (-)-epicatechin 5-sulfate: 1.15 ± 0.20μmol/24 h	2.3 ± 0.8; 4.1 ± 0.9;		
			3′-O-methyl-(-)-epicatechin 4′-sulfate; 3′-O-methyl-(-)-epicatechin 5-sulfate; 3′-O-methyl-(-)-epicatechin 7-sulfate; 4′-O-methyl- (-)-epicatechin 5-sulfate 4′-O-methyl-(-)-epicatechin 7-sulfate	49 ± 14 nM; 153 ± 43 nM; 40 ± 10 nM; 18 ± 6 nM; 13 ± 4 nM	3.6 ± 0.3; 3.8 ± 0.2; 3.8 ± 0.2; 3.8 ± 0.3; 3.8 ± 0.2	269 ± 74 nM/h; 679 ± 160 nM/h; 222 ± 59 nM/h; 94 ± 19 nM/h; 70 ± 22 nM/h	1.67 ± 0.62 μmol/24 h; 14.1 ± 3.88 μmol/24 h; 2.33 ± 0.68 μmol/24 h; 1.37 ± 0.34 μmol/24 h; 0.73 ± 0.23 μmol/24 h	2.5 ± 0.6, 2.1 ± 0.6; 2.1 ± 0.6; 2.3 ± 0.5; 2.0 ± 0.8		

Given the above, Figure 1 shows that in both humans and rats a large proportion of (-)-epicatechin (approximately 90%) is absorbed in different conjugated metabolites (glucuronides, sulfates, methylates) that are produced in the intestinal mucous lining. Through portal circulation, these compounds are transported to the liver where they undergo different biotransformation reactions of phase I and II, which produce new O-sulfated, O-glucuronidated and O-methylated forms of (-)-epicatechin. Afterwards, these new forms can be distributed among tissues or can be excreted from the body via bile or urine. The fraction of polymeric procyanidins unabsorbed in the upper gastrointestinal tract can reach the colon and become available for the local microbiota. Gut flora can produce several low–molecular-weight metabolites that can then be efficiently absorbed [16,20,21,31].

Regarding the absorption of procyanidin B2, it has been reported that it is less efficiently and less rapidly absorbed than epicatechin (5%–10% of the absorbed concentration of (-)-epicatechin), and, therefore, its human and rat plasma concentration is lower (10–40 nM/L) [27,32].

Using liquid chromatography and mass spectrometry, studies of kinetic absorption in humans and rats revealed that procyanidin B2 reachs a maximum plasma concentration at 30 to 60 min after administration of the pure compound [27]. Another study showed that human plasma level of procyanidin B2 reached the maximum at about 2 h after the consumption of a flavanol-rich cocoa [26].

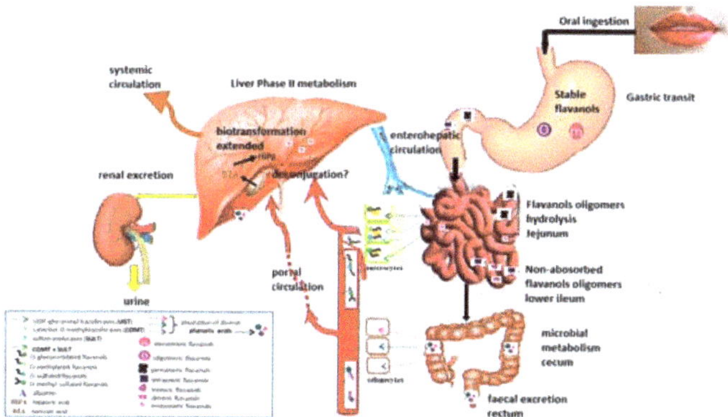

Figure 1. Cocoa flavanols bioavailability. Depolymerization of cocoa procyanidins in the stomach is negligible after ingestion of cacao-derived food products. Thus, most of them reach the small intestine unchanged. Once in the upper intestine, the flavanol monomers and oligomers undergo extensive metabolism (mainly phase II reactions: catechol-o-methytransferase (COMT), sulfotransferase (SULT) and uridine 5 diphosphate glucuronilsyltransferase (UGT)) within the enterocyte (jejunum) that gives rise to a range of O-methylated, O-glucuronidated, and O-sulfated flavanol derivatives. After absorption, the conjugated metabolites are bound to albumin and transported to the liver via the portal vein. Inside the hepatocytes, cocoa flavanols experience extended phase II biotransformations. The resulting metabolites can take 3 different pathways: reach other tissues through systemic circulation or get back to the duodenum through the bile (enterohepatic circulation) or be excreted in the urine. The fraction of the ingested cocoa procyanidins that are not absorbed in the small intestine can be metabolized by colonic microflora (lower part of the ileum and the cecum) into several phenolic acids (such as phenyl propionic acid, phenyl acetic acid and benzoic acid derivatives). These compounds may further be metabolized in the liver and undergo renal excretion, although some may enter other tissues.

However, a subsequent study showed that the maximal concentrations (Cmax) for total (^{14}C) in blood were not attained until 5 to 6 h after oral administration (10.5 and 21 mg/kg) in rats [23]. Therefore, it was suggested that much of the radioactivity was absorbed from the distal part of the small intestine and/or the colon, whereas plasma concentrations of procyanidin B2 detected 30 min and 2 h after oral administration corresponded to the absorption in the proximal part of the small intestine.

Considering the low levels of procyanidin dimer B2 detected in human plasma and that little was known about its colonic metabolism, the catabolism by human faecal microbiota of (-)-epicatechin and procyanidin B2 was compared using an in vitro culture model. Results showed that from 10 phenolic acid catabolites common to both substrates, solely five phenolic catabolites were unique to procyanidin B2 [23]. Although full characterization and further investigation of these catabolites is needed, it has been suggested that they might be of interest with regard to potential biological effects.

Another topic of discussion regarding procyanidin B2 absorption is its possible biotransformations at the gastrointestinal tract. An in vitro study where human gastric conditions were simulated (gastric juice (pH 2.0) at 37 °C for up to 3.5), showed that oligomeric procyanidins (dimer to hexamer) decompose essentially to epicatechin monomeric and dimeric units. It was also suggested that the latter were the major components for absorption via the small intestine [19].

In a following study, the perfusion of isolated small intestine with cocoa procyanidin dimers B2 and B5 (50 mM) showed that both forms are transferred to the serosal side of enterocytes in a lesser extent than the monomer subunits (<1%). Instead, it was observed that unconjugated (-)-epicatechin was the most abundant bioavailable form of procyanidin B2 in plasma (95.8%). These observations provide an explanation for the high ratio of epicatechin to catechin observed [26]. In addition, the latter confirmed the presence of modest concentrations of procyanidin dimers (<1%) in human plasma after the intake of a flavanol-rich cocoa beverage [26]. Small amounts of methylated B2 dimer have also been detected in plasma (3.2%) after ex vivo perfusion of a rat small intestine [19].

Using an in situ rat small intestinal perfusion model it was shown that the presence of tetrameric procyanidins enhanced the absorption of procyanidin B2. These results are consistent with the findings, where elevated concentrations of procyanidin B2 was detected when fed in combination with high-degree of polymerization (DP) oligomers (>DP8) [32,33].

This suggests that further studies are needed to fully understand the synergy between procyanidins with different degrees of polymerization, particularly when considering that these coexist naturally in foodstuffs.

Hepatic glucuronidation, sulfation and methylation of procyanidin B2 have also been assessed using mice microsomal incubations. Unlike (-)-epicatechin, it has been shown that procyanidin B2 remains mostly unmetabolized. Only a small percentage was converted to four minor glucuronide products, although formation mechanisms have not yet been determined due to their low concentrations. This confirms that most of the biotransformations experienced by procyanidin B2 take place in the colon [22,23,34].

A wide range of phenolic acid microbial metabolites (high and low molecular weight) derived from (-)-epicatechin and procyanidin B2 biotransformations have been detected in urine samples collected after consumption of cocoa in humans and rats. It is noteworthy that the variations in the urinary excretion profiles in humans and rats may be influenced by the differences in the ingested dose of cocoa and to the different microbiota present in the intestine of each species [34].

After cocoa consumption, the major microbial metabolites found in human urine samples were caffeic acid, ferulic acid, 3-hydroxyphenylacetic acid, vanillic acid, 3-hydroxybenzoic acid, hippuric acid, 4-hydroxyhippuric acid, (-)-epicatechin and procyanidin B2. On the other hand, the major metabolites in rat urine samples were 3,4-dihydroxyphenylpropionic acid, cumaric acid, 3-hydroxyphenylacetic acid, protocatechuic acid, vanillic acid and (-)-epicatechin [34].

There are few studies regarding the colonic metabolism of phenolic compounds, and even fewer regarding oligomeric flavanols. However, it is now known that colonic microbiota has a large catalytic potential for enzymatic degradation of flavonoids, which results in a huge array of new metabolites with several biological and health-promoting properties [20,23,35,36].

4. Cacao Flavanols and Their Health Effects

After ingestion, flavanols can undergo significant modifications that result in several bioactive molecules with beneficial effects in chronic diseases related to metabolic disorders and oxidative stress.

The mechanisms that have been proposed to explain the biological actions of flavanols are based on their capacity to act as antioxidants and to interact with signaling proteins, enzymes, DNA and membranes. According to the concentrations achieved in their target tissues, their mechanisms have been classified as direct (high concentration) or indirect (low concentration).

4.1. Direct Mechanisms

Until now, the most studied direct effects of cacao flavanols are related to their antioxidant capacity. It is well documented that the latter depends on their aromatic rings with hydroxyl substituents, which give flavanols an adequate configuration to act as electron donors (e^-) and thus stabilize free radicals [18].

On the other hand, the degree of polymerization, partition coefficient and number and distribution of hydroxyl groups will influence the type of interactions that occur between flavanols and the cell membrane. For example, flavanols can partition in the hydrophobic core of membranes or form hydrogen bonds with the polar headgroups of membrane lipids [37].

Given the above, it has been proposed that flavanols may protect the integrity and function of the cell membrane by modulating changes in its fluidity and permeability produced by molecules with oxidation potential [37,38].

It is well known that when membrane fluidity decreases it is more prone to be oxidized. Instead, when fluidity increases membrane lipids are less exposed to oxidation. The effects of cocoa procyanidins on bilayer fluidity and susceptibility to oxidation have been studied using predominantly Jurkat T cells and liposomes. Cocoa derived dimers showed to protect Jurkat T cells from AMVN (2,2'-azobis (2,4-dimethylvaleronitrile)-mediated oxidation and to increase membrane fluidity, measured by a decrease in 1,6-diphenyl-1,3,5-hexatriene (DPH) fluorescence polarization. It was proposed that this effect could be mediated through complex interactions of dimers with membrane proteins, rather than lipids [37].

This was confirmed when the interaction of flavanols and procyanidins (dimers to hexamers) with liposomes (composed of phosphatidylcholine and phosphatidylserin) did not influence its membrane fluidity or lipid lateral phase separation [39].

Membrane lipid oxidation induces the formation of pores that allow the leakage of certain molecules. The increase in membrane permeability due to lipid oxidation has also been studied in liposomes oxidized with AMVN or ferrous iron. An in vitro study showed that preincubation with procyanidins significantly reduces the effect of ferrous iron on liposome permeability [37].

Taking into consideration flavanols' interactions with cell membrane, lipid peroxidation has been widely used to study flavanols' effects against oxidative stress. Breaking initiation and propagation reactions are considered as the most important antioxidant strategy of flavanols for inhibiting or retarding lipid oxidation. During initiation, free radicals (generally HO• and •O_2^-) substract a hydrogen (H) atom from membrane polyunsaturated fatty acids (PUFA) methylene groups [40]. The unpaired electron on the carbon is stabilized by a molecular rearrangement of the double bonds to form a conjugated diene, which then combines with oxygen to form a peroxyl radical (LOO•). The latter has the potential to extend the damage by reacting with other polyunsaturated fatty acids to produce lipoperoxides (LOOH) that can subtract hydrogen atoms from another polyunsaturated fatty acids (PUFA) (propagation reaction). This chain reaction generates irreversible structural and functional damages in cell membranes [18,40].

Additionally, there has been a growing interest in the ability of flavanols to chelate redox-active metals (iron and copper). In biological systems, oxidative stress breaks iron homeostasis and increases its intracellular concentration, promoting free radical-producing reactions and increasing DNA oxidative damage [41,42].

In the presence of hydrogen peroxide (H_2O_2), redox active metal ions such as Fe^{2+} or Cu^+ that are covalently bound to the nucleotide bases of DNA react with it to form highly reactive hydroxyl radical (•OH). The latter abstracts a hydrogen atom from the deoxyribose sugar backbone, which in turn promotes the phosphodiester backbone cleavage and strand scission. Together, nucleotide bases damage (oxidation) and strand breakage have been associated to genetic mutations, cancer and cell death. Flavanols metal chelating properties reside in the presence of a catechol group (B ring) and hydroxyl substituents, since they are centers of high affinity for metal ions. However, there have been observed differences in the magnitude of their chelating activity depending on modifications in their chemical structure [41,42].

Mechanisms underlying flavanols direct effects have been mainly assessed by in vitro studies. However, in vivo studies (biological systems) have been useful for evaluating in vitro evidence.

The beneficial effect of (-)-epicatechin on lipid peroxidation was evaluated in ApoE knockout rats. Administration of epicatechin (64 mg/kg body weight) during 20 weeks significantly reduced aortic F2-isoprostanes, vascular superoxide and endothelin-1 production ($p < 0.05$ versus control ApoE($-/-$) mice) [43,44].

Considering that direct effects require the presence of high concentrations of flavanols, it is thought that in living organisms these effects can be observed just in the gastrointestinal tract, since their absorption has shown to be limited [18].

Cocoa polyphenols are expected to activate Nrf2, which induces the transcription of antioxidant enzymes such as glutathione peroxidase, superoxide dismutase, and heme oxygenase 1, thus blocking the production of reactive oxygen species (ROS) and nitric oxide synthase (NOS), and attenuating oxidative stress, as well as a number of cellular kinases, including the mitogenactivated protein kinases (MAPKs) [45,46]. In Zucker diabetic fatty (ZDF) rats, the ingestion of a cocoa-rich diet (10%) for 9 weeks attenuated hyperglycemia, improved insulin sensitivity, and increased β-cell mass and function. At molecular level, cocoa intake prevented β-cell apoptosis by increasing antiapoptotic proteins (Bcl-xL) and decreasing proapoptotic proteins (Bax and caspase-3 activity) [47].

4.2. Indirect Mechanisms

A major limitation for the direct effects of flavanols in vivo conditions is their relatively low bioavailability. Therefore, it has been suggested that in living organisms the main effects of flavanols are mediated through modifications of enzymatic activities (induction or inhibition), receptors-ligand binding, regulation of protein synthesis and activities, transcription factors binding to their specific sites in DNA, among others (indirect mechanisms) [13,15,48]. In the particular case of the metabolic syndrome, indirect mechanisms have been the most studied. Both in vivo and in vitro models have been useful to link physiological mechanisms with health effects.

Figure 2 summarizes the main mechanisms underlying the effects of cocoa (-)-epicatechin and procyanidin B2 on specific risk factors associated with the development of metabolic syndrome.

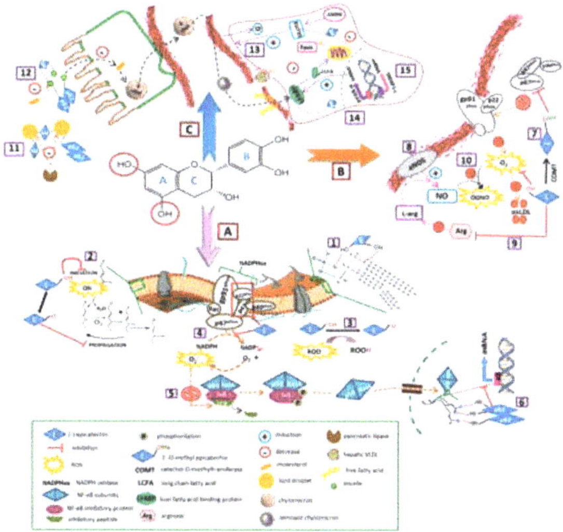

Figure 2. Mechanisms underlying the effects of cocoa (-)-epicatechin and procyanidin B2 on some of the major risk factors for developing metabolic syndrome. A and B mechanisms involved in flavanols antioxidant and anti-inflammatory effects. ① The presence of both hydrophobic and hydrophilic domains in flavanol molecules allow them to be adsorbed on the polar head of membrane lipids and/or to interact with the hydrophobic chains of lipids inside the bilayer in order to modify membrane fluidity and permeability. ② When inserted into the lipid bilayer, flavanols are in close proximity to scavenge free oxygen radicals (such as HO⁻) and lipid soluble radicals (L⁻, LOO⁻) derived from lipid peroxidation. ③ Scavenging free radicals is considered as one of the most important antioxidant mechanisms of flavanols, due to their OH groups (one-electron donation) and their aromatic structures (stabilization by resonance of the resultant radicals). ④ A high dose of epicatechin can prevent upregulation of nicotinamide dinucleotide phosphate NADPH oxidase subunits p47phox and p22phox, the increased enzyme complex activity and generation of O_2^-. ⑤ A decrease in cell reactive oxygen species (ROS) decreases the redox-sensitive release of the LC8 inhibitory peptide, preventing IκBα phosphorylation and degradation and the release of the active NF-κB complex. ⑥ Inside the nucleus, procyanidin B2 can mimic the guanine pairs in the κB DNA sequence and establish hydrogen bonds similar to those that specifically interact with the arginine residues of both p50 and RelA. This inhibits the interaction of NF-κB with κB sites in gene promoters and dependent gene transcription. ⑦ 3'-O-Methyl-(-)-epicatechin has shown to be an inhibitor of endothelial NADPH oxidase by blocking the translocation and interaction of p47phox/p67phox/p40phox with gp91phox and p22phox (transmembranal subunits). Thus, epicatechin improves bioavailability and bioactivity of NO in the arterial vascular endothelium. ⑧ Epicatechin can also increase circulating NO pool via eNOS activation. ⑨ (-)-epicatechin has shown to down-regulate endothelial cell arginase expression and activity. This leads to an increase in vascular L-arginine pool and substrate supply for the eNOS-catalyzed NO synthesis. ⑩ An adequate supply of L-arginine avoids eNOS uncoupling and formation of large quantities of O_2^-, which can scavenge NO to generate peroxynitrite or enhance the production of oxLDL. C. Mechanisms involved in flavanols hypolipidemic and hypoglycemic effects. ⑪ (-)-epicatechin and related flavanol oligomers may suppress triglyceride intestinal absorption by blocking the interaction between pancreatic lipase and the surface of emulsified lipid droplets. ⑫ Monomeric and oligomeric flavanols can lower plasma cholesterol concentrations by decreasing its solubility in intestinal micelles. ⑬ Cocoa flavanols could ameliorate hyperglycaemia by promoting translocation of GLUT4 in insulin-sensitive tissues via activation of AMPK signaling pathways. ⑭ Cocoa flavanols could also attenuate Non-alcoholic steatohepatitis (NASH) by increasing intracellular trafficking of (Long Chain Fatty Acid (LCFA) via liver fatty acid binding protein (LFABP) mRNA and protein expression. ⑮ In addition, cocoa flavanols may decrease fatty acid synthesis by down-regulation of the *Fasn* gene and protein expression.

4.2.1. Effects on Inflammation and Oxidation

Oxidative stress and inflammation are factors with great potential to exacerbate the progression of the metabolic syndrome. Visceral fat accumulation (central obesity) is closely related to an increase in ROS production and to the expression of inflammation-related genes (TNF-α, CRP, IL-6, IL-18, NF-$\kappa\beta$) that increase the risk of developing non-communicable chronic diseases [49,50].

Together, the antioxidant capacity of flavanols and the inverse association of cocoa and chocolate intake and the development of cardiovascular diseases have suggested that these compounds have the potential to reduce oxidative stress and inflammatory processes. Cocoa polyphenols diffuse into the cell and can inhibit MAPKs, thus blocking inflammatory transcription factors, such as NF-kB and AP-1. Together these signals repress the expression of inflammatory genes of many proinflammatory mediators, eg. TNF-α, IL-6, IL-8, IL-1, MCP-1, NO [45].

Studies in animal models have reported that supplementation of high-fat diets with cocoa decreases the plasma concentrations of inflammatory mediators (such as IL-6 and the monocyte chemoattractant protein-1 (MCP-1)), as well as the expression of genes encoding pro-inflammatory molecules (*IL-6*, *IL-12b* and *NOS 2*) in white adipose tissue [51,52].

NADPH Oxidase and Endotheline 1

Given the importance of nitric oxide (NO) on the regulation of vascular homeostasis, the enzymes and transcription factors involved in the decrease in NO synthesis and bioavailability have been proposed as targets for the action of cacao flavanols. Such is the case of NADPH oxidase (Nox), an enzymatic complex that catalyzes the formation of ROS (mainly O_2^- and de H_2O_2), reducing the bioavailability of NO and favoring endothelial dysfunction.

On this regard, an in vitro study with cultures of human umbilical vein endothelial cells (HUVEC) demonstrated that the metabolite 3′-O-methyl (-)-epicatechin had the capacity to inhibit Nox, due to its structural similarities with apocynin (first known inhibitor of Nox) [26,53].

It has been determined that the presence of a mono methylated catechol ring is essential for the inhibitory effect of apocynin, 3′-O-methyl (-)-epicatechin and 4′-O-methyl (-)-epicatechin on Nox activity [54,55]. The latter was corroborated when incubation of 3′-O-methyl (-)-epicatechin in the presence of 3,5-dinitrocatechol (DNC), an inhibitor of catechol-O-methyltransferase (COMT), showed an increase in the production of O_2^- and a decrease in the concentration of NO, due to its reaction with O_2^- to form peroxynitrite ($ONOO^-$) [53].

It has been suggested that epicatechin mono-O-methylated metabolites inhibit NADPH oxidase by affecting the assembly of the multi-protein complex (membrane-linked components and cytosolic proteins).

Oxidative stress also plays an important role in the pathogenesis of hypertension (HTA), which constitutes one of the main risk factors in the development and rapid progression of atherosclerosis. In a murine model of arterial hypertension (DOCA-salt HTA), the effect of two different doses of (-)-epicatechin on the activity of NADPH oxidase were studied. Results show that the chronic administration (5 weeks) of 10mg/kg of epicatechin significantly reduces the activity of this enzymatic complex, plasma markers of oxidative stress, and vascular production of O_2^-.

The proposed mechanism by which (−)-epicatechin inhibits Nox activity involves a decrease in the expression of its cytoplasmic subunit ($p47_{phox}$), which acts as an adaptor protein that facilitates activation of gp91phox (membrane-bound subunit) [56].

Moreover, it has been shown that pretreatment with (-)-epicatechin suppresses the NADPH-oxidase-mediated generation of O_2^- elicited by oxidized low-density lipoproteins (LDL) or angiotensin II in endothelial cells [54].

Previous in vitro studies studied the effect of other flavonoids on the vascular bioavailability of NO and the activity of endothelin-1 (ET-1). The latter is a powerful vasoconstrictor produced in the blood vessel walls that increases the production of O_2^- through the endothelin receptor A (ETA)/NADPH oxidase pathway. Vascular dilatation is critically impaired by an upregulated ETA in association with a reducing nitric oxide bioavailability. In a study of vascular endothelial cultures, it was reported that green tea epigallocatechin-3-gallate (EGCG) stimulates the production of NO through the activation of the Fyn/PI3K/Akt/nitric oxide synthase and down-regulation of ET-1 gene and protein expression [57].

Phosphorylation and activation of Akt kinase (Ser^{473}) and AMPK by EGCG are implicated in the phosphorylation and inhibition of FOXO1 (Thr^{24}), which results in nuclear exclusion of this transcription factor and in its dissociation from hET-1 promoter [56,57].

Polyphenol compounds have demonstrated the capacity to inhibit endogenous antioxidant enzymes (superoxide dismutase (SOD), catalase (CAT), glutathione peroxidase (GPx)) [58]. The flavanols rich in cocoa revert N^{G}-N-L-arginina methyl ester, which is an inhibitor of NO synthase, so that increases the production of NO (whose bioavailability is directly related to oxidative stress, because when it increases it is not converted to peroxynitrites) oxidants that reduce NO, having as a direct consequence vasoconstriction. The flavanols of cacao, due to their antioxidant properties, inhibit the production of peroxynitrites, increasing the bioavailability of NO and finally lowering blood pressure; these last effects are also produced by inhibiting the enzyme angiotensin converting enzyme, a key enzyme in the control of blood pressure, which inhibits the renin-angiotensin-aldosterone system [8].

Endothelial Nitric Oxide Synthase (eNOS) and Arginase

Inhibition of Nox is not the only mechanism proposed to explain cacao flavanols' effect on increased bioavailability of NO. It has been reported that (-)-epicatechin can induce endothelial NO synthase (eNOS) activation through several mechanisms. eNOS is an homodimer binded by two calmodulins (CaMs) that present N-terminal oxigenase and C-terminal reductase domains. The first domain has binding sites for cofactors such as Zn, tetrahydrobioperin, heme group and L-arginin; while the second has biding sites for NADPH, flavin adenine dinucleotide (FAD) and Flavin mononucleotide (FMN) [18]. NO production is mediated by diverse NOS activities and its reaction with superoxide anion, on which its production and bioavailability depend. Exposure of vascular cells to high nanomolar levels or low micromolar concentrations of flavanols triggers a cellular response that indicates a potential existence of cellular surface aceceptors/effectors for epicatechin, which can mediate the subsequent activation of eNOS. This is associated with calcium homeostasis, both kinase II dependent and calmodulin-independent or increased through phosphorylation of serine residues [8].

L-arginine is a key substrate for eNOS endothelial production of NO. Limited availability of this amino acid changes the functional profile of eNos and instead of oxidizing L-arginine, the enzyme reduces molecular oxygen to O_2^{-}. The latter reduces NO bioavailability by rapidly reacting with it to form $ONOO^{-}$ [59]. Vascular arginase competes with eNOS for their common substrate L-arginine. During inflammation and cellular oxidation, the increase in the concentration of arginase is linked to a decrease in the availability of L-arginine, a decreased synthesis of NO, an increase in ROS production, and endothelial dysfunction.

Given the above, an increase in the expression and activity of arginase has been related to the progression of atherosclerosis, hypertension and cardiovascular diseases. An in vitro study showed that (-)-epicatechin and its structurally related metabolites lower arginase-2 expression and arginase activity (dose-dependent manner) in HUVEC. In accordance with these results, the consumption of a high-flavanol cocoa drink (985 mg/serving) showed a decreased arginase activity in human erythrocytes. Further in vivo evidence showed that high-flavanol containing cocoa-based diet (4% cocoa) lowers renal arginase-2 activity in rats [6,60].

Nuclear Factor κB (NF-κB) and Tumoral Necrosis Factor (TNF-α)

In view of the strong association between oxidative stress, chronic inflammation and metabolic syndrome, nuclear factor κB (NF-κB) has emerged as an important target to reduce chronic inflammatory response and development of non-communicable diseases. NF-κB is a ubiquitous heterodimeric protein (including c-Rel, RelB, RelA (p65), p50/p105 and p52/p100) that regulates the expression of a large family of genes that encode proteins involved in inflammation, innate immune response and regulation of cellular survival. It is also considered a useful marker of oxidative stress, since it is activated by ROS and pro-inflammatory cytokines (IL-1, IL-6, TNF-α). In most cells, NF-kB remains inactive in the cytoplasm when inhibitory IκB proteins bind to dimers of RelA, c-Rel, and p50. This interaction blocks the ability of NF-kB to act as a transcription factor. On the other hand, high levels of superoxide-derived ROS and proinflammatory cytokines (such as TNF-α and IL-1) lead to the activation of a specific IκB-kinase (IKK) complex that targets IκB for ubiquitination and proteasomal degradation (by phosphorylation on S32 and S36). The latter releases an active NF-κB that can translocate to the nucleus and bind to specific κB sites in select gene promoters to activate the transcription of genes involved in inflammation [48,61–63].

Procyanidin B2 and (-)-epicatechin can interfere in different levels of the NF-κB activation pathway, which plays a central role in the development of inflammation and regulator of the adhesion and expression molecules of cytokines. A decrease in cell oxidants through free radicals scavenging and inhibition of NADPH oxidase activity are two well known mechanisms that modulate the activation of NF-κB. Other mechanism that has been proposed is a previous inhibition of TNF-α. An in vitro study showed that pre-incubation of Jurkat T cells with increasing doses of procyanidin B1 and B2 reduces NF-κB activation mediated by TNF-α treatment. The expression of *IL-2* (NF-kB regulated gene) was also evaluated by measuring its release to the media. A decrease in its production, after treatment with procyanidin B1 and B2, suggested an inhibition of the formation of DNA/NF-κB complex. Evidence supported by a molecular model has shown that procyanidin B2 can interact with NF-kB proteins (RelA and p50) and prevent their binding to the DNA kB sites. The presence of certain OH groups in the B2 dimer allows the conformation of a folded structure that mimics the guanine pairs in the κB DNA sequence, which interact (forming hydrogen bonds) with the arginine residues of both p50 (Arg 54 and Arg 56) and RelA (Arg 33 and Arg 35) [34,48,62].

All of the above show that procyanidin B2 has the capacity to inhibit the union of NF-κB to DNA and thus decrease the expression of genes involved in oxidation and inflammation processes of the metabolic syndrome. Curcumin and resveratrol (found mainly in grapes and wines) have also been studied for their potential inhibitory effect on NF-κB activation and translocation to the nucleus in TNF-α-stimulated adipocytes. Evidence showed that when cells were co-incubated with TNF-α and either curcumin or resveratrol, degradation of IκB was inhibited, as well as NF-κB nuclear translocation. In accordance with these observations, it was demonstrated that both curcumin and resveratrol were able to down-regulate *TNF-α*, *IL-1β*, *IL-6*, and *COX-2* gene expression in a dose-dependent manner. Moreover, a significant reduction in secreted cytokine levels was also observed [64].

Considering the association between NF-κB activation and up-regulation of endothelial cell adhesion molecules expression, the effect of an acute consumption of a cocoa beverage (40 g of cocoa powder) on NF-κB activation was evaluated in human peripheral blood mononuclear cells (PBMC).

In this clinical trial, cocoa flavanols showed to reduce phosphorylation of p65, which is the transcriptional active subunit of NF-kB. A reduced activation of NF-kB resulted in lower concentrations of both inflammation markers: intercellular adhesion molecule (ICAM-1) and E-selectin. However, vascular cell adhesion protein-1 (VCAM-1) concentrations were not modified after the intervention. These results are in accordance with a previous study, where the consumption of dark chocolate decreased plasma levels of ICAM-1 but not of VCAM-1 [34].

Oxidized Low-Density Lipoproteins (LDLox)

An increase in the oxidation of LDL particles has been associated with low NO bioavailability and high production of ROS in the arterial wall.

The presence of oxidized LDL (oxLDL) constitutes a crucial factor in the development of atherosclerosis and vascular diseases. The latter is supported by evidence that shows that oxidized LDL promotes the formation of fatty streak lesions that give rise to the formation and progression of atherosclerotic lesions by increasing foam cell formation and by inducing inflammatory cytokines, chemokines and adhesion molecules production [65].

The increase in vascular oxLDL concentration has also been associated with an increase in proteosomal degradation of eNOS, thus inverting the eNOS/iNOS ratio. Epicatechin has been shown to protect endothelial cells against both oxLDL-mediated cytotoxicity and loss of eNOS protein [66].

Pretreatment of bovine aortic endothelial cells (BAEC) with (-)-epicatechin prevented oxLDL-elicited down-regulation of eNOS protein and partially the up-regulation of iNOS protein, thus shifting the eNOS/iNOS ratio toward preferencial expression of eNOS protein [67].

Furthermore, oxLDL-provoked oxidative stress in endothelial cells renders cellular proteins vulnerable to oxidative modification, which includes formation of protein-bound carbonyls and tyrosine-nitrated proteins with functional alterations. It has been observed that these modified proteins are mainly localized in the cytosol with highest concentrations at the border zone between cytosol and nucleus. In HUVEC, (-)-epicatechin (10 μM) showed a protective action on protein carbonyl and tyrosine-nitrated proteins formation elicited by oxLDL. Since elevated protein nitration is a consequence of increased iNOS-dependent nitrite formation, these results corroborate that (-)-epicatechin reduces the induction iNOS by oxLDL [67].

Table 2 summarizes the most relevant in vivo and in vitro studies demonstrating the antioxidant and anti-inflammatory effects and underlying mechanisms of cocoa and its main flavanols and procyanidins.

Table 2. Flavanols: antioxidant and anti-inflammatory effects.

Type of Study	Product/Compound	Dose/Duration	Intervention	Target	Outcome (S)	Reference
2-year-old male Wistar rats ($n = 48$)	(-)-epicatechin	2 and 10 mg/kg bw intragastric administration, during 5 weeks	DOCA-salt induced hypertension vs. DOCA-salt EPI2 and DOCA-salt EPI10	Vascular Nox activity Protein expression of Nox p47^{phox} and p22^{phox} subunits	DOCA-salt–EPI10 ↓ Nox activity in aortic rings by suppression of protein over-expression of p47^{phox} and p22^{phox} subunits and ↓ in ET-1 plasma levels Both DOCA-salt–EPI2 and EPI10 restored impaired endothelial function due to an ↑ in eNOS phosphorylation and a ↓ in O_2^{-} vascular content	[56]
Double blind study with crossover-design in healthy volunteers ($n = 10$)	High-flavanol cocoa beverage (98 mg total flavanols: 183 mg epicatechin and 215 mg dimers) Low-flavanol cocoa beverage (80.4 mg total flavanols: 19.8 mg epicatechin and 23.1 mg dimers)	54 g/200 mL of high or low-flavanol cocoa beverage	High-flavol cocoa (HFC) vs. low-flavanol cocoa (LFC)	Erythrocyte arginase activity	Ingestion of a high-flavanol cocoa beverage resulted in the highest decrease in erythrocyte arginase activity after 24 h (HFC: 3.0± 0.4; $p < 0.05$ vs. LFC: 3.5 ± 0.5 µmol urea mg protein-1 h-1)	[60]

Table 2. *Cont.*

Type of Study	Product/Compound	Dose/Duration	Intervention	Target	Outcome (S)	Reference
Jurkat T cells culture	Procyanidin A1, procyanidin A2, procyanidin B1 and procyanidin B2	Cells (1×10^6 cells/ml) were pre-incubated with 2.5–50 µM A1, A2, B1 or B2 for 24 h	Effect of preincubation of Jurkat T cells (further incubation with or without the addition of either TNF-α or PMA)	NF-κB-DNA binding	Pre-incubation (24 h) with B1 or B2 procyanidins (50 µM) ↓ NF-κB-Luc activity (34–52%) and ↓ by 80 and 85% IL-2 release in Jurkat cells subsequently treated with TNF-α or PMA A concentration-dependent (5-50µM) inhibition of NF-κB-DNA binding was observed in cells pre-incubated with B1 or B2 procyanidins	[62]
HCAEC (human coronary artery endothelial cells) culture	(-)-epicatechin			eNOS activation	At 100 nM, B1 and B2 caused a 29–38% and 38–47% inhibition of either p50 or RelA binding to its DNA consensus sequence	[68]
		Incubation of 0.1 nM-100 µmol/L during 10 minutes	Identification of epicatechin intracellular signaling pathways on eNOS-NO production			[69]
Obesity mice 2 months old C57BL			Inflammatory status: TNF αand IL-6		Epicatechin (1 µM/L) induced eNOS activation via Ser1177 and Ser633 phosphorylation and Thr495 de-phosphorylation Epicatechin (1 µM/L) activated eNOS via Akt phosphorylation (induction of Ser1177 phosphorylation) Epicatechin stimulated dissociation of eNOS from Cav-1 and therefore stimulated its activation	
		1 mg epicatechin/kG Body weight 15 days			TNF α level decreased by 50 % while IL-6 decreased by 30%	

Table 2. *Cont.*

Type of Study	Product/Compound	Dose/Duration	Intervention	Target	Outcome (S)	Reference
RAEC, BAEC and human umbilical endothelial cells (HUVEC) cultures	(-)-epicatechin	20 μM incubation for 24 h	Protective effects of (-)-epicatechin against oxLDL protein damage	NADPH oxidase (NOX) activity and oxLDL protein damage	Pretreatment of BAEC and RAEC with epicate-chin prevented oxLDL-elicited downregulation of eNOS protein and par-tially the upregulation of iNOS protein In BAEC and HUVEC incubated with oxLDL, (-)-epicatechin showed a potent O_2^- scavenging activity and a strong inhibition of its production (10 μM) Pretreatment of HUVEC with (-)-epicatechin su-ppressed the formation of all 3 types of modified proteins (protein carbo-nyls and tyrosine-nitrated proteins) in a dose-dependent manner (com-plete inhibition at 10 μM)	[67]
HUVEC culture	(-)-epicatechin, its metabolites (3′-O-methyl epicatechin, 4′-O- methyl epicatechin) and pB2	0.1-100 μM incubation for 24 h	Effect of pB2, epicatechin and its metabolites on NADPH oxidase activity	NADPH oxidase (NOX) activity and O_2^- generation	All 4 compouds (10 μM) inhibited O_2^- re-lease in Angiotensin-II estimulated HUVEC, after 24 h preincubation Methylated epicatechin metabolites proved to be Nox inhibitors (100 μM) without O_2^- scavenging activity Epicatechin showed O_2^- scavenging activity (100 μM) dependent on the duration of preincuba-tion, but did not affect NOX oxidation pB2 showed both inhibitory Nox and O_2^- scavenging activities	[54]

Table 2. *Cont.*

Type of Study	Product/Compound	Dose/Duration	Intervention	Target	Outcome (S)	Reference
Sprague–Dawley male rats ($n = 10$)	Cocoa powder (11 mg epicatechin/g and 43 mg procyanidins/g)	Purified egg white protein-based diet containing 40 g cocoa/kg diet, during 28 days	Diet 0% cocoa vs. Diet 4% cocoa	Renal arginase activity	4% cocoa supplementation ↓ renal arginase activity, compared with control group (0.13 ± 0.02 vs. 0.18 ± 0.02 U/mg protein)	[60]
HUVEC culture	(-)-epicatechin flavanol metabolite mixture (2.6 μM total flavanols: 0.1 μM epicatechin and 2.15 μM epi-catechin metaboli-tes found in human plasma 2 h after high-flavanol cocoa beverage consumption)	mix: 0.4, 2.6 and 7.8 μM epicatechin: 1, 3 and 10 μM 48 hour incubation	Comparison between different concentrations of flavanol mix and epicatechin	Arginase-2 (Arg-2) mRNA expression and activity	Flavanol mix and epicatechin signifi-cantly ↓ Arg-2 mRNA expression in HUVEC, at 24 h in a dose-dependent manner Cells incubated with flavanol mix and epica-techin exhibited ↓ Arg-2 activity, at 48 h in a dose-dependent manner	[60]
Randomized, crossover clinical trial in healthy volunteers ($n = 18$)	Cocoa powder	40 g cocoa powder (28.2mg epicatechin and 25.5 mg pB2/40 g) with 250 mL whole milk or water, during 3 weeks	Cocoa powder with milk (CM) vs. cocoa powder with water (CW)	NF-κB activation and protein expression of adhesion molecules (sICAM-1, sVCAM-1 and sE-selectin) in PBMC (periphe-ral blood mono-nuclear cells)	CW significantly ↓NF-κB activation (determined by protein expression) after 6 h of ingestion, compared with CM Both CM and CW ↓ serum [sICAM-1] after intervention but only CW ↓ [sE-selectin]	[34]
Human hepatoma HepG2,	(-)-Epicatechin (EC) and cocoa phenolic extract (CPE)	10 μM EC or 1 μg//mL CPE were added to the cells for 24 h;	Comparison between epicatechin and polyphenol extract	Nrf2; GPx, GX and CAT	Antioxidant exnzymes were regulating and Nfr2 has been stimulated.	[46]

bw: body weight; DOCA-salt: deoxycorticosterone acetate and sodium chloride; EPI: (-)-epicatechin; Nox: NADPH oxidase; ET-1: endothelin-1; eNOS: endothelial nitric oxide synthase; O_2^-: superoxide; NO: nitric oxide; oxLDL: oxidized LDL; RAEC: rat aortic endothelial cells; BAEC: bovine aortic endothelial cells; iNOS: inducible nitric oxide synthase; pB2: procyanidin B2; NF-κB: nuclear factor κB; PMA: phorbol myristate acetate; TNF-α: tumor necrosis factor-alpha; Cav-1: caveolin-1. ↑ increased; ↓decreased.

4.2.2. Effects of Flavanols on Lipid Metabolism Disorders

According to the criteria for the diagnosis of the metabolic syndrome, the main lipid metabolism disorders include elevated plasma triglyceride (TG) levels and low concentrations of high-density lipoproteins (HDL-c).

Nevertheless, metabolic syndrome dyslipidemia also involves an increase in total cholesterol (TC), remnant chylomicrons, low-density lipoproteins (LDL-c) and very low-density lipoproteins (VLDL-c).

An overview of current clinical trials and in vitro studies evaluating cocoa and derived products consumption on metabolic syndrome lipid profile is summarized in Table 3.

Epidemiological evidence shows a direct relationship between the intake of polyphenol-rich vegetables and a lower predisposition to develop dyslipidemia and cardiovascular diseases. Observational studies and clinical assays in animals and humans have demonstrated the positive effects of cacao flavanols on oxidation, inflammation and endothelial function. It has also been highlighted that these compounds are able to improve the lipid profile that predisposes subjects to the development of atherosclerosis and cardiovascular diseases [70,71].

Meta-analysis of clinical assays have demonstrated that short term consumption of cacao-derived products (cocoa and dark chocolate) has benefits on the lipid profile of patients with some type of cardiovascular disease or with metabolic risk factors. Most of the studies are consistent when showing a decrease in the plasma levels of total (TC) and LDL-c. However, with regard to the increase in HDL levels, results are heterogeneous [70,71].

Other studies on animals and humans (healthy or with CV risk) have also reported a significant decrease in the plasma levels of TG, TC and LDL-c, as well as an increase in the HDL-c levels after chronic and acute consumption of cocoa powder or dark chocolate [27,72].

Additionally, supplementation with cocoa has shown a significant increase in the plasma concentrations of adiponectin, which has been related to a decrease in hepatic triglyceride content of mice and rats fed a high-fat diet [51,73].

An animal study evaluated the potential mechanisms for the hypocholesterolemic effect of a polyphenol extract from cocoa powder and a mixture of catechin (0.024%) and epicatechin (0.058%), after the ingestion of a high-cholesterol diet for 4 weeks. Results showed that the polyphenol extract group had significantly lower plasma cholesterol concentrations, and had significantly greater fecal cholesterol and total bile acids excretion than the mix (catechin/epicatechin) and the control group [74].

Table 3. Flavanols: hypolipidemic and hypoglycemic effects.

Type OF Study	Product/Compound	Dose/Duration	Intervention	Target	Outcome (s)	Reference
Comparative, double-blind study in normo and mild hyper cholesterolemic japanese subjects ($n = 160$)	Low PFT cocoa powder (64.5 mg epicatechin and 36.3 mg pB2/g) middle PFT cocoa powder (96.7 mg epicatechin and 54.4 mg pB2/g) high PFT cocoa powder (129 mg epicatechin and 72.5 mg pB2/g)	Consumption of 13 g low PFT cocoa; or 19.5 g middle PFT cocoa; or 26 g high PFT cocoa, during 4 weeks	Intake of low PFT cocoa powder vs. middle PFT cocoa vs. high PFT cocoa in normo and mild hyper cholesterolemic subjects	Serum LDL, HDL and oxLDL	Consumption of 3 cocoa doses in subjects with LDL ≥3.23 mmol/L, resulted in significantly ↓ serum [LDL], after 4 wk Consumption of 3 cocoa doses (normo and mild hypercholesterolemic subjects) resulted in ↑ serum [HDL], compared with baseline after 4 wk Plasma oxLDL levels were significantly ↓ after 4 wk consumption of 3 cocoa doses (normo and mild hyper cholesterole-mic subjects)	[72]
Randomized, placebo-controlled, double blind, crossover study in DM 2 subjects ($n = 12$)	High (16.6 mg epicatechin) and low (<2 mg epicatechin) polyphenol content chocolate	45 g high or low polyphenol chocolate, during 8 week	High polyphenol chococolate (HPC) intake vs. low polyphenol chococolate (LPC) intake	Serum c-HDL, c-LDL, TG, HbA1c, fasting glucose and insulin and C- reactive protein	Consumption of HPC and LPC improved lipid profile through ↑ HDL/↓ LDL No changes were observed in fasting glucose or HbA1c levels in none of the 2 treat-ment groups Insulin levels showed an ↑ after LPC intake No changes were observed in C-reactive protein levels in none of the 2 treatment groups	[75]
Cross-sectional study in 4098 patients from NHLBI	Chocolate	-	Association between self-reported chococlate consumption and prevalence of metabolic syndrome (MS) in adult population	ATP-III criteria for clinical diagnosis of metabolic syndrome	Higher intake of chocolate was associated with ↓ prevalence of coronary heart disease and ↓ glycemia From the lowest to the highest levels of choco-late consumption, the prevalence of MS odd ratios were: + women: 1.0 (0/wk); 1.26 (<1/wk); 1.15 (1–4/wk) and 0.9 (+5/wk) + men: 1.0 (0/wk); 1.13 (<1/wk); 1.02 (1-4/wk) and 1.21 (+5/wk) the highest odd ratios of obesity prevalence were observed with higher chocolate consumption	[71]

Table 3. *Cont.*

Type OF Study	Product/Compound	Dose/Duration	Intervention	Target	Outcome (s)	Reference
9-week-old male Sprague–Dawley rats ($n = 40$)	Cacao procya-nidins (CP) extracted from cacao liquor (CLPr: 79.3% total polyphenols; 5.9% epicatechin; and 4% PB2)	High-cholesterol diet (HCD: 1% cholesterol and 15% fat) supplemented with 0.5 or 1.0% of CLPr	HCD with 0.5% CP vs. HCD with 1.0% CP	Plasma and liver cholesterol liver and feces TG	Both CP groups (0.5 and 1%) inhibited drastic elevation of plasma TC levels Liver cholesterol and TG levels were significantly ↓ in HCD groups supplemented with both CP doses (more marked effects in 1% CP)	[76]
12-week-old female Sprague–Dawley rats ($n = 56$)	Cocoa powder	C1: methionine-choline deficient diet (MCD) + 28 d of 12.5% cocoa supplementation C2: MCD diet + 56 d of cocoa supplementation C3: 80 d of MCD + cocoa supp. C4: 108 d of MCD + cocoa supplementation	C1 and C2 were selected to test NASH treatmet effects of cocoa supplementation C3 and C4 were used to test if cocoa supple-mentation could prevent NASH development	mRNA and protein expression of LFABP serum TG, glucose and superoxide levels	All cocoa supplemented groups showed ↓ serum TG and glucose levels C3 group had the ↓ superoxide levels, compared to C1, janeC2 and C4 groups C1 had the ↑ mRNA and protein expression levels of LFABP	[77]
3-week-old female diabetic obese mice ($n = 44$)	Cacao liquor procyanidins (72.32% total polyphenols; 5.89% epi-catechin and 3.93% PB2)	Supplementation with 0.5 or 1 % CLPr, during 3 weeks	Dietary supplementation with 0% CLPr vs. supplementation with 0.5 or 1.0% CLPr	Plasma glucose (hyperglycemia) and renal function	Levels of blood glucose were significantly ↓ in mice fed 1% CLPr At the end of the study, group supplemented with 1% CLPr had ↓ levels of BUN and creatinine and suppressed membrane lipoxidation in kidney (↓ 4-hidroxy-2-nonenal antibody levels)	[78]
Streptozotocin-diabetic male Wistar rats ($n = 80$: 200–300 g)	Cocoa beans extract (CE)	Supplementation with 1, 2 or 3% CE (1 g CE/100 g diet)	Streptozotocin-diabetic rats + normal diet vs. diabetic induced rats + 1, 2 or 3% CE	Serum CT, HDL, LDL, TG and glucose	All 3 diabetic groups treated with CE showed significant ↓ in body weight gain and serum TG levels Diabetic groups treated with 1 and 3% CE exhibited significant ↓ in plasma glucose levels Diabetic group treated with 1% CE had the ↓ serum levels of CT and LDL	[79]

Table 3. *Cont.*

Type OF Study	Product/Compound	Dose/Duration	Intervention	Target	Outcome (s)	Reference
Male Sprague–Dawley rats (n= 90)	Polyphenol-rich cocoa extract (CE)	Intragastric administration of 1, 2 or 3% CE (1 mL/100 g bw) during 4 weeks	Assessment of CE effectiveness in reducing hyperglyce-mia in diabetic-induced rats	Plasma glucose levels and body weight gain	A significant body weight reduction was observed ($p < 0.05$) in diabetic-induced rats treated with 1 and 2% CE Diabetic-induced rats treated with 3% CE showed the most significant ↓ in glucose levels	[80] $
Glucose-responsive pancreatic cell lines (BRIN-BD11)		Polyphenol-rich cocoa extract (CE)	Incubation with CE at 2, 1, 0.5, 0.1 and 0.05 mg/mL	Evaluation of different concentrations of CE on insulin secretion	Insulin-release from rat pancreatic β-cells	[81]
Randomized, crossover trial in patients with hypertension (HTA-I) and impaired glucose tolerance (IGT) ($n = 19$)		Flavanol-rich dark chocolate bar (FRDC: 110.9 mg epicatechin/bar)	100 g of dark chocolate bar, during 15 days	Flavanol rich dark chocolate bar vs. flavanol free chocolate bar	Fasting glucose and insulin sensitivity Serum C-reactive protein (CRP) Serum lipid profile (HDL, CT, LDL and TG)	
Male C57BL/6 4-week-old mice ($n = 36$)	cacao liquor procyanidin extract (CLPr: 6.12% epicatechin and 3.60% PB2)	Supplementation with 0.5 or 2% CLPr, during 13 weeks	High fat diet (HFD) vs. HFD + 0.5% (HF-0.5) or 2% (HF-2) CLPr	Glucose parameters mRNA and protein expression of UCP-1, UCP-2, GLUT-4 and AMPKα	At week 7, fasting glucose levels in HF-2 group were significantly lower At week 11, OGTT^e showed ↓ glucose levels in HF-2 group (0 and 15 minutes after glucose load) At the end of the study, HF-0.5 and HF-2 completely suppressed HF diet-induced hyper-glycemia, hyper-insulinemia (↓ HOMA-IR) and hypercholestero-lemia, compared to control group CLPr supplementation promoted AMPKα phosphorilation (BAT, WAT, liver and skeletal muscle), which enhanced GLUT-4 translocation to plasma membrane in BAT and skeletal muscle in a dose-dependent manner CLPr ↑ UCP-1 and UCP-2 gene and protein expression in BAT and WAT, respectively	[82]

PFT: total polyphenols; OGTT: oral glucose tolerance test; oxLDL: oxidized LDL; HbA1c: glycated hemoglobin; NHLBI: National Heart, Lung, and Blood Institute; NASH: non-alcoholic steatohepatitis; LFAP: liver fatty acid binding protein; bw: body weight.↑ increased; ↓decreased.

In vivo polyphenol extract effects were also assessed in vitro. It was shown that micellar solubility of cholesterol was significantly lower for procyanidin B2 (dimer), B5 (dimer), C1 (trimer) and A2 (tetramer) (main components of polyphenol extract) compared to catechin and epicatechin. These results indicate that oligomeric procyanidins from cocoa powder are the main active components responsible for the inhibition of cholesterol and bile acids intestinal absorption, through the decrease in micellar cholesterol solubility [75].

It has also been proposed that flavanols can act synergistically with tea polyphenols to increase their overall lipid-lowering effect through the inhibition of pancreatic lipase activity [83].

Chocolate intake has also been evaluated on the prevalence of the metabolic syndrome. The National Heart, Lung and Blood Institute (NHLBI) family heart study examined the association between self-reported chocolate consumption and the prevalence of MS in an adult US population. Results showed a lower prevalence of metabolic syndrome in groups with higher chocolate intake per week (>5 times/week) [71]. Table 3 shows that even when there is a big difference in the food matrix (powder or chocolate) or extracts, as well as in the concentrations tested; in addition if they are tested in humans, animals (obese or diabetic) or cell culture, consensus is found in the observed effects, that is, reduction of total cholesterol levels, triacylglycerols, both in serum and liver; in some cases reduction in glucose levels, lipoxidation in the kidney, and increase in insulin secretion were shown. In addition to observing these in humans, but without changes in the CRP. The inolucrated mechanisms, although already mentioned previously, the flavanols promote the phosphorylation of AMPK, in adipose tissue, by which the translocation of GLT4 is improved, in addition to increases in the gene and protein expression of UCP1 and UCP2 [76–82].

The hypocholesterolemic effects of apple procyanidins have also been studied. In rats fed a purified diet containing 0.5% cholesterol, the supplementation with apple procyanidins (AP: 0.2%, 0.5% and 1%) showed a decrease in liver and serum TG and TC levels in a dose-dependent manner compared with the control group. Additionally, the levels of HDL-c were significantly higher in the groups supplemented with AP 0.5% and 1% compared to the control group

In the liver, AP at levels of 0.5% and 1% significantly lowered cholesterol levels. This effect was associated with the modulation of hepatic cholesterol 7alpha-hydroxylase (CYP7) activity and an increased fecal excretion of acidic steroids [84].

A similar study evaluated the anti-obesity effects of apple and tea procyanidins. Supplementation of a high-fat diet (HF) with 1% of tea or apple procyanidins showed lower total white adipose tissue levels compared to the control group (HF alone). This result was in accordance with a decrease in serum leptin levels by dietary procyanidins (tea and apple). However, no changes were detected in serum adiponectin and insulin levels.

On the other hand, serum and hepatic TG levels were lowered by dietary procyanidins compared to the control group. In contrast, tea and apple procyanidins reduced serum TC levels, but showed no effect on liver TC levels.

It was suggested that these effects were mediated by a combination of an agonist-like action of PPARα and a moderate antagonist-like action of PPAR-γ, which in turn promoted a decrease in hepatic fatty acid synthase (FAS) activity and an increase in the peroxisomal β-oxidase activity [84].

A reduction on intrahepatic TG and cholesterol levels have been attributed in part to the modulation in the expression of genes related to their metabolism. In vitro, the acute administration of grape seed procyanidins decreased the expression of 3-hydroxy-3 methyl-glutaryl-CoA reductase (HMG-CoA-reductase), the limiting enzyme in the biosynthesis of cholesterol.

In the liver, lipid metabolism disorders promote an increase in fat infiltration and lipotoxicity that result in the development of non-alcoholic steatohepatitis (NASH), one of the most severe comorbidities of the metabolic syndrome [65,77].

An experimental study evaluated the preventive effects that cocoa supplementation may have on the development of NASH induced by a high-fat and choline deficient diet. Results showed that cocoa supplementation reduces the degree of hepatic steatosis, hepatic fibrosis and portal inflammation.

It was suggested that these effects could be mediated in part by up-regulation of gene and protein expression of the liver fatty-acid binding protein (LFABP) in rats with NASH. LFABP prevents non-alcoholic fatty liver disease (NAFLD) development by shuttling long-chain fatty acids towards the mitochondria for β-oxidation and possibly by acting as an antioxidant when intracellular antioxidants (CAT, SOD, GPx) are insufficient [77].

Effects of catechin, epicatechin and procyanidinB2, as well as extract of cocoa rich in flavanols on lipid metabolism disorders have also been studied. In an animal study, the effects of supplementing and hyperlipidic diet (21.2% of fat), 10 mg/Kg body weight of the pure compounds and 100 mg of the extract during eight weeks were evaluated. Results showed that flavanols supplementation reduced the accumulation of intrahepatic TG and fatty acids, thus inhibiting the development of steatosis. The proposed underlying mechanisms were the down-regulation of hepatic lipogenic genes expression (fatty acid synthase, PPAR-γ, and Cd36), as well as an up-regulation of lipolytic genes expression (PPARα, PGC1α, SIRT1) [85]. These results agree with those observed in vivo with vaticanol C (a resveratrol tetramer), which showed to be a more potent activator for PPARα than resveratrol. Vaticanol C also showed to induce PPARα-dependent genes, such as hepatic fatty acid binding protein-1 (FABP1) and acyl-CoA oxidase-1 (Acox1) [86].

Naringin is an active flavanone glycoside present in cocoa although at low concentrations that showed a significant lipid-lowering effect in a mouse model fed with a high-fat diet for 20 weeks. Supplementation with this compound (0.2 g/kg) improved dyslipidemia and hepatic steatosis by reducing serum and liver cholesterol levels. However, it did not reduce liver TG levels. Additionally, in white adipose tissue, naringin treatment reversed hypertrophic adipocytes. Among the mechanisms underlying naringin hypolipidemic effects were an up-regulation of four genes involved in fatty acid oxidation (*PPARα, CPT-1a, ACOX, UCP2*), and a down-regulation of three genes encoding enzymes involved in fatty acid synthesis (*SERBP-1c, FAS, ACC*). Stearoyl-CoA desaturase-1 (SCD-1), a very important lipogenic enzyme responsible for triglyceride metabolism, showed no alteration after naringin treatment. The latter explains why naringin was not able to reduce liver TG levels [87].

4.2.3. Effects on Hyperglycemia and Insulin Resistance

Besides the antioxidant, anti-inflammatory and hypolipidemic effects demonstrated by cacao flavanols, publications have emphasized their potential benefits in reducing hyperglycemia, insulin resistance and diabetes (Table 3). The latter are closely related to dyslipidemia, accumulation of abdominal fat and to the pathogenesis of the metabolic syndrome.

The effect of flavanol-rich dark chocolate intake (110.9 mg epicatechin/100 g) on endothelial function, insulin sensitivity, β-cell function and blood pressure was evaluated in hypertensive patients with impaired glucose tolerance. After 15 days of dark chocolate intake, an increase in vasodilation was observed (by increasing NO availability), insulin sensitivity and β-cell function, as well as a decrease in BP and hyperglycemia [81,88].

In vivo studies have shown that the administration of cacao liquor rich in procyanidins (CLPr) to diabetic and obese mice decreased hyperglycemia in a dose-dependent manner. The mechanisms proposed for this effect involved an increase in the translocation of GLUT4 towards the cell membrane, an increase in the phosphorylation of AMPK and the up-regulation of *UCP-2* gene expression in the skeletal muscle [78,82]. These results agree with those obtained by Ruzaidi et al, in which the administration of a cocoa extract in a DM 2 animal model, resulted in hypoglycemic (reduction of plasma glucose levels and insulin mimicking activity) and hypolipidemic effects [79,80].

In a further study, the effect of supplementing a standard diet with different concentrations of CLPr was evaluated in obese female rats with DM2. Results showed that groups fed a standard diet containing 0.5% and 1% CLPr had lower glycemic levels from the first week of the treatment as compared to the control group. It is noteworthy that this effect was dose-dependent [78]. A similar study evaluated if supplementation with CLPr was able to attenuate the development of obesity, insulin resistance and hyperglycemia induced by a high-fat diet. At the end of the experiment, it was observed that different doses of CLPr decreased the levels of fasting plasma glucose as compared to the group fed with a high-fat diet without supplementation. Oral glucose tolerance was examined in order to evaluate the effects of CLPr on postprandial hyperglycemia. Results indicated that supplementation with 2% CLPr suppressed postprandial hyperglycemia and hyperinsulinemia [82].

In this study, it was also confirmed that the molecular mechanisms involved in CLPr hypoglycemic effects were the activation of AMPK and an increase in GLUT4 translocation. Finally, the effect of the CLPr supplementation on the regulation of thermogenesis and energy metabolism was also studied [82].

Results showed that both concentrations of CLPr (0.5 AND 2%) increased energy expenditure through up-regulation of gene and protein expression of UCP1 (expressed in brown adipose tissue) and UCP2 (expressed in white adipose tissue and liver) [82]. These results agree with those obtained [87], in which the effect of supplementing a high-fat diet with naringin (0.2 g/kg of body weight) was evaluated in a murine model of diet-induced obesity.

Results showed a significant decrease in fasting plasma glucose and insulin levels, which indicated a relevant improvement in insulin sensitivity among different tissues.

In animal models of obesity, low plasma adiponectin levels and high plasma TNF-α levels were observed. Supplementation with naringin decreased plasma TNF-α levels and increased those of adiponectin [87]. Besides, it was observed that through the activation of the AMPK pathway, naringin can down-regulate glycogenic enzymes gene expression (PEPCK and G6Pase), and thus increase glucose uptake by an insulin-independent pathway [87]. Treatment of rat hepatic cell cultures with epigallocatechin-3-gallate has been shown to induce insulin receptor and insulin receptor substrate (IRS-1) tyrosine kinase activity and to down-regulate phosphoenlopyruvate carboxykinase gene expression (a key enzyme in gluconeogenesis) [89].

5. Conclusions

In summary, there is evidence from in vitro assays, animal experiments and human intervention trials, which demonstrate that cacao flavanols have the potential to modulate several risk factors associated with the development of the metabolic syndrome.

Considering that flavanols' bioactivity in vivo depends on their absorption and metabolism, there has been a great advance in the analysis of cocoa flavanols metabolites in plasma and urine (mainly from (-)-epicatechin) after the administration of pure compounds or the consumption of cocoa or dark chocolate.

However, further studies on cellular uptake of O-methylated, O-glucuronidated, and O-sulphated forms are required in order to provide information on which metabolites are the most able to enter cells. The latter will be useful to predict their localizations in different body tissues.

On the other hand, since cocoa is a complex mixture of monomeric and oligomeric flavanols, future experiments should be focused on the possible interactions between flavanols in the gastrointestinal milieu and their synergic effects on their bioavailability. Moreover, it would be important to study the possible effects that the presence of different macronutrients (proteins, lipids and sugars) in cocoa-derived products could have on flavanols absorption.

Additionally, we have shown some of the potential molecular targets of cacao flavanols and the proposed mechanisms responsible for their alleviating effects on metabolic syndrome development.

Further studies on cocoa flavanols' effects should include their major metabolites (at a physiologically achievable dose), in order to reveal if they are linked to the biological effects and underlying mechanisms of cocoa products.

Together, all this information will serve as the basis for a careful review to provide meaningful dietary recommendations for the consumption of cocoa flavanols.

Funding: This work was financially supported through the project SIP-IPN 201905333; CONACYT-CB-2013-01 No. 220732- I010/532/2014.

Acknowledgments: The author thanks the support received from EDI-IPN; COFAA-IPN; SNI-CONACYT.

Conflicts of Interest: The author declare no conflict of interest.

References

1. Rangel-Fajardo, M.A.; Zavaleta-Mancera, H.A.; Córdova-Téllez, L.; López-Andrade, A.P.; Delgado-Alvarado, A.; Vidales-Fernández, I.; Villegas-Monter, Á. Anatomía e histoquímica de la semilla del cacao (*Theobroma cacao* L.) criollo mexicano. Anatomy and histochemistry of the mexican cacao (*Theobroma cacao* L.) seed. *Rev. Fitotecnia Mex* **2012**, *35*, 189–197.

2. Colombo, M.L.; Pinorini-Godly, M.T.; Conti, A. Botany and Pharmacognosy of the Cacao Tree. In *Chocolate and Health*; Conti, A., Paoletti, R., Poli, A., Visioli, F., Eds.; Springer: Berlin/Heidelberg, Garmany, 2012; pp. 41–62.

3. Martín, M.A.; Ramos, S. Cocoa polyphenols in oxidative stress: Potential health implications. *J. Funct. Foods* **2016**, *27*, 570–588. [CrossRef]

4. Redovniković, I.R.; Delonga, K.; Mazor, S.; Dragović-Uzelac, V.; Carić, M.; Vorkapić-Furač, J. Polyphenolic content and composition and antioxidative activity of different Cocoa liquors. *Czech J. Food Sci.* **2009**, *27*, 330–337. [CrossRef]

5. Bauer, S.R.; Ding, E.L.; Smit, L.A. Cocoa Consumption, Cocoa Flavonoids, and Effects on Cardiovascular Risk Factors: An Evidence-Based Review. *Curr. Cardiovasc. Risk. Rep.* **2011**, *5*, 120–127. [CrossRef]

6. Sudano, I.; Flammer, A.J.; Roas, S.; Enseleit, F.; Ruschitzka, F.; Corti, R. Cocoa, blood pressure, and vascular function. *Curr. Hypertens. Rep.* **2012**, *14*. [CrossRef] [PubMed]

7. Loffredo, L.; Violi, F. Polyphenolic antioxidants and health. *Choc. Heal.* **2013**, 77–85. [CrossRef]

8. Wang, Y.; Feltham, B.A.; Suh, M.; Jones, P.J.H. Cocoa flavanols and blood pressure reduction: Is there enough evidence to support a health claim in the United States? *Trends Food Sci. Technol.* **2019**, *83*, 203–210. [CrossRef]

9. Corella, D.; Ordovas, J.M. Metabolic syndrome pathophysiology: The role of adipose tissue. *Nutr. Metab. Cardiovasc. Dis.* **2007**, *17*, 125–139. [CrossRef]

10. Ritchie, S.A.; Connell, J.M.C. The link between abdominal obesity, metabolic syndrome and cardiovascular disease. *Nutr. Metab Cardiovasc. Dis.* **2007**, *17*, 319–326. [CrossRef]

11. Srikanthan, K.; Feyh, A.; Visweshwar, H.; Shapiro, J.I.; Sodhi, K. Systematic review of metabolic syndrome biomarkers: A panel for early detection, management, and risk stratification in the West Virginian population. *Int. J. Med. Sci.* **2016**, *13*, 25–38. [CrossRef] [PubMed]

12. Alberti, K.G.M.; Zimmet, P.; Shaw, J. Metabolic syndrome—A new world-wide definition. A Consensus Statement from the International Diabetes Federation. *Diabet. Med.* **2006**, *2010*, 469–480.

13. Trivedi, R.; Patel, P.N.; Jani, H.J.; Trivedi, B.A. Nutrigenomics: From molecular nutrition to prevention of disease. *Biotechnol. Indian J.* **2011**, *5*, 112–116.

14. Espín, J.C.; García-Conesa, M.T.; Tomás-Barberán, F.A. Nutraceuticals: Facts and fiction. *Phytochemistry* **2007**, *68*, 2986–3008. [CrossRef] [PubMed]

15. Strat, K.M.; Rowley, T.J.; Smithson, A.T.; Tessem, J.S.; Hulver, M.W.; Liu, D.; Davy, B.M.; Davy, K.P.; Neilsonc, A.P. ScienceDirect Mechanisms by which cocoa flavanols improve metabolic syndrome and related disorders ☆. *J. Nutr. Biochem.* **2016**, *35*, 1–21. [CrossRef]

16. Rusconi, M.; Conti, A. Theobroma cacao L., the Food of the Gods: A scientific approach beyond myths and claims. *Pharmacol. Res.* **2010**, *61*, 5–13. [CrossRef]

17. Braicu, C.; Irimie, A.; Berindan; Pilecki, V.; Balacescu, O.; Neagoe, I. The Relationships Between Biological Activities and Structure of Flavan-3-Ols. *Int. J. Mol. Sci.* **2011**, *12*, 9342–9353. [CrossRef]

18. Fraga, C.G.; Galleano, M.; Verstraeten, S.V.; Oteiza, P.I. Basic biochemical mechanisms behind the health benefits of polyphenols. *Mol. Aspects Med.* **2010**, *31*, 435–445. [CrossRef] [PubMed]

19. Spencer, J.P.E.; Chaudry, F.; Pannala, A.S.; Srai, S.K.; Debnam, E.; Rice-Evans, C. Decomposition of cocoa procyanidins in the gastric milieu. *Biochem. Biophys. Res. Commun.* **2000**, *272*, 236–241. [CrossRef] [PubMed]

20. Bennett, R.N.; Scalbert, A.; Rémésy, C.; Rios, L.Y.; Williamson, G.; Lazarus, S.A. Cocoa procyanidins are stable during gastric transit in humans. *Am. J. Clin. Nutr.* **2018**, *76*, 1106–1110.

21. Hackman, R.M.; Polagruto, J.A.; Zhu, Q.Y.; Sun, B.; Fujii, H.; Keen, C.L. Flavanols: Digestion, absorption and bioactivity. *Phytochem. Rev.* **2008**, *7*, 195–208. [CrossRef]

22. Kwik-Uribe, C.; Bektash, R.M. Cocoa flavanols: Measurement, bioavailability and bioactivity. *Asia. Pac. J. Clin. Nutr.* **2008**, *17*, 280–283.

23. Stoupi, S.; Williamson, G.; Drynan, J.W.; Barron, D.; Clifford, M.N. A comparison of the in vitro biotransformation of (-)-epicatechin and procyanidin B2 by human faecal microbiota. *Mol. Nutr. Food Res.* **2010**, *54*, 747–759. [CrossRef]

24. Camps-Bossacoma, M.; Pérez-Cano, F.J.; Franch, À.; Untersmayr, E.; Castell, M. Effect of a cocoa diet on the small intestine and gut-associated lymphoid tissue composition in an oral sensitization model in rats. *J. Nutr. Biochem.* **2017**, *42*, 182–193. [CrossRef] [PubMed]

25. Arola-Arnal, A.; Muguerza, B.; Margalef, M.; Bravo, F.I.; Iglesias-Carres, L.; Pons, Z. Flavanol plasma bioavailability is affected by metabolic syndrome in rats. *Food Chem.* **2017**, *231*, 287–294.

26. Holt, R.R.; Lazarus , S.A.; Sullards, M.C.; Zhu, Q.Y.; Schramm, D.D.; Hammerstone, J.F.; Fraga, C.G.; Schmitz, H.H.; Keen, C.L. Procyanidin dimer B2 [epicatechin-(4β-8)-epicatechin] in human plasma after the consumption of a flavanol-rich cocoa. *Am. J. Clin. Nutr.* **2018**, *76*, 798–804.

27. Baba, S.; Osakabe, N.; Natsume, M.; Muto, Y.; Takizawa, T.; Terao, J. Absorption and urinary excretion of (-)-epicatechin after administration of different levels of cocoa powder or (-)-epicatechin in rats. *J. Agric. Food Chem.* **2001**, *49*, 6050–6056. [CrossRef]

28. Takizawa, T.; Baba, S.; Terao, J.; Osakabe, N.; Muto, Y.; Natsume, M. In Vivo Comparison of the Bioavailability of (+)-Catechin, (−)-Epicatechin and Their Mixture in Orally Administered Rats. *J. Nutr.* **2018**, *131*, 2885–2891.

29. Ottaviani, J.I.; Heiss, C.; Spencer, J.P.E.; Kelm, M.; Schroeter, H. Recommending flavanols and procyanidins for cardiovascular health: Revisited. *Mol. Aspects Med.* **2018**, *61*, 63–75. [CrossRef]

30. Actis-Goretta, L.; Lévèques, A.; Giuffrida, F.; Romanov-Michailidis, F.; Viton, F.; Barron, D.; Duenas-Paton, M.; Gonzalez-Manzano, S.; Santos-Buelga, C.; Williamson, G.; et al. Elucidation of (-)-epicatechin metabolites after ingestion of chocolate by healthy humans. *Free Radic Biol. Med.* **2012**, *53*, 787–795. [CrossRef] [PubMed]

31. Aron, P.M.; Kennedy, J.A. Flavan-3-ols: Nature, occurrence and biological activity. *Mol. Nutr. Food Res.* **2008**, *52*, 79–104. [CrossRef]

32. Ohtake, Y.; Moriichi, N.; Shoji, T.; Masumoto, S.; Akiyama, H.; Kanda, T.; Goda, Y. Apple Procyanidin Oligomers Absorption in Rats after Oral Administration: Analysis of Procyanidins in Plasma Using the Porter Method and High-Performance Liquid Chromatography/Tandem Mass Spectrometry. *J. Agric Food Chem.* **2006**, *54*, 884–892.

33. Appeldoorn, M.M.; Vincken, J.-P.; Gruppen, H.; Hollman, P.C.H. Procyanidin Dimers A1, A2, and B2 Are Absorbed without Conjugation or Methylation from the Small Intestine of Rats. *J. Nutr.* **2009**, *139*, 1469–1473. [CrossRef] [PubMed]

34. Llorente-Cortés, V.; Urpi-Sarda, M.; Sacanella, E.; Camino-López, S.; Vázquez-Agell, M.; Chiva-Blanch, G.; Tobias, E.; Rourae, E.; Andres-Lacueva, C.; Lamuela-Raventos, R.M.; et al. Cocoa consumption reduces NF-κB activation in peripheral blood mononuclear cells in humans. *Nutr. Metab. Cardiovasc. Dis.* **2011**, *23*, 257–263.

35. Aherne, S.A.; O 'brien, N.M. Dietary Flavonols: Chemistry, Food Content, and Metabolism chemistry and structure of the flavonoids. *Nutrition* **2002**, *18*, 75–81. [CrossRef]

36. Williamson, G.; Clifford, M.N. Role of the small intestine, colon and microbiota in determining the metabolic fate of polyphenols. *Biochem. Pharmacol.* **2017**, *139*, 24–39. [CrossRef]

37. Verstraeten, S.V.; Oteiza, P.I.; Fraga, C.G. Membrane effects of Cocoa procyanidins in liposomes and Jurkat T cells. *Biol. Res.* **2004**, *37*, 293–300. [CrossRef] [PubMed]

38. Oteiza, P.I.; Erlejman, A.G.; Verstraeten, S.V.; Keen, C.L.; Fraga, C.G. Flavonoid-membrane interactions: A protective role of flavonoids at the membrane surface? *Clin. Dev. Immunol.* **2005**, *12*, 19–25. [CrossRef] [PubMed]

39. Verstraeten, S.V.; Keen, C.L.; Schmitz, H.H.; Fraga, C.G.; Oteiza, P.I. Flavan-3-ols and procyanidins protect liposomes against lipid oxidation and disruption of the bilayer structure. *Free Radic Biol. Med.* **2003**, *34*, 84–92. [CrossRef]

40. Repetto, M.; Semprine, J.; Boveris, A. Lipid Peroxidation: Chemical Mechanism, Biological Implications and Analytical Determination. *InTechOpen* **2012**. [CrossRef]

41. Perron, N.R.; Brumaghim, J.L. A review of the antioxidant mechanisms of polyphenol compounds related to iron binding. *Cell Biochem. Biophys.* **2009**, *53*, 75–100. [CrossRef]

42. Cherrak, S.A.; Mokhtari-Soulimane, N.; Berroukeche, F.; Bensenane, B.; Cherbonnel, A.; Merzouk, H.; Elhabiri, M. In vitro antioxidant versus metal ion chelating properties of flavonoids: A structure-activity investigation. *PLoS ONE* **2016**, *11*, 1–21. [CrossRef] [PubMed]

43. Loke, W.M.; Proudfoot, J.M.; Hodgson, J.M.; McKinley, A.J.; Hime, N.; Magat, M.; Stocker, R.; Croft, K.D. Specific dietary polyphenols attenuate atherosclerosis in apolipoprotein e-knockout mice by alleviating inflammation and endothelial dysfunction. *Arterioscler. Thromb. Vasc. Biol.* **2010**, *30*, 749–757. [CrossRef]

44. Cheng, Y.-C.; Sheen, J.-M.; Hu, W.L.; Hung, Y.-C. Polyphenols and Oxidative Stress in Atherosclerosis-Related Ischemic Heart Disease and Stroke. *Oxid. Med. Cell Longev.* **2017**, *2017*, 1–16. [CrossRef]

45. Ali, F.; Ismail, A.; Kersten, S. Molecular mechanisms underlying the potential antiobesity-related diseases effect of cocoa polyphenols. *Mol. Nutr. Food Res.* **2014**, *58*, 33–48. [CrossRef]

46. Cordero-Herrera, I.; Martín, M.A.; Goya, L.; Ramos, S. Cocoa flavonoids protect hepatic cells against high-glucose-induced oxidative stress: Relevance of MAPKs. *Mol. Nutr. Food Res.* **2015**, *59*, 597–609. [CrossRef]

47. Goya, L.; Martin, M.Á.; Ramos, S. Antidiabetic actions of cocoa flavanols. *Mol. Nutr. Food Res.* **2016**, *60*, 1756–1769.

48. Fraga, C.G.; Oteiza, P.I. Dietary flavonoids: Role of (-)-epicatechin and related procyanidins in cell signaling. *Free Radic Biol. Med.* **2011**, *51*, 813–823. [CrossRef]

49. Van Guilder, G.P.; Hoetzer, G.L.; Greiner, J.J.; Stauffer, B.L.; DeSouza, C.A. Influence of metabolic syndrome on biomarkers of oxidative stress and inflammation in obese adults. *Obesity* **2006**, *14*, 2127–2231. [CrossRef] [PubMed]

50. Catrysse, L.; van Loo, G. Inflammation and the Metabolic Syndrome: The Tissue-Specific Functions of NF-κB. *Trends. Cell Biol.* **2017**, *27*, 417–429. [CrossRef]

51. Gu, Y.; Yu, S.; Lambert, J.D. Dietary cocoa ameliorates obesity-related inflammation in high fat-fed mice. *Eur. J. Nutr.* **2014**, *53*, 149–158. [CrossRef]

52. Telle-Hansen, V.H.; Christensen, J.J.; Ulven, S.M.; Holven, K.B. Does dietary fat affect inflammatory markers in overweight and obese individuals?—A review of randomized controlled trials from 2010 to 2016. *Genes Nutr.* **2017**, *12*, 1–18. [CrossRef] [PubMed]

53. Schewe, T.; Steffen, Y.; Sies, H. How do dietary flavanols improve vascular function? A position paper. *Arch. Biochem. Biophys.* **2008**, *476*, 102–106. [CrossRef] [PubMed]

54. Steffen, Y.; Gruber, C.; Schewe, T.; Sies, H. Mono-O-methylated flavanols and other flavonoids as inhibitors of endothelial NADPH oxidase. *Arch. Biochem. Biophys.* **2008**, *469*, 209–219. [CrossRef]

55. Rassaf, T.; Kelm, M. Cocoa flavanols and the nitric oxide-pathway: targeting endothelial dysfunction by dietary intervention. *Drug Discov. Today Dis. Mech.* **2008**, *5*, 273–278. [CrossRef]

56. Gómez-Guzmán, M.; Jiménez, R.; Sánchez, M.; Zarzuelo, M.J.; Galindo, P.; Quintela, A.M.; López-Sepulveda, R.; Romero, M.; Tamargo, J.; Vargas, F.; et al. Epicatechin lowers blood pressure, restores endothelial function, and decreases oxidative stress and endothelin-1 and NADPH oxidase activity in DOCA-salt hypertension. *Free Radic. Biol. Med.* **2011**, *52*, 70–79.

57. Reiter, C.E.N.; Kim, J.A.; Quon, M.J. Green tea polyphenol epigallocatechin gallate reduces endothelin-1 expression and secretion in vascular endothelial cells: Roles for AMP-activated protein kinase, Akt, and FOXO1. *Endocrinology* **2010**, *151*, 103–114. [CrossRef] [PubMed]

58. Yahfoufi, N.; Alsadi, N.; Jambi, M.; Matar, C. The Immunomodulatory and Anti-Inflammatory Role of Polyphenols. *Nutrients* **2018**, *10*, 1618. [CrossRef] [PubMed]

59. Heiss, C.; Lauer, T.; Dejam, A.; Kleinbongard, P.; Hamada, S.; Rassaf, T.; Matern, S.; Feelisch, M.; Kelm, M. Plasma nitroso compounds are decreased in patients with endothelial dysfunction. *J. Am. Coll. Cardiol.* **2006**, *47*, 573–579. [CrossRef]

60. Schnorr, O.; Brossette, T.; Momma, T.Y.; Kleinbongard, P.; Keen, C.L.; Schroeter, H.; Sies, H. Cocoa flavanols lower vascular arginase activity in human endothelial cells in vitro and in erythrocytes in vivo. *Arch. Biochem. Biophys.* **2008**, *476*, 211–215. [CrossRef]

61. Hajer, G.R.; Van Haeften, T.W.; Visseren, F.L.J. Adipose tissue dysfunction in obesity, diabetes, and vascular diseases. *Eur. Heart J.* **2008**, *29*, 2959–2971. [CrossRef] [PubMed]

62. Mackenzie, G.G.; Delfino, J.M.; Keen, C.L.; Fraga, C.G.; Oteiza, P.I. Dimeric procyanidins are inhibitors of NF-κB-DNA binding. *Biochem. Pharmacol.* **2009**, *78*, 1252–1262. [CrossRef]

63. Fuster, J.J.; Ouchi, N.; Gokce, N.; Walsh, K. Obesity-induced changes in adipose tissue microenvironment and their impact on cardiovascular disease. *Circ. Res.* **2016**, *118*, 1786–1807. [CrossRef] [PubMed]

64. Gonzales, A.M.; Orlando, R.A. Curcumin and resveratrol inhibit nuclear factor-kappaB-mediated cytokine expression in adipocytes. *Nutr. Metab.* **2008**, *5*, 1–13. [CrossRef] [PubMed]

65. Grattagliano, I.; Palmieri, V.O.; Portincasa, P.; Moschetta, A.; Palasciano, G. Oxidative stress-induced risk factors associated with the metabolic syndrome: a unifying hypothesis. *J. Nutr. Biochem.* **2008**, *19*, 491–504. [CrossRef]

66. Steffen, Y.; Schewe, T.; Sies, H. Epicatechin protects endothelial cells against oxidized LDL and maintains NO synthase. *Biochem. Biophys. Res. Commun.* **2005**, *331*, 1277–1283. [CrossRef]
67. Steffen, Y.; Jung, T.; Klotz, L.O.; Schewe, T.; Grune, T.; Sies, H. Protein modification elicited by oxidized low-density lipoprotein (LDL) in endothelial cells: Protection by (-)-epicatechin. *Free Radic Biol. Med.* **2007**, *42*, 955–970. [CrossRef]
68. Ramirez-Sanchez, I.; Maya, L.; Ceballos, G.; Villarreal, F. (-)-Epicatechin activation of endothelial cell endothelial nitric oxide synthase, nitric oxide, and related signaling pathways. *Hypertension* **2010**, *55*, 1398–1405. [CrossRef] [PubMed]
69. Mendoza-Lorenzo, P.; Ceballos, G.; Villarreal, F.; Varela, C.E.; Rodriguez, A.; Romero-Valdovinos, M.; Mendoza-Loren, P.; Ramirez-Sanchez, I. Browning effects of (-)-epicatechin on adipocytes and white adipose tissue. *Eur. J. Pharmacol.* **2017**, *811*, 48–59.
70. Jia, L.; Liu, X.; Bai, Y.Y.; Li, S.H.; Sun, K.; He, C. Short-term effect of cocoa product consumption on lipid profile: A meta-analysis of randomized controlled trials 1–3. *Am. J. Clin. Nutr.* **2010**, *2*, 218–225. [CrossRef]
71. Tokede, O.A.; Ellison, C.R.; Pankow, J.S.; North, K.E.; Hunt, S.C.; Kraja, A.T.; Arnett, D.K.; Djoussé, L. Chocolate consumption and prevalence of metabolic syndrome in the NHLBI Family Heart Study. *e-SPEN J.* **2012**, *7*, e139–e143. [CrossRef]
72. Bladé, C.; Arola, L.; Salvadó, M.J. Hypolipidemic effects of proanthocyanidins and their underlying biochemical and molecular mechanisms. *Mol. Nutr. Food Res.* **2010**, *54*, 37–59. [CrossRef] [PubMed]
73. Rabadan-Chávez, G.; Reyes-Maldonado, E.; Quevedo-Corona, L.; Jaramillo-Flores, M.E.; Miliar, G.A. Cocoa powder, cocoa extract and epicatechin attenuate hypercaloric diet-induced obesity through enhanced β-oxidation and energy expenditure in white adipose tissue. *J. Funct. Foods* **2015**, *20*, 54–67. [CrossRef]
74. Natsume, M.; Kanegae, M.; Yasuda, A.; Sasaki, K.; Nagaoka, S.; Baba, S.; Natsume, M. Cacao procyanidins reduce plasma cholesterol and increase fecal steroid excretion in rats fed a high-cholesterol diet. *BioFactors* **2009**, *33*, 211–223.
75. Mellor, D.D.; Sathyapalan, T.; Kilpatrick, E.S.; Beckett, S.; Atkin, S.L. High-cocoa polyphenol-rich chocolate improves HDL cholesterol in Type 2 diabetes patients. *Diabet. Med.* **2010**, *27*, 1318–1321. [CrossRef] [PubMed]
76. Osakabe, N.; Yamagishi, M. Procyanidins in Theobroma cacao Reduce Plasma Cholesterol Levels in High Cholesterol-Fed Rats. *J. Clin. Biochem. Nutr.* **2009**, *45*, 131–136. [CrossRef] [PubMed]
77. Janevski, M.; Antonas, K.N.; Sullivan-Gunn, M.J.; McGlynn, M.A.; Lewandowski, P.A. The effect of cocoa supplementation on hepatic steatosis, reactive oxygen species and LFABP in a rat model of NASH. *Comp. Hepatol.* **2011**, *10*, 10. [CrossRef] [PubMed]
78. Tomaru, M.; Takano, H.; Osakabe, N.; Yasuda, A.; Inoue, K.; Yanagisawa, R.; Ohwatari, T.; Uematsu, H. Dietary supplementation with cacao liquor proanthocyanidins prevents elevation of blood glucose levels in diabetic obese mice. *Nutrition* **2007**, *23*, 351–355. [CrossRef] [PubMed]
79. Faizul, H.A.; Nawalyah, A.G.; Hamid, M.; Ruzaidi, A.; Amin, I. The effect of Malaysian cocoa extract on glucose levels and lipid profiles in diabetic rats. *J. Ethnopharmacol.* **2005**, *98*, 55–60.
80. León, L.; De la Rosa, D.; Gracia, A.; Barranco, D.; Rallo, L. Fatty acid composition of advanced olive. *J. Food Sci. Agric.* **2008**, *1926*, 1921–1926. [CrossRef]
81. Grassi, D.; Lippi, C.; Desideri, G.; Necozione, S.; Ferri, C. Short-term administration of dark chocolate is followed by a significant increase in insulin sensitivity and a decrease in blood pressure in healthy persons. *Am. J. Clin. Nutr.* **2018**, *81*, 611–614. [CrossRef]
82. Yamashita, Y.; Okabe, M.; Natsume, M.; Ashida, H. Prevention mechanisms of glucose intolerance and obesity by cacao liquor procyanidin extract in high-fat diet-fed C57BL/6 mice. *Arch. Biochem. Biophys.* **2012**, *527*, 95–104. [CrossRef]
83. Gondoin, A.; Grussu, D.; Stewart, D.; McDougall, G.J. White and green tea polyphenols inhibit pancreatic lipase in vitro. *Food Res. Int. [Internet].* **2010**, *43*, 1537–1544. [CrossRef]
84. Osada, K.; Suzuki, T.; Kawakami, Y.; Senda, M.; Kasai, A.; Sami, M.; Ohta, Y.; Kanda, T.; Ikeda, M. Dose-dependent hypocholesterolemic actions of dietary apple polyphenol in rats fed cholesterol. *Lipids* **2006**, *41*, 133–139. [CrossRef]
85. Rabadán-Chávez, G.M.; Miliar Garcia, A.; Paniagua Castro, N.; Escalona Cardoso, G.; Quevedo-Corona, L.; Reyes-Maldonado, E.; Jaramillo Flores, M.E. Modulating the expression of genes associated with hepatic lipid metabolism, lipoperoxidation and inflammation by cocoa, cocoa extract and cocoa flavanols related to hepatic steatosis induced by a hypercaloric diet. *Food Res. Int.* **2016**, *89*, 937–945. [CrossRef]

86. Tsukamoto, T.; Nakata, R.; Tamura, E.; Kosuge, Y.; Kariya, A.; Katsukawa, M.; Mishima, S.; Ito, T.; Iinuma, M.; Akao, Y.; et al. Vaticanol C, a resveratrol tetramer, activates PPARalpha and PPARbeta/delta in vitro and in vivo. *Nutr. Metab.* **2010**, *7*, 46. [CrossRef] [PubMed]
87. Jiang, H.; Zhou, X.-Y.; Gao, D.-M.; Zhou, N.-J.; Xie, J.; Chen, J. Naringin ameliorates metabolic syndrome by activating AMP-activated protein kinase in mice fed a high-fat diet. *Arch. Biochem. Biophys.* **2011**, *518*, 61–70.
88. Najim, N.I.; Shah, S.A.; Kazi, A.N.; Dharani, A.M.; Shah, S.R.; Jangda, M.A. Use of dark chocolate for diabetic patients: a review of the literature and current evidence. *J. Commu. Hosp. Intern. Med. Perspect.* **2017**, *7*, 218–221.
89. Reygaert, W. An Update on the Health Benefits of Green Tea. *Beverages* **2017**, *3*, 6. [CrossRef]

Article

Theobromine Improves Working Memory by Activating the CaMKII/CREB/BDNF Pathway in Rats

Rafiad Islam [1,2], Kentaro Matsuzaki [1,*], Eri Sumiyoshi [1], Md Emon Hossain [1,3], Michio Hashimoto [1], Masanori Katakura [1,4], Naotoshi Sugimoto [1,5,*] and Osamu Shido [1]

[1] Department of Environmental Physiology, Faculty of Medicine, Shimane University, 89-1 Enya-cho, Izumo 693-8501, Japan; rafiad@med.shimane-u.ac.jp (R.I.); erisumi@med.shimane-u.ac.jp (E.S.); emon@uab.edu (M.E.H.); michio1@med.shimane-u.ac.jp (M.H.); mkatakur@josai.ac.jp (M.K.); o-shido@med.shimane-u.ac.jp (O.S.)

[2] Department of Biotechnology and Genetic Engineering, Mawlana Bhashani Science and Technology University, Tangail 1902, Bangladesh

[3] Department of Biochemistry and Molecular Genetics, University of Alabama at Birmingham, Birmingham AL 35294, USA

[4] Department of Nutritional Physiology, Faculty of Pharmaceutical Sciences, Josai University, 1-1 Keyakidai, Sakado, Saitama 350-0295, Japan

[5] Department of Physiology, Graduate School of Medical Science, Kanazawa University, 13-1 Takara-machi, Kanazawa 920-8640, Japan

* Correspondence: matuzaki@med.shimane-u.ac.jp (K.M.); ns@med.kanazawa-u.ac.jp (N.S.); Tel.: +81-853-20-2114 (K.M.); +81-76-365-2314 (N.S.); Fax: +81-853-20-2110 (K.M.); +81-76-262-1866 (N.S.)

Received: 22 February 2019; Accepted: 17 April 2019; Published: 20 April 2019

Abstract: Theobromine (TB) is a primary methylxanthine found in cacao beans. cAMP-response element-binding protein (CREB) is a transcription factor, which is involved in different brain processes that bring about cellular changes in response to discrete sets of instructions, including the induction of brain-derived neurotropic factor (BDNF). Ca^{2+}/calmodulin-dependent protein kinase II (CaMKII) has been strongly implicated in the memory formation of different species as a key regulator of gene expression. Here we investigated whether TB acts on the CaMKII/CREB/BDNF pathway in a way that might improve the cognitive and learning function in rats. Male Wistar rats (5 weeks old) were divided into two groups. For 73 days, the control rats (CN rats) were fed a normal diet, while the TB-fed rats (TB rats) received the same food, but with a 0.05% TB supplement. To assess the effects of TB on cognitive and learning ability in rats: The radial arm maze task, novel object recognition test, and Y-maze test were used. Then, the brain was removed and the medial prefrontal cortex (mPFC) was isolated for Western Blot, real-time PCR and enzyme-linked immunosorbent assay. Phosphorylated CaMKII (p-CaMKII), phosphorylated CREB (p-CREB), and BDNF level in the mPFC were measured. In all the behavior tests, working memory seemed to be improved by TB ingestion. In addition, p-CaMKII and p-CREB levels were significantly elevated in the mPFC of TB rats in comparison to those of CN rats. We also found that cortical BDNF protein and mRNA levels in TB rats were significantly greater than those in CN rats. These results suggest that orally supplemented TB upregulates the CaMKII/CREB/BDNF pathway in the mPFC, which may then improve working memory in rats.

Keywords: theobromine; cacao; working memory; behavior; CaMKII; CREB; BDNF

---

1. Introduction

Coffee, cocoa, and chocolate are among the most frequently consumed substances in the world [1]. Coffee has various beneficial effects on human health, as it appears to be cardio-protective, neuroprotective, hepatoprotective, and nephroprotective. It is now well known that coffee (caffeine) suppresses the activities of nuclear factors κB (NF-κB), Akt, and ERK [2]. Like coffee, the use of the seeds from the cacao tree (Theobroma cacao) to create beverages, dates back to the early formative period of Mesoamerican history (2000–1000 BC). In recent years, there has been a notable interest in the neuroprotective effects of flavonoids, with evidence emerging that they may lead to improvements in memory and learning by improving neuronal functioning, while also promoting neuronal protection and regeneration [3]. Cacao beans, a very popular food worldwide, contains many flavonoids, which have pleotropic effects in cognition and neuroprotection [4,5]. Theobromine (TB) is a primary methylxanthine found in products made from cacao beans, which generally contain approximately 1% TB [6]. From recent literature, both scientific and popular, TB has been implicated in the health benefits of cacao intake. We know that TB traverses the blood-brain barrier (BBB) [7], and that this might induce effects on the brain, alter the cellular redox environment, modulate neuronal signaling pathways, and influence gene expression, as well as protein activity, perhaps in a manner similar to other flavonoids [8].

In our previous study, we confirmed that mice fed TB performed better on learning tasks and TB acted as a phosphodiesterase (PDE) inhibitor [9,10], by enhancing the cAMP/cAMP-response element-binding protein (CREB)/brain-derived neurotrophic factor (BDNF) pathway [7]. We also confirmed that TB supplemented chow-inhibited mTOR signaling in the brain and liver [11]. In the current experiment, we want to assess the effects of TB on rats, using a prolonged feeding period of 73 days, and to check whether different types of memory functions are affected by a variety of behavior experiments. We have checked the rats using the radial arm maze (RAM) task, Y-maze test, and novel object recognition (NOR) test for assessing the memory function, especially working memory. Working memory is the cognitive capacity to actively and temporarily maintain information for the purpose of task execution [12]. The dorsolateral prefrontal cortex in primates, which is homologous to the medial PFC (mPFC) in rodents [13,14], is essential for working memory, as evidenced by numerous lesion studies [15], electrophysiological recordings [16], and brain imaging investigations [17,18]. Many signaling pathways control the memory process, such as (i) cAMP-dependent protein kinase (PKA), (ii) Ca^{2+}/calmodulin-dependent protein kinases (CaMKs), (iii) mitogen-activated protein kinases. These pathways converge to signal to CREB, a transcription factor, which is associated with memory and synaptic plasticity that binds to the promoter regions of many genes [19–21]. Remarkably, many genes have been altered following CREB activation [22,23], including key proteins involved in neuronal plasticity, such as BDNF [24]. This protein mediates neuronal development and synaptic function [24], which is critical for the differentiation and survival of neurons during development [25]. Studies have shown that Ca^{2+}, acting as an important messenger via CaMKs, triggers phosphorylation of CREB [26]. This phosphorylated CREB (p-CREB) activates BDNF transcription by binding to a cAMP response element within the gene [24].

In this study, we found that TB fed rats appear to have improved working memories. As with the cAMP/CREB/BDNF pathway, which was established in our previous study for motor learning in mice, it appears that yet another novel molecular pathway, CaMKII/CREB/BDNF, may be responsible for working memory improvement in rats.

2. Materials and Methods

2.1. Animals

Forty four male Wistar rats (5 weeks old, 120~140 g body weight) purchased from Japan SLC Inc. (Shizuoka, Japan) were maintained at an ambient temperature of 24 ± 0.1 °C and relative humidity of 45% ± 5% under a 12:12-h light–dark cycle (light on at 7:00 h), with food and water ad libitum. All the animal experiments were performed in accordance with the Guidelines for Animal Experimentation of the Shimane University Faculty of Medicine, compiled from the Guidelines for Animal Experimentation of the Japanese Association for Laboratory Animal Science. The Committee on the Ethics of Animal Experiments of the Shimane University approved the protocol for this study.

2.2. Feeding and Experiment Schedules

Figures 1 and 2 summarize the feeding and schedules for experiment 1 and 2. All of the rats had free access to a standard chow (CRF-1, Oriental Yeast Co. Ltd., Tokyo, Japan) for 10 days after admission. On day 0, the rats were divided into two groups. In both experiments, the control rats (CN rats), was fed the CRF-1 chow and the second group (TB rats), was fed the standard CRF-1 chow supplemented with 0.05% (W/W) of TB (Oriental Yeast Co. Ltd.) over the entire experiment time.

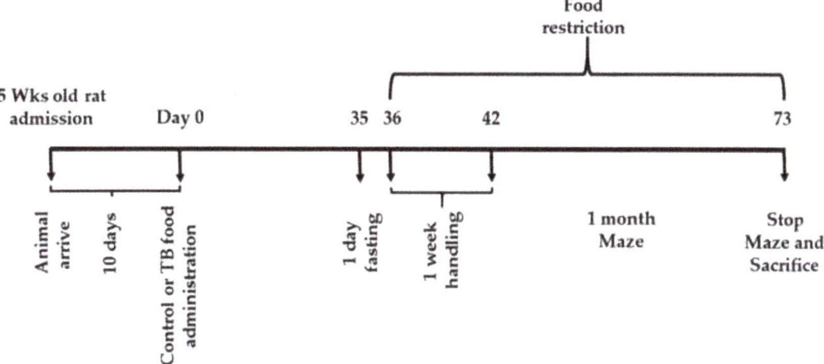

Figure 1. Schedule for Experiment 1 (radial arm maze (RAM) task).

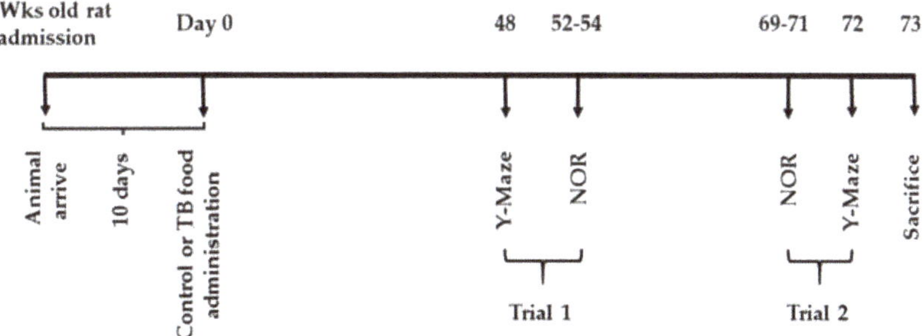

Figure 2. Schedule for Experiment 2 (Y-maze test and novel object recognition (NOR) test).

2.2.1. Behavioral Tests

RAM Task (Experiment 1)

CN rats (n = 12) and TB rats (n = 12) were behaviorally tested for their learning-related cognitive abilities by determining their ability to complete a task in a RAM, as described previously [27,28]. Here we used an eight arm radial maze (Toyo Sangyo, Toyama, Japan) for RAM task. Four weeks after the start of TB administration, rats were transferred to a regimen of food deprivation to keep their body weight at 80–85% of their free feeding weight, and each rat was handled for 3 minutes every day for a total of 5 consecutive days with constant monitoring of body weight. Their mean body weight was approximately 290 g at the beginning of the behavioral testing. Then, they were familiarized with the radial maze apparatus, across the entire surface on which reward pellets (Dustless Precision Pellets, Bio Serv®, Flemington, NJ, USA) were scattered. Then, the rats were trained to acquire a reward at the end of each of four arms of the eight arm radial maze. Three parameters of memory function were examined: Reference memory error (RME), which was determined by the number of entries into unbaited arms, working memory error (WME), which was estimated by the number of repeated entries into arms that had already been visited within a trial, and latency, which was determined by total time needed to finish each trial. A lower number of RMEs and WMEs suggested better spatial learning ability.

After the RAM task, the rats were anaesthetized using isoflurane and brains were rapidly separated from the skull, and the hippocampus were bilaterally collected. Afterwards, olfactory bulbs were removed and a coronal section was made on ice at +4.70 to +2.20 mm from bregma, according to a brain atlas [29]. The mPFC, containing the prelimbic, infralimbic, and anterior cingulate cortices, was immediately dissected from a coronal section. mPFC and hippocampal samples for Enzyme-linked immunosorbent assay (ELISA) were immediately frozen on liquid nitrogen and stored at −80 °C until use.

Y-Maze Test and NOR Test (Experiment 2)

It was clear that the procedural memory requirements and the stresses (both behavioral and metabolic) related to experimental procedures, including food deprivation, may non-specifically affect animal performance in the eight arm radial maze. Thus, to further assess the specificity of working memory improvement related to TB supplementation, we chose to test spontaneous alternation in the Y-maze test, which is devoid of all these procedural aspects, as it is based on the natural tendency of rats to explore novel environments. Another behavior test, the NOR test, in which spontaneous behavior is studied, where no artificial stimuli, food deprivation, reinforcement, and/or prior special training are required.

CN rats (n = 10) and TB rats (n = 10) were used for the Y-maze test and NOR test. Body weight was measured three times in the experimental time period. The Y-maze was performed twice: On day 48 and day 72. The NOR test was also performed twice, over days 52 to 54 and again from day 69 to day 71. All the experiments were done during the light phase from 10 a.m. to 6 p.m. Generally, the rats were placed in the experimental room at least 1 h prior to any test. In between subjects being tested, the apparatus was thoroughly cleaned with water wetted paper towels and a 70% ethanol solution.

Behavioral assessments were performed on rats to determine a spontaneous alteration of behavior using a Y-maze test, as this behavior is considered to reflect the strength of short-term memory [30–32]. The Y-maze was made of gray plastic with three arms (40 cm × 3 cm × 10 cm) extending from a central platform at an angle of 120°. Each rat was placed in the center of the arms and allowed to move freely around the three arms of the maze during an 8 min session. Arm entry was defined as the entry of four paws into one arm [33,34]. Alteration was defined over multiple sets. The percentage of spontaneous alternations were calculated as the ratio of the actual to possible alternations (defined as the total number of arm entries minus 2) multiplied by 100, as shown in the following equation: Alteration (%) = [(number of alternations)/(total arm entries-2)] × 100 [35]. For example, if the following sequence of arm entries was observed: ABACBCAABABC, the animal would have exhibited twelve arm entries, and three correct spontaneous alternations. The alteration percent here would be 40%. The behavior was recorded by a video camera located above the Y-maze, and correct spontaneous alterations were calculated at a later time by a trained observer.

Training and testing in the NOR task were carried out in an arena (70 × 70 × 30 cm) built of plywood as described by Ennaceur and Delacour [36]. The objects were made of plastic and were chosen after determining, in preliminary experiments with other animals, that they were equally preferred. Shapes, colors, and textures were different among these objects. Exploration of the objects was defined as sniffing or touching with the nose toward the objects at a distance of less than 1 cm; however, sitting on the object was not considered [37]. The circumstances where the rats explored the objects for <4 s were excluded. The animals were handled daily for the week that preceded the testing. Then, the animals were habituated to the NOR apparatus by placing them in it for 10 min per day to freely explore 1 day before the training (the habituation phase). On the training day, two identical objects (A1 and A2) were placed in the apparatus and the animals were allowed to explore them freely for 5 min (the familiarization phase). At one hour and 24 h later, in the test phase, one of the objects was randomly replaced by a novel object (named B, and C, respectively) and the rats were reintroduced into the apparatus for an additional 5 min period of free exploration [38–40]. To avoid confounds by lingering olfactory stimuli and preferences, the objects and the arena were cleaned with 70% ethanol after testing each animal. The Discrimination Index (DI), being the ability to distinguish the novel from the familiar object, was calculated: [novel object(s)/(novel object(s) + familiar object(s)) × 100%]. Thus, an index > 50% indicates novel object preference, <50% reveals a familiar object preference, and 50% implies no preference [41].

After the Y-maze test and NOR test, the rats were anaesthetized using isoflurane and the mPFCs, hippocampus, and plasma were sampled as described above. mPFC and hippocampal samples for ELISA were immediately frozen on liquid nitrogen and stored at −80 °C until use. mPFCs samples for western blot were immediately frozen on liquid nitrogen and stored at −80 °C until use. mPFCs samples for real time PCR were collected in RNAlatter reagent (Applied Biosystems, Warrington, UK), and stored at −30 °C until use. Whereas, plasma was used for liver and kidney function test as described below.

2.3. Liver and Kidney Function Test

Plasma was tested for the quantitative determination of ten parameters: glutamic oxaloacetic transaminase/aspartate transaminase (GOT/AST), glutamate–pyruvate transaminase/alanine transaminase (GPT/ALT), γ-glutamyl transpeptidase (GGT), triglyceride (TG), total cholesterol (T-Cho) and creatinine (Cre-2), albumin (Alb), total protein (T-Pro), urea acid (UA), blood urea nitrogen (BUN). These liver and kidney function tests were carried out using the automated biochemical analyzer Spotchem EZ SP-4430 (Arkray, Kyoto, Japan), and the Spotchem EZ Reagent Strip KENSHIN-2 (Arkray, Kyoto, Japan), Spotchem EZ Reagent Strip Kidney-3 (Arkray, Kyoto, Japan) were used.

2.4. ELISA

The mPFCs were homogenized from rats of experiment 1 and 2 with Tris-buffer (pH 7.4) and centrifuged at 800× *g* for 15 min at 4 °C to remove tissue debris. Protein assays were performed using the Pierce BCA Protein Assay Kit (Thermo Fisher Scientific, Waltham MA, USA) to determine protein concentration. Homogenized samples were analyzed by ELISA, as described previously [42]. Briefly, equal amount of protein were analyzed with the BDNF Emax® ImmunoAssay System (Promega, WI, USA) according to the manufacturer's protocol. Absorbance, at 450 nm, was measured by a plate reader (DTX880, Beckman Coulter, CA, USA) and BDNF concentrations were calculated using SoftMax pro software (Molecular Devices, LLC, San Jose, CA, USA).

2.5. Western Blot Analysis

The mPFCs were extracted from rats of experiment 2 with a lysis buffer composed of 1 mM EDTA, 1% SDS, 1x complete protease inhibitor cocktail (Roche Diagnostics, Schweiz), 1x phosphatase inhibitor cocktail (Fujifilm Wako Pure Chemical Corporation, Osaka, Japan) and 20 mM Tris-HCl (pH 7.4). The lysates were sonicated and centrifuged at 14,000 rpm for 20 min at 4 °C to obtain the supernatant as the cell extract. Then, the lysates were analyzed by Western blotting as described previously [43–45]. Briefly, lysates were separated using a 10–12.5% SDS-PAGE system and transferred onto PVDF membranes (Immobilon-P, Merck Millipore, Burlington, MA, USA). Then, membranes were incubated with antibodies of monoclonal rabbit anti-p-CREB (Ser133) (1:1000, Cell signaling, Danvers, MA, USA), monoclonal mouse anti-CREB (1:1000, Cell signaling, Danvers, MA, USA), monoclonal rabbit anti-p-CaMKII (Thr^{286}/Thr^{287}) (1:1000, Cayman Chemical, MI, USA) and polyclonal rabbit anti-CaMKII beta (1:1000, GeneTex, CA, USA). HRP-conjugated anti-rabbit IgG (1:2000, Cell Signaling, Danvers, MA, USA) and HRP-conjugated anti-mouse IgG (1:2000, Cell Signaling, Danvers, MA, USA) were used as the secondary antibody. Immunoblots were incubated and visualized with the ECL detection kit (Amersham ECL Prime, GE Healthcare, Little Chalfont, Buckinghamshire, UK) and visualized with an image analyzer (LAS-4000, FUJI FILM, Tokyo, Japan). Then, membranes were stripped and reprobed with monoclonal rabbit anti-β-actin antibody as a loading control (1:2000, Cell Signaling, Danvers, MA, USA).

2.6. Real Time PCR

The total RNA from the mPFC was purified using a NucleoSpin®RNA isolation kit (TAKARA, Shiga, Japan) and reverse transcribed using a reverse transcription kit (TAKARA, Shiga, Japan). Real time PCR was carried out, as described previously, using the QuantiTect SYBR Green PCR Kit (Qiagen, Hilden, Germany) [46]. The quantification of mRNA was calculated using a cDNA sample as a calibrator. The quantified value of each sample was normalized with that of the β-actin value of the same sample, which was amplified simultaneously with the target gene. Supplementary Table S1 shows the list of primer sequences used for real time PCR.

The PCR conditions were as follows: initial activation at 95 °C for 10 min, then 40 amplification cycles of denaturation at 95 °C for 15 s, followed by annealing and extension at 60 °C for 1 min.

2.7. Sample Size of Each Tested Assay

In RAM task (Experiment 1), 12 animals were used per group. In Y-maze (Experiment 2), 8 animals and in NOR test (Experiment 2), 5 animals were used. For ELISA of BDNF (Experiment 1), we used 5 animal from each group. For real time PCR, ELISA of BDNF, western blot analysis of p-CERB and p-CaMKII (Experiment 2), we used 6 animal per group.

2.8. Statistical Analysis

The data are expressed as the mean ± S.E.M. The Stat View statistical package (Version 5.0) was used for statistical analysis. A two-way analysis of variance (ANOVA) with Fisher's PLSD post-hoc tests to determine significant differences in the various pairwise comparisons were used to analyze the RAM task results. Students' t-tests were used to examine the differences between the two experimental groups in the Y-maze test, NOR test and biological analyses. The probability level of $p < 0.05$ was considered as statistically significant.

3. Results

3.1. Experiment 1 (RAM Task)

3.1.1. Body Weight

First, we measured the body weights at three different time point during the experiment. The body weights of TB rats ($n = 12$) did not differ from those of CN rats ($n = 12$) on day 0 ($p = 0.32$), day 45 ($p = 0.29$) or day 73 ($p = 0.12$) (Table 1).

Table 1. Body weights for control rats (CN rats) and theobromine-fed rats (TB rats) in experiment 1.

	Body Weight (g)		p Value
	CN Rats	**TB Rats**	
Initial day (day 0)	239.2 ± 7.13	229.9 ± 5.20	0.32
Half way (day 45)	360.6 ± 3.95	353.6 ± 4.98	0.29
Final (day 73)	384.9 ± 3.10	375.5 ± 4.65	0.12

Values are the mean ± S.E.M. ($n = 12$ for each group).

3.1.2. Effects of TB on Memory Related Learning Abilities on the RAM Task

The effects of TB supplementation on reference and working memory-related learning abilities were determined by examining the changes in the mean number of WMEs and RMEs, with the data averaged over a block of five trials (Figure 3A,B). Randomized two factor (block and group) ANOVA was used to analyze the possible impact of TB, and revealed a significant main effect in both blocks of trials [F (4,88) = 17.583, $p < 0.0001$] and groups [F (1,88) = 27.265, $p < 0.0001$] on the number of WMEs, with a significant block x group interaction [F (4,88) = 3.131, $p < 0.05$]. However, even without a significant main group effect, and block x group interaction on the number of RMEs, a significant main effect of blocks of trials [F (4,88) = 11.774, $p < 0.0001$)] were observed. Another metric, latency was also determined by examining changes in the total mean time, with the data averaged over a block of five trials (Figure 3C). We found a significant main effect in both the blocks of trials [F (4,88) = 27.882, $p < 0.0001$] and the groups [F (1,88) = 21.558, $p < 0.0001$] on the number of total times, with a significant block x group interaction [F (4,88) = 2.815, $p < 0.05$].

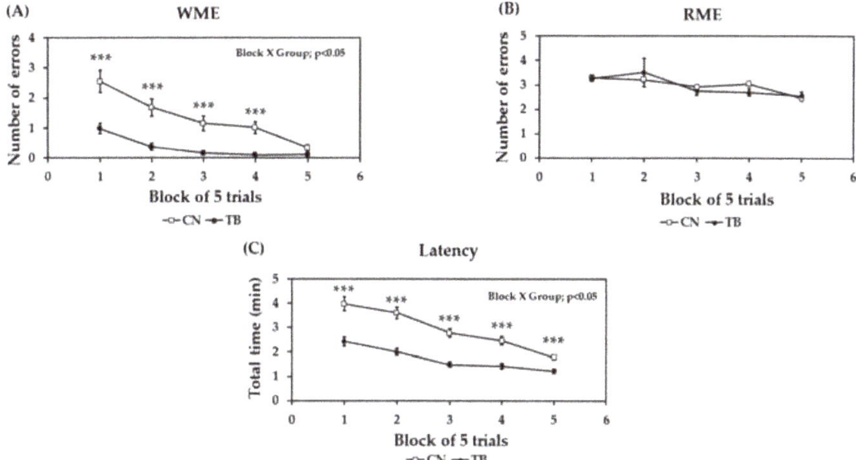

Figure 3. The effect of theobromine (TB) on (A) working memory errors (WME); (B) reference memory errors (RME); and (C) latency assessed by the radial arm maze task. Results are expressed as the mean ± S.E.M. (*n* = 12 for each group). *** $p < 0.001$; significant difference between groups.

3.1.3. BDNF Protein Expression Level in the mPFC

BDNF protein expression level in the mPFC of TB rats (*n* = 5) was significantly higher than that of CN rats (*n* = 5) (Figure 4; $p = 0.009$).

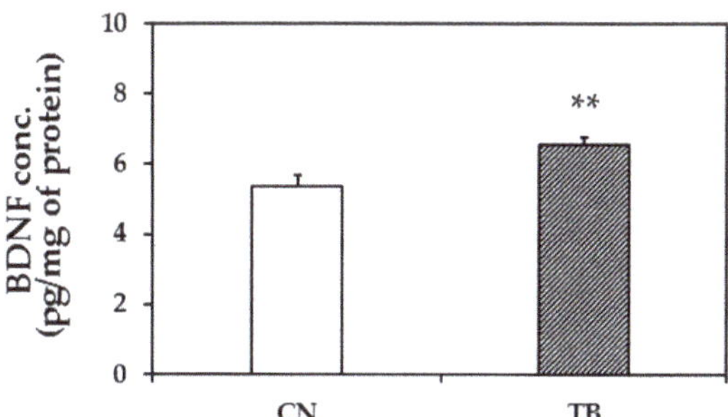

Figure 4. The effect of theobromine (TB) on the brain-derived neurotrophic factor (BDNF) protein level in the medial prefrontal cortex (mPFC). BDNF protein level in the mPFC of TB-fed rats (TB rats) was significantly higher than that of control rats (CN rats). Hippocampal BDNF protein levels did not differ between groups (Figure S1). Values are the mean ± S.E.M. (*n* = 5 for each group). ** $p < 0.01$; significant difference between groups.

3.2. Experiment 2 (Y-Maze Test and NOR Test)

3.2.1. Body Weight, Liver Function, and Kidney Function Tests

The body weights of TB rats (n = 10) did not differ from those of CN rats (n = 10) on day 0 (p = 0.47), day 45 (p = 0.30), or day 73 (p = 0.21), respectively (Table 2). The biochemical parameters for liver and kidney functions of TB rats (n = 10) also did not differ from those of CN rats (n = 10) rats after day 73 (Table 3). These results indicated that TB did not affect the feeding behavior and normal liver or kidney functions.

Table 2. Body weights for control rats (CN rats) and theobromine-fed rats (TB rats) in experiment 2.

	Body Weight (g)		*p* Value
	CN Rats	**TB Rats**	
Initial day (day 0)	183.1 ± 2.03	185.5 ± 2.52	0.47
Half way (day 45)	416.0 ± 5.67	428.3 ± 9.92	0.30
Final (day 73)	463.0 ± 7.89	481.0 ± 11.3	0.21

Values are the mean ± S.E.M. (n = 10 for each group).

Table 3. Biochemical parameters for liver function and kidney function in experiment 2 for control rats (CN rats) and theobromine-fed rats (TB rats).

Liver Function Test				Kidney Function Test			
Biochemical Parameters	CN Rats	TB Rats	*p* Value	Biochemical Parameters	CN Rats	TB Rats	*p* Value
GOT/AST (IU/L)	67.4 ± 8.43	63.7 ± 8.14	0.76	T-Pro (g/dl)	5.4 ± 0.10	5.2 ± 0.12	0.09
GPT/ALT (IU/L)	25.1 ± 1.74	23.7 ± 2.23	0.63	Alb (g/dl)	3.1 ± 0.05	3.0 ± 0.07	0.50
GGT (IU/L)	2.7 ± 0.21	3.3 ± 0.21	0.06	BUN (mg/dl)	16.6 ± 0.40	16.3 ± 0.50	0.64
T-Cho (mg/dl)	79.0 ± 4.94	90.6 ± 5.74	0.14	UA (mg/dl)	1.2 ± 0.06	1.0 ± 0.09	0.11
TG (mg/dl)	95.0 ± 10.97	102.6 ± 13.44	0.67	Cre-2 (mg/dl)	0.3 ± 0.02	0.3 ± 0.02	0.75

GOT/AST, glutamic oxaloacetic transaminase/aspartate transaminase; GPT/ALT, glutamate–pyruvate transaminase/alanine transaminase; GGT, γ-glutamyl transpeptidase; T-Cho, total cholesterol; TG, triglyceride; T-Pro, total protein; Alb, albumin; BUN, blood urea nitrogen; UA, urea acid; Cre-2, creatinine. Values are the mean ± S.E.M. (n = 10 for each group).

3.2.2. Effects of TB Supplementation on Working Memory Improvement in Y-Maze Test

The exploration rates were not affected by TB supplementation, as the total number of arm entries between groups did not show any significant difference (Figure 5A). TB rats displayed spontaneous alternations of 68.9 ± 1.72 and 72.5 ± 2.41 of choices, which was significantly higher than the 60.2 ± 2.59 and 57.8 ± 3.04 of values for the control group (Figure 5B, p = 0.014, p = 0.002), as measured across two trials, respectively.

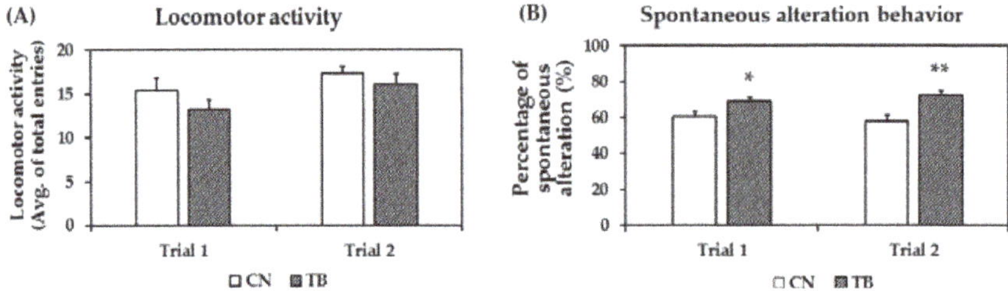

Figure 5. The effect of theobromine (TB) on locomotor activity and spontaneous alteration behavior in rats assessed by the Y-maze test. (**A**) Locomotor activity of TB-fed rats (TB rats) and control rats (CN rats). There were no significant differences between groups in both trials. (**B**) TB rats had a significantly higher number of alteration in both trials compared to CN rats. Values are the mean ± S.E.M. (n = 8 for each group). * p < 0.05, ** p < 0.01; significant difference between groups.

3.2.3. Effects of TB Supplementation on Memory Function in the NOR Test

During the training session (the familiarization phase), the rats spent a similar period of time exploring two identical objects. Figure 6A shows the DIs for the identical objects (A1 and A2) between the two groups were statistically not significant in both trials ($p = 0.478$ for trial 1, $p = 0.627$ for trial 2).

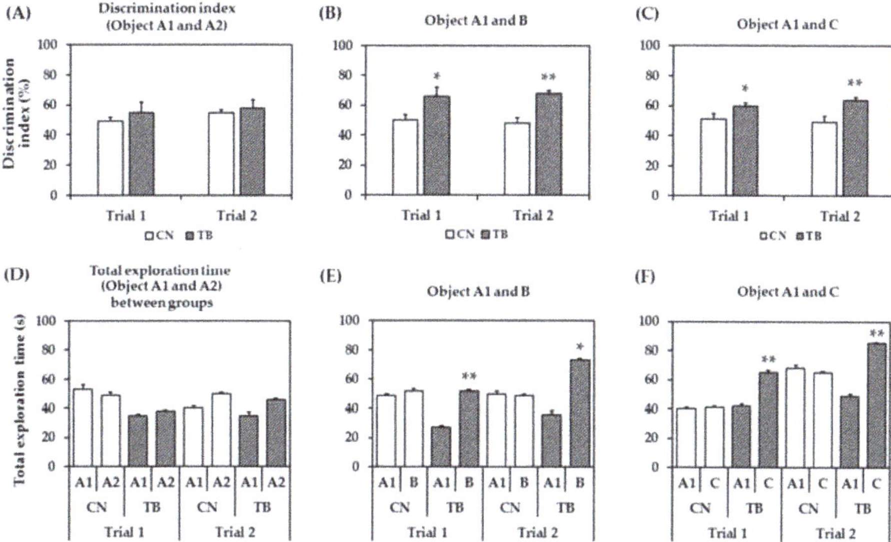

Figure 6. The effect of theobromine (TB) on memory performance assessed by the novel object recognition (NOR) test. The discrimination index (DI) for exploring the object A1, A2 (**A**); object A1, B (**B**); and object A1, C (**C**). Both DIs (Object A1, B and Object A1, C) were significantly greater in TB-fed rats (TB rats) in 2 trials. Values are the mean ± S.E.M. ($n = 5$ for each group). * $p < 0.05$, ** $p < 0.01$; significant difference between groups. Total exploration time for same object (object A1, A2) (**D**); total exploration time for each object (object A1, B) (**E**); and total exploration time for each object (object A1, C) (**F**). TB rats stayed significantly longer time at the new object B and C in 2 trials. Values are the mean ± S.E.M. ($n = 5$ for each group). * $p < 0.05$, ** $p < 0.01$; significant difference between A1 and B or C.

Figure 6B shows that the DIs for object A1 and object B (after 1 h object A2 was replaced by object B) in test phase was significantly increased between groups in both trials ($p = 0.043$ for trial 1, $p = 0.005$ for trial 2).

In Figure 6C, the DIs for object A1 and novel object C (after 24 h object B was changed to new object C) were significantly higher in TB rats in both trials ($p = 0.041$ for trial 1, $p = 0.012$ for trial 2). The results suggest that TB rats has a > 50% DI, implying a higher novel object preference.

To help determine whether explorative and/or anxiety-like behaviors are affected, it is quite important to assess the total exploration time spent at an object. Total exploration time for the same object did not show significant differences between groups in both trials (Figure 6D, $p = 0.744$ for trial 1 and $p = 0.509$ for trial 2), whereas the total exploration times regarding objects A1 and B (Figure 6E, $p = 0.006$ for 1st trial and $p = 0.037$ for 2nd trial), and objects A1 and C (Figure 6F, $p = 0.036$ for trial 1 and $p = 0.005$ for trial 2) were significantly different between groups in both trials. However, the CN rats did not show that much of preference to novel object. Though, the average time spent near the novel object increased numerically in the control rats, however, it did not reach significance.

3.2.4. p-CaMKII, p-CREB and BDNF mRNA and Protein Expression in the mPFC

The expression level of p-CaMKII in the mPFC was significantly higher in TB rats than that of CN rats (Figure 7A, $p = 0.005$). The expression level of p-CREB in the mPFC was also significantly higher in TB rats than that of CN rats (Figure 7B, $p = 0.023$). BDNF mRNA and protein expression levels in the mPFC were also significantly higher in TB rats than in their CN counterparts (Figure 8A, $p = 0.030$ and Figure 8B, $p = 0.015$).

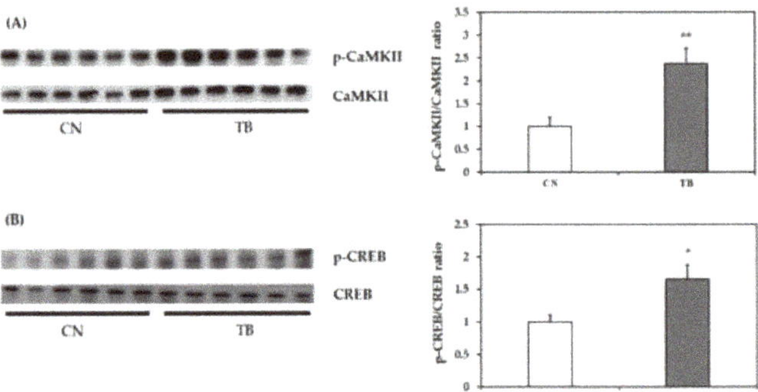

Figure 7. Phospho-CaMKII (p-CaMKII) and phospho-CREB (p-CREB) levels in the medial prefrontal cortex (mPFC) of control rats (CN rats) and theobromine (TB)-fed rats (TB rats). TB enhanced (**A**) p-CaMKII; and (**B**) p-CREB levels in the mPFC. Values are the mean ± S.E.M. ($n = 6$ per group). * $p < 0.05$, ** $p < 0.01$; significant difference between groups.

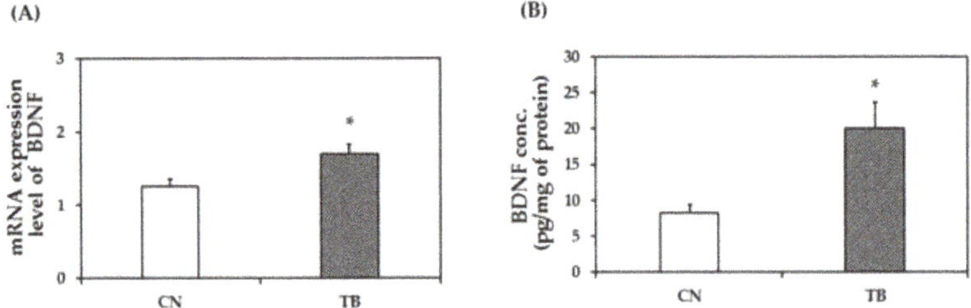

Figure 8. The effect of theobromine (TB) on brain derived neurotrophic factor (BDNF) mRNA and protein levels. (**A**) Relative mRNA expression and (**B**) protein level in TB-fed rat (TB rats) of BDNF were significantly higher than those of control rats (CN rats) in the prefrontal cortex. Hippocampal BDNF protein levels did not differ between groups (Figure S2). Values are the mean ± S.E.M. ($n = 6$ for each group). * $p < 0.05$; significant difference between groups.

4. Discussion

The results of our present study demonstrated that the oral administration of TB influenced the signaling pathway in the mPFC, including those for CaMKII, CREB and BDNF, and concurrently improved the working memory function of rats in the RAM task, the Y-maze test, and also in the NOR test. Moreover, long-term TB supplementation did not show any adverse effects on body weight, food, water intake, liver, and kidney functions in rats (Tables 1–3).

To exert material effects against cognitive disorders, TB must be taken up in the brain from the blood by crossing over the BBB. Our previous data showed that TB was detectable in rat plasma and the brain after 30 and 40 days of TB ingestion, and gradually increased in a time-dependent manner [11]. The concentration of TB in rat brain and plasma after 40 days of TB ingestion might have been sufficient to produce significant pharmacological effects, as described in [4,7,47].

Previously we showed that a 30-day orally administrated TB enhances motor learning in mice [7]. In the present study, we prolonged TB administration to 73 days, and found it intriguing that the TB fed group of rats made fewer WMEs and lower latency in the RAM task, as the number of trials increased. In order to have the food efficiently with minimal effort when a baited arm is visited for food, the rat has to avoid re-entry, and this learning strategy involves working memory. The results of the decreased WMEs and shorter latency suggest that the administration of TB significantly improved the short-term/working memory (Figure 3A). However, in our experimental setting, it could not affect the long-term memory, as RME scores between the two groups were not significantly different (Figure 3B). Concurrently, TB, compared to control animals, produced the learning-related memory significantly at a faster pace (shorter time) (Figure 3C). Oral administration of TB also triggered a significant increase in the spatial working memory, as indicated by a higher spontaneous alteration ratio in the Y-maze test [48]. TB not only ameliorated the RAM task- and Y-maze-determined working memory, but also significantly contributed to the enhancement of memory examined in NOR test, which was determined with an Inter Trial Interval (ITI) at 1h, and 24h, respectively. This was confirmed by increases in the DI and exploration time at both 1h and 24h of the post-familiarization test, which respectively can be referred to as short-term and long-term memory. With regards to this, Ennaceur et al. (1997) and Quillfeldt et al. (2016) also reported that object recognition in the NOR task is usually more employed to examine working memory [49,50], while recognition tasks tested the 'delays' of more than 6h allow one to determine the long-term memory [49,50]. Therefore, TB supplementation is significant in the betterment of learning-related memory cognition.

Many mechanisms have been proposed in relation to the location of memory. The working memory, which relates to the 'temporary operation and storage of information', is mainly stored in the prefrontal cortex of the brain [12,51]. Albeit, there are some reports that working memory is also stored in the hippocampus, at least to some extent [51]. On the other hand, long-term memory formation mainly occurs in the hippocampus [52,53]. However, the neocortex [54] and perirhinal cortex [55,56] of the medial temporal lobe are anatomically interlinked in the formation of long-term memory [57]. The long-term memory could be better described to occur in both the hippocampus and cortex regions, while the former is used for the formation and/or storage of new memory trace, and the latter is needed for long term storage [58–62]. Thus suggesting that the exact location of the neural circuitry of the memory is yet to be clearly elucidated.

The enhancement of both working- and long-term memory is controlled at the molecular level in neurons [63]. Whereas, working memory involves modifications of pre-existing neurochemicals/proteins, and long-term memory requires the synthesis of new mRNAs and proteins [64,65]. Furthermore, memory encompasses cholinergic, noradrenergic, dopaminergic systems [66], and most importantly glutamatergic system is critical to the learning and memory, and related plasticity [67,68]. *N*-methyl-D-aspartate receptor (NMDAR) is a voltage sensitive glutamate receptor. Glutamate-NMDAR interaction causes activation of Ca^{2+}-calmodulin cascade, including CaMKII [69]. Functional maintenance of neuronal circuitry depends on different neurotrophic factors like nerve growth factor (NGF), glial cell line-derived neurotrophic factor (GDNF), BDNF, etc. [70–72]. Besides the BDNF, many other signaling proteins e.g., CaMKII, MAPK, PKC, PKA, PI3K/Akt [73–77], and transcription factors, such as CREB [78], are important in memory-related neural plasticity. For example, CaMKII is a key synaptic signaling molecule that facilitates learning and memory processes by mediating a wide variety of intercellular signals [79,80]. CaMKII contributes to long term potentiation (LTP) and hence long-term memory [81].

Many studies have suggested that CaMKII phosphorylates the transcription factor CREB and transforms it into active form p-CREB [73,75,77]. p-CREB then initiates transcription and translation of proteins/receptors required for neuronal plasticity. This could be confirmed by inhibiting p-CREB and subsequent inhibition of new protein synthesis and, therefore, impair memory cognition [75]. In the current study, TB supplementation significantly upregulated CaMKII, as indicated by increased ratios of p-CaMKII/CaMKII in the cortical tissues (Figure 7A). TB augmented the levels of p-CREB concomitantly, as compared to those of the control rats (Figure 7B). Therefore, we speculate that TB-induced an increase in the levels of p-CaMKII, and p-CREB contributed to the improvement of neuronal plasticity, hence, the learning and memory of TB rats. BDNF has been implicated in LTP that occurs in the hippocampus and other brain regions. LTP plays a prime role in learning and memory [82,83]. Improvements of working memory were accompanied with an increased level of BDNF proteins in the frontal cortex [84,85]. Consistently, TB-fed rats, in the current study, had also higher BDNF levels in the mPFC. Moreover, the improvements in learning and memory were positively correlated with the levels of memory-related substrates—p-CaMKII, p-CREB, and BDNF (Table S2). Therefore, it is conceivable that TB-instigated increases in the levels of p-CaMKII, p-CREB, and BDNF improved the learning and the memory of the rats. TB supplementation did not have an effect on ERK1/2 pathway, as indicated by no change in the level of p-c-Raf, p-MEK1/2, p-90RSK; and p-MSK1 in the mPFC (Figure S3). However, the cause remains to be clarified.

There are accumulating shreds of evidences that the deficit in working memory has been implicated in several neurodegenerative (Parkinson's disease, Alzheimer's disease, and ageing) and/or neurodevelopmental disorders, such as attention deficit hyperactivity disorder (ADHD), Schizophrenia [86–89]. ADHD is associated with loss of function of the mPFC, which plays a role in regulating complex cognitive, emotional, and behavioral activities [90]. Interestingly, Yabuki et al. reported that deficits in working memory in ADHD model rats and spontaneously hypertensive rats (SHR), were closely associated with dysfunction of CaMKII in the mPFC, but not in the hippocampus [91]. Since the accumulation of TB in the cortex has been clearly detected [7,11], neurodegenerative and/or neurodevelopmental disorders, associated with these regions, might be beneficent from the oral administration of TB. Moreover, this study demonstrated that chronic oral administration of TB did not cause any adverse health effects in rats. TB may be considered as a safe functional food to include in the daily diet.

TB can be metabolized from caffeine, and both caffeine and TB may work through a similar pathway. Caffeine has some pharmacological effects in the central nervous system and beneficial effects on memory and/or cognitive functions in humans and rodents [92–96]. Since TB can be metabolized from caffeine, these components may work through similar mechanisms. For instance, caffeine and TB can pass through BBB and block cell surface adenosine receptors, which distributed widely throughout cortical regions [97]. TB and caffeine also act as PDE inhibitor that increases in intracellular cAMP level [7,11,97]. Moreover, these reagents are known to affect Ca^{2+} release from intracellular stores of the brain [97]. Although, exact differences between caffeine and TB on memory and/or cognitive functions have not been examined in this study, a comparison study for these reagents may be required in the future.

5. Conclusions

This study demonstrated that the oral administration of TB for 73 days resulted in the upregulation of p-CaMKII and p-CREB in the mPFC, and also found both BDNF mRNA expression level, as well as protein level upregulation in the mPFC of TB rats. These results clearly suggest that TB supplementation may facilitate the CaMKII/CREB/BDNF pathway in the mPFC. We also clearly observed a significant improvement in working memory in TB rats. These observations are also firmly supported by previous findings concerning the role of the CaMKII/CREB/BDNF pathway in working memory and learning in rats.

Supplementary Materials: The following are available online at http://www.mdpi.com/2072-6643/11/4/888/s1, In supplementary file 1 (Table S1: Primer sequence for real time PCR), In supplementary file 2 (Figure S1: The effect of theobromine (TB) on the brain-derived neurotrophic factor (BDNF) protein level in the hippocampus (Experiment 1), Figure S2: The effect of theobromine (TB) on the brain-derived neurotrophic factor (BDNF) protein level in the hippocampus (Experiment 2), Figure S3: ERK1/2 pathway proteins levels in the medial prefrontal cortex (mPFC), Table S2: Correlation between biological markers and behavioral measurements).

Author Contributions: R.I., K.M., N.S., and O.S. conceived the study; R.I., K.M., M.E.H., and E.S. performed the animal rearing and sample collection; R.I. carried out the behavioral analysis; R.I., K.M., and M.E.H. participated in biological analysis; R.I. and E.S. performed statistical analysis; R.I. prepared the figures and drafted the manuscript; all authors read, edited, and revised the manuscript and approved the final manuscript.

Funding: This study was supported by a Ministry of Education, Culture, Sports, Science, and Technology of Japan Grant-in-Aid for Scientific Research (B) 17H01963 and 25282021, and Support funds of Shimane University for young researchers.

Acknowledgments: We thank Shahdat Hossain, Sheikh Md. Abdullah and Abdullah Al Mamun for their helpful advice. We would also like to acknowledge the technical expertise of the Department of Laboratory Medicine, Department of Psychology and the Interdisciplinary Center for Science Research Organization for Research and Academic Information, Shimane University.

Conflicts of Interest: The authors declare no conflict of interest.

References

1. Fredholm, B.B.; Bättig, K.; Holmén, J.; Nehlig, A.; Zvartau, E.E. Actions of caffeine in the brain with special reference to factors that contribute to its widespread use. *Pharmacol. Rev.* **1999**, *51*, 83–133.
2. Miwa, S.; Sugimoto, N.; Yamamoto, N.; Shirai, T.; Nishida, H.; Hayashi, K.; Kimura, H.; Takeuchi, A.; Igarashi, K.; Yachie, A.; et al. Caffeine induces apoptosis of osteosarcoma cells by inhibiting AKT/mTOR/S6K, NF-κB and MAPK pathways. *Anticancer Res.* **2012**, *32*, 3643–3649.
3. Spencer, J.P.E. Flavonoids and brain health: Multiple effects underpinned by common mechanisms. *Genes Nutr.* **2009**, *4*, 243–250. [CrossRef]
4. Pura Naik, J. Improved high-performance liquid chromatography method to determine theobromine and caffeine in cocoa and cocoa products. *J. Agric. Food Chem.* **2001**, *49*, 3579–3583. [CrossRef]
5. Sokolov, A.N.; Pavlova, M.A.; Klosterhalfen, S.; Enck, P. Chocolate and the brain: Neurobiological impact of cocoa flavanols on cognition and behavior. *Neurosci. Biobehav. Rev.* **2013**, *37*, 2445–2453. [CrossRef] [PubMed]
6. Shively, C.A.; Tarka, S.M. Methylxanthine composition and consumption patterns of cocoa and chocolate products. *Prog. Clin. Biol. Res.* **1984**, *158*, 149–178. [CrossRef] [PubMed]
7. Yoneda, M.; Sugimoto, N.; Katakura, M.; Matsuzaki, K.; Tanigami, H.; Yachie, A.; Ohno-Shosaku, T.; Shido, O. Theobromine up-regulates cerebral brain-derived neurotrophic factor and facilitates motor learning in mice. *J. Nutr. Biochem.* **2017**, *39*, 110–116. [CrossRef]
8. Williams, R.J.; Spencer, J.P.E.; Rice-Evans, C. Flavonoids: Antioxidants or signalling molecules? *Free Radic. Biol. Med.* **2004**, *36*, 838–849. [CrossRef] [PubMed]
9. Sugimoto, N.; Miwa, S.; Ohno-Shosaku, T.; Tsuchiya, H.; Hitomi, Y.; Nakamura, H.; Tomita, K.; Yachie, A.; Koizumi, S. Activation of tumor suppressor protein PTEN and induction of apoptosis are involved in cAMP-mediated inhibition of cell number in B92 glial cells. *Neurosci. Lett.* **2011**, *497*, 55–59. [CrossRef] [PubMed]
10. Sugimoto, N.; Miwa, S.; Hitomi, Y.; Nakamura, H.; Tsuchiya, H.; Yachie, A. Theobromine, the primary methylxanthine found in Theobroma cacao, prevents malignant glioblastoma proliferation by negatively regulating phosphodiesterase-4, extracellular signal-regulated kinase, Akt/mammalian target of rapamycin kinase, and nuclear fact. *Nutr. Cancer* **2014**, *66*, 419–423. [CrossRef] [PubMed]
11. Sugimoto, N.; Katakura, M.; Matsuzaki, K.; Sumiyoshi, E.; Yachie, A.; Shido, O. Chronic administration of theobromine inhibits mTOR signal in rats. *Basic Clin. Pharmacol. Toxicol.* **2018**, 1–7. [CrossRef]
12. Khan, Z.U.; Muly, E.C. Molecular mechanisms of working memory. *Behav. Brain Res.* **2011**, *219*, 329–341. [CrossRef] [PubMed]
13. Kolb, B.; Tees, R.C. *The Cerebral Cortex of the Rat*; The MIT Press: Cambridge, MA, USA, 1990.
14. Guldin, W.O.; Pritzel, M.; Markowitsch, H.J. Prefrontal cortex of the mouse defined as cortical projection area of the thalamic mediodorsal nucleus. *Brain Behav. Evol.* **1981**, *19*, 93–107. [CrossRef]
15. Jacobsen, C.F. Functions of frontal association area in primates. *Arch. Neurol. Psychiatry* **1935**, *33*, 558–569. [CrossRef]
16. Fuster, J.M.; Alexander, G.E. Neuron activity related to short-term memory. *Science* **1971**, *173*, 652–654. [CrossRef] [PubMed]
17. Cohen, J.D.; Perlstein, W.M.; Braver, T.S.; Nystrom, L.E.; Noll, D.C.; Jonides, J.; Smith, E.E. Temporal dynamics of brain activation during a working memory task. *Nature* **1997**, *386*, 604–608. [CrossRef]
18. Goldman-Rakic, P.S. Cellular basis of working memory. *Neuron* **1995**, *14*, 477–485. [CrossRef]

19. Impey, S.; McCorkle, S.R.; Cha-Molstad, H.; Dwyer, J.M.; Yochum, G.S.; Boss, J.M.; McWeeney, S.; Dunn, J.J.; Mandel, G.; Goodman, R.H. Defining the CREB regulon: A genome-wide analysis of transcription factor regulatory regions. *Cell* **2004**, *119*, 1041–1054. [CrossRef]

20. Impey, S.; Smith, D.M.; Obrietan, K.; Donahue, R.; Wade, C.; Storm, D.R. Stimulation of cAMP response element (CRE)-mediated transcription during contextual learning. *Nat. Neurosci.* **1998**, *1*, 595–601. [CrossRef] [PubMed]

21. Barco, A.; Bailey, C.H.; Kandel, E.R. Common molecular mechanisms in explicit and implicit memory. *J. Neurochem.* **2006**, *97*, 1520–1533. [CrossRef] [PubMed]

22. Zhang, X.; Odom, D.T.; Koo, S.-H.; Conkright, M.D.; Canettieri, G.; Best, J.; Chen, H.; Jenner, R.; Herbolsheimer, E.; Jacobsen, E.; et al. Genome-wide analysis of cAMP-response element binding protein occupancy, phosphorylation, and target gene activation in human tissues. *Proc. Natl. Acad. Sci. USA* **2005**, *102*, 4459–4464. [CrossRef] [PubMed]

23. Shieh, P.B.; Hu, S.C.; Bobb, K.; Timmusk, T.; Ghosh, A. Identification of a signaling pathway involved in calcium regulation of BDNF expression. *Neuron* **1998**, *20*, 727–740. [CrossRef]

24. Tao, X.; Finkbeiner, S.; Arnold, D.B.; Shaywitz, A.J.; Greenberg, M.E. Ca^{2+} influx regulates BDNF transcription by a CREB family transcription factor-dependent mechanism. *Neuron* **1998**, *20*, 709–726. [CrossRef]

25. Gordon, T. The role of neurotrophic factors in nerve regeneration. *Neurosurg. Focus* **2009**, *26*, E3. [CrossRef] [PubMed]

26. Sheng, M.; Thompson, M.A.; Greenberg, M.E. CREB: A Ca(2+)-regulated transcription factor phosphorylated by calmodulin-dependent kinases. *Science* **1991**, *252*, 1427–1430. [CrossRef] [PubMed]

27. Hashimoto, M.; Hossain, S.; Agdul, H.; Shido, O. Docosahexaenoic acid-induced amelioration on impairment of memory learning in amyloid beta-infused rats relates to the decreases of amyloid beta and cholesterol levels in detergent-insoluble membrane fractions. *Biochim. Biophys. Acta* **2005**, *1738*, 91–98. [CrossRef] [PubMed]

28. Mamun, A.; Hashimoto, M.; Katakura, M.; Matsuzaki, K.; Hossain, S.; Arai, H.; Shido, O. Neuroprotective Effect of Madecassoside Evaluated Using Amyloid B1-42-Mediated in Vitro and in Vivo alzheimer's Disease Models. *Int. J. Indig. Med. Plants* **2014**, *47*, 1669–1682.

29. Paxinos, G.; Watson, C. *The Rat Brain in Stereotaxic Coordinates*; Academic Press: San Diego, CA, USA, 1998; ISBN 0125476191.

30. Maurice, T.; Hiramatsu, M.; Itoh, J.; Kameyama, T.; Hasegawa, T.; Nabeshima, T. Behavioral evidence for a modulating role of sigma ligands in memory processes. I. Attenuation of dizocilpine (MK-801)-induced amnesia. *Brain Res.* **1994**, *647*, 44–56. [CrossRef]

31. Kameyama, T.; Ukai, M.; Shinkai, N. Ameliorative effects of tachykinins on scopolamine-induced impairment of spontaneous alternation performance in mice. *Methods Find. Exp. Clin. Pharmacol.* **1998**, *20*, 555–560. [CrossRef]

32. Ukai, M.; Shinkai, N.; Kameyama, T. Involvement of dopamine receptors in beneficial effects of tachykinins on scopolamine-induced impairment of alternation performance in mice. *Eur. J. Pharmacol.* **1998**, *350*, 39–45. [CrossRef]

33. Dossat, A.M.; Jourdi, H.; Wright, K.N.; Strong, C.E.; Sarkar, A.; Kabbaj, M. Viral-mediated Zif268 expression in the prefrontal cortex protects against gonadectomy-induced working memory, long-term memory, and social interaction deficits in male rats. *Neuroscience* **2017**, *340*, 243–257. [CrossRef] [PubMed]

34. Hidaka, N.; Suemaru, K.; Li, B.; Araki, H. Effects of repeated electroconvulsive seizures on spontaneous alternation behavior and locomotor activity in rats. *Biol. Pharm. Bull.* **2008**, *31*, 1928–1932. [CrossRef]

35. Nakamura, K.; Ito, M.; Liu, Y.; Seki, T.; Suzuki, T.; Arai, H. Effects of single and repeated electroconvulsive stimulation on hippocampal cell proliferation and spontaneous behaviors in the rat. *Brain Res.* **2013**, *1491*, 88–97. [CrossRef]

36. Ennaceur, A.; Delacour, J. A new one-trial test for neurobiological studies of memory in rats. 1: Behavioral data. *Behav. Brain Res.* **1988**, *31*, 47–59. [CrossRef]

37. Clark, R.E.; Zola, S.M.; Squire, L.R. Impaired recognition memory in rats after damage to the hippocampus. *J. Neurosci.* **2000**, *20*, 8853–8860. [CrossRef] [PubMed]

38. Gomez, J.L.; Lewis, M.J.; Sebastian, V.; Serrano, P.; Luine, V.N. Alcohol administration blocks stress-induced impairments in memory and anxiety, and alters hippocampal neurotransmitter receptor expression in male rats. *Horm. Behav.* **2013**, *63*, 659–666. [CrossRef]

39. Reichel, C.M.; Schwendt, M.; McGinty, J.F.; Olive, M.F.; See, R.E. Loss of object recognition memory produced by extended access to methamphetamine self-administration is reversed by positive allosteric modulation of metabotropic glutamate receptor 5. *Neuropsychopharmacology* **2011**, *36*, 782–792. [CrossRef] [PubMed]

40. Davis, S.; Renaudineau, S.; Poirier, R.; Poucet, B.; Save, E.; Laroche, S. The formation and stability of recognition memory: What happens upon recall? *Front. Behav. Neurosci.* **2010**, *4*, 177. [CrossRef]

41. Hammond, R.S.; Tull, L.E.; Stackman, R.W. On the delay-dependent involvement of the hippocampus in object recognition memory. *Neurobiol. Learn. Mem.* **2004**, *82*, 26–34. [CrossRef]

42. Matsuzaki, K.; Katakura, M.; Sugimoto, N.; Hara, T.; Hashimoto, M.; Shido, O. Neural progenitor cell proliferation in the hypothalamus is involved in acquired heat tolerance in long-term heat-acclimated rats. *PLoS ONE* **2017**, *12*, e0178787. [CrossRef]

43. Matsuzaki, K.; Katakura, M.; Sugimoto, N.; Hara, T.; Hashimoto, M.; Shido, O. β-amyloid infusion into lateral ventricle alters behavioral thermoregulation and attenuates acquired heat tolerance in rats. *Temperature* **2015**, *2*, 418–424. [CrossRef]

44. Matsuzaki, K.; Yamakuni, T.; Hashimoto, M.; Haque, A.M.; Shido, O.; Mimaki, Y.; Sashida, Y.; Ohizumi, Y. Nobiletin restoring beta-amyloid-impaired CREB phosphorylation rescues memory deterioration in Alzheimer's disease model rats. *Neurosci. Lett.* **2006**, *400*, 230–234. [CrossRef] [PubMed]

45. Hossain, M.E.; Matsuzaki, K.; Katakura, M.; Sugimoto, N.; Mamun, A.; Islam, R.; Hashimoto, M.; Shido, O. Direct exposure to mild heat promotes proliferation and neuronal differentiation of neural stem/progenitor cells in vitro. *PLoS ONE* **2017**, *12*, e0190356. [CrossRef] [PubMed]

46. Li, G.H.; Katakura, M.; Maruyama, M.; Enhkjargal, B.; Matsuzaki, K.; Hashimoto, M.; Shido, O. Changes of noradrenaline-induced contractility and gene expression in aorta of rats acclimated to heat in two different modes. *Eur. J. Appl. Physiol.* **2008**, *104*, 29–40. [CrossRef] [PubMed]

47. Martínez-Pinilla, E.; Oñatibia-Astibia, A.; Franco, R. The relevance of theobromine for the beneficial effects of cocoa consumption. *Front. Pharmacol.* **2015**, *6*, 30. [CrossRef]

48. Sarter, M.; Bodewitz, G.; Stephens, D.N. Attenuation of scopolamine-induced impairment of spontaneous alternation behaviour by antagonist but not inverse agonist and agonist β-carbolines. *Psychopharmacology* **1988**. [CrossRef]

49. Ennaceur, A.; Neave, N.; Aggleton, J.P. Spontaneous object recognition and object location memory in rats: The effects of lesions in the cingulate cortices, the medial prefrontal cortex, the cingulum bundle and the fornix. *Exp. Brain Res.* **1997**, *113*, 509–519. [CrossRef] [PubMed]

50. Quillfeldt, J.A. Behavioral methods to study learning and memory in rats. In *Rodent Model as Tools in Ethical Biomedical Research*; Springer: Cham, Switzerland, 2016; pp. 271–311.

51. Yoon, T.; Okada, J.; Jung, M.W.; Kim, J.J. Prefrontal cortex and hippocampus subserve different components of working memory in rats. *Learn. Mem.* **2008**, *15*, 97–105. [CrossRef]

52. Han, S.; Bhattarai, J. Phasic and tonic type A γ-Aminobutyric acid receptor mediated effect of Withania somnifera on mice hippocampal CA1 pyramidal Neurons. *J. Ayurveda Integr. Med.* **2014**, *5*, 216. [CrossRef]

53. Ravichandran, V.; Kim, M.; Han, S.; Cha, Y. Stachys sieboldii Extract Supplementation Attenuates Memory Deficits by Modulating BDNF-CREB and Its Downstream Molecules, in Animal Models of Memory Impairment. *Nutrients* **2018**, *10*, 917. [CrossRef]

54. Bontempi, B.; Laurent-Demir, C.; Destrade, C.; Jaffard, R. Time-dependent reorganization of brain circuitry underlying long-term memory storage. *Nature* **1999**, *400*, 671. [CrossRef]

55. Mumby, D.G.; Glenn, M.J. Anterograde and retrograde memory for object discriminations and places in rats with perirhinal cortex lesions. *Behav. Brain Res.* **2000**, *114*, 119–134. [CrossRef]

56. Ramos, J.M.J. Long-term spatial memory in rats with hippocampal lesions. *Eur. J. Neurosci.* **2000**, *12*, 3375–3384. [CrossRef] [PubMed]

57. Witter, M.P.; Naber, P.A.; Van Haeften, T.; Machielsen, W.C.M.; Rombouts, S.A.R.B.; Barkhof, F.; Scheltens, P.; da Silva, F.H. Cortico-hippocampal communication by way of parallel parahippocampal-subicular pathways. *Hippocampus* **2000**, *10*, 398–410. [CrossRef]

58. Graff, J.; Woldemichael, B.T.; Berchtold, D.; Dewarrat, G.; Mansuy, I.M. Dynamic histone marks in the hippocampus and cortex facilitate memory consolidation. *Nat. Commun.* **2012**, *3*, 991. [CrossRef] [PubMed]

59. Frankland, P.W.; Bontempi, B.; Talton, L.E.; Kaczmarek, L.; Silva, A.J. The involvement of the anterior cingulate cortex in remote contextual fear memory. *Science* **2004**, *304*, 881–883. [CrossRef] [PubMed]

60. Jung, M.W.; Baeg, E.H.; Kim, M.J.; Kim, Y.B.; Kim, J.J. Plasticity and memory in the prefrontal cortex. *Rev. Neurosci.* **2008**, *19*, 29–46. [CrossRef]

61. Nieuwenhuis, I.L.C.; Takashima, A. The role of the ventromedial prefrontal cortex in memory consolidation. *Behav. Brain Res.* **2011**, *218*, 325–334. [CrossRef]

62. Squire, L.R.; Alvarez, P. Retrograde amnesia and memory consolidation: A neurobiological perspective. *Curr. Opin. Neurobiol.* **1995**, *5*, 169–177. [CrossRef]

63. Carew, T.J. Molecular enhancement of memory formation. *Neuron* **1996**, *16*, 5–8. [CrossRef]

64. Martin, K.C.; Barad, M.; Kandel, E.R. Local protein synthesis and its role in synapse-specific plasticity. *Curr. Opin. Neurobiol.* **2000**, *10*, 587–592. [CrossRef]

65. Kelleher, R.J., III; Govindarajan, A.; Tonegawa, S. Translational regulatory mechanisms in persistent forms of synaptic plasticity. *Neuron* **2004**, *44*, 59–73. [CrossRef] [PubMed]

66. Briand, L.A.; Gritton, H.; Howe, W.M.; Young, D.A.; Sarter, M. Modulators in concert for cognition: Modulator interactions in the prefrontal cortex. *Prog. Neurobiol.* **2007**, *83*, 69–91. [CrossRef]

67. Dudai, Y. Molecular bases of long-term memories: A question of persistence. *Curr. Opin. Neurobiol.* **2002**, *12*, 211–216. [CrossRef]

68. Lamprecht, R.; LeDoux, J. Structural plasticity and memory. *Nat. Rev. Neurosci.* **2004**, *5*, 45–54. [CrossRef] [PubMed]

69. Park, H.; Poo, M. Neurotrophin regulation of neural circuit development and function. *Nat. Rev. Neurosci.* **2013**, *14*, 7. [CrossRef] [PubMed]

70. Henderson, C.E. Role of neurotrophic factors in neuronal development. *Curr. Opin. Neurobiol.* **1996**, *6*, 64–70. [CrossRef]

71. Levi-Montalcini, R. The nerve growth factor 35 years later. *Science* **1987**, *237*, 1154–1162. [CrossRef]

72. Davies, A.M. The neurotrophic hypothesis: Where does it stand? *Philos. Trans. R. Soc. Lond. Ser. B Biol. Sci.* **1996**, *351*, 389–394.

73. Miyamoto, E. Molecular mechanism of neuronal plasticity: Induction and maintenance of long-term potentiation in the hippocampus. *J. Pharmacol. Sci.* **2006**, *100*, 433–442. [CrossRef]

74. Leinninger, G.M.; Backus, C.; Uhler, M.D.; Lentz, S.I.; Feldman, E.L. Phosphatidylinositol 3-kinase and Akt effectors mediate insulin-like growth factor-I neuroprotection in dorsal root ganglia neurons. *FASEB J. Off. Publ. Fed. Am. Soc. Exp. Biol.* **2004**, *18*, 1544–1546. [CrossRef] [PubMed]

75. Vitolo, O.V.; Sant'Angelo, A.; Costanzo, V.; Battaglia, F.; Arancio, O.; Shelanski, M. Amyloid beta -peptide inhibition of the PKA/CREB pathway and long-term potentiation: Reversibility by drugs that enhance cAMP signaling. *Proc. Natl. Acad. Sci. USA* **2002**, *99*, 13217–13221. [CrossRef] [PubMed]

76. Zhao, L.; Brinton, R.D. Vasopressin-induced cytoplasmic and nuclear calcium signaling in embryonic cortical astrocytes: Dynamics of calcium and calcium-dependent kinase translocation. *J. Neurosci.* **2003**, *23*, 4228–4239. [CrossRef] [PubMed]

77. Yan, X.; Liu, J.; Ye, Z.; Huang, J.; He, F.; Xiao, W.; Hu, X.; Luo, Z. CaMKII-Mediated CREB phosphorylation is involved in Ca^{2+}-Induced BDNF mRNA transcription and neurite outgrowth promoted by electrical stimulation. *PLoS ONE* **2016**, *11*, e0162784. [CrossRef]

78. Kandel, E.R. The molecular biology of memory storage: A dialogue between genes and synapses. *Science* **2001**, *294*, 1030–1038. [CrossRef] [PubMed]

79. Schulman, H.; Hanson, P.I. Multifunctional Ca^{2+}/calmodulin-dependent protein kinase. *Neurochem. Res.* **1993**, *18*, 65–77. [CrossRef] [PubMed]

80. Kennedy, M.B. Signal transduction molecules at the glutamatergic postsynaptic membrane. *Brain Res. Brain Res. Rev.* **1998**, *26*, 243–257. [CrossRef]

81. Lisman, J.; Yasuda, R.; Raghavachari, S. Mechanisms of CaMKII action in long-term potentiation. *Nat. Rev. Neurosci.* **2012**, *13*, 169–182. [CrossRef]

82. Korte, M.; Kang, H.; Bonhoeffer, T.; Schuman, E. A role for BDNF in the late-phase of hippocampal long-term potentiation. *Neuropharmacology* **1998**, *37*, 553–559. [CrossRef]

83. Lessmann, V. Neurotrophin-dependent modulation of glutamatergic synaptic transmission in the mammalian CNS. *Gen. Pharmacol.* **1998**, *31*, 667–674. [CrossRef]

84. Bimonte, H.A.; Nelson, M.E.; Granholm, A.-C.E. Age-related deficits as working memory load increases: Relationships with growth factors. *Neurobiol. Aging* **2003**, *24*, 37–48. [CrossRef]

85. Bimonte-Nelson, H.A.; Hunter, C.L.; Nelson, M.E.; Granholm, A.-C.E. Frontal cortex BDNF levels correlate with working memory in an animal model of Down syndrome. *Behav. Brain Res.* **2003**, *139*, 47–57. [CrossRef]

86. Arnsten, A.F.T.; Ramos, B.P.; Birnbaum, S.G.; Taylor, J.R. Protein kinase A as a therapeutic target for memory disorders: Rationale and challenges. *Trends Mol. Med.* **2005**, *11*, 121–128. [CrossRef]

87. Germano, C.; Kinsella, G.J. Working memory and learning in early Alzheimer's disease. *Neuropsychol. Rev.* **2005**, *15*, 1–10. [CrossRef]

88. Goldman-Rakic, P.S.; Castner, S.A.; Svensson, T.H.; Siever, L.J.; Williams, G.V. Targeting the dopamine D1 receptor in schizophrenia: Insights for cognitive dysfunction. *Psychopharmacology* **2004**, *174*, 3–16. [CrossRef] [PubMed]

89. Morris, R.G.; Baddeley, A.D. Primary and working memory functioning in Alzheimer-type dementia. *J. Clin. Exp. Neuropsychol.* **1988**, *10*, 279–296. [CrossRef]

90. Castellanos, F.X.; Tannock, R. Neuroscience of attention-deficit/hyperactivity disorder: The search for endophenotypes. *Nat. Rev. Neurosci.* **2002**, *3*, 617–628. [CrossRef] [PubMed]

91. Yabuki, Y.; Shioda, N.; Maeda, T.; Hiraide, S.; Togashi, H.; Fukunaga, K. Aberrant CaMKII activity in the medial prefrontal cortex is associated with cognitive dysfunction in ADHD model rats. *Brain Res.* **2014**, *1557*, 90–100. [CrossRef] [PubMed]

92. Shabir, A.; Hooton, A.; Tallis, J.; F Higgins, M. The Influence of Caffeine Expectancies on Sport, Exercise, and Cognitive Performance. *Nutrients* **2018**, *10*, 1528. [CrossRef]

93. Ritchie, K.; Carriere, I.; de Mendonca, A.; Portet, F.; Dartigues, J.F.; Rouaud, O.; Barberger-Gateau, P.; Ancelin, M.L. The neuroprotective effects of caffeine: A prospective population study (the Three City Study). *Neurology* **2007**, *69*, 536–545. [CrossRef] [PubMed]

94. van Gelder, B.M.; Buijsse, B.; Tijhuis, M.; Kalmijn, S.; Giampaoli, S.; Nissinen, A.; Kromhout, D. Coffee consumption is inversely associated with cognitive decline in elderly European men: The FINE Study. *Eur. J. Clin. Nutr.* **2007**, *61*, 226–232. [CrossRef] [PubMed]

95. Santos, C.; Lunet, N.; Azevedo, A.; de Mendonca, A.; Ritchie, K.; Barros, H. Caffeine intake is associated with a lower risk of cognitive decline: A cohort study from Portugal. *J. Alzheimers. Dis.* **2010**, *20* (Suppl. 1), S175–S185. [CrossRef] [PubMed]

96. Alzoubi, K.H.; Mhaidat, N.M.; Obaid, E.A.; Khabour, O.F. Caffeine Prevents Memory Impairment Induced by Hyperhomocysteinemia. *J. Mol. Neurosci.* **2018**, *66*, 222–228. [CrossRef]

97. Chen, X.; Ghribi, O.; Geiger, J.D. Caffeine protects against disruptions of the blood-brain barrier in animal models of Alzheimer's and Parkinson's diseases. *J. Alzheimers. Dis.* **2010**, *20* (Suppl. 1), S127–S141. [CrossRef] [PubMed]

Article

Association of Chocolate Consumption with Hearing Loss and Tinnitus in Middle-Aged People Based on the Korean National Health and Nutrition Examination Survey 2012–2013

Sang-Yeon Lee [1], Gucheol Jung [2], Myoung-jin Jang [2], Myung-Whan Suh [1,3], Jun ho Lee [1,3], Seung-Ha Oh [1,3] and Moo Kyun Park [1,3,*]

[1] Department of Otorhinolaryngology-Head and Neck Surgery, Seoul National University Hospital, Seoul 03080, Korea; maru4843@hanmail.net (S.-Y.L.); drmung@naver.com (M.-W.S.); junlee@snu.ac.kr (J.h.L.); shaoh@snu.ac.kr (S.-H.O.)
[2] Medical Research Collaborating Center, Seoul National University Hospital, Seoul 03080, Korea; 9E887@snuh.org (G.J.); mjjang2014@naver.com (M.-j.-J.)
[3] Sensory Organ Research Institute, Seoul National University Medical Research Center, Seoul 03080, Korea
* Correspondence: entpmk@gmail.com; Tel.: +82-2-2072-2446; Fax: +82-2-745-2387

Received: 2 March 2019; Accepted: 22 March 2019; Published: 30 March 2019

Abstract: Chocolate, which is produced from cocoa, exerts antioxidant and anti-inflammatory effects that ameliorate neurodegenerative diseases. We hypothesized that chocolate consumption would protect against hearing loss and tinnitus. We evaluated the hearing and tinnitus data, as well as the chocolate consumption, of middle-aged participants (40–64 years of age) of the 2012–2013 Korean National Health and Nutrition Examination Survey. All of the subjects underwent a medical interview, physical examination, audiological evaluation, tinnitus questionnaire, and nutrition examination. A total of 3575 subjects 40–64 years of age were enrolled. The rate of any hearing loss (unilateral or bilateral) in the subjects who consumed chocolate (26.78% (338/1262)) was significantly lower than that in those who did not (35.97% (832/2313)) ($p < 0.001$). Chocolate consumption was independently associated with low odds of any hearing loss (adjusted odds ratio = 0.83, 95% confidence interval = 0.70 to 0.98, $p = 0.03$). Moreover, the severity of hearing loss was inversely correlated with the frequency of chocolate consumption. In contrast to chocolate, there was no association between hearing loss and the consumption of sweet products without cocoa. Chocolate consumption was also not associated with tinnitus or tinnitus-related annoyance. Our results suggest that a chocolate-based diet may protect middle-aged people from hearing loss.

Keywords: chocolate; hearing loss; tinnitus; cohort study

---

1. Introduction

Hearing loss, a highly prevalent sensorineural disorder, imposes a major economic and social burden [1]. The incidence of hearing loss is approximately 20% when mild and unilateral hearing losses are included [2]. Hearing loss typically hampers communication and relationships, thereby resulting in social isolation [3]. In addition, hearing loss increases the risk of depression symptoms and deterioration of the quality of life [4]. Hearing loss is an important risk factor for neurodegenerative dementia [5,6]. Auditory rehabilitation using, for example, hearing aids and cochlear implants, restores auditory function, but the protective effect of certain foodstuffs against hearing loss is unclear.

Tinnitus is a common otologic symptom, often called a "phantom sound" as it is a conscious auditory perception without a corresponding physical source [7]. The prevalence of tinnitus ranges from 12% to 30% per 100,000 people [8,9]. Tinnitus can be bothersome and negatively affect sleep, concentration, emotions, and social enjoyment [10]. Tinnitus is associated particularly with hearing loss, and so correction of any cochlear pathology would ameliorate tinnitus and tinnitus-related distress. However, the selection of a treatment for tinnitus is supported by limited evidence [11]. Moreover, neither medical therapy nor dietary supplements substantially reduce tinnitus-related distress [12].

Patients and clinicians have concerns regarding the effect of diet on hearing loss and tinnitus [13,14]. Therefore, efforts to identify foods that protect hearing are needed. The health-promoting effects of cocoa have attracted the attention of researchers, health-conscious consumers, and manufacturers of cocoa products. Chocolate, which is produced from cocoa, is consumed for pleasure worldwide and is a source of health-promoting compounds [15]. Chocolate contains abundant polyphenols, the antioxidant and anti-inflammatory effects of which might protect against audiological impairment. Recent animal studies demonstrated that polyphenols attenuate oxidative stress and inflammation in the cochlea [16,17]. However, no human study has investigated the association between chocolate intake and hearing loss and/or tinnitus.

Based on the therapeutic effect of chocolate [18], we hypothesized that chocolate consumption would decrease the burden of hearing loss and tinnitus. We investigated the effect of chocolate consumption on hearing loss and tinnitus in middle-aged people in a large Korean cohort.

2. Materials and Methods

2.1. Study Population

This cohort study used data, from the Korean National Health and Nutrition Examination Survey (KNHANES), representative of the health and nutritional status of Koreans. We examined the association of chocolate consumption with hearing loss and tinnitus in participants 40 to 65 years of age. All subjects gave their informed consent for inclusion before they participated in the study. This study was approved by the Institutional Review Board of Seoul National University Hospital (1902-046-1008).

A total of 16,076 individuals participated in the 2012–2013 KNHANES. Subjects with external or middle ear pathologies, a history of congenital hearing loss, or retrocochlear lesions were excluded from the study. Of the remaining 5673 middle-aged participants (40–64 years of age), 2098 were excluded because they did not receive a hearing threshold test or respond to a tinnitus-related questionnaire (n = 887), did not provide their chocolate consumption frequency (n = 874), or had missing values for confounders (n = 337). Finally, 3575 participants from the 2012–2013 KNHANES aged 40 to 65 years were enrolled in the study (Figure 1).

2.2. Audiological Assessment

A structured physical examination, to exclude middle and external ear problems, was conducted by a physician. The pathologies of perforation or retraction of the tympanic membrane, otitis media with effusion, and cholesteatoma were identified by ear endoscopy. All subjects underwent pure-tone audiometry for six different octave frequencies (0.5, 1, 2, 3, 4, and 6 kHz) in a soundproof room. The mean hearing threshold was calculated as the average of the hearing thresholds at 0.5, 1, 2, and 4 kHz. The mean high-tone hearing threshold was determined using the average of the hearing thresholds at 3, 4, and 6 kHz. To exclude subjects with systemic diseases, blood samples were collected and analyzed at the Neodin Medical Institute in Seoul, Korea.

With regard to tinnitus, all subjects were also interviewed about the presence of tinnitus and tinnitus-related annoyance. The severity of tinnitus was classified as follows: "no," "slightly annoying", and "very annoying and difficult to sleep". In this study, a response of "slightly annoying" or "very annoying and difficult to sleep" was defined as tinnitus-related annoyance. The parameters of mean hearing threshold according to frequency, presence of tinnitus, and tinnitus-related annoyance were used to evaluate the association with chocolate.

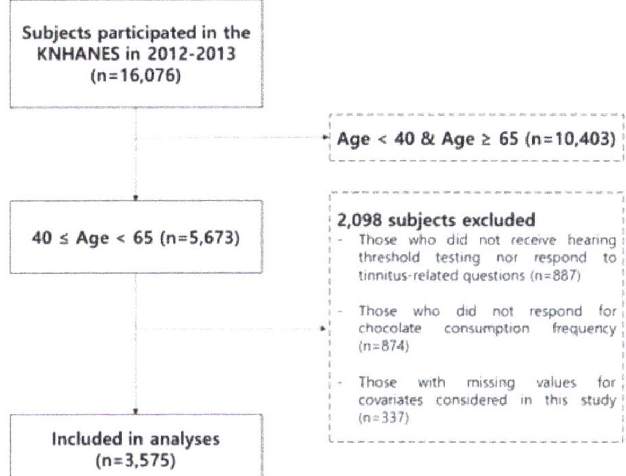

Figure 1. Schematic illustration of the selection of subjects.

2.3. Potential Confounders

The following potential confounders were adjusted for: sleep duration, stress severity, income, current smoking habits, alcohol consumption, noise exposure, and medical conditions. The information obtained included sleep duration (<6, 6–7, 7–8, or ≥8 h), rate of perceived stress, income (<25%, 25–50%, 50–75%, or >75% of the equalized household income per month), current smoking habits, exposure to indoor second-hand smoke (at work or at home), alcohol consumption (social drinker, heavy drinker, or problem drinker), and difficulties in controlling alcohol use. The duration of occupational exposure to noise and earphone and headphone use were also measured. In addition, the health status (presence of hypertension, diabetes, dyslipidemia, anemia, kidney failure, thyroid disorder, and menopause) of the subjects was evaluated. Finally, we used sweet products without cocoa (including ice cream, cake, and cookies) as a confounder in the evaluation of the non-specific effect of pleasant sensations.

2.4. Assessment of Chocolate Consumption

As in previous study [19], chocolate consumption was assessed using a food-frequency questionnaire (FFQ). Participants were asked to indicate how frequently they consumed chocolate in the previous year, from 2012 to 2013; there were 10 possible responses (never or seldom, once per month, two to three times per month, once per week, two to four times per week, five to six times per week, once per day, twice per day, three times per day, and no response). In addition, average chocolate intake was categorized into rarely (less than once per month), one quarter of a chocolate tablet, one half of a chocolate tablet, one chocolate tablet, and no response. Information on the frequency and quantity of chocolate consumed over the past year was collected by trained dietitians after the health interview (Supplementary Table S1).

2.5. Statistical Analysis

All statistical analyses were performed using SAS software (version 9.2; SAS Institute, Cary, NC, USA). The subjects' demographic and clinical characteristics are presented as medians (interquartile ranges) or numbers (proportions). The comparison between groups, performed by Fisher exact test (binary covariates), chi-squared test (more than three categories), or Wilcoxon rank sum test (continuous covariates), was deemed appropriate. The association between chocolate consumption and hearing loss and tinnitus was evaluated by means of multivariate logistic regression models with adjustment for the following potential confounders: age; sex; sleep duration; perceived stress; current smoking habits; exposure to indoor second-hand smoke; heavy drinking; drinking-related problem; duration of earphone use; occupational exposure to noise; menopause; and history of hypertension, diabetes mellitus, dyslipidemia, anemia, kidney failure, and thyroid disorder. The Spearman correlation rank order was used to evaluate the correlation between the frequency of chocolate consumption and the severity of hearing loss. In the multivariate models of tinnitus or tinnitus-related annoyance, we controlled for unilateral or bilateral hearing loss as well as the above potential confounders.

3. Results

The rate of any hearing loss (unilateral or bilateral hearing loss) was significantly lower in the subjects who consumed chocolate (26.78% (338/1262)) than in those who did not (35.97% (832/2313)) ($p < 0.001$, Table 1). In addition, chocolate consumption decreased the risk of bilateral hearing loss (13.31% (168/1262) vs. 20.32% (470/2313), $p < 0.001$) and high-tone hearing loss (51.58% (651/1262) vs. 63.60% (1,471/2313), $p < 0.001$), respectively.

Table 1. Characteristics of the 40–64-year-old subjects according to chocolate consumption.

	Chocolate Consumption			*p*-Value [1]
	Total (*n* = 3575)	More Than Once (*n* = 1262)	None (*n* = 2313)	
Hearing loss [2], *n* (%)				
Unilateral or bilateral	1170 (32.73%)	338 (26.78%)	832 (35.97%)	<0.0001
Bilateral	638 (17.85%)	168 (13.31%)	470 (20.32%)	<0.0001
High-tone hearing loss	2122 (59.36%)	651 (51.58%)	1471 (63.60%)	<0.001
Tinnitus, *n* (%)	811 (22.69%)	266 (21.08%)	545 (23.56%)	0.0947
Tinnitus-related annoyance, *n* (%)	259 (7.24%)	78 (6.18%)	181 (7.83%)	0.0790
Age (year), median (interquartile range, IQR)	52 (45, 58)	50 (44, 56)	53 (47, 59)	<0.0001
Male, *n* (%)	1415 (39.58%)	493 (39.06%)	922 (39.86%)	0.6677
Monthly household income [3], median (IQR)	333.33 (200, 518.33)	400 (250, 591.67)	310.67 (184.17, 500)	<0.0001
Use of earphones, *n* (%)	186 (5.20%)	90 (7.13%)	96 (4.15%)	0.0002
Duration of earphone use (min), median (IQR)				
Total	0 (0, 0)	0 (0, 0)	0 (0, 0)	0.0002
User of earphones	60 (30, 60)	30 (20, 60)	60 (30, 90)	0.0617
Occupational exposure to noise, *n* (%)	587 (16.42%)	186 (14.74%)	401 (17.34%)	0.0473
Duration of occupational exposure to noise (months), median (IQR)				
Total	0 (0, 0)	0 (0, 0)	0 (0, 0)	0.0448
Occupational exposure to noise	96 (36, 216)	96 (36, 216)	108 (36, 216)	0.8920
Sleep duration (hours), *n* (%)				0.0295
<6	504 (14.1%)	163 (12.92%)	341 (14.74%)	
6–7	1,039 (29.06%)	377 (29.87%)	662 (28.62%)	
7–8	1,099 (30.74%)	418 (33.12%)	681 (29.44%)	
≥8	933 (26.1%)	304 (24.09%)	629 (27.19%)	
High perceived stress, *n* (%)	770 (21.54%)	274 (21.71%)	496 (21.44%)	0.8648
Exposure to indoor second-hand smoke				
At work, *n* (%)	1162 (32.5%)	418 (33.12%)	744 (32.17%)	0.5753
At home, *n* (%)	349 (9.76%)	123 (9.75%)	226 (9.77%)	>0.9999
Current smoking, *n* (%)	653 (18.27%)	203 (16.09%)	450 (19.46%)	0.0128
Heavy drinking [4], *n* (%)	665 (18.60%)	193 (15.29%)	472 (20.41%)	0.0002
Difficulties in controlling alcohol use, *n* (%)	294 (8.22%)	87 (6.89%)	207 (8.95%)	0.0354
Having drinking-related problem in life, *n* (%)	165 (4.62%)	56 (4.44%)	109 (4.71%)	0.7393
Menopause (females)				<0.0001
Yes	1209 (55.97%)	372 (48.37%)	837 (60.17%)	
No	951 (44.03%)	397 (51.63%)	554 (39.83%)	
Hypertension, *n* (%)	683 (19.10%)	196 (15.53%)	487 (21.05%)	<0.0001
Diabetes mellitus, *n* (%)	247 (6.91%)	60 (4.75%)	187 (8.08%)	0.0001
Anemia, *n* (%)	284 (7.94%)	115 (9.11%)	169 (7.31%)	0.0606
Kidney failure, *n* (%)	13 (0.36%)	6 (0.48%)	7 (0.30%)	0.4004
Thyroid disorder, *n* (%)	85 (2.38%)	35 (2.77%)	50 (2.16%)	0.2527
Dyslipidemia, *n* (%)	331 (9.26%)	87 (6.89%)	244 (10.55%)	0.0003

[1] *p*-values by Fisher exact test (binary covariates), chi-squared test (more than three categories) and Wilcoxon rank sum test (continuous covariates). [2] Hearing loss ≥20 dB for four frequency average of pure-tone thresholds at 500, 1000, 2000, and 4000 Hz. [3] Monthly household income (10,000 Korean won). [4] Heavy drinking defined as more than three drinks per average drinking session more than twice a week.

In a multivariate logistic regression analysis, compared to chocolate non-consumers, the subjects who consumed chocolate had lower odds of any hearing loss (adjusted odds ratio (OR) = 0.83, 95% confidence interval (CI) = 0.70 to 0.98, $p = 0.03$), bilateral hearing loss (adjusted OR = 0.79, 95% CI = 0.64 to 0.98, $p = 0.03$), and high-tone hearing loss (adjusted OR = 0.78, 95% CI = 0.66 to 0.91, $p = 0.02$) (Table 2).

Table 2. Odds ratios (OR) and 95% confidence intervals (CI) for hearing loss and tinnitus according to chocolate consumption.

	n = 3575	Univariate Analysis		Multivariate Analysis	
		OR (95% CI)	*p*-Value	OR (95% CI)	*p*-Value
Hearing loss (unilateral or bilateral)	1170: 2405	0.651 (0.560, 0.757)	<0.0001	0.829 (0.701, 0.980) [1]	0.0285
Hearing loss (bilateral)	638: 2937	0.602 (0.497, 0.729)	<0.0001	0.791 (0.641, 0.976) [1]	0.0287
High-tone hearing loss	2122: 1453	0.610 (0.531, 0.701)	<0.0001	0.777 (0.661, 0.912) [1]	0.0021
Tinnitus	811: 2764	0.866 (0.734, 1.023)	0.0902	0.911 (0.767, 1.081) [2]	0.2847
Tinnitus-related annoyance	259: 3316	0.776 (0.590, 1.021)	0.0705	0.886 (0.668, 1.176) [2]	0.4036

[1] Adjusted for: age; sex; perceived stress; exposure to indoor second-hand smoke; current smoking habits; heavy drinking; drinking-related problems; menopause; histories of hypertension, diabetes mellitus, anemia, kidney failure, thyroid disorder, and dyslipidemia; income level; sleep duration; duration of occupational exposure to noise; and earphone and headphone use time. [2] Adjusted for all covariates used in the hearing-loss model in addition to hearing loss (unilateral or bilateral).

In contrast to chocolate, neither hearing loss nor bilateral hearing loss was associated with the consumption of sweet products without cocoa, according to multivariable logistic regression model. Furthermore, chocolate was significantly associated with a decrease of hearing loss after adjusting for cofounders, including consumption of sweet products without cocoa (Table 3).

Table 3. Odds ratios and 95% confidence intervals for hearing loss and tinnitus according to consumption of chocolate and sweet products without cocoa.

	Multivariable Analysis		
	Consumption Per Week (Reference = No)	OR (95% CI)	*p*-Value
Hearing loss (unilateral or bilateral)	Chocolate	0.835 (0.703, 0.992) [1]	0.0406
	Cookie	0.989 (0.973, 1.006) [1]	0.1912
	Ice cream	1.055 (0.967, 1.151) [1]	0.2308
	Cake	0.975 (0.861, 1.105) [1]	0.6931
Hearing loss (bilateral)	Chocolate	0.766 (0.617, 0.951) [1]	0.0156
	Cookie	0.989 (0.968, 1.010) [1]	0.3043
	Ice cream	1.146 (1.039, 1.265) [1]	0.0063
	Cake	1.041 (0.901, 1.203) [1]	0.5826
Tinnitus	Chocolate	0.920 (0.772, 1.097) [2]	0.3536
	Cookie	1.001 (0.985, 1.018) [2]	0.8901
	Ice cream	1.009 (0.922, 1.104) [2]	0.8478
	Cake	0.932 (0.814, 1.067) [2]	0.3081
Tinnitus-related annoyance	Chocolate	0.876 (0.655, 1.172) [2]	0.3735
	Cookie	0.987 (0.956, 1.020) [2]	0.4431
	Ice cream	1.120 (0.990, 1.267) [2]	0.0721
	Cake	0.973 (0.787, 1.204) [2]	0.8028

[1] Variables included in multivariable models: age; sex; perceived stress; exposure to indoor second-hand smoke; current smoking habits; heavy drinking; drinking-related problem; menopause; history of hypertension, diabetes mellitus, anemia, kidney failure, thyroid disorder, and dyslipidemia; income level; sleep duration; duration of occupational exposure to noise; and earphone and headphone use time; chocolate consumption; cookie consumption; ice cream consumption; cake consumption. [2] Variables included in multivariable models: all covariates used in the hearing-loss model as well as hearing loss (unilateral or bilateral).

Moreover, Spearman's rank order correlation analyses of the data revealed a significant inverse correlation between the mean hearing threshold and the frequency of chocolate consumption per week ($\rho = -0.117$, 95% CI = -0.15 to -0.08, $p < 0.001$) (Figure 2A). In addition, the mean threshold of high-tone frequencies was negatively correlated with the frequency of chocolate consumption per week ($\rho = -0.121$, 95% CI = -0.15 to -0.09, $p < 0.001$) (Figure 2B).

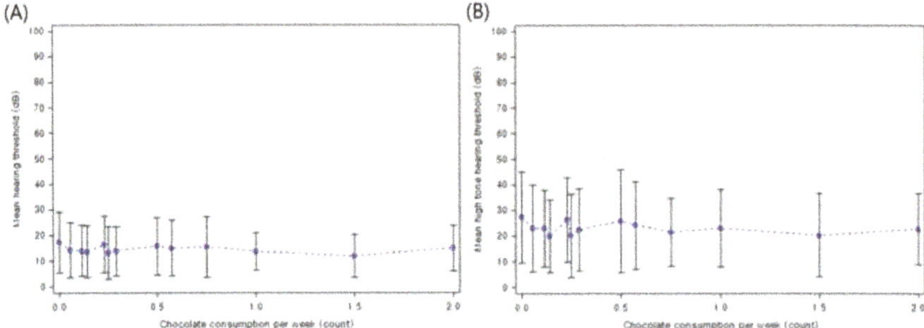

Figure 2. Scatter plot of the severity of hearing loss and the frequency of chocolate consumption per week. Chocolate consumption was inversely correlated with the (**A**) mean hearing threshold (average of 0.5, 1, 2, and 4 kHz) and (**B**) mean high-tone hearing threshold (average of 3, 4, and 6 kHz).

Neither the rate of tinnitus nor the rate of tinnitus-related annoyance differed significantly according to chocolate consumption. In addition, subjects who consumed chocolate had 9% and 11% lower odds of tinnitus (adjusted OR = 0.91, 95% CI = 0.74 to 1.10) and tinnitus-related annoyance (adjusted OR = 0.89, 95% CI = 0.67 to 1.18), respectively (Table 2). However, the association between chocolate consumption and tinnitus and tinnitus-related annoyance was not significant.

4. Discussion

To our knowledge, this is the first study to explore the effect of chocolate on hearing loss and tinnitus in a large cohort of middle-aged people. The rate of hearing loss was significantly lower in the subjects who consumed chocolate than in those who did not. Additionally, there was an inverse correlation between the severity of hearing loss and the frequency of chocolate consumption. These results support those of previous animal studies, which reported that various compounds in chocolate protect against hearing loss [20]. Specifically, there was no association between hearing loss and the consumption of sweet products without cocoa. Contrary to our hypothesis, chocolate consumption was not associated with tinnitus or tinnitus-related annoyance.

The cochlea is susceptible to oxidative stress because of its high energy requirements and lack of an adequate collateral blood supply [21]. Thus, the cochlea is likely to be an initial target organ of ischemia. Several clinical risk factors, such as noise exposure and smoking, are associated with oxidative stress in the cochlea [22]. We adjusted for possible confounders but were unable to perform strict matching of confounders due to the high drop-out rate. Our data indicate that chocolate consumption is independently associated with a decreased risk of hearing loss.

Although we could not evaluate the direct effect of chocolate on hearing loss by quantifying cocoa, our results suggest that chocolate may prevent hearing loss by a mechanisms other than the pleasant sensation of chocolate. Those who did not consume chocolate had higher risks of hypertension and dyslipidemia than did those who consumed chocolate. Chocolate exerts antioxidant and anti-inflammatory effects and has been shown to have therapeutic benefits for patients with cardiovascular diseases. Specifically, cocoa, a major ingredient of chocolate, attenuates vascular risks by reducing blood pressure and improving endothelium-dependent vasodilation [23]. In line with this, several investigations demonstrated the causal relationship between vascular risk factors and hearing loss [24,25]. Thus, our results suggest that chocolate decreases the rates of hypertension and dyslipidemia, which enables the preservation of hearing loss. In addition to vascular risk factors, chocolate products provide therapeutic benefits to patients with neurological, intestinal, and neurodegenerative diseases [18].

Chocolate is an important dietary source of flavonoids, a subclass of polyphenols [26]. Polyphenols exert antioxidant and anti-inflammatory effects [20,27]. A recent animal study demonstrated that polyphenols ameliorated oxidative stress inside the cochlea by downregulating the apoptotic signaling pathway [16]. Notably, the level of oxidative stress in the cochlea increased with age; this effect was attenuated by polyphenols. Consistent with this, polyphenols exhibited significant radical scavenging activity in individuals exposed to noise [17] and significantly improved the auditory thresholds in rats [20]. Moreover, chocolate-mediated nitric oxide (NO) release enhances blood circulation in the inner ear and decreases the levels of inflammatory markers; e.g., C-reactive protein, COX-2, and atherogenesis [18,28]. The biological activities of chocolate vary according to the processing strategy used in its production, but chocolate produced using a gentle processing technique may protect against hearing loss, particularly in middle-aged people.

Unexpectedly, chocolate intake was not associated with tinnitus or tinnitus-related annoyance. In humans there is a relationship between the hearing-loss frequency and the tinnitus pitch, suggesting a strong relationship between auditory deafferentation and tinnitus. However, the association between hearing loss and tinnitus may not be straightforward. For example, subjective tinnitus was reported by approximately 19% of individuals with normal hearing in a cohort study in Korea [14], suggesting that subjective tinnitus can develop in the absence of hearing loss. The persistence of tinnitus after cochlear nerve section suggests a central origin [29]. Although the pathophysiological mechanisms of tinnitus are unclear, it may be caused by maladaptive cortical plasticity between auditory and non-auditory regions [11]. Thus, chocolate may improve cochlear rheology, such as microcirculation and vasodilation, but does not significantly ameliorate tinnitus or tinnitus-related annoyance.

The lack of an association between chocolate consumption and tinnitus may be attributable to strict adjustment for comorbidities, e.g., hearing loss, stress, and sleep [30]. Moreover, systemic inflammatory mediators that predispose individuals to tinnitus are likely to be associated with the mechanisms of the protective effect of chocolate [31]. Specifically, chocolate restores imbalances in lipid and glucose levels, which are linked to tinnitus [32,33]. Therefore, other as-yet-unknown effects of chocolate are likely to contribute to the development of tinnitus. This is in line with the borderline significance detected in univariate regression analyses, which was lost after adjustment for confounders.

This study had several limitations that should be noted. First, we could not assess the causality of the relationship between chocolate consumption and hearing loss due to the cross-sectional nature of the study. Moreover, no information on chocolate consumption was available for participants over 65 years of age. Given the protective effect of chocolate on age-related hearing loss, the association might have been significant if older persons had been included. Second, the processing technique used in the production of chocolate affects its total polyphenol content [15]. The data regarding chocolate were obtained using subjective questionnaires; thus, chocolate's functional properties could not be evaluated. The levels of polyphenols, such as epicatechins and catechins, in plasma are correlated with those of markers of antioxidant activity in humans [34]. Third, the type of chocolate, dose, and duration of consumption were not controlled, and milk chocolate and chocolate drinks reportedly do not exert a significant effect on health [15]. To establish causality, a randomized controlled study of the functional properties of chocolate is needed. Lastly, we were unable to compare the effect of chocolate according to laterality. There is some evidence suggesting that the genotype influences the phenotype, particularly in the laterality, of self-reported tinnitus; the discrepancy between unilateral and bilateral tinnitus was more evident in men than women [35]. Although the effect of chocolate on hearing loss and tinnitus was similar irrespective of sex as documented by interaction and multivariate analyses in this study (Supplementary Table S2), this finding awaits further confirmation.

Nonetheless, the present study had several strengths. First, a large cohort from a nationally representative database was analyzed. Second, this is the first study of the relationship between chocolate intake and hearing loss. Our results will enhance nutritional support for patients with hearing loss.

5. Conclusions

Our results suggest that chocolate plays an otoprotective role against hearing loss but not tinnitus in middle-aged people.

Supplementary Materials: The following are available online at http://www.mdpi.com/2072-6643/11/4/746/s1, Table S1: Assessment of chocolate consumption Food-Frequency Questionnaire (FFQ); Table S2: Subgroup analysis regarding effect of chocolate on hearing loss and tinnitus according to sex.

Author Contributions: Conceptualization, M.K.P.; investigation, M.K.P., M.-j.J., S.-Y.L., and G.J.; methodology, M.K.P., M.-j.J., S.-Y.L., G.J., S.-H.O., J.h.L., and M.-W.S.; data analysis, M.-j.J. and G.J.; original draft preparation, M.K.P. and S.-Y.L.; review and editing, M.K.P., M.-j.J., S.-Y.L., G.J., S.-H.O., J.H.L., and M.-W.S.; supervision, S.-H.O., J.h.L., and M.-W.S.; project administration, S.-Y.L. and G.J.

Funding: This research was supported by the Basic Science Research Program through the National Research Foundation of Korea (NRF) and funded by the Ministry of Science, ICT and Future Planning (NRF-2017R1D1A 1B03034832).

Conflicts of Interest: The authors declare no conflicts of interest.

References

1.	Ramsey, T.; Svider, P.F.; Folbe, A.J. Health burden and socioeconomic disparities from hearing loss: A global perspective. *Otol. Neurotol.* **2018**, *39*, 12–16. [CrossRef]

2.	Stevens, G.; Flaxman, S.; Brunskill, E.; Mascarenhas, M.; Mathers, C.D.; Finucane, M. Global Burden of Disease Hearing Loss Expert Group. Global and regional hearing impairment prevalence: An analysis of 42 studies in 29 countries. *Eur. J. Public Health* **2013**, *23*, 146–152. [CrossRef]

3.	Crealey, G.E.; O'Neill, C. Hearing loss, mental well-being and healthcare use: Results from the Health Survey for England (HSE). *J. Public Health* **2018**. [CrossRef]

4.	Leverton, T. Depression in older adults: Hearing loss is an important factor. *BMJ* **2019**, *364*, l160. [CrossRef]

5.	Peracino, A. Hearing loss and dementia in the aging population. *Audiol. Neurootol.* **2014**, *19*, 6–9. [CrossRef] [PubMed]

6.	Lin, F.R.; Metter, E.J.; O'Brien, R.J.; Resnick, S.M.; Zonderman, A.B.; Ferrucci, L. Hearing loss and incident dementia. *Arch Neurol.* **2011**, *68*, 214–220. [CrossRef] [PubMed]

7.	Lee, S.Y.; Nam, D.W.; Koo, J.W.; De Ridder, D.; Vanneste, S.; Song, J.J. No auditory experience, no tinnitus: Lessons from subjects with congenital- and acquired single-sided deafness. *Hear. Res.* **2017**, *354*, 9–15. [CrossRef] [PubMed]

8.	Kim, H.J.; Lee, H.J.; An, S.Y.; Sim, S.; Park, B.; Kim, S.W.; Lee, J.S.; Hong, S.K.; Choi, H.G. Analysis of the prevalence and associated risk factors of tinnitus in adults. *PLoS ONE* **2015**, *10*, e0127578. [CrossRef]

9.	McCormack, A.; Edmondson-Jones, M.; Somerset, S.; Hall, D. A systematic review of the reporting of tinnitus prevalence and severity. *Hear. Res.* **2016**, *337*, 70–79. [CrossRef] [PubMed]

10.	Aazh, H.; Danesh, A.A.; Moore, B.C.J. Parental Mental Health In Childhood As A Risk Factor For Anxiety And Depression Among People Seeking Help For Tinnitus And Hyperacusis. *J. Am. Acad. Audiol.* **2018**. [CrossRef]

11.	Bauer, C.A. Tinnitus. *N. Engl. J. Med.* **2018**, *378*, 1224–1231. [CrossRef] [PubMed]

12.	Tunkel, D.E.; Bauer, C.A.; Sun, G.H.; Rosenfeld, R.M.; Chandrasekhar, S.S.; Cunningham, E.R., Jr.; Archer, S.M.; Blakley, B.W.; Carter, J.M.; Granieri, E.C.; et al. Clinical practice guideline: Tinnitus. *Otolaryngol. Head Neck Surg.* **2014**, *151*, S1–S40. [CrossRef] [PubMed]

13.	Jung, S.Y.; Kim, S.H.; Yeo, S.G. Association of nutritional factors with hearing loss. *Nutrients* **2019**, *11*, 307. [CrossRef]

14.	Lee, D.Y.; Kim, Y.H. Relationship between diet and tinnitus: Korea national health and nutrition examination survey. *Clin. Exp. Otorhinolaryngol.* **2018**, *11*, 158–165. [CrossRef]

15.	Di Mattia, C.D.; Sacchetti, G.; Mastrocola, D.; Serafini, M. From cocoa to chocolate: The impact of processing on in vitro antioxidant activity and the effects of chocolate on antioxidant markers in vivo. *Front. Immunol.* **2017**, *8*, 1207. [CrossRef]

16.	Sanchez-Rodriguez, C.; Cuadrado, E.; Riestra-Ayora, J.; Sanz-Fernandez, R. Polyphenols protect against age-associated apoptosis in female rat cochleae. *Biogerontology* **2018**, *19*, 159–169. [CrossRef] [PubMed]

17. Chang, M.Y.; Han, S.Y.; Shin, H.C.; Byun, J.Y.; Rah, Y.C.; Park, M.K. Protective effect of a purified polyphenolic extract from Ecklonia cava against noise-induced hearing loss: Prevention of temporary threshold shift. *Int. J. Pediatr. Otorhinolaryngol.* **2016**, *87*, 178–184. [CrossRef] [PubMed]

18. Magrone, T.; Russo, M.A.; Jirillo, E. Cocoa and dark chocolate polyphenols: From biology to clinical applications. *Front. Immunol.* **2017**, *8*, 677. [CrossRef] [PubMed]

19. Lee, S.Y.; Jung, G.; Jang, M.J.; Suh, M.W.; Lee, J.H.; Oh, S.H.; Park, M.K. Association of coffee consumption with hearing and tinnitus based on a national population-based survey. *Nutrients* **2018**, *10*, 1429. [CrossRef]

20. Sanz-Fernandez, R.; Sanchez-Rodriguez, C.; Granizo, J.J.; Durio-Calero, E.; Martin-Sanz, E. Accuracy of auditory steady state and auditory brainstem responses to detect the preventive effect of polyphenols on age-related hearing loss in Sprague-Dawley rats. *Eur. Arch. Otorhinolaryngol.* **2016**, *273*, 341–347. [CrossRef]

21. Møller, A.R. *Hearing: Anatomy, Physiology, and Disorders of the Auditory System*; Plural Publishing: San Diego, CA, USA, 2012.

22. Agrawal, Y.; Platz, E.A.; Niparko, J.K. Risk factors for hearing loss in US adults: Data from the national health and nutrition examination survey, 1999 to 2002. *Otol. Neurotol.* **2009**, *30*, 139–145. [CrossRef] [PubMed]

23. Grassi, D.; Necozione, S.; Lippi, C.; Croce, G.; Valeri, L.; Pasqualetti, P.; Desideri, G.; Blumberg, J.B.; Ferri, C. Cocoa reduces blood pressure and insulin resistance and improves endothelium-dependent vasodilation in hypertensives. *Hypertension* **2005**, *46*, 398–405. [CrossRef] [PubMed]

24. Bener, A.; Al-Hamaq, A.; Abdulhadi, K.; Salahaldin, A.H.; Gansan, L. Interaction between diabetes mellitus and hypertension on risk of hearing loss in highly endogamous population. *Diabetes Metab. Syndr.* **2017**, *11*, S45–S51. [CrossRef] [PubMed]

25. Lin, B.M.; Curhan, S.G.; Wang, M.; Eavey, R.; Stankovic, K.M.; Curhan, G.C. Hypertension, diuretic use, and risk of hearing loss. *Am. J. Med.* **2016**, *129*, 416–422. [CrossRef] [PubMed]

26. Gianfredi, V.; Salvatori, T.; Nucci, D.; Villarini, M.; Moretti, M. Can chocolate consumption reduce cardio-cerebrovascular risk? A systematic review and meta-analysis. *Nutrition* **2018**, *46*, 103–114. [CrossRef] [PubMed]

27. Williamson, G. Bioavailability and health effects of cocoa polyphenols. *Inflammopharmacology* **2009**, *17*, 111. [CrossRef]

28. Vauzour, D.; Rodriguez-Mateos, A.; Corona, G.; Oruna-Concha, M.J.; Spencer, J.P. Polyphenols and human health: Prevention of disease and mechanisms of action. *Nutrients* **2010**, *2*, 1106–1131. [CrossRef] [PubMed]

29. Baguley, D.M. Mechanisms of tinnitus. *Br. Med. Bull.* **2002**, *63*, 195–212. [CrossRef]

30. Zheng, Y.; Stiles, L.; Chien, Y.T.; Darlington, C.L.; Smith, P.F. The effects of acute stress-induced sleep disturbance on acoustic trauma-induced tinnitus in rats. *Biomed. Res. Int.* **2014**, *2014*, 724195. [CrossRef] [PubMed]

31. Gibrin, P.C.; Melo, J.J.; Marchiori, L.L. Prevalence of tinnitus complaints and probable association with hearing loss, diabetes mellitus and hypertension in elderly. *Codas* **2013**, *25*, 176–180. [CrossRef]

32. McCormack, A.; Edmondson-Jones, M.; Mellor, D.; Dawes, P.; Munro, K.J.; Moore, D.R.; Fortnum, H. Association of dietary factors with presence and severity of tinnitus in a middle-aged UK population. *PLoS ONE* **2014**, *9*, e114711. [CrossRef] [PubMed]

33. Steinberg, F.M.; Bearden, M.M.; Keen, C.L. Cocoa and chocolate flavonoids: Implications for cardiovascular health. *J. Am. Diet. Assoc.* **2003**, *103*, 215–223. [CrossRef] [PubMed]

34. Li, A.-N.; Li, S.; Zhang, Y.-J.; Xu, X.-R.; Chen, Y.-M.; Li, H.-B. Resources and biological activities of natural polyphenols. *Nutrients* **2014**, *6*, 6020–6047. [CrossRef] [PubMed]

35. Maas, I.L.; Bruggemann, P.; Requena, T.; Bulla, J.; Edvall, N.K.; Hjelmborg, J.V.B.; Szczepek, A.J.; Canlon, B.; Mazurek, B.; Lopez-Escamez, J.A.; et al. Genetic susceptibility to bilateral tinnitus in a Swedish twin cohort. *Genet. Med.* **2017**, *19*, 1007–1012. [CrossRef] [PubMed]

 *nutrients*

Review

Traceability of Functional Volatile Compounds Generated on Inoculated Cocoa Fermentation and Its Potential Health Benefits

Jatziri Mota-Gutierrez [1], **Letricia Barbosa-Pereira** [1,2], **Ilario Ferrocino** [1] and **Luca Cocolin** [1,*]

[1] Department of Agricultural, Forestry, and Food Science, University of Turin, Largo Paolo Braccini 2, 10095 Grugliasco, Torino, Italy; jatziri.motagutierrez@unito.it (J.M.-G.); letricia.barbosapereira@unito.it (L.B.-P.); ilario.ferrocino@unito.it (I.F.)

[2] Department of Analytical Chemistry, Nutrition and Food Science, Faculty of Pharmacy, University of Santiago de Compostela, 15782 Santigo de Compostela, Spain

* Correspondence: lucasimone.cocolin@unito.it; Tel.: +39-011-670-8553

Received: 1 March 2019; Accepted: 17 April 2019; Published: 19 April 2019

Abstract: Microbial communities are responsible for the unique functional properties of chocolate. During microbial growth, several antimicrobial and antioxidant metabolites are produced and can influence human wellbeing. In the last decades, the use of starter cultures in cocoa fermentation has been pushed to improve nutritional value, quality, and the overall product safety. However, it must be noted that unpredictable changes in cocoa flavor have been reported between the different strains from the same species used as a starter, causing a loss of desirable notes and flavors. Thus, the importance of an accurate selection of the starter cultures based on the biogenic effect to complement and optimize chocolate quality has become a major interest for the chocolate industry. This paper aimed to review the microbial communities identified from spontaneous cocoa fermentations and focused on the yeast starter strains used in cocoa beans and their sensorial and flavor profile. The potential compounds that could have health-promoting benefits like limonene, benzaldehyde, 2-phenylethanol, 2-methylbutanal, phenylacetaldehyde, and 2-phenylethyl acetate were also evaluated as their presence remained constant after roasting. Further research is needed to highlight the future perspectives of microbial volatile compounds as biomarkers to warrant food quality and safety.

Keywords: fermentation; functional volatile compounds; starter culture; yeast; roasting; chocolate; cocoa beans

---

1. Introduction

Certainly, people have been changing their food consumption patterns and lifestyle over the last decade [1]. To counteract unhealthy food choices, functional food has emerged as a strategy to increase the consciousness of the relationship between diet and disease/health to consumers. The objective of a successful functional food is to target a specific group of consumers and to meet their health demands without compromising flavor, taste, and color. In this context, the most used bioactive compounds in the food industry include alkaloids, anthocyanins, carotenoids, flavonoids, glucosinolates, isoflavones, phenolic acids, hydrolysate proteins, tannins, and phytochemical terpenes [2].

Volatile organic compounds (VOCs) are organic molecules that include esters, alcohols, aldehydes, ketones, phenols, terpenes, etc. These VOCs are synthesized naturally by a broad number of plants or microorganisms (as secondary metabolites) to enable interactions with their environment. In addition, it has been demonstrated that VOCs provide health benefits to consumers [3]. Health benefits provided by the microbial communities can be either direct or indirect. The difference between these two concepts is the ingestion of a live microorganism (direct) or the ingestion of microbial metabolites (indirect or biogenic effect) [4]. Undeniably, a biogenic effect is commonly observed in fermented foods such as chocolate.

The production of microbial metabolites in cocoa beans begins during fermentation. In this process, microorganisms, encompassing bacteria and yeasts, serve to confer taste, texture, and desirable aromas to the final product. An effective cocoa fermentation develops when a correct microbial succession of yeasts, lactic acid bacteria (LAB), and acetic acid bacteria (AAB) takes place [5,6]. The success of these dynamics is due to the nutrient content of the cocoa pulp that is used as an optimal substrate for the microbial growth, and yeasts are considered the first microorganisms growing at the beginning of the fermentation process, producing ethanol, organic acids and VOCs, that contribute as precursors of chocolate flavor [7]. For those reasons, yeasts have been widely used as starter cultures in cocoa beans with the aim to enrich the sensorial quality of chocolate. However, the modulation of the remarkable complexity of microbial communities in cocoa beans to obtain an optimal flavor fingerprinting as well as understanding the metabolic and regulatory networks concerning the production of secondary metabolites are still not clear. In this context, the present review aims to describe the development of the microbes in fermented cocoa beans, and to evaluate the individual capacities of yeast species to form aroma compounds to enhance flavor perception and nutritional or healthy values. More importantly, it assesses the most frequently identified VOCs during the three different steps of chocolate elaboration, including fermentation, roasting, and the final product, chocolate (Figure 1). It is important to clarify that the VOCs identified in the fermentation and final product were the most frequently identified VOCs in inoculated cocoa beans with yeast, while the most frequently identified VOCs during roasting were assessed from non-inoculated cocoa beans.

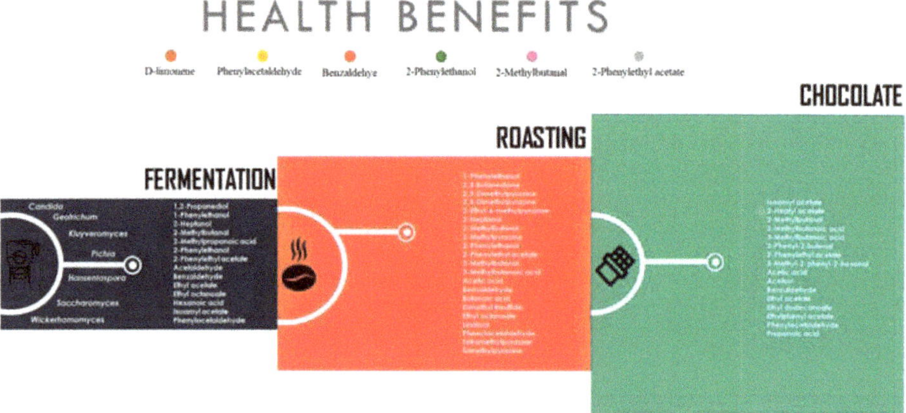

Figure 1. Tracking volatile compounds from chocolate.

2. Microbial Composition of Fermented Cocoa Beans

The fermentation step is considered a key stage that influences the flavor potential of cocoa beans. Existing scientific data shows the complexity of the composition of the microbial population on fermented cocoa beans, that varies depending on: plant variety; environmental conditions; post-harvesting processing; type of fermentation; and agricultural practices [6,8–10]. The bacteria population often present during cocoa fermentation are mainly composed by LAB mostly belonging to the *Lactobacillus* and *Leuconostoc* genera, as well as AAB such as members of the genus *Acetobacter* [11–24]. In addition, some species belonging to *Bacillus* have also been rarely isolated from fermented cocoa beans. Despite the lower complexity of the bacteria population in the cocoa-fermented system, several yeasts have been identified including species belonging to *Candida, Debaromyces, Geotrichum, Hanseniaspora, Kluyveromyces, Pichia, Saccharomyces, Rhodotorula, Saccharomycopsis*, and *Wickerhamomyces* [6,12,14,15,17,18,20–31]. Besides yeast species, filamentous fungi belonging to *Aspergillus, Mucor, Neurospora, Penicillium*, and *Rhizopus* are also often reported [6,23]. Interestingly, there are some discrepancies between the relative abundance of microbial communities reported in fermented cocoa beans from different origins and fermented from different types of fermentations (box, heap). Nonetheless, a recent study suggested that not only the most abundant microbial species could affect the production of organic molecules, but also rare species were involved [6].

2.1. Yeasts Species Used as Starter during Cocoa Fermentation

Some yeasts and fungi are considered a safe source of ingredients and additives for food processing because they have a positive image with consumers [32]. The interest in yeasts as starter cultures has arisen in recent years especially in relation to the addition of *Saccharomyces, Pichia, Kluyveromyces, Candida,* and *Torulaspora* to fermented cocoa beans [6,25,28,30,31,33–37]. Yeasts that are being used to ferment cocoa beans are shown in Table 1. It should be noted that starter cultures used to drive cocoa fermentation processes have been applied only in a few cocoa-producing countries such as Brazil, Malaysia, Indonesia, and Cameroon [6,25,28,30,31,33–36,38]. The importance of standardized cocoa fermentation process has become controversial, considering that the environmental conditions are difficult to control in most of the cocoa-producing countries. Therefore, the choice of selecting cocoa fermenting starters, discriminated based on origin and microbial communities, should be a logical choice from the available options.

Table 1. Functional yeasts used as starters in cocoa fermentation.

Genera/Species	Year	Country	Type of Cocoa Bean	Type of Fermentation	Amount	VOCs F	VOCs C	Sensorial Analysis	Reference
Kluyveromyces marxianus MMIII-41	2008	Brazil	NM	Plastic basket	45 kg	-	-	+	Leal et al., 2008 [28]
Saccharomyces cerevisiae UFLA CA11	2014	Brazil	PH16	Wooden box	60 kg	+	-	-	Ramos et al., 2014 [33]
Saccharomyces cerevisiae UFLA CA11	2014	Brazil	PS1030	Wooden box	60 kg	+	-	-	Ramos et al., 2014 [33]
Saccharomyces cerevisiae UFLA CA11	2014	Brazil	FA13	Wooden box	60 kg	+	-	-	Ramos et al., 2014 [33]
Saccharomyces cerevisiae UFLA CA11	2014	Brazil	PS1319	Wooden box	60 kg	+	-	-	Ramos et al., 2014 [33]
Saccharomyces cerevisiae, Pichia kluyveri and *Hanseniaspora uvarum*	2015	Brazil	PS1319	Wooden box	100 kg	-	-	+	Batista et al., 2015 [30]
Candida sp.	2015	Malaysia	NM	Basket	5 kg	-	-	-	Mahazar et al., 2015 [36]
Saccharomyces cerevisiae H19	2015	Malaysia	NM	Basket	50 kg	-	+	+	Meersman et al., 2016 [37]
Saccharomyces cerevisiae H28	2015	Malaysia	NM	Basket	50 kg	-	+	+	Mersman et al., 2016 [37]
Saccharomyces cerevisiae H37	2015	Malaysia	NM	Basket	50 kg	-	+	+	Mersman et al., 2016 [37]
Saccharomyces cerevisae var. chevalieri	2015	Indonesia	Forastero	Plastic bags	NM	-	-	-	Cempaka et al., 2014 [35]
Saccharomyces cerevisiae	2016	Brazil	CCN51	Wooden box	100 kg	-	+	+	Menezes et al., 2016 [34]
Saccharomyces cerevisiae	2016	Brazil	CEPEC2004	Wooden box	100 kg	-	+	+	Menezes et al., 2016 [34]
Saccharomyces cerevisiae	2016	Brazil	FA13	Wooden box	100 kg	-	+	+	Menezes et al., 2016 [34]
Saccharomyces cerevisiae	2016	Brazil	PS1030	Wooden box	100 kg	-	+	+	Menezes et al., 2016 [34]
Torulaspora delbrueckii	2017	Brazil	PS1319	Wooden box	300 kg	-	+	+	Visintin et al., 2017 [31]
T. delbrueckii	2017	Brazil	SJ02	Wooden box	300 kg	-	+	+	Visintin et al., 2017 [31]
S. cerevisiae and T. delbrueckii	2017	Brazil	PS1319	Wooden box	300 kg	-	+	+	Visintin et al., 2017 [31]
Pichia kudriavzevii LPB06	2017	Brazil	NM	Lab scale	400 g	+	-	-	Pereira et al., 2017 [25]
Pichia kudriavzevii LPB07	2017	Brazil	NM	Lab scale	400 g	+	-	-	Pereira et al., 2017 [25]
Saccharomyces cerevisiae	2018	Cameroon	Forastero	Wooden box	200 kg	+	-	-	Mota-Gutierrez et al., 2018 [6]
Saccharomyces cerevisiae	2018	Cameroon	Forastero	Heap	100 kg	+	-	-	Mota-Gutierrez et al., 2018 [6]
Saccharomyces cerevisiae and T. delbrueckii	2018	Cameroon	Forastero	Wooden box	200 kg	+	-	-	Mota-Gutierrez et al., 2018 [6]
Saccharomyces cerevisiae and T. delbrueckii	2018	Cameroon	Forastero	Heap	100 kg	+	-	-	Mota-Gutierrez et al., 2018 [6]

Abbreviations: NM, not mentioned; F, Fermented cocoa volatile compounds profile; C, Chocolate volatile compounds profile from inoculated cocoa beans; PH16 (Porto hibrido/Sao Jose da Vitoria, Brazil), PS1030 (Porto Seguro/Urucuca, Brazil), FA13 (Angola/Itahuípe Brazil), PS1319 (Bahia, Brazil), CCN51 (Ecuador), CEPEC2004 (Ilhéus/Bahia, Brazil), SJ02 (Bahia, Brazil), Witches broom- resistant varieties; +, Presence; -, Absence; VOCs, volatile organic compounds.

The most frequently used yeast culture in fermented cocoa beans is *Saccharomyces cerevisiae*. This yeast has the capability to assimilate and ferment reducing sugars and citric acid, produce aroma substances and killer-like toxins, and it has a high pectinolytic activity and can prevent microbial pathogen growth [5,7,10,39–42]. Despite the well-known *Saccharomyces*, non-*Saccharomyces* yeast (*Kluyveromyces, Hanseniaspora, Pichia*, and *Torulaspora)* have also shown a relevant pectinolytic activity and increased the aroma complexity in wine [43,44]. However, these species exhibit a lower ethanol yield, and sugar consumption compared to *S. cerevisiae* [45]. Regardless of this characteristic, several studies have used mixed yeast cultures to inoculate fermented cocoa beans (Table 1) [6,31,46]. However, the combination of different yeast species often results in unpredictable compounds produced and/or different microbial communities, which can affect both the chemical and ecological population of fermented cocoa beans. The unpredictable changes, specifically from the ecological point of view, might be explained by the antagonistic ability of some yeast, such as *S. cerevisiae*, to inhibit the growth of non-*Saccharomyces* species (*Hanseniaspora guilliermondii, Torulaspora delbrueckii, Kluveromyces marxianus*, and *Lachancea thermotolerants)* by the production of antimicrobial peptides with a 4.0, 4.5, and 6.0 kDa [47]. Therefore, the selection of starter cultures to produce chocolate plays an important role not only in the modulation of the microbial communities, but rather to achieve optimal sensorial properties, such as cocoa, malty, and fruity flavors [6,31,48]. To this regard, future research is needed to elucidate the variability at the strain level that contributes an added value to the cocoa fermentation [25].

2.2. Quality Evaluation of the Chocolate Produce from Inoculated Cocoa Beans

Contrasting findings on the sensory analysis of chocolate produced from cocoa beans inoculated with yeasts has been recently assessed (Table 1) [28,30,31,34,38]. In detail, the consumer panel from Brazil and Malaysia described chocolates inoculated with *K. marxianus* (Brazil), *S. cerevisiae* (Malayzia and Brazil), *T. delbrueckii*, and a mixed culture of *S. cerevisiae* and *T. delbrueckii* (Brazil) with better desirable notes, flavor attributes, and global acceptability compared with chocolate produced from spontaneous cocoa bean fermentation [28,31,38]. In contrast, coffee and sour attributes with a worse acceptance were described from the chocolate produced in Brazil inoculated with a mixture of three yeast starters (*S. cerevisiae, P. kluyveri*, and *H. uvarum*) during cocoa fermentation [30]. Interestingly, the chocolate produced from different cocoa varieties originated from Brazil inoculated with *S. cerevisiae* during fermentation were clearly discriminated based on the perceptible attributes of each variety [31,34]. However, the lack of the small number of published studies regarding the sensory analysis of chocolate produced from inoculated cocoa beans with yeast species during fermentation from different countries to improve sensorial attributes are not conclusive.

3. Changes in the Nutrient Composition from Fermented to Roasted Cocoa Beans

The transformation of the nutrient content of cocoa beans during fermentation plays an important role in the development of selected attributes in the final product (chocolate). Fats, proteins, and carbohydrates are the main macronutrients found in cocoa seeds (Table 2) [49–53]. Beans also contain amines that are already present in the unfermented dried cocoa and as expected their amount increased after fermentation and decreased after thermal cocoa processing [54,55]. The first step in processing cocoa beans is to ferment amino acids and oligopeptides, and reduce sugars (Table 2). This step is crucial for the development of the quality cocoa flavor that depends on the balance of organic compounds. In general, the biochemical processes involved over the fermentation and roasting of cocoa beans comprise the hydrolysis of sucrose and proteins, oxidation and hydrolysis of phenolic compounds, biosynthesis of alkaloids, amino acids, release of alcohols (that are also oxidized into acetic and lactic acid), and the breakdown of fatty acids [6,49,50,56,57].

Table 2. Nutritional composition of cocoa beans expressed as g/kg.

Source	Origin	Variety	Genetic Material	Carbohydrates				Lipids	Proteins
				Sucrose	Fructose	Glucose	Total Carbohydrates		
Afoakwa et al., 2013 [56]	Ghana	NM	Unfermented				155.00	552.00	216.00
Efraim et al., 2010 [50]	Brazil	Forastero	Unfermented					548.20	238.80
Afoakwa et al., 2013 [56]	Ghana	NM	Fermented				210.00	534.00	188.00
Efraim et al., 2010 [50]	Brazil	Forastero	Fermented					556.00	169.90
Redgwell et al., 2003 [52]	Ghana	NM	Dry cocoa beans	1.58	4.18	0.62			
Redgwell et al., 2003 [52]	Ivory Coast	NM	Dry cocoa beans	1.55	2.80	0.80			
Redgwell et al., 2003 [52]	Ecuador	NM	Dry cocoa beans	4.83	1.72	0.84			
Gu et al., 2013 [53]	Papua New Guinea	Trinitario	Roasted						458.60
Gu et al., 2013 [53]	Indonesia	Trinitario	Roasted						498.50
Gu et al., 2013 [53]	China	Trinitario	Roasted					392.40	
Gu et al., 2013 [53]	China	Trinitario	Roasted					434.40	
Redgwell et al., 2003 [52]	Ghana	NM	Roasted	1.41	0.60	0.05			
Redgwell et al., 2003 [52]	Ivory Coast	NM	Roasted	2.03	0.44	0.05			134.40
Redgwell et al., 2003 [52]	Ecuador	NM	Roasted	6.24	0.61	0.11			181.70

Abbreviations: NM, not mentioned.

3.1. Composition of Volatile Compounds from Cocoa Beans

More than 600 different VOCs have been identified in chocolate flavor. Substances such as aliphatic esters, polyphenols, unsaturated aromatic carbonyls, diketopiperazines, pyrazines, and theobromine are developed, and these compounds provide the characteristic chocolate flavor [49].

3.1.1. VOCs Associated with Inoculated Cocoa Beans

A total of twenty VOCs profiles from inoculated cocoa beans with yeast starters ($n = 10$) and from chocolate produced from inoculated cocoa beans also with yeasts ($n = 10$) has been recently reported from ten different cocoa varieties using eleven different yeast strains over the world (Table 1). The identified VOCs from five different cocoa varieties inoculated with different yeasts during fermentation originated from Cameroon and Brazil were used to create a list of all the identified compounds. The data were treated as dummy variables indicating whether the VOCs were identified and only the most frequently reported were used to increase our knowledge of the probable VOCs formed when cocoa beans are inoculated with yeasts [6,25,33]. As expected, esters, alcohols, and aldehydes were the three major VOCs groups characterized in fermented cocoa beans inoculated with yeast (Figure 2).

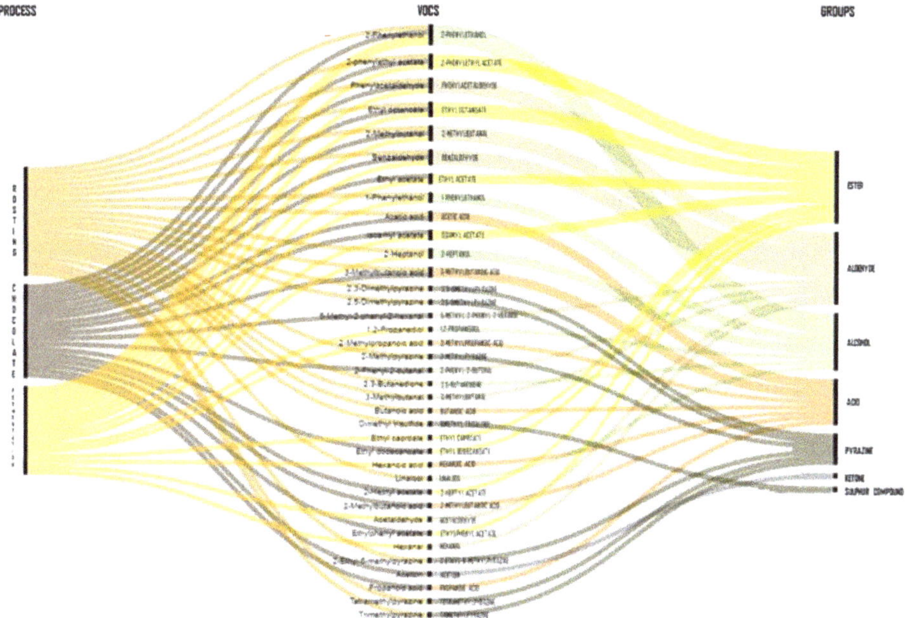

Figure 2. Most identified and abundant volatile compounds in fermented and roasted cocoa beans and chocolate.

In detail, the most predominant VOCs among the three studies were ethyl acetate, benzaldehyde, hexanoic acid, and the key aromatic markers for chocolate (2-heptanol, 2-phenylethanol, 2-phenylethyl acetate, and phenylacetaldehyde) [48], while the most abundant compounds at the end of the fermentation were ethyl octanoate, 1-butanol, 1-pentanol, phenylacetaldehyde, ethyl acetate, isoamyl acetate, limonene, and acetic acid (Table 3) [6,25,33]. It is important to highlight that recently, it has been demonstrated that the volatilome profile of cocoa beans fermented in boxes increased the production of alcohols and esters compared to heap fermentations [6]. However, not only the type of fermentation could influence the volatilome profile. It has been shown that the effect of the yeast starter on different cocoa varieties also influences the relative percentage of VOCs, such as 2-phenylethanol and ethyl acetate [33]. Concerning the dynamics of VOCs, it has been reported that the concentrations of limonene-epoxide and 1-butanol decreased over the fermentation time, while ethyl acetate, limonene, benzaldehyde, benzyl alcohol, acetoin, 3-methyl-1-butanol, acetic acid, and the key-aroma markers (phenylacetaldehyde, 2-heptanol and, 2-phenylethanol) increased (Table 3). The development of VOCs during cocoa fermentation and the appropriate selection of starter culture play a crucial role especially for consumers that follow a raw-food diet. Fermented cocoa beans are a suitable food product for this new trend towards raw foods and desirable attributes should also be met after fermentation [58].

Despite the development of VOCs in inoculated cocoa fermentations, the volatilome profile of chocolate produced from cocoa beans originated from Brazil and Malaysia, inoculated with *S. cerevisiae* and *T. delbrueckii*, and a mixed culture of these two yeasts at the beginning of the fermentation has been assessed (Table 1) [31,34,38]. Interesting observations can be made regarding the most frequently identified VOCs in the fermented cocoa beans inoculated with yeasts and the chocolate produced also from inoculated cocoa beans with yeasts, which support the idea that some VOCs produced during fermentation can remain after processing (Figure 2). Remarkably, acetic acid was the most abundant VOC during fermentation and remained the most abundant VOC in chocolate followed by acetoin and 2-phenylethanol (Table 3) [31,34,37].

Several limitations were noted during the collection of the reported VOCs in both the inoculated fermented cocoa beans and the chocolate produced from different inoculated cocoa beans. The first limitation was related to the incongruency of the total number of VOCs and terpenoids reported, whereas some studies have not reported any terpenoids and the total number of VOCs identified vary from 34 to 72 compounds [6,25,31,33,34,37]. Second, studies that identified VOCs in inoculated fermented cocoa beans and chocolate are limited. Although there are no studies that have been tracking the presence of VOCs over the whole chocolate process, this review provides us with an idea of which VOCs are only formed during the fermentation of cocoa beans inoculated with yeast species and could probably remain in the end product. It is worth noting that future research in the identification of VOCs may further increase our knowledge on the role of yeasts, particularly if they increase the production of esters, aldehydes, and terpenoids. This could heighten the positive impacts of yeasts during cocoa fermentation.

Table 3. Concentration ranges (µg/kg) of volatile compounds of raw, fermented, and roasted cocoa beans and chocolate.

Volatile Aroma Compounds	Raw Beans [6,33]			End of Fermentation [6,33]			Roasting [59–64]			Chocolate [31,34,37]		
Aldehydes												
2-Methylbutanal	0.70	-	1.24	0.49	-	1.46	111.00	-	4500.00	0.21	-	38.30
Acetaldehyde	0.02	-	0.85	0.00	-	0.18	285.00	-	285.00	0.60	-	41.70
Benzaldehyde	0.21	-	0.55	0.59	-	0.75	28.00	-	895.00	2.77	-	53.50
Decanal	0.03	-	0.06	0.02	-	0.04				1.00	-	1.00
Dodecanal	0.00	-	0.02	0.00	-	0.01				0.10	-	0.50
Furfural	0.00	-	0.24	0.00	-	0.25	26.00	-	87.00			
Hexanal	0.02	-	3.65	0.01	-	6.55						
Nonanal	0.14	-	0.19	0.09	-	0.14	46.00	-	46.00	0.05	-	1.52
Phenylacetaldehyde	4.06	-	6.09	3.49	-	12.37	60.00	-	5500.00	0.06	-	0.15
(E)-2-Undecenal	0.00	-	0.01	0.00	-	0.05						
2-Phenyl-2-butenal	0.00	-	0.00	0.00	-	0.05						
Alcohols												
(Z)-3-Hexen-1-ol	0.00	-	37.65	0.01	-	0.02						
1,2-Propanediol	0.00	-	0.00	0.07	-	0.35				1.10	-	1.70
1-Butanol	3.20	-	33.26	0.91	-	10.50						
1-Decanol	0.01	-	0.01	0.01	-	0.01						
1-Dodecanol	0.02	-	0.17	0.05	-	0.38						
1-Heptadecanol	0.03	-	0.10	0.06	-	0.21						
1-Heptanol	0.04	-	0.05	0.00	-	0.00				0.03	-	0.05
1-Hexanol	0.21	-	0.43	0.15	-	0.22						
1-Octanol	0.06	-	0.09	0.09	-	0.17						
1-Octen-3-ol	0.03	-	0.05	0.00	-	0.18						
1-Pentanol	0.13	-	0.83	0.07	-	0.14						
1-Phenylethanol	0.29	-	0.55	0.22	-	0.34						
1-Propanol	0.00	-	1.01	0.02	-	1.02						
2,3-Butanediol	0.00	-	9.60	0.00	-	2.07	62.00	-	356.00	35.40	-	65.35
2-Ethyl-1-hexanol	0.31	-	0.49	0.14	-	0.34				0.37	-	0.71
Furfuryl alcohol	0.00	-	0.00	0.00	-	10.71				0.49	-	0.90
2-Heptanol	0.35	-	0.54	0.00	-	8.97	32.00	-	1070.00	0.00	-	0.00
2-Hexanol	0.42	-	1.13	0.07	-	0.18						
2-Methyl-1-butanol	0.00	-	3.36	0.00	-	2.75				0.10	-	3.70
2-Methyl-1-propanol	0.00	-	0.22	0.00	-	10.33						

Table 3. *Cont.*

Volatile Aroma Compounds	Raw Beans [6,33]			End of Fermentation [6,33]			Roasting [59–64]			Chocolate [31,34,37]		
2-Nonanol	0.04	–	0.06	0.16	–	0.78				1.00	–	1.00
2-Pentanol	25.70	–	47.70	1.52	–	4.32				0.47	–	0.47
2-Phenylethanol	0.31	–	0.55	0.00	–	6.87	63.00	–	7500.00	3.60	–	142.00
3-Methyl-1-butanol	1.09	–	1.30	0.88	–	1.86	27.00	–	238.00	0.10	–	27.10
3-Methyl-1-pentanol	0.63	–	7.64	0.00	–	3.08						
Benzyl alcohol	0.04	–	0.05	0.03	–	0.07	104.00	–	104.00	0.20	–	0.23
Ethanol	2.17	–	3.89	1.25	–	3.81	124.00	–	124.00	4.06	–	6.71
Isobutanol	0.10	–	1.54	0.06	–	0.14						
Methanol	0.00	–	15.74	0.00	–	24.41	9068.00	–	9068.00			
(E)-3-Hexen-1-ol				0.00	–	43.75						
Acids												
2-Methylpropanoic acid	0.00	–	0.00	0.00	–	0.60	79.00	–	79.00	7.70	–	48.80
3-Methylbutanoic acid	0.05	–	0.10	3.51	–	9.20	86.00	–	9700.00	0.10	–	48.10
Acetic acid	0.68	–	1.30	4.33	–	28.40	5.60	–	330000.00	734.00	–	2555.70
Butanoic acid	0.00	–	7.36	0.00	–	13.10	21.00	–	570.00	1.30	–	2555.70
Decanoic acid	0.00	–	1.32	0.00	–	0.00						
Heptanoic acid	0.00	–	9.79	0.00	–	0.09	31.00	–	31.00			
Hexanoic acid	0.16	–	2.71	0.00	–	0.50	116.00	–	116.00	0.40	–	1.47
Nonanoic acid	0.00	–	10.28	0.00	–	0.00				0.10	–	0.10
Octanoic acid	0.03	–	0.06	0.11	–	0.27						
Ketones												
2-Heptanone	0.66	–	1.28	0.88	–	3.61	85.00	–	140.00	1.10	–	5.20
2-Pentanone	1.55	–	9.73	1.01	–	2.23						
2-Undecanone	0.04	–	0.05	0.00	–	0.03				1.00	–	1.00
Acetoin	0.38	–	0.47	1.23	–	5.98	14.00	–	1143.00	1.99	–	505.20
Acetophenone	1.17	–	3.06	0.81	–	2.31	14.00	–	225.00			
Esters												
1,2-Propanediol diacetate	6.50	–	8.11	1.21	–	2.53						
Isoamyl acetate	0.00	–	56.50	0.00	–	17.65						
2,3-Butanediol diacetate	0.15	–	0.30	0.03	–	1.20						
2-Pentanol acetate	1.42	–	2.55	1.78	–	3.93						
Diethyl malate	0.00	–	0.00	0.18	–	0.44						
Diethyl succinate	0.06	–	11.65	0.00	–	0.93						
Ethyl acetate	0.00	–	18.45	0.00	–	22.82	66.00	–	66.00	1.40	–	28.90

Table 3. *Cont.*

Volatile Aroma Compounds	Raw Beans [6,33]			End of Fermentation [6,33]			Roasting [59–64]			Chocolate [31,34,37]		
Ethyl benzoate	0.00	-	0.02	0.13	-	0.24	2.10	-	2.10			
Ethyl butanoate	0.26	-	3.99	0.06	-	4.18						
Ethyl caproate	0.17	-	0.22	0.43	-	0.94						
Ethyl dodecanoate	0.00	-	1.64	0.38	-	1.77	24.00	-	24.00			
Ethyl octanoate	0.00	-	0.03	0.00	-	74.29	3.30	-	143.00	0.09	-	19.10
Ethyl pyruvate	0.00	-	0.88	1.78	-	20.88						
Ethyl-o-toluate	0.00	-	0.00	0.33	-	0.62						
Furfuryl acetate	1.59	-	27.04	0.13	-	3.57						
Hexyl acetate	0.00	-	0.01	0.00	-	0.04						
Isoamyl benzoate	0.10	-	0.19	0.02	-	0.56						
Isobutyl acetate	0.14	-	1.97	0.06	-	1.98						
Methyl octanoate	0.10	-	0.15	0.00	-	0.00						
Mono-ethyl succinate	0.00	-	1.98	0.00	-	0.52						
Hexyl butanoate	0.00	-	0.05	0.00	-	0.00						
Phenyl acetate	0.00	-	0.54	0.00	-	0.14						
α-Phenylethyl acetate	0.00	-	0.45	0.17	-	0.89	34.00	-	930.00	2.60	-	37.10
Propyl acetate	0.00	-	0.09	0.00	-	1.34						
β-Phenylethyl acetate	0.03	-	0.12	0.73	-	1.69						
Terpenes												
Carveol	0.01	-	0.05	0.00	-	0.00						
(Z)-Linalool oxide pyranoid	0.10	-	0.15	0.06	-	0.17						
(Z)-Linalool oxide furanoid	0.03	-	0.10	0.00	-	0.10	21.00	-	21.00			
Nerylacetone	0.02	-	0.04	0.00	-	0.00						
Limonene	6.65	-	12.37	6.43	-	30.60						
Geraniol	0.00	-	0.31	0.00	-	0.00						
Limonene epoxide	0.29	-	0.90	0.00	-	0.02						
Sabinene	0.05	-	0.17	0.05	-	0.30						
α-Caryophyllene	0.08	-	0.09	0.07	-	0.18						
α-Citral	0.03	-	0.10	0.02	-	0.15						
α-Limonene diepoxide	0.00	-	0.02	0.00	-	0.02						
β-Caryophyllene	0.01	-	0.02	0.01	-	0.03						
β-Citronellol	0.00	-	0.00	0.00	-	0.41						
β-Myrcene	1.98	-	2.32	0.96	-	3.14	66.00	-	66.00			
(E)-β-ocimene	0.08	-	0.34	0.06	-	0.51						

Table 3. *Cont.*

Volatile Aroma Compounds	Raw Beans [6,33]			End of Fermentation [6,33]			Roasting [59–64]			Chocolate [31,34,37]
Lactones										
Δ-Decalactone				0.00	-	0.20				
Other compounds										
1,1-Diethoxyethane	0.06	-	21.65	0.12	-	5.83				
o-Guaiacol	0.00	-	0.01	0.02	-	0.62	230.00	-	230.00	
Phenol	0.02	-	0.03	0.02	-	0.37	7.00	-	7.00	
trans-Methyl dihydrojasmonate	0.02	-	0.04	0.02	-	0.04				

Values are expressed as concentration ranges (µg/kg). Not statistical analysis was applied due to unbalanced sample size. Different color showed decrease (light blue) or increase (light green) of selected VOC concentrations.

3.1.2. Dynamics of VOCs during Roasting

Roasting of cocoa beans is used to diminish moisture and acidity by reducing concentrations of volatile acids such as acetic acid and water [49]. However, the degree of this reduction depends on the time/temperature conditions used [59]. Several chemical reactions such as Maillard and Strecker reactions play an important role during roasting to develop the characteristic aroma and flavor of chocolate [55]. These reactions reduce sugars and amino acids to produce mainly heterocyclic groups such as aldehydes and pyrazines. Indeed, roasting has been shown to be a more effective amine generator than fermentation and it has been observed that the fermentation process supplied precursors for Strecker aldehyde formation. Overall, these reactions also depend on temperature and pH, in which higher temperatures increase amine generation [49,55,59].

Enormous progress is currently being made in the identification of VOCs during roasting [48,59–65]. In detail, a total of 243 VOCs has recently been reported from three different cocoa varieties originating from ten different countries (Table 4). The most frequently identified and abundant VOCs in roasted cocoa beans are acetic acid, 3-methylbutanoic acid, benzaldehyde, and the key aromatic compounds (2-heptanol, 2-phenylethanol, phenylacetaldehyde, and 2-methylbutanal, Table 3) [59–64]. Interestingly, we observed that the key aromatic compounds (2-phenylethyl acetate, phenylacetaldehyde, and 2-heptanol), benzaldehyde, acetic acid, as well as trimethylpirazine and 3-methylbutanal, formed during inoculated fermentations, were still present after the roasting process [48,59–65].

Table 4. Overview of the volatile organic compounds of roasted cocoa beans from different origins under different roasting conditions.

Source	Country	Variety	Equipment	Roasting conditions	
				Temperature (°C)	Time (min)
Bonhevi et al., 2005 [60]	Ghana, Cameroon, Ivory Coast, Brazil and Ecuador	NM	GC-MS	130	48
Ramli et al., 2006 [59]	Malaysia	NM	GC-MSD	150	30
Frauendorfer and Shieberle, 2008 [63]	Grenada	Criollo	HRGC-MS	95	14
Huang and Barringer, 2011 [61]	Ecuador	NM	SIFT-MS	150	30
Van Durme et al., 2016 [62]	Ghana and Tanzania	NM	HS-SPME-GC-MS	150	30
Magagna et al., 2018 [65]	Mexico	NM	HS-SPME-GCxGC-MS	100–130	20–40
Tan and Kerr, 2018 [64]	United States of America	Forastero	GC-MS and ANN-based-e-nose	135	0–40
Magagna et al., 2017 [48]	Ecuador and Mexico	Trinitario hybrids	GCxGC-MS, GCx2GC-MS/FID	nm	nm

Abbreviations: nm: Not mentioned, SIFT-MS: Selected ion flow tube-mass spectrometry, GC-MS: Gas chromatograph-mass spectrometer, ANN: Artificial neural network, GC-MSD: Gas chromatography-Mass selective detector, HRGC-MS: High-resolution gas chromatography-mass spectrometry, HS-SPME: Head-space solid-phase micro-extraction, FID: Flame ionization detector.

In terms of the concentration changes in VOCs during roasting, it has been shown that the key odorants formed during fermentation 2-heptanol, 2-phenylethyl acetate, 2-phenylethanol, butanoic acid, and ethyl 2-methylbutanoate remained nearly constant during the roasting process, while the formation of pyrazines, a by-product of Maillard reaction, mainly occurs during roasting [48,63,64]. It should also be noted that the loss and development of limonene, ethyl acetate, benzaldehyde, and 2-methylbutanal after thermal processing remains unclear.

4. Synthesis of VOCs by Fungal Communities and their Potential Health Benefits

Research on microbial flavor generation has tremendously increased over the last two decades and special attention was given to understand the microbial processes or microbial strategies to produce flavor compounds [25,27,66–69]. Interestingly, VOCs have been traditionally used and added to food products more for pleasure and consumers' acceptability than for nutritional reasons. However, microorganisms and their metabolites produced have been also exploited for their tremendous potential to provide health benefits in humans. In fact, it has been recently pointed out the potential health benefits contributed mainly by VOCs in plant foods [3].

Volatile organic compounds can be synthesized by biological process (microorganisms during fermentation), chemical reactions (synthetic and semi-synthetic) or plant extracts, depending on the type of compound that needs to be synthesized. Concerning biological processes, the microbial metabolism includes the transformation of natural precursor (sugars, organic acids, amino acids, and fatty acids) to a wide range of flavor molecules such as aliphatics, aromatics, terpenes, lactones, O-heterocycles, and S- and N-containing compounds [68]. This review focuses on the formation of six VOCs (2-phenylethanol, phenylacetaldehyde, 2-methylbutanal, benzaldehyde, limonene, and 2-phenylethyl acetate) that are formed during inoculated cocoa fermentations with yeasts, and remain present after roasting, and the potential health benefits of these VOCs. Overall, a total of 36 fungi have been described as producers of the selected VOCs, as shown in Figure 3. In detail, *S. cerevisiae*, *P. anomala*, *H. uvarum*, *H. guilliermondii*, and *Galactomyces geotrichum* have been demonstrated to produce the majority of the selected VOCs, while a high variation between species of *Candida* and *Pichia* has been reported (Figure 3).

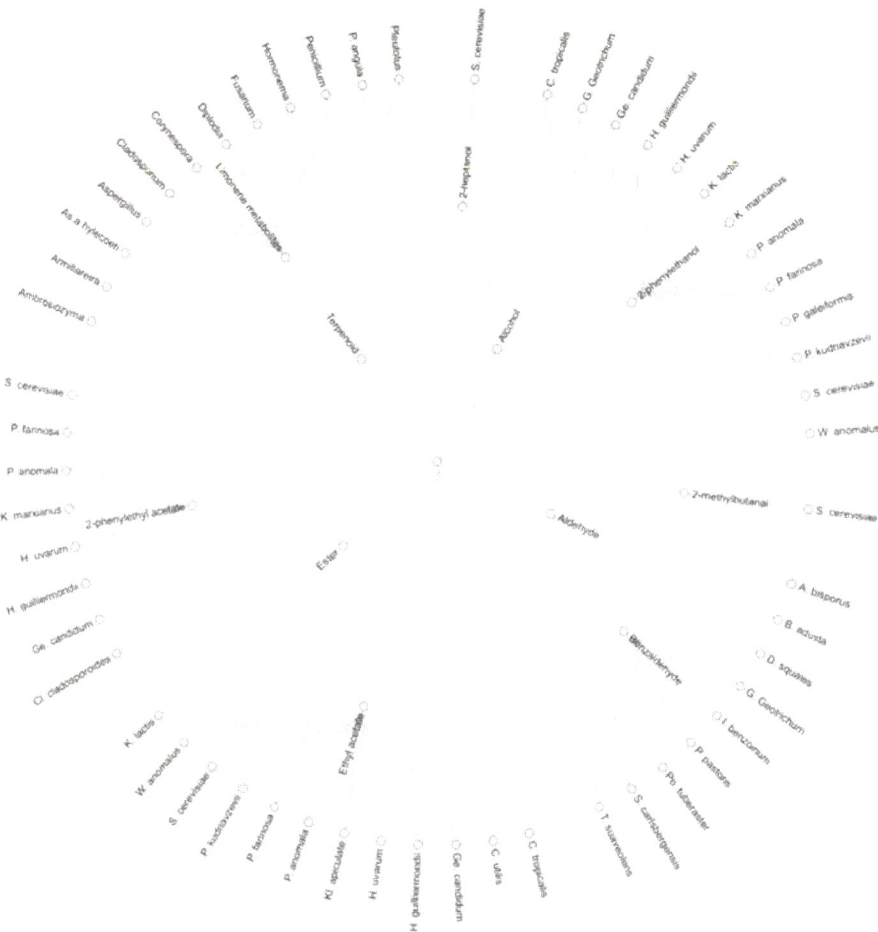

Figure 3. Yeast producer of selected key aromatic compounds in cocoa beans. Abbreviations: S: *Saccharomyces*, P: *Pichia*, C: *Candida*, G: *Galactomyces*, Ge: *Geotrichum*, H: *Hanseniaspora*, K: *Kluyveromyces*, W: *Wickerhamomyces*, A: *Agaricus*, B: *Bjerkandera*, D: *Dichomitus*, I: *Ischnoderma*, Po: *Polyporus*, T: *Trametes*. Kl: *Kloeckera*, Cl: *Cladosporium*, As: *Ascoide*.

Fusel alcohols are generally synthetized by the yeast's Ehrlich pathway by the conversion of reducing sugars; this pathway contains a three-enzyme cascade that converts valine, leucine, and isoleucine into their corresponding alcohols [68]. The microbial production of L-phenylalanine to 2-phenylethanol (a rose like odor) involves the transamination of the amino acid to phenylpyruvate, decarboxylation to phenylacetaldehyde, and reduction to alcohol by yeast species [27,66,70–74], while the synthesis of secondary alcohols, such as 2-heptanol can be obtained from 2-heptanone (Table 5) [75]. Regarding the potential health benefits, 2-phenylethanol has been demonstrated to inhibit the growth of Gram-negative bacteria and filamentous fungi [76,77].

Table 5. Summary table of the yeast producer of selected key aromatic compounds in cocoa beans.

Group	VOCs	Microorganism	Reference
	2-heptanol	*Saccharomyces cerevisiae*	Cappaert and Laroche, 2004 [75]
Alcohol	2-phenylethanol	*Candida tropicalis*	Koné et al., 2016 [27]
		Galactomyces geotrichum	Koné et al., 2016 [27]
		Geotrichum candidum	Janssens et al., 1992 [66]
		Hanseniaspora guilliermondii	Moreira et al., 2005 [73]
		Hanseniaspora uvarum	Moreira et al., 2005 [73]
		Kluyveromyces lactis	Janssens et al., 1992 [66], Fabre et al., 1997 [74]
		Kluyveromyces marxianus	Janssens et al., 1992 [66], Whittmann et al., 2002 [72], Etschman et al., 2005 [71], Fabre et al., 1997 [74]
		Pichia anomala	Janssens et al., 1992 [66]
		Pichia farinosa	Janssens et al., 1992 [66]
		Pichia galeiformis	Koné et al., 2016 [27]
		Pichia kudriavzevii	Koné et al., 2016 [27]
		Saccharomyces cerevisiae	Kim et al., 2014 [70], Koné et al., 2016 [27], Schwan and Wheals, 2004 [7], Moreira et al., 2005 [73], Fabre et al., 1997 [74]
		Wickerhamomyces anomalus	Koné et al., 2016 [27]
Aldehydes	2-methylbutanal	*Saccharomyces cerevisiae*	Janssens et al., 1992 [66], Larroy et al., 2002 [78]
	Benzaldehyde	*Agaricus bisporus*	Janssens et al., 1992 [66]
		Bjerkandera adusta	Lapadatescu et al., 1997 [79]
		Dichomitus squales	Lapadatescu et al., 1997 [79]
		Galactomyces geotrichum	Koné et al., 2016 [27]
		Ischnoderma benzoinum	Lapadatescu et al., 1997 [79]
		Pichia pastoris	Berger, 2007 [68]
		Polyporus tuberaster	Kawabe and Morita, 1994 [80]
		Saccharomyces carlsbergensis	Pal et al., 2009 [81]
	Phenylacetaldehyde	*Kluyveromyces marxianus*	Etschman et al., 2005 [71]
		Acetobacter	Berger, 2007 [68]
Ester	Ethyl acetate	*Candida tropicalis*	Koné et al., 2016 [27]
		Candida utilis	Janssens et al., 1992 [66]
		Geotrichum candidum	Janssens et al., 1992 [66]
		Hanseniaspora guilliermondii	Rojas et al., 2001 [82]
		Hanseniaspora uvarum	Rojas et al., 2001 [82]
		Kloeckera apiculate	Schwan and Wheals, 2004 [7]
		Pichia anomala	Janssens et al., 1992 [66], Rojas et al., 2001 [82]
		Pichia farinosa	Janssens et al., 1992 [66]
		Pichia kudriavzevii	Koné et al., 2016 [27], Pereira et al., 2017 [25]
		Saccharomyces cerevisiae	Janssens et al., 1992 [66], Koné et al., 2016 [27], Rojas et al., 2001 [82], Schwan and Wheals, 2004 [7]
		Wickerhamomyces anomalus	Koné et al., 2016 [27]
		Kluyveromyces lactis	Van Laere et al., 2008 [83]
	2-Phenylethyl acetate	*Cladosporium cladosporoides*	Janssens et al., 1992 [66]
		Geotrichum candidum	Janssens et al., 1992 [66]
		Hanseniaspora guilliermondii	Rojas et al., 2001 [82], Moreira et al., 2005 [73]
		Hanseniaspora uvarum	Rojas et al., 2001 [82]
		Kluyveromyces marxianus	Janssens et al., 1992 [66], Whittmann et al., 2002 [72], Etschman et al., 2005 [71]
		Pichia anomala	Janssens et al., 1992 [66], Rojas et al., 2001 [82]
		Pichia farinosa	Janssens et al., 1992 [66]
		Saccharomyces cerevisiae	Kone et al., 2016 [27], Rojas et al., 2001 [82]

Table 5. *Cont.*

Group	VOCs	Microorganism	Reference
	Limonene	*Ascoidea hylecoeti*	Janssens et al., 1992 [66]
	Limonene metabolites (terpineol, verbenol)	*Armillareira, Aspergillus Cladosporium*	Duetz et al., 2003 [84], Janssens et al., 1992 [66]
Terpenoid	Limonene metabolites (limonene-1,2-epoxide)	*Corynespora Diplodia*	Duetz et al., 2003 [84]
	Limonene metabolites (verbenone)	*Hormonema*	Berger, 2007 [68]
	Limonene metabolites (carvone, carveol)	*Penicillium Pleutotus*	Janssens et al., 1992 [66], Duetz et al., 2003 [84]
	Limonene metabolites	*Pichia angula Ambrosiozyma Fusarium*	Janssens et al., 1992 [66] Berger, 2007 [68]

Besides 2-phenylethanol, other VOCs such as benzaldehyde and its derivates have been used as preservatives [85,86]. However, this compound is also able to induce antitumor activity in human cells [85–88] and antioxidant activity [53,89]. Concerning the conversion of benzyl alcohol or L-phenylalanine into benzaldehyde, this conversion has been attributed not only to yeasts but also to the basidiomycetes' activity (Table 5) [27,66,79–81]. In general, aldehydes can be produced by the oxidation of alcohols such as 2-methylbutanol, 3-methylbutanol, and 2-methyl-1-propanol derived from short-chain aliphatic aldehydes such as acetaldehyde, 2-methyl-1-propanal, 2-methylbutanal, and 3-methylbutanal efficiently produced by the metabolism of yeast [66,78].

Regarding the biosynthesis and conversion of monoterpenes, it has been associated with the basidiomycetes' metabolism. Limonene is produced by plants as a defense for pathogens, and this transforms into other monoterpenoids such as carvone, terpineol, perillyl alcohol, limonene epoxide, and verbenone, which can be associated with the activity of several fungal species (Table 5) [66,68,84]. Interestingly, recent *in vivo* and *in vitro* studies have reported anticarcinogenic and antinociceptive activity of limonene [90–96], and this compound has also been used as a preservative [97].

Last but not least, the well-known ester 2-phenylethyl acetate is recognized for its antimicrobial activity [98]. In general, the biotransformation of esters includes a more complex catabolic reaction, and it comprises the esterification of amino acids or short-chain aliphatic fatty acid and terpenyl alcohol into the desired flavor ester. The transformation of 2-phenylethyl acetate is usually metabolized from amino acids, such as phenylalanine and/or phenylpyruvic acid also from yeast species (Table 5) [66,73,82]. Besides 2-phenylethyl acetate, ethyl acetate is formed from the esterification of leucine, isoleucine or valine, and a natural aliphatic alcohol has been attributed to the activity of yeast (Table 5) [7,25,27,73,82,83].

Overall, the potential health effect of the selected VOCs synthesized by chemical reactions or biological processes have been linked to prevent or delay diseases or the growth of undesirable microorganisms. In summary, it has been reported from *in vivo* and *in vitro* studies the anticarcinogenic and the antinociceptive activity of limonene, [90–96], the antitumor activity of benzaldehyde [85–88], and antioxidant activity of benzaldehyde and its derivates [99,100]. In addition, 2-phenylethanol [76,77,101], 2-phenylethyl acetate [98], limonene [97], benzaldehydes, and derivates [89,99,100,102] have been widely used as preservatives.

While most studies have focused on describing the capacity of VOCs to prevent, slow or inhibit the growth of microorganisms, tumors, or cells to provide health benefits, recent literature has demonstrated the capacity of these compounds to stimulate communication with the limbic system of the brain via neurons through oral routes and olfactory receptors in the nose, which changed mood and emotions by creating a sedative effect for the reduction of stress and anxiety, and finally by reducing the pain perception [103]. On the other hand, dysfunction of the chemosensory activities were highly related to differences in dietary behaviors, including loss of appetite, unintended weight loss, malnutrition, and well-known psychiatric and neurological disorders [104–109]. More important is the fact that this loss has been reported to affect the general population and it remains undiagnosed in some patients [104,110]. In this regard, 2-phenylethanol has been used to counteract the olfactory dysfunction due to multiple etiologies [111–117]. However, the mechanism of action of the improvements of the smell progresses and the association of chemosensory function with dietary and health outcomes remains unclear. There is no doubt that individuals with this dysfunction, highly observed in neurological diseases such as Parkinson's, are more likely to experience a hazardous event and are the major concern for public health. Considering the positive effect of the single compounds, also produced by microbial communities during cocoa fermentation, this review hypothesizes that the consumption of chocolate produced from inoculated cocoa beans with yeasts could provide a positive health effect to consumers. However, more comprehensive studies are required to confirm the potential effect of VOCs from chocolate in human health.

In terms of international legal regulations, according to the Join FAO/WHO expert committee on food additives, all the VOCs proposed in this review are categorized as flavoring agents and do not represent a safety concern since they are predictably metabolized efficiently into innocuous products and their estimated daily intake are below the threshold for daily human intake [118].

5. Conclusions

Microbial communities in, on, and around our food are essential for exploring the interaction between the food system and its inter-connectedness with human health. Tracking the production of functional compounds produced by microbes will serve to improve the formation of desirable compounds. Future perspectives on the selection of the best candidate starter cultures possessing genes coding for oral usage to acquire desirable compounds and the mechanism underlying flavor perception linked to nutritional or health values need to be assessed. The findings of the present review and future analyses of VOCs may help to inform researchers, policy makers, the chocolate industry, and the general public to explore yeasts as proper producers of important VOCs to improve quality and health.

Author Contributions: Conceptualization, J.M.-G., L.C.; Validation, L.C., I.F., L.B.-P.; Investigation, J.M.-G., L.B.-P.; Writing-original Draft Preparation, J.M.-G.; Writing; Review and Editing, L.C., I.F., L.B.-P.; Supervision, L.C.

Funding: This research received no external funding.

Conflicts of Interest: The authors declare no conflict of interest.

References

1. Schmieder, R.; Edwards, R. Quality control and preprocessing of metagenomic datasets. *Bioinformatics* **2011**, *27*, 863–864. [CrossRef]
2. Sun-Waterhouse, D.; Wadhwa, S.S. Industry-relevant approaches for minimising the bitterness of bioactive compounds in functional foods: A Review. *Food Bioprocess Technol.* **2013**, *6*, 607–627. [CrossRef]
3. Ayseli, M.T.; Ipek Ayseli, Y. Flavors of the future: Health benefits of flavor precursors and volatile compounds in plant foods. *Trends Food Sci. Technol.* **2016**, *48*, 69–77. [CrossRef]
4. Gobbetti, M.; Di Cagno, R.; de Angelis, M. Functional microorganisms for functional food quality. *Crit. Rev. Food Sci. Nutr.* **2010**, *50*, 716–727. [CrossRef]
5. Nielsen, D.S.; Teniola, O.D.; Ban-Koffi, L.; Owusu, M.; Andersson, T.S.; Holzapfel, W.H. The microbiology of Ghanaian cocoa fermentations analysed using culture-dependent and culture-independent methods. *Int. J. Food Microbiol.* **2007**, *114*, 168–186. [CrossRef]
6. Mota-Gutierrez, J.; Botta, C.; Ferrocino, I.; Giordano, M.; Bertolino, M.; Dolci, P.; Cannoni, M.; Cocolin, L. Dynamics and biodiversity of bacterial and yeast communities during the fermentation of cocoa beans. *Appl. Environ. Microbiol.* **2018**, AEM.01164-18. [CrossRef]
7. Schwan, R.F.; Wheals, A.E. The microbiology of cocoa fermentation and its role in chocolate quality. *Crit. Rev. Food Sci. Nutr.* **2004**, *44*, 205–221. [CrossRef]

8. Camu, N.; De Winter, T.; Verbrugghe, K.; Cleenwerck, I.; Vandamme, P.; Takrama, J.S.; Vancanneyt, M.; De Vuyst, L. Dynamics and biodiversity of populations of lactic acid bacteria and acetic acid bacteria involved in spontaneous heap fermentation of cocoa beans in Ghana. *Appl. Environ. Microbiol.* **2007**, *73*, 1809–1824. [CrossRef]

9. Camu, N.; De Winter, T.; Van Schoor, A.; De Bruyne, K.; Vandamme, P.; Takrama, J.S.; Addo, S.K.; De Vuyst, L. Influence of turning and environmental contamination on the dynamics of populations of lactic acid and acetic acid bacteria involved in spontaneous cocoa bean heap fermentation in Ghana. *Appl. Environ. Microbiol.* **2008**, *74*, 86–98. [CrossRef]

10. Jespersen, L.; Nielsen, D.S.; Hønholt, S.; Jakobsen, M. Occurrence and diversity of yeasts involved in fermentation of West African cocoa beans. *FEMS Yeast Res.* **2005**, *5*, 441–453. [CrossRef]

11. Papalexandratou, Z.; Vrancken, G.; De Bruyne, K.; Vandamme, P.; De Vuyst, L. Spontaneous organic cocoa bean box fermentations in Brazil are characterized by a restricted species diversity of lactic acid bacteria and acetic acid bacteria. *Food Microbiol.* **2011**, *28*, 1326–1338. [CrossRef] [PubMed]

12. Papalexandratou, Z.; De Vuyst, L. Assessment of the yeast species composition of cocoa bean fermentations in different cocoa-producing regions using denaturing gradient gel electrophoresis. *FEMS Yeast Res.* **2011**, *11*, 564–574. [CrossRef]

13. Kostinek, M.; Ban-Koffi, L.; Ottah-Atikpo, M.; Teniola, D.; Schillinger, U.; Holzapfel, W.H.; Franz, C.M.A.P. Diversity of predominant lactic acid bacteria associated with cocoa fermentation in Nigeria. *Curr. Microbiol.* **2008**, *56*, 306–314. [CrossRef] [PubMed]

14. Miescher Schwenninger, S.; Freimüller Leischtfeld, S.; Gantenbein-Demarchi, C. High-throughput identification of the microbial biodiversity of cocoa bean fermentation by MALDI-TOF MS. *Lett. Appl. Microbiol.* **2016**, *63*, 347–355. [CrossRef]

15. Ho, V.T.T.; Zhao, J.; Fleet, G. The effect of lactic acid bacteria on cocoa bean fermentation. *Int. J. Food Microbiol.* **2015**, *205*, 54–67. [CrossRef] [PubMed]

16. García-Alamilla, P.; Lagunes-Gálvez, L.M.; Barajas-Fernández, J.; García-Alamilla, R.; Garcia-Armisen, T.; Papalexandratou, Z.; Hendryckx, H.; Camu, N.; Vrancken, G.; De Vuyst, L.; et al. Diversity of the total bacterial community associated with Ghanaian and Brazilian cocoa bean fermentation samples as revealed by a 16 S rRNA gene clone library. *Appl. Microbiol. Biotechnol.* **2010**, *87*, 2281–2292. [CrossRef]

17. Papalexandratou, Z.; Falony, G.; Romanens, E.; Jimenez, J.C.; Amores, F.; Daniel, H.; De Vuyst, L. Species diversity, community dynamics, and metabolite kinetics of the microbiota associated with traditional Ecuadorian spontaneous cocoa bean fermentations. *Appl. Environ. Microbiol.* **2011**, *77*, 7698–7714. [CrossRef]

18. Papalexandratou, Z.; Lefeber, T.; Bahrim, B.; Seng, O.; Daniel, H.; De Vuyst, L. *Acetobacter pasteurianus* predominate during well-performed Malaysian cocoa bean box fermentations, underlining the importance of these microbial species for a successful cocoa bean fermentation process. *Food Microbiol.* **2013**, *35*, 73–85. [CrossRef]

19. Bortolini, C.; Patrone, V.; Puglisi, E.; Morelli, L. Detailed analyses of the bacterial populations in processed cocoa beans of different geographic origin, subject to varied fermentation conditions. *Int. J. Food Microbiol.* **2016**, *236*, 98–106. [CrossRef]

20. Miguel, M.G.D.C.P.; de Castro Reis, L.V.; Efraim, P.; Santos, C.; Lima, N.; Schwan, R.F. Cocoa fermentation: Microbial identification by MALDI-TOF MS, and sensory evaluation of produced chocolate. *LWT—Food Sci. Technol.* **2017**, *77*. [CrossRef]

21. de Melo Pereira, G.V.; Magalhães, K.T.; de Almeida, E.G.; da Silva Coelho, I.; Schwan, R.F. Spontaneous cocoa bean fermentation carried out in a novel-design stainless steel tank: Influence on the dynamics of microbial populations and physical-chemical properties. *Int. J. Food Microbiol.* **2013**, *161*, 121–133. [CrossRef]

22. Illeghems, K.; De Vuyst, L.; Papalexandratou, Z.; Weckx, S. Phylogenetic analysis of a spontaneous cocoa bean fermentation metagenome reveals new insights into its bacterial and fungal community diversity. *PLoS ONE* **2012**, *7*. [CrossRef]

23. da Veiga Moreira, I.M.; Gabriela, M.; Miguel, P.; Ferreira, W.; Ribeiro, D.; Freitas, R. Microbial succession and the dynamics of metabolites and sugars during the fermentation of three different cocoa (*Theobroma cacao* L.) hybrids. *Food Res. Int.* **2013**, *54*, 9–17. [CrossRef]

24. Hamdouche, Y.; Guehi, T.; Durand, N.; Kedjebo, K.B.D.; Montet, D.; Meile, J.C. Dynamics of microbial ecology during cocoa fermentation and drying: Towards the identification of molecular markers. *Food Control* **2015**, *48*, 117–122. [CrossRef]

25. Pereira, G.V.M.; Alvarez, J.P.; de Neto, D.P.C.; Soccol, V.T.; Tanobe, V.O.A.; Rogez, H.; Góes-Neto, A.; Soccol, C.R. Great intraspecies diversity of *Pichia kudriavzevii* in cocoa fermentation highlights the importance of yeast strain selection for flavor modulation of cocoa beans. *LWT—Food Sci. Technol.* **2017**, *84*, 290–297. [CrossRef]

26. Arana-Sánchez, A.; Segura-García, L.E.; Kirchmayr, M.; Orozco-Ávila, I.; Lugo-Cervantes, E.; Gschaedler-Mathis, A. Identification of predominant yeasts associated with artisan Mexican cocoa fermentations using culture-dependent and culture-independent approaches. *World J. Microbiol. Biotechnol.* **2015**, *31*, 359–369. [CrossRef]

27. Koné, M.K.; Guéhi, S.T.; Durand, N.; Ban-kof, L.; Berthiot, L.; Tachon, A.F.; Brou, K.; Boulanger, R.; Montet, D. Contribution of predominant yeasts to the occurrence of aroma compounds during cocoa bean fermentation. *Food Res. Int.* **2016**, *89*, 910–917. [CrossRef]

28. Leal, G.A.; Gomes, L.H.; Efraim, P.; De Almeida Tavares, F.C.; Figueira, A. Fermentation of cacao (*Theobroma cacao* L.) seeds with a hybrid *Kluyveromyces marxianus* strain improved product quality attributes. *FEMS Yeast Res.* **2008**, *8*, 788–798. [CrossRef] [PubMed]

29. Crafack, M.; Mikkelsen, M.B.; Saerens, S.; Knudsen, M.; Blennow, A.; Lowor, S.; Takrama, J.; Swiegers, J.H.; Petersen, G.B.; Heimdal, H.; et al. Influencing cocoa flavour using *Pichia kluyveri* and *Kluyveromyces marxianus* in a defined mixed starter culture for cocoa fermentation. *Int. J. Food Microbiol.* **2013**, *167*, 103–116. [CrossRef] [PubMed]

30. Batista, N.N.; Ramos, C.L.; Ribeiro, D.D.; Pinheiro, A.C.M.; Schwan, R.F. Dynamic behavior of *Saccharomyces cerevisiae*, *Pichia kluyveri* and *Hanseniaspora uvarum* during spontaneous and inoculated cocoa fermentations and their effect on sensory characteristics of chocolate. *LWT—Food Sci. Technol.* **2015**, *63*, 221–227. [CrossRef]

31. Visintin, S.; Ramos, L.; Batista, N.; Dolci, P.; Schwan, F.; Cocolin, L. Impact of *Saccharomyces cerevisiae* and *Torulaspora delbrueckii* starter cultures on cocoa beans fermentation. *Int. J. Food Microbiol.* **2017**, *257*, 31–40. [CrossRef] [PubMed]

32. Fleet, G.H. Yeasts in foods and beverages: Impact on product quality and safety. *Curr. Opin. Biotechnol.* **2007**, *18*, 170–175. [CrossRef]

33. Ramos, C.L.; Dias, D.R.; Miguel, M.G.D.C.P.; Schwan, R.F. Impact of different cocoa hybrids (*Theobroma cacao* L.) and *S. cerevisiae* UFLA CA11 inoculation on microbial communities and volatile compounds of cocoa fermentation. *Food Res. Int.* **2014**, *64*, 908–918. [CrossRef] [PubMed]

34. Menezes, A.G.T.; Batista, N.N.; Ramos, C.L.; e Silva, A.R.D.A.; Efraim, P.; Pinheiro, A.C.M.; Schwan, R.F. Investigation of chocolate produced from four different Brazilian varieties of cocoa (*Theobroma cacao* L.) inoculated with *Saccharomyces cerevisiae*. *Food Res. Int.* **2016**, *81*, 83–90. [CrossRef]

35. Cempaka, L.; Aliwarga, L.; Purwo, S.; Kresnowati, M.T.A.P. Dynamics of cocoa bean pulp degradation during cocoa bean fermentation: Effects of yeast starter culture addition. *J. Math. Fundam. Sci.* **2014**, *46*, 14–25. [CrossRef]

36. Mahazar, N.H.; Sufian, N.F.; Meor Hussin, A.S.; Norhayati, H.; Mathawan, M.; Rukayadi, Y. *Candida* sp. as a starter culture for cocoa (*Theobroma cacao* L.) beans fermentation. *Int. Food Res. J.* **2015**, *22*, 1783–1787.

37. Meersman, E.; Steensels, J.; Struyf, N.; Paulus, T.; Saels, V.; Mathawan, M.; Allegaert, L.; Vrancken, G.; Verstrepen, K.J. Tuning chocolate flavor through development of thermotolerant *Saccharomyces cerevisiae* starter cultures with increased acetate ester. *Appl. Environ. Microbiol.* **2016**, *82*, 732–746. [CrossRef]

38. Meersman, E.; Steensels, J.; Paulus, T.; Struyf, N.; Saels, V.; Mathawan, M.; Koffi, J.; Vrancken, G.; Verstrepen, K.J. Breeding strategy to generate robust yeast starter cultures for cocoa pulp fermentations. *Appl. Environ. Microbiol.* **2015**, *81*, 6166–6176. [CrossRef] [PubMed]

39. Daniel, H.M.; Vrancken, G.; Takrama, J.F.; Camu, N.; De Vos, P.; De Vuyst, L. Yeast diversity of Ghanaian cocoa bean heap fermentations. *FEMS Yeast Res.* **2009**, *9*, 774–783. [CrossRef] [PubMed]

40. de Melo Pereira, G.V.; Miguel, M.G.D.C.P.; Ramos, C.L.; Schwan, R.F. Microbiological and physicochemical characterization of small-scale cocoa fermentations and screening of yeast and bacterial strains to develop a defined starter culture. *Appl. Environ. Microbiol.* **2014**, *78*, 5395–5405. [CrossRef]

41. Samagaci, L.; Ouattara, H.; Niamké, S.; Lemaire, M. *Pichia kudrazevii* and *Candida nitrativorans* are the most well-adapted and relevant yeast species fermenting cocoa in Agneby-Tiassa, a local Ivorian cocoa producing region. *Food Res. Int.* **2016**, *89*, 773–780. [CrossRef] [PubMed]

42. Branco, P.; Francisco, D.; Chambon, C.; Hébraud, M.; Arneborg, N.; Almeida, M.G.; Caldeira, J.; Albergaria, H. Identification of novel GAPDH-derived antimicrobial peptides secreted by *Saccharomyces cerevisiae* and involved in wine microbial interactions. *Appl. Microbiol. Biotechnol.* **2014**, *98*, 843–853. [CrossRef]

43. Jayani, R.S.; Saxena, S.; Gupta, R. Microbial pectinolytic enzymes: A review. *Process Biochem.* **2005**, *40*, 2931–2944. [CrossRef]

44. Whitener, M.E.B.; Stanstrup, J.; Carlin, S.; Divol, B.; Du Toit, M.; Vrhovsek, U. Effect of non-*Saccharomyces* yeasts on the volatile chemical profile of Shiraz wine. *Aust. J. Grape Wine Res.* **2017**, *23*, 179–192. [CrossRef]

45. Contreras, A.; Hidalgo, C.; Henschke, P.A.; Chambers, P.J.; Curtin, C.; Varela, C. Evaluation of non-*Saccharomyces* yeasts for the reduction of alcohol content in wine. *Appl. Environ. Microbiol.* **2014**, *80*, 1670–1678. [CrossRef] [PubMed]

46. Batista, N.N.; Ramos, C.L.; Dias, D.R. The impact of yeast starter cultures on the microbial communities and volatile compounds in cocoa fermentation and the resulting sensory attributes of chocolate. *J. Food Sci. Technol.* **2016**, *53*, 1101–1110. [CrossRef] [PubMed]

47. Albergaria, H.; Francisco, D.; Gori, K.; Arneborg, N.; Gírio, F. *Saccharomyces cerevisiae* CCMI 885 secretes peptides that inhibit the growth of some non-*Saccharomyces* wine-related strains. *Appl. Microbiol. Biotechnol.* **2010**, *86*, 965–972. [CrossRef]

48. Magagna, F.; Guglielmetti, A.; Liberto, E.; Reichenbach, S.E.; Allegrucci, E.; Gobino, G.; Bicchi, C.; Cordero, C. Comprehensive chemical fingerprinting of high-quality cocoa at early stages of processing: Effectiveness of combined untargeted and targeted approaches for classification and discrimination. *J. Agric. Food Chem.* **2017**, *65*, 6329–6341. [CrossRef] [PubMed]

49. Afoakwa, E.O.; Paterson, A.; Fowler, M.; Ryan, A.; Ohene, E.; Paterson, A.; Fowler, M.; Ryan, A.; Fowler, M.; Ryan, A. Flavor formation and character in cocoa and chocolate: A critical review. *Crit. Rev. Food Sci. Nutr.* **2008**, *48*, 840–857. [CrossRef] [PubMed]

50. Efraim, P.; Pezoa-García, N.H.; Calil, D.; Jardim, P.; Nishikawa, A.; Haddad, R.; Eberlin, M.N. Influência da fermentação e secagem de amêndoas de cacau no teor de compostos fenólicos e na aceitação sensorial. *Ciência Tecnol. Aliment.* **2010**, *30*, 142–150. [CrossRef]

51. Caligiani, A.; Cirlini, M.; Palla, G.; Ravaglia, R.; Arlorio, M. GC-MS detection of chiral markers in cocoa beans of different quality and geographic origin. *Chirality Pharmacol. Biol. Chem. Consequences Mol. Asymmetry* **2007**, *334*, 329–334. [CrossRef]

52. Redgwell, R.J.; Trovato, V.; Curti, D. Cocoa bean carbohydrates: Roasting-induced changes and polymer interactions. *Food Chem.* **2003**, *80*, 511–516. [CrossRef]

53. Gu, F.; Tan, L.; Wu, H.; Fang, Y.; Xu, F.; Chu, Z.; Wang, Q. Comparison of cocoa beans from China, Indonesia and Papua New Guinea. *Foods* **2013**, *2*, 183–197. [CrossRef] [PubMed]

54. Pätzold, R.; Brückner, H.; Oracz, J.; Nebesny, E.; Sukha, D.A.; Butler, D.R.; Umaharan, P.; Boult, E.; Miescher Schwenninger, S.; Freimüller Leischtfeld, S.; et al. Assessment methodology to predict quality of cocoa beans for export. *J. Agric. Food Chem.* **2008**, *56*, 773–780. [CrossRef]

55. Granvogl, M.; Bugan, S.; Schieberle, P. Formation of amines and aldehydes from parent amino acids during thermal processing of cocoa and model systems: New insights into pathways of the strecker reaction. *J. Agric. Food Chem.* **2006**, *54*, 1730–1739. [CrossRef]

56. Afoakwa, E.O.; Quao, J.; Takrama, J.; Budu, A.S. Chemical composition and physical quality characteristics of Ghanaian cocoa beans as affected by pulp pre-conditioning and fermentation. *J. Food Sci. Technol.* **2013**, *50*, 1097–1105. [CrossRef]

57. Afoakwa, E.O.; Kongor, J.E.; Takrama, J.F.; Budu, A.S. Changes in acidification, sugars and mineral composition of cocoa pulp during fermentation of pulp pre-conditioned cocoa (*Theobroma cacao*) beans. *Int. Food Res. J.* **2013**, *20*, 1215–1222.

58. Scott-Thomas, C. Raw Food on the Rise. Available online: https://www.foodnavigator.com/Article/2015/06/09/Raw-food-on-the-rise (accessed on 15 January 2019).

59. Ramli, N.; Hassan, O.; Said, M.; Samsudin, W.; Idris, N.A. Influence of roasting conditions on volatile flavor of roasted Malaysian cocoa beans. *J. Food Process. Preserv.* **2006**, *30*, 280–298. [CrossRef]

60. Bonhevi, J.S. Investigation of aromatic compounds in roasted cocoa powder. *Eur. Food Res. Tecnhonol.* **2005**, *221*, 19–29. [CrossRef]

61. Huang, Y.; Barringer, S.A. Monitoring of cocoa volatiles produced during roasting by selected ion flow tube-mass spectrometry (SIFT-MS). *J. Food Sci.* **2011**, *76*, 279–286. [CrossRef]

62. Van Durme, J.; Ingels, I.; De Winne, A. Online roasting hyphenated with gas chromatography-mass spectrometry as an innovative approach for assessment of cocoa fermentation quality and aroma formation potential. *Food Chem.* **2016**, *205*, 66–72. [CrossRef]

63. Frauendorfer, F.; Schieberle, P. Changes in key aroma compounds of criollo cocoa. *J. Agric. Food Chem.* **2008**, *56*, 10244–10251. [CrossRef]

64. Tan, J.; Kerr, W.L. Determining degree of roasting in cocoa beans by artificial neural network (ANN)-based electronic nose system and gas chromatography/mass spectrometry (GC/MS). *J. Sci. Food Agric.* **2018**, *98*, 3851–3859. [CrossRef]

65. Magagna, F.; Liberto, E.; Reichenbach, S.E.; Tao, Q.; Carretta, A.; Cobelli, L.; Giardina, M.; Bicchi, C.; Cordero, C. Advanced fingerprinting of high-quality cocoa: Challenges in transferring methods from thermal to differential-flow modulated comprehensive two dimensional gas chromatography. *J. Chromatogr. A* **2018**, *1536*, 122–136. [CrossRef]

66. Janssens, L.; De Pooter, H.L.; Schamp, N.M.; Vandamme, E.J. Production of flavours by microorganisms. *Process Biochem.* **1992**, *27*, 195–215. [CrossRef]

67. Hagedorn, S.; Kaphammer, B. Microbial biocatalysis in the generation of flavor and fragrance chemicals. *Annu. Rev. Microbiol.* **1994**, *46*, 1561–1563. [CrossRef] [PubMed]

68. Berger, R.G. *Flavours and Fragrances: Chemistry, Bioprocessing and Sustainability*; Springer: Berlin, Germany, 2007; ISBN 9783540493389.

69. Schrader, J.; Etschmann, M.M.W.; Sell, D.; Hilmer, J.-M.; Rabenhorst, J. Applied biocatalysis for the synthesis of natural flavour compounds: Curent industrial processes and future prospects. *Biotechnol. Lett.* **2004**, 463–472. [CrossRef]

70. Kim, B.; Cho, B.R.; Hahn, J.S. Metabolic engineering of *Saccharomyces cerevisiae* for the production of 2-phenylethanol via Ehrlich pathway. *Biotechnol. Bioeng.* **2014**, *111*, 115–124. [CrossRef]

71. Etschmann, M.M.W.; Sell, D.; Schrader, J. Production of 2-phenylethanol and 2-phenylethylacetate from L-phenylalanine by coupling whole-cell biocatalysis with organophilic pervaporation. *Biotechnol. Bioeng.* **2005**, *92*, 624–634. [CrossRef]

72. Wittmann, C.; Hans, M.; Bluemke, W. Metabolic physiology of aroma-producing *Kluyveromyces marxianus*. *Yeast* **2002**, *19*, 1351–1363. [CrossRef]

73. Moreira, N.; Mendes, F.; Hogg, T.; Vasconcelos, I. Alcohols, esters and heavy sulphur compounds production by pure and mixed cultures of apiculate wine yeasts. *Int. J. Food Microbiol.* **2005**, *103*, 285–294. [CrossRef] [PubMed]

74. Fabre, C.E.; Blanc, P.J.; Goma, G. Screening of yeasts producing 2-phenylethylalcohol. *Biotechnol. Tech.* **1997**, *11*, 523–525. [CrossRef]

75. Cappaert, L.; Larroche, C. Oxidation of a mixture of 2-(R) and 2-(S)-heptanol to 2-heptanone by *Saccharomyces cerevisiae* in a biphasic system. *Biocatal. Biotransform.* **2004**, *22*, 291–296. [CrossRef]

76. Chen, H.; Fink, G.R. Feedback control of morphogenesis in fungi by aromatic alcohols. *Genes Dev.* **2006**, *20*, 1150–1161. [CrossRef]

77. Fraud, S.; Rees, E.L.; Mahenthiralingam, E.; Russell, A.D.; Maillard, J.Y. Aromatic alcohols and their effect on Gram-negative bacteria, cocci and mycobacteria (1). *J. Antimicrob. Chemother.* **2003**, *51*, 1435–1436. [CrossRef]

78. Larroy, C.; Parés, X.; Biosca, J.A. Characterization of a *Saccharomyces cerevisiae* NADP(H)-dependent alcohol dehydrogenase (ADHVII), a member of the cinnamyl alcohol dehydrogenase family. *Eur. J. Biochem.* **2002**, *269*, 5738–5745. [CrossRef] [PubMed]

79. Lapadatescu, C.; Feron, G.; Vergoignan, C.; Djian, A.; Durand, A.; Bonnarme, P. Influence of cell immobilization on the production of benzaldehyde and benzyl alcohol by the white-rot fungi *Bjerkandera adusta*, *Ischnoderma benzoinum* and *Dichomitus squalens*. *Appl. Microbiol. Biotechnol.* **1997**, *47*, 708–714. [CrossRef]

80. Kawabe, T.; Morita, H. Production of benzaldehyde and benzyl alcohol by the mushroom *Polyporus tuberaster* K2606. *J. Agric. Food Chem.* **1994**, *42*, 2556–2560. [CrossRef]

81. Pal, S.; Park, D.H.; Plapp, B.V. Activity of yeast alcohol dehydrogenases on benzyl alcohols and benzaldehydes. Characterization of ADH1 from *Saccharomyces carlsbergensis* and transition state analysis. *Chem. Biol. Interact.* **2009**, *178*, 16–23. [CrossRef] [PubMed]

82. Rojas, V.; Gil, J.V.; Piñaga, F.; Manzanares, P. Studies on acetate ester production by non-*Saccharomyces* wine yeasts. *Int. J. Food Microbiol.* **2001**, *70*, 283–289. [CrossRef]

83. Van Laere, S.D.M.; Saerens, S.M.G.; Verstrepen, K.J.; Van Dijck, P.; Thevelein, J.M.; Delvaux, F.R. Flavour formation in fungi: Characterisation of KlAtf, the *Kluyveromyces lactis* orthologue of the *Saccharomyces cerevisiae* alcohol acetyltransferases Atf1 and Atf2. *Appl. Microbiol. Biotechnol.* **2008**, *78*, 783–792. [CrossRef]

84. Duetz, W.A.; Bouwmeester, H.; Beilen, J.B.; Witholt, B. Biotransformation of limonene by bacteria, fungi, yeasts, and plants. *Appl. Microbiol. Biotechnol.* **2003**, *61*, 269–277. [CrossRef]

85. Ariyoshi-Kishino, K.; Hashimoto, K.; Amano, O.; Saitoh, J.; Kochi, M.; Sakagami, H. Tumor-specific cytotoxicity and type of cell death induced by benzaldehyde. *Anticancer Res.* **2010**, *30*, 5069–5076.

86. Stringer, T.; Therrien, B.; Hendricks, D.T.; Guzgay, H.; Smith, G.S. Mono- and dinuclear (η6-arene) ruthenium (II) benzaldehyde thiosemicarbazone complexes: Synthesis, characterization and cytotoxicity. *Inorg. Chem. Commun.* **2011**, *14*, 956–960. [CrossRef]

87. Su, W.; Zhou, Q.; Huang, Y.; Huang, Q.; Huo, L.; Xiao, Q.; Huang, S.; Huang, C.; Chen, R.; Qian, Q.; et al. Synthesis, crystal and electronic structure, anticancer activity of ruthenium (II) arene complexes with thiosemicarbazones. *Appl. Organomet. Chem.* **2013**, *27*, 307–312. [CrossRef]

88. Liu, Y.; Sakagami, H.; Hashimoto, K.; Kikuchi, H.; Amano, O.; Ishibara, M.; Yu, G. Tumor-specific cytotoxicity and type of cell death induced by beta-cyclodextrin benzaldehyde inclusion compound. *Anticancer Res.* **2008**, *28*, 229–236. [PubMed]

89. Alamri, A.; El-Newehy, M.H.; Al-Deyab, S.S. Biocidal polymers: Synthesis and antimicrobial properties of benzaldehyde derivatives immobilized onto amine-terminated polyacrylonitrile. *Chem. Cent. J.* **2012**, *6*, 1–13. [CrossRef]

90. Rabi, T.; Bishayee, A. d-Limonene sensitizes docetaxel-induced cytotoxicity in human prostate cancer cells: Generation of reactive oxygen species and induction of apoptosis. *J. Carcinog.* **2009**, *8*, 9. [CrossRef]

91. Vigushin, D.M.; Poon, G.K.; Boddy, A.; English, J.; Halbert, G.W.; Pagonis, C.; Jarman, M.; Coombes, R.C. Phase I and pharmacokinetic study of d-limonene in patients with advanced cancer. Cancer Research Campaign Phase I/II Clinical Trials Committee. *Cancer Chemother. Pharmacol.* **1998**, *42*, 111–117. [CrossRef] [PubMed]

92. Igimi, H.; Watanabe, D.; Yamamoto, F.; Asakawa, S.; Toraishi, K.; Shimura, H. A useful cholesterol solvent for medical dissolution of gallstones. *Gastroenterol. Jpn.* **1992**, *27*, 536–545. [CrossRef] [PubMed]

93. De Almeida, A.A.C.; Costa, J.P.; De Carvalho, R.B.F.; De Sousa, D.P.; De Freitas, R.M. Evaluation of acute toxicity of a natural compound (+)-limonene epoxide and its anxiolytic-like action. *Brain Res.* **2012**, *1448*, 56–62. [CrossRef]

94. do Amaral, J.F.; Silva, M.I.G.; de Aquino Neto, M.R.A.; Neto, P.F.T.; Moura, B.A.; de Melo, C.T.V.; de Sousa, F.C.F. Antinociceptive effect of the monoterpene R-(+)-limonene in mice. *Biol. Pharm. Bull.* **2007**, *30*, 1217–1220. [CrossRef]

95. Giri, R.K.; Parija, T.; Das, B.R. D-limonene chemoprevention of hepatocarcinogenesis in AKR mice: Inhibition of c-jun and c-myc. *Oncol. Rep.* **1999**, *6*, 1123–1127. [CrossRef]

96. Kaji, I.; Tatsuta, M.; Lishi, H.; Baba, M.; Inoue, A.; Kasugai, H. Inhibition by d-limonene of experimental hepatocarcinogenesis in sprague-dawley rats does not involve P21ras plasma membrane association. *Int. J. Cancer* **2001**, *93*, 441–444. [CrossRef]

97. Vuuren, S.V.; Viljoen, A.M. Antimicrobial activity of limonene enantiomers and 1,8-cineole alone and in combinatio. *Flavours Fragr. J.* **2007**, *22*, 540–544. [CrossRef]

98. Jirovetz, L.; Buchbauer, G.; Schmidt, E.; Denkova, Z.; Slavchev, A.; Stoyanova, A.; Geissler, M. Purity, antimicrobial activities and olfactory evaluations of 2-phenylethanol and some derivatives. *J. Essent. Oil Res.* **2008**, *20*, 82–85. [CrossRef]

99. Ullah, I.; Khan, A.L.; Ali, L.; Khan, A.R.; Waqas, M.; Hussain, J.; Lee, I.J.; Shin, J.H. Benzaldehyde as an insecticidal, antimicrobial, and antioxidant compound produced by *Photorhabdus temperata* M1021. *J. Microbiol.* **2015**, *53*, 127–133. [CrossRef]

100. Wang, J.; Liu, H.; Zhao, J.; Gao, H.; Zhou, L.; Liu, Z.; Chen, Y.; Sui, P. Antimicrobial and antioxidant activities of the root bark essential oil of *Periploca sepium* and its main component 2-hydroxy-4-methoxybenzaldehyde. *Molecules* **2010**, *15*, 5807–5817. [CrossRef] [PubMed]

101. Lingappa, B.T.; Prasad, M.; Lingappa, Y.; Hunt, D.F.; Biemann, K. Phenethyl alcohol and tryptophol: Autoantibiotics produced by the fungus *Candida albicans*. *Science* **1969**, *163*, 192–194. [CrossRef]

102. Wang, Q.; Ke, L.; Xue, C.; Luo, W.; Chen, Q. Inhibitory kinetics of p-substituted benzaldehydes on polyphenol oxidase from the fifth instar of *Pieris rapae* L. *Tsinghua Sci. Technol.* **2007**, *12*, 400–404. [CrossRef]

103. Wilkinson, S.M.; Love, S.B.; Westcombe, A.M.; Gambles, M.A.; Burgess, C.C.; Cargill, A.; Young, T.; Maher, E.J.; Ramirez, A.J. Effectiveness of aromatherapy massage in the management of anxiety and depression in patients with cancer: A multicenter randomized controlled trial. *J. Clin. Oncol.* **2007**, *25*, 532–539. [CrossRef]
104. Hoffman, H.J.; Rawal, S.; Li, C.; Duffy, V.B. New chemosensory component in the U.S. National Health and Nutrition Examination Survey (NHANES): First-year results for measured olfactory dysfunction. *Rev. Endocr. Metab. Disord.* **2017**, *17*, 221–240. [CrossRef]
105. Naka, A.; Riedl, M.; Luger, A.; Hummel, T.; Mueller, C.A. Clinical significance of smell and taste disorders in patients with diabetes mellitus. *Eur. Arch. Oto-Rhino-Laryngol.* **2010**, *267*, 547–550. [CrossRef]
106. Aschenbrenner, K.; Scholze, N.; Joraschky, P.; Hummel, T. Gustatory and olfactory sensitivity in patients with anorexia and bulimia in the course of treatment. *J. Psychiatr. Res.* **2008**, *43*, 129–137. [CrossRef] [PubMed]
107. Skrandies, W.; Zschieschang, R. Olfactory and gustatory functions and its relation to body weight. *Physiol. Behav.* **2015**, *142*, 1–4. [CrossRef]
108. Clepce, M.; Reich, K.; Gossler, A.; Kornhuber, J.; Thuerauf, N. Olfactory abnormalities in anxiety disorders. *Neurosci. Lett.* **2012**, *511*, 43–46. [CrossRef]
109. Kershaw, J.C.; Mattes, R.D. Nutrition and taste and smell dysfunction. *World J. Otorhinolaryngol.-Head Neck Surg.* **2018**, *4*, 3–10. [CrossRef]
110. Pekala, K.; Chandra, R.K.; Turner, J.H. Efficacy of olfactory training in patients with olfactory loss: A systematic review and meta-analysis. *Int. Forum Allergy Rhinol.* **2016**, *6*, 299–307. [CrossRef]
111. Jiang, R.S.; Twu, C.W.; Liang, K.L. The effect of olfactory training on the odor threshold in patients with traumatic anosmia. *Am. J. Rhinol. Allergy* **2017**, *31*, 317–322. [CrossRef] [PubMed]
112. Hummel, T.; Reden, K.R.J.; Hähner, A.; Weidenbecher, M.; Hüttenbrink, K.B. Effects of olfactory training in patients with olfactory loss. *Laryngoscope* **2009**, *119*, 496–499. [CrossRef] [PubMed]
113. Haehner, A.; Tosch, C.; Wolz, M.; Klingelhoefer, L.; Fauser, M.; Storch, A.; Reichmann, H.; Hummel, T. Olfactory training in patients with Parkinson's disease. *PLoS ONE* **2013**, *8*. [CrossRef]
114. Geißler, K.; Reimann, H.; Gudziol, H.; Bitter, T.; Guntinas-Lichius, O. Olfactory training for patients with olfactory loss after upper respiratory tract infections. *Eur. Arch. Oto-Rhino-Laryngol.* **2014**, *271*, 1557–1562. [CrossRef] [PubMed]
115. Damm, M.; Pikart, L.K.; Reimann, H.; Burkert, S.; Göktas, Ö.; Haxel, B.; Frey, S.; Charalampakis, I.; Beule, A.; Renner, B.; Hummel, T.; et al. Olfactory training is helpful in postinfectious olfactory loss: A randomized, controlled, multicenter study. *Laryngoscope* **2014**, *124*, 826–831. [CrossRef]
116. Kollndorfer, K.; Kowalczyk, K.; Hoche, E.; Mueller, C.A.; Pollak, M.; Trattnig, S.; Schöpf, V. Recovery of olfactory function induces neuroplasticity effects in patients with smell loss. *Neural Plast.* **2014**, *2014*. [CrossRef] [PubMed]
117. Konstantinidis, I.; Tsakiropoulou, E.; Bekiaridou, P.; Kazantzidou, C.; Constantinidis, J. Use of olfactory training in post-traumatic and postinfectious olfactory dysfunction. *Laryngoscope* **2013**, *123*. [CrossRef] [PubMed]
118. Evaluation of Certain Food Additives: Fifty-Ninth Report of the Joint FAO/WHO Expert Committee on Food Additives. Available online: https://apps.who.int/iris/bitstream/handle/10665/42601/WHO_TRS_913.pdf?sequence=1&isAllowed=y (accessed on 10 January 2019).

Article

Simulated Gastrointestinal Digestion of Cocoa: Detection of Resistant Peptides and In Silico/In Vitro Prediction of Their Ace Inhibitory Activity

Angela Marseglia, Luca Dellafiora, Barbara Prandi, Veronica Lolli, Stefano Sforza, Pietro Cozzini, Tullia Tedeschi, Gianni Galaverna and Augusta Caligiani *

Department of Food and Drug, University of Parma, 43124 Parma PR, Italy; angela.marseglia@unipr.it (A.M.); luca.dellafiora@unipr.it (L.D.); barbara.prandi@unipr.it (B.P.); veronica.lolli@unipr.it (V.L.); stefano.sforza@unipr.it (S.S.); pietro.cozzini@unipr.it (P.C.); tullia.tedeschi@unipr.it (T.T.); gianni.galaverna@unipr.it (G.G.)
* Correspondence: augusta.caligiani@unipr.it; Tel.: +39-0521-905407

Received: 14 March 2019; Accepted: 17 April 2019; Published: 30 April 2019

Abstract: In this study we investigated the oligopeptide pattern in fermented cocoa beans and derived products after simulated gastrointestinal digestion. Peptides in digested cocoa samples were identified based on the mass fragmentation and on the software analysis of vicilin and 21 KDa cocoa seed protein sequences, the most abundant cocoa proteins. Quantification was carried out by liquid chromatography/electrospray ionisation mass spectrometry (LC/ESI-MS) using an internal standard. Sixty five peptides were identified in the digested samples, including three pyroglutamyl derivatives. The in vitro angiotensin-converting enzyme (ACE)-inhibitory activity of cocoa digests were tested, demonstrating a high inhibition activity, especially for digestates of cocoa beans. The peptides identified were screened for their potential ACE inhibitory activity through an in silico approach, and about 20 di-, three- and tetra-peptides actually present in our samples were predicted as active. Two of the potentially active peptides were chemically synthesized and then assessed for their inhibitory activity by using the ACE in vitro assay. These peptides demonstrated an ACE inhibitory activity, however, that was too weak to explain alone the high activity of cocoa digestates, suggesting a synergic effect of all cocoa peptides. As a whole, results showed that an average chocolate portion (30 g) ensures an amount of peptides after digestion that, assuming complete absorption, could reach almost a complete inhibition of ACE.

Keywords: cocoa; oligopeptides; simulated gastrointestinal digestion; angiotensin-converting enzyme (ACE) inhibitory activity

---

1. Introduction

Cocoa beans, from the fruit of the cocoa tree (*Theobroma cacao* L.), are transformed into chocolate and other cocoa products by a complex process involving fermentation, drying and roasting. By the 1600s and 1700s, chocolate and cocoa were viewed not just as a beverage with a pleasurable taste, but also as a food to treat a number of disorders [1]. Possible health benefits of chocolate have been reported for many years but it is only recently that some of these claims are being more clearly identified and studied. For example, the antioxidant and health-promoting properties of cocoa and cocoa-related products have been thoroughly investigated and various health claims for cocoa polyphenolics have been proposed [2]. Recently, the European Food Safety Authority (EFSA) issued a positive opinion on cocoa flavanols and maintenance of endothelium-dependent vasodilation, which contributes to normal blood flow [3]. While the polyphenols and antioxidant activity of cocoa have been extensively studied, little is known about the potential health effects of other cocoa components such as peptides/proteins [4]. Peptides are currently considered important bioactive constituents of food, however the potential biological activities of oligopeptides found in cocoa are under-investigated in cocoa literature. Biologically active or functional peptides are food-derived peptides that exert, beyond their nutritional value, a physiological effect in the body [5]. Dietary proteins provide a rich source of bioactive peptides, which are hidden in a latent state within the native protein, requiring enzymatic proteolysis for their release. Bioactive peptides can be produced during in vivo gastrointestinal digestion and/or food processing and they have been reported in a wide range of animal and vegetable proteins, such as bovine and human milk, fish, meat, soybean and cereals [6]. In vitro and in vivo studies demonstrated several biological functions attributed to bioactive peptides, such as antimicrobial, immunomodulatory, enhancement of mineral absorption, antithrombotic, antihypertensive, opioid and antioxidant activities [7].

In cocoa, peptides are naturally formed during cocoa beans' fermentation, as previously reported [8–11], and together with amino acids they are considered important flavour precursors [12–14]. Peptides in cocoa derive from the two major protein fractions, globulins, consisting of a 66 kDa vicilin-like storage protein [15], and albumin consisting of a 21 kDa protein with trypsin inhibitory properties [16]. Cocoa proteins during natural cocoa fermentation are cleaved to hydrophilic and hydrophobic peptides as well as amino acids through autolysis by two endogenous enzymes, aspartic endoprotease and carboxypeptidase activated by microbial metabolites as acetic acid [17]. Besides generating the characteristic chocolate aroma, the potential biological activities of oligopeptides found in cocoa are of interest. Two papers report the physiological effects (antioxidant, angiotensin-converting enzyme inhibitors and hypoglycaemic activities) of cocoa autolysates containing peptides and amino acids [18,19] and, more recently, the antioxidant properties of cocoa protein enzymatic hydrolysates [20]. A patent for the production of angiotensin-converting enzyme (ACE) inhibitory peptides from cocoa was also developed [21], demonstrating the growing interest in the topic. However, the identification of peptides responsible for the activities is largely unknown, together with the resistance of peptides through the gastrointestinal tract, a pre-requisite for their bioavailability. In fact, before testing any systemic biological activity of food peptides mixtures, it is of utmost importance to assess their bioavailability, and as a first step their resistance to gastrointestinal digestion. Moreover, gastrointestinal digestion might even form new peptides, with more biological activity. However, the information about the peptide composition of cocoa after ingestion is, at the moment, lacking in the literature. Therefore, the aim of this study was the identification of the peptides released after in vitro simulated gastrointestinal digestion of cocoa beans and derived products, using a physiological digestion model. Moreover, the potential ACE inhibitory activity of cocoa peptides found after digestion was evaluated by an in silico/in vitro combined approach.

2. Materials and Methods

2.1. Cocoa Samples

The release of oligopeptides by in vitro gastrointestinal digestion model was investigated for different typologies of cocoa samples. Two samples of well fermented cocoa beans of Forastero varieties of different geographical origins (Congo and Dominican Republic) were kindly provided by Barry Callebaut, Belgium. To highlight the eventual different release of peptides due to the cocoa process, one intermediate product of the cocoa processing chain (cocoa paste) and one end product (dark chocolate bar, 40% of cocoa mass) were also included in the experimental plan.

2.2. Extraction of Peptides from Non-Digested Samples

Peptides were extracted according to the method previously described [8]. A total of 10 g of finely grinded cocoa sample was suspended in 45 mL of 0.1 N HCl. (L,L)-phenylalanylphenylalanine (Phe-Phe) was added as an internal standard (2.25 mL of a 1 mM solution). The suspension was homogeneized for 1.5 min by Ultra Turrax T50 at 4000 rpm (Janke and Hunkel Labortechnik, Germany) and then centrifuged at 4000 rpm for 30 min at 4 °C by an ALC 4237R centrifuge. The solution was filtered through paper filters (pore dimensions 15–20 µm) and then extracted four times with 50 mL of ethyl ether. The solution was filtered again with a Millipore 47 mm Steril Aseptic system through 0.45 µm HVLP millipore filters. A total of 1.5 mL of the resulting solution were mixed with 0.5 mL of a formic acid solution (0.1%). The solution was diafiltered through Sartorius Vivaspin 2 filters (nominal molecular cut-off 10 KDa) by using an Amicon Micropartition system MPS-1. The filtrate was dried under nitrogen, redissolved in water with formic acid (0.1% *v/v*), and analyzed by ultra-high performance liquid chromatography/electrospray ionisation mass spectrometry (UPLC/ESI-MS).

2.3. Simulated In Vitro Gastro-Intestinal Digestion

Following the procedure described by Minekus et al. [22], 2.5 g of finely ground cocoa were digested. Briefly, three main steps were carried out: salivary phase, gastric phase and intestinal phase. 3.5 mL of simulated salivary fluid (15.1 mM KCl, 3.7 mM KH_2PO_4, 13.6 mM $NaHCO_3$, 0.15 mM $MgCl_2$ and 0.06 mM $(NH_4)_2CO_3$) were added to the sample, together with 0.5 mL of amylase solution (1500 U/mL), 25 µl of calcium chloride (300 mM) and 0.975 mL of distilled water. Samples were briefly stirred using a vortex and incubated for 2 min at 37 °C on a reciprocating shaker (Stuart Scientific, Staffordshire, UK). Then, 7.5 mL of simulated gastric fluid (6.9 mM KCl, 0.9 mM KH_2PO_4, 25 mM $NaHCO_3$, 47.2 mM NaCl, 0.1 mM $MgCl_2$ and 0.5 mM $(NH_4)_2CO_3$) were added to the samples, together with 1.6 mL of pepsin solution (25,000 U/mL), 5 µL of calcium chloride (300 mM), 0.2 mL of 1 M hydrochloric acid and 0.695 mL of distilled water. The pH was adjusted to 3 with 1 M HCl and the samples briefly stirred using a vortex and incubated for 2 h at 37 °C on a reciprocating shaker (Stuart Scientific, Staffordshire, UK). Finally, 11 mL of simulated intestinal fluid (6.8 mM KCl, 0.8 mM KH_2PO_4, 85 mM $NaHCO_3$, 38.4 mM NaCl and 0.33 mM $MgCl_2$) were added, together with 5 mL of pancreatin solution (800 U/mL), 2.5 mL of bile solution (75 mg/mL), 400 µL of calcium chloride (300 mM), 150 µL of 1 M sodium hydroxide and 1.31 mL of distilled water. The pH was adjusted to 7 using NaOH 1 M and the samples briefly stirred using a vortex and incubated for 2 h at 37 °C on a reciprocating shaker (Stuart Scientific, Staffordshire, UK). To stop the digestion and inactivate enzymes, samples were heated at 100 °C for 10 min.

After cooling, 1 mL of the internal standard Phe-Phe (1 mM) is added and the solution acidified by HCl (pH 1–2). Samples are centrifuged for 30 min at 4000 rpm at 4 °C to precipitate insoluble proteins and undigested compounds, then the aqueous phase is separated and subjected to two extractions with diethyl ether to remove lipids and then treated as previously reported for undigested samples.

2.4. Ultra-High Performance Liquid Chromatography/Electrospray Ionisation Mass Spectrometry (UHPLC/ESI-MS) Analysis Conditions

Peptides were analyzed by a UHPLC/ESI-MS system (ACQUITY Ultra Performance LC, WATERS, Milford, MA, USA) in the following conditions. Eluent A: water with 0.1% (*v/v*) formic acid and 0.2% (*v/v*) acetonitrile; eluent B acetonitrile with 0.1% (*v/v*) formic acid; gradient elution was performed according to the following steps: 0–7 min isocratic 100% A, 7–50 min linear gradient from 100% A to 50% A, 50–52 min isocratic 50% A, 53–58 min from 50% A to 0% A and reconditioning. Column: AQUITY UPLC BEH C18 (1.7 µm, 2.1 mm × 150 mm). Column temperature was 35 °C. Injection volume was 2 µL; flow rate: was 0.2 mL/min. MS conditions: ESI, positive ions, single quadrupole analyzer. Capillary voltage: 3.2 kV; cone voltage: 30 V; source temperature: 150 °C; desolvation temperature: 300 °C; cone gas flow (N2): 100 L/h; desolvation gas (N2): 650 L/h; acquisition: 100:2000 *m/z*. All data were acquired and processed by the software MassLynx 4.0 (Waters, Milford, MA, USA).

2.5. Peptides Identification by High-Performance Liquid Chromatography/Tandem Mass Spectrometry (HPLC/MS-MS)

Low-resolution mass spectrometry (LRMS) analysis was performed on digested cocoa samples, in order to identify the amino acid sequences of the peptides. LRMS analysis was performed in positive mode accordingly to Prandi et al. [23]. Prior to LC-MS analysis, samples were centrifuged at 6708× g, 4 °C for 10 min to precipitate insoluble compounds. Chromatographic separation was achieved using a reverse phase column (Aeris Peptide 1.7 µm XB-C18, 150 mm × 2.10 mm, Phenomenex, Torrance, CA, USA) in an UHPLC system (Dionex Ultimate 3000, Thermo Scientific, Waltham, MA, USA). Eluent A was water with 0.1% (*v/v*) formic acid and 0.2% (*v/v*) acetonitrile, eluent B was acetonitrile with 0.1% (*v/v*) formic acid and 0.2% water. Flow was maintained at 0.2 mL/min and the gradient applied was: 0–7 min, 100% A; 7–50 min, from 100% A to 50% A; 50–52.6 min, 50% A; 52.6–53 min, from 50% A to 0% A; 53–58.2 min, 0% A; 58.2–59 min, from 0% A to 100% A; 59–72 min, 100% A. Total run time: 72 min; column temperature: 35 °C; sample temperature: 18 °C; injection volume: 2 µL. Detection was achieved using a triple quadrupole TSQ Vantage (Thermo Scientific, Waltham, MA, USA) using the following parameters: positive ion mode, acquisition time: 7–58.2 min (7 min of solvent delay were applied at the beginning of the chromatographic run), acquisition range: 100–1500 *m/z*; micro scans: 1; scan time: 0.50; Q1 PW: 0.70; spray voltage: 3200 V; capillary temperature: 250 °C; vaporizer temperature: 250 °C; sheath gas flow: 22 units. Different collision energies (CE) were applied depending on the mass and charge of the ion to be fragmented. Peptides fragments were also compared with cocoa peptide sequences reported by D'Souza et al. [11].

2.6. Peptides Quantification

Peptides from digested and undigested samples were quantified by comparison to the internal standard (Phe-Phe), assuming a response factor equal to 1. For the correct integration of peaks, the extract ion chromatogram (XIC) technique was applied. The oligopeptides were semi-quantified by measuring the ratio between the XIC peptide area and the relative XIC area of Phe-Phe, as previously described [24]. Data obtained were expressed as mg of peptide respect to cocoa sample assuming that the specific response factor for each peptides is equal to 1, which is certainly not the case. With this limitation in mind, the quantitative data were used mostly for comparative purposes and to highlight trends and differences among samples.

2.7. Determination of In Vitro Angiotensin-Converting Enzyme (ACE) Inhibitory Activity of Cocoa Digestates

The percentage of ACE inhibitory activity for digested cocoa samples was determined by using the methods of Cushman et al. [25] with some modifications as reported by Dellafiora et al. [26] and according to the following equation:

$$I\% = [(ACEmax - Bmax) - (ACEmin - Bmin)]/(ACEmax - Bmax) \times 100 \tag{1}$$

where ACEmax is the maximum activity of ACE (in the absence of the cocoa peptides), ACEmin is the minimal activity of ACE (in the presence of the peptides), Bmax is the control blank of ACE and Bmin is the control blank of sample/pure peptide.

The following solutions were prepared: sodium borate buffer (0.1 M, NaBB) with NaCl (300 mM), pH 8.3; potassium phosphate buffer (0.01 M, KPB) with NaCl (500 mM), pH 7; 5 mM hippurylhistidyl-leucine (HHL) in NaBB buffer; and ACE 0.1 U/mL in KPB + 5% glycerol (g/mL). The experiment was carried out at 37 °C in a thermostatic bath with the following parameters: maximum activity of ACE (ACEmax) = 200 µL of HHL + 80 µL of digestion blank + 20 µL of ACE; control blank of ACE (Bmax) = 200 µL of HHL + 80 µL of NaBB + 20 µL of KPB; minimal activity of ACE (ACEmin) = 200 µL of HHL + 80 µL of sample (i.e., fraction or pure peptide) + 20 µL of ACE; control blank of sample (i.e., fraction or pure peptide) (Bmin) = 200 µL of HHL + 80 µL of sample (i.e., fraction or pure peptide) + 20 µL of KPB. After 60 min of incubation, the reaction was quenched with 250 µL of HCl (1 N).

Because the digested cocoa extract is a complex mixture containing several compounds, each of which could possess a potential ACE inhibitory activity, solutions containing compounds representative of the main cocoa molecular classes present in the extracts (theobromine for methylxanthines, epicatechin for polyphenols and free amino acids) and simulating the real concentrations contained in the cocoa extract were also prepared. The analysis of HHL and hyppuric acid (HA) was performed by UPLC-ESI-MS in the following conditions: eluent A: H_2O (0.2% CH_3CN and 0.1% HCOOH); eluent B CH_3CN (0.1% HCOOH); gradient elution was performed according to the following steps: 2 min isocratic 100% A, 2–6 min linear gradient from 100% A to 0% A and reconditioning. Column: AQUITY UPLC BEH C18 (1.7 µm, 2.1 mm × 150 mm). Flow rate: 0.25 mL/min. MS conditions: ESI, negative ions, single quadrupole analyzer. Capillary voltage: 2 kV; cone voltage: 30 V; source temperature: 150 °C; desolvation temperature: 300 °C; cone gas flow (N2): 100 L/h; desolvation gas (N2): 650 L/h; acquisition: 80:1000 *m/z*. ACE activity was determined by calculating the ratio of HA and HHL, utilizing as quantification ions *m/z* 178 and 428 respectively.

Antihypertensive activity was determined on cocoa digestate NaBB solutions, containing 50 mg cocoa mass/mL, corresponding to an approximate range of peptides of 10–30 µg/mL (quantitative amount are reported in Table 1).

2.8. Prediction of ACE Inhibitory Activity of Cocoa Digestate Peptides by Computational Procedures

2.8.1. Pharmachopore Models

The anatomy of the open and closed ACE binding sites was investigated by using the Flapsite tool of FLAP software (Fingerprint for Ligand And Protein; http://www.moldiscovery.com, Hertfordshire, UK) [27], and the GRID molecular interaction fields (MIFs) was used to investigate the corresponding pharmacophoric space. The DRY probe was used to describe the potential hydrophobic interactions, while the sp2 carbonyl oxygen (O) and the neutral flat amino (N1) probes were used to describe the hydrogen bond donor and acceptor capacity of the target, respectively. All images were obtained using the software PyMol version 1.7 (http://www.pymol.org, Schrodinger, LLC, New York, NY, USA).

2.8.2. Molecular Modelling

The models for both C- and N-domains of ACE were derived from the Protein Data Bank (http://www.rcsb.org) structures having PDB codes 4APH and 4BZS, respectively. Protein structures and ligands were processed by using the software Sybyl, version 8.1 (www.tripos.com, Certara USA, Inc., Princeton, NJ, USA). All atoms were checked for atom- and bond-type assignments. Amino- and carboxyl-terminal groups were set as protonated and deprotonated, respectively. Hydrogen atoms were computationally added to the protein and energy-minimized using the Powell algorithm whit a coverage gradient of ≤0.5 kcal (mol Å)–1 and a maximum of 1500 cycles.

Anatomy of the pocket. The two catalytic domains of ACE originated from tandem gene duplication (ref) and maintain the same 3D organization with 51% of sequence identity (according to global alignment by using the Needleman–Wunsch algorithm; http://www.ebi.ac.uk/Tools/psa/emboss_needle). Both domains hold a huge pocket with similar shape which crosses the entire protein body. However, analyses were focused on catalytic sites retracing the mode of action of inhibitory drugs (Yates et al., 2014). The regions lining the catalytic site maintain the same organization in both domains and both pockets share a prevalently hydrophobic environment, albeit they differ for 7 amino acid substitutions.

Table 1. Peptides identified in cocoa and derived products before and after digestion, semi-quantitative amounts (mg/kg cocoa) and reported bioactivity from BIOPEP database. Quantitative amounts are calculated utilizing the ratio of peptide area vs. internal standard area (phe-phe), assuming that all response factors are equal to 1.

Retention Time (min)	MH$^+$	Sequence	Cocoa Protein (Vicilin, V; Albumin, 21k)	Bioactivity	Cocoa Bean (Congo) before Digestion	Cocoa Bean (Congo) after Digestion	Cocoa Bean (Dominican Rep.) before Digestion	Cocoa Bean (Dominican Rep.) after Digestion	Cocoa Paste before Digestion	Cocoa Paste after Digestio	Chocolate before Digestion	Chocolate after Digestion
10.62	203.2	AI	v	Angiotensin-converting enzyme (ACE) inhibitor	35.8	6.0	6.8	2.2	5.3	5.9	1.7	4.1
12.99	403.4	RLD	21k		2.2	0.0	11.7	1.4	0.5	0.3	0.1	0.3
13.02	295	FE			7.8	5.0	41.4	21.6	2.2	4.0	0.8	3.1
13.13	485.3				4.0	0.2	43.7	0.3	2.8	0.2	0.6	0.0
13.23	302	SPV			4.9	3.8	29.6	21.1	0.4	2.6	0.2	1.5
13.26	290.2	GTI	v		0.2	1.6	1.5	0.2	0.2	1.9	0.0	2.1
13.35	237.2	MS			1.4	1.3	19.4	8.8	1.0	1.6	0.4	1.0
13.42	281.2	DF		ACE inhibitor	3.1	4.4	28.4	14.1	1.6	4.2	0.6	3.4
13.5	223.2	FG	V	ACE inhibitor (FG)	14.2	5.6	51.4	27.3	3.5	4.2	1.2	2.8
13.74	229.3	PI (L)	21k/v	dipeptidyl peptidase IV inhibitor (PI, PL); ACE inhibitor (PL)	4.0	7.1	20.3	12.7	2.3	12.0	0.8	13.3
13.81	231.2	L (I) V	v/21k	glucose uptake stimulating peptide (IV, LV); dipeptidyl peptidase IV inhibitor (LV)	13.7	17.4	2.0	0.0	2.0	27.0	0.7	30.2
14	276.2	QE		dipeptidyl peptidase IV inhibitor	9.1	6.1	28.2	8.9	2.6	1.7	1.1	1.3
14.35	761.4	RRSDLD	21k		0.0	0.3	0.3	0.4	0.0	0.3	0.0	0.3
14.35	231.2	VI (L)	v/21k	dipeptidyl peptidase IV inhibitor (VI, VL), glucose uptake stimulating peptide (VL)	8.2	7.1	32.7	26.1	1.9	6.1	0.7	5.3
14.45	223.2	GF		ACE inhibitor, dipeptidyl peptidase IV inhibitor	3.5	4.0	0.0	4.2	0.3	5.8	0.1	6.7
14.64	237.2	AF		dipeptidyl peptidase IV inhibitor; ACE inhibitor	17.0	4.3	33.7	14.6	2.9	4.4	0.9	2.8
14.83	231.2	L (I) V	v/21k	glucose uptake stimulating peptide (IV, LV); dipeptidyl peptidase IV inhibitor (LV)	8.1	12.0	17.7	11.8	1.9	8.7	0.7	7.4
14.9	634.3	VSTDVN	21k		0.3	0.3	11.8	5.1	0.3	0.9	0.0	0.5
14.95	229	PI (L)		dipeptidyl peptidase IV inhibitor (PI, PL); ACE inhibitor (PL)	11.8	7.4	23.4	18.9	3.5	8.2	1.3	8.8
15.1	431	unk			7.5	9.2	52.5	28.4	2.5	2.7	0.7	1.2
15.24	487.3	ANSPV	21k		2.6	2.7	27.5	25.9	1.5	4.9	0.4	2.9

Table 1. *Cont.*

Retention Time (min)	MH$^+$	Sequence	Cocoa Protein (Vicilin, V; Albumin, 21k)	Bioactivity	Cocoa Bean (Congo) before Digestion	Cocoa Bean (Congo) after Digestion	Cocoa Bean (Dominican Rep.) before Digestion	Cocoa Bean (Dominican Rep.) after Digestion	Cocoa Paste before Digestion	Cocoa Paste after Digestio	Chocolate before Digestion	Chocolate after Digestion
15.32	295	EF		CaMPDE inhibitor; Renin inhibitor (HYPOTENSIVE)	6.3	4.4	23.1	11.9	1.3	2.8	0.3	2.7
15.44	838.4	DEEGNFK	v		0.1	0.0	2.4	0.0	0.1	0.0	0.0	0.0
15.77	231.2	VI (L)	v/21	dipeptidyl peptidase IV inhibitor (VI, VL), glucose uptake stimulating peptide (VL)	44.6	12.5	17.7	11.8	6.7	11.6	2.3	6.8
15.8	488.3	GAGGGGL	v		4.8	3.6	28.4	12.5	0.6	0.8	0.1	0.5
16.06	296.2	YN		dipeptidyl peptidase IV inhibitor, ACE inhibitor	6.9	7.3	14.8	13.1	1.7	3.7	0.7	1.8
16.25	263.3	FP		dipeptidyl peptidase IV inhibitor; ACE inhibitor	3.8	5.3	10.7	7.5	0.8	3.1	0.3	3.5
16.37	379.9 (757.4)	ASKDQPL	v		1.3	0.2	4.4	0.7	1.4	0.6	0.3	0.3
16.8	265.3	FV	v/21k		7.6	6.6	30.2	18.7	2.3	4.7	1.0	3.0
17.16	245.2	II IL LL LI		ACE inhibitor (IL), glucose uptake stimulating peptide; dipeptidyl peptidase IV inhibitor	5.3	7.1	9.1	6.8	0.8	7.2	0.2	7.8
17.3	276.1	AW	21k	ACE inhibitor; antioxidant; dipeptidyl peptidase IV inhibitor	8.6	5.7	29.1	13.9	1.1	1.9	0.1	0.8
17.86	360.3	VLE	V		0.1	4.4	0.6	0.0	0.5	0.8	0.1	0.7
18.1	265.2	VF	V	ACE inhibitor; dipeptidyl peptidase IV inhibitor	19.6	10.7	56.6	27.8	3.8	6.8	1.2	4.6
18.19	393.3	FLN/SSIS	V/21k		2.3	0.6	17.6	8.1	1.2	1.4	0.3	0.5
18.19	245.2	II IL LL LI		ACE inhibitor (IL), glucose uptake stimulating peptide; dipeptidyl peptidase IV inhibitor	4.1	19.0	0.0	10.8	0.8	7.5	0.3	7.5
18.22	710.4	DEEGNF	V		1.1	0.1	20.8	0.6	0.8	0.2	0.1	0.3
18.57	243.2	Pyroglu-LEU			9.3	13.9	29.8	35.2	2.7	6.8	1.1	5.6
18.69	245	II IL LL LI		ACE inhibitor (IL), glucose uptake stimulating peptide; dipeptidyl peptidase IV inhibitor	19.6	15.8	0.0	9.9	4.1	11.8	1.3	8.5
18.93	690.3	NGKGTIT	V		0.2	0.1	5.5	2.0	0.1	0.1	0.0	0.1
19.05	328	VPI			35.7	25.5	133.3	72.4	1.9	7.7	0.6	5.6
19.53	243.2	Pyroglu-ILE			6.5	7.9	24.2	17.2	2.5	4.4	1.0	3.9
19.7	245.2	II IL LL LI		ACE inhibitor (IL), glucose uptake stimulating peptide; dipeptidyl peptidase IV inhibitor	7.2	9.3	20.2	12.8	1.8	9.8	0.5	8.8
19.85	534.3	PGDVF	V		0.0	0.0	17.5	0.0	0.6	0.1	0.2	0.0
19.89	933.6	DSKDDVVR	21k		0.1	0.0	2.8	0.0	0.1	0.1	0.0	0.1
20.16	564.3	RRSF	V		0.6	0.0	5.5	1.1	0.2	0.1	0.1	0.2

Table 1. *Cont.*

Retention Time (min)	MH+	Sequence	Cocoa Protein (Vicilin, V; Albumin, 21k)	Bioactivity	Cocoa Bean (Congo) before Digestion	Cocoa Bean (Congo) after Digestion	Cocoa Bean (Dominican Rep.) before Digestion	Cocoa Bean (Dominican Rep.) after Digestion	Cocoa Paste before Digestion	Cocoa Paste after Digestio	Chocolate before Digestion	Chocolate after Digestion
20.22	279.3	L (I) F		ACE inhibitor	7.1	7.7	27.3	13.3	1.6	4.3	0.8	3.2
20.28	563.3	DEEGN	V		3.9	0.0	21.8	0.0	0.8	0.3	0.1	0.3
20.32	360.3	EVL	V		2.6	0.0	4.4	3.4	0.3	0.3	0.1	0.2
20.58	862.5	SSISGAGGGGL	21K		0.3	0.1	12.5	3.8	0.3	0.7	0.1	0.4
21	437.3	GDVF	V		0.6	0.2	5.5	0.0	0.6	0.1	0.2	0.2
21.16	279.3	L (I) F		ACE inhibitor	24.5	18.3	74.7	60.0	4.9	19.1	1.8	11.8
21.19	600.4	KDQPL	V		0.8	0.0	29.4	0.0	0.9	0.0	0.2	0.2
21.31	277	Pyroglu-PHE			8.8	9.0	24.1	22.0	3.2	4.2	1.3	2.2
21.4	380.2	DVF	V		8.0	0.1	25.0	1.9	1.6	0.3	0.6	0.2
22.09	279.3	FL (I)		dipeptidyl peptidase IV inhibitor (FL)	3.6	3.3	30.2	0.0	2.3	4.2	0.7	3.2
22.41	279.3	FL (I)	v		9.4	2.2	30.2	0.0	2.3	3.2	0.7	1.9
23	621.5	SPGDVF	V		0.4	0.0	37.7	0.5	1.6	0.1	0.5	0.0
23.96	408.3	IEF	21K		1.9	0.0	17.9	0.4	1.8	0.1	0.5	0.0
25.69	603.3 (1204)	SNADSKDDVVR	21k		0.0	2.3	0.3	0.0	0.0	2.2	0.0	4.1
26.31	789.6	TVWRLD	21K		0.0	0.6	0.2	0.1	0.0	0.0	0.0	0.3
27.59	601.3	NNKPE			0.0	3.6	0.7	0.0	0.0	1.3	0.0	1.7
27.92	747.5	NGTPVIF	21K		0.6	0.0	12.7	0.4	0.1	0.0	0.0	0.1
28.13	533.1 (1063.5)	DEEGNFKIL	V		0.0	0.0	2.3	0.4	0.2	0.1	0.0	0.0
28.54	902.6	APLSPGDVF	V		0.0	0.0	1.2	0.0	0.3	0.0	0.1	0.0
29.76	820.5	DNEWAW	21K		0.1	0.0	2.2	0.0	0.0	0.1	0.0	0.2
				Total peptides	427.9	313.2	1348.7	655.0	100.0	242.3	33.1	203.2

2.8.3. Docking Simulations and Rescoring Procedure

The coupling of GOLD, to perform docking simulations, and HINT (Hydrophatic INTeraction) software, as re-scoring function, has been already proved to be effectively able to evaluate the bioactivity of small molecules [28–31] including peptides [32]. The docking simulations of compounds were performed with the GOLD version 5.1 (CCDC, Cambridge, UK; http://www.ccd.cam.ac.uk). All crystallographic waters and ligands were removed and 25 poses for each compound were generated. No constraints were set up, and the explorable space was defined in a radius of 10 Å from the centroid of the catalitic site. For each GOLD docking search, a maximum number of 100,000 operations were performed on a population of 100 individuals with a selection pressure of 1.1. Operator weights for crossover, mutation and migration were set to 95, 95, and 10, respectively. The number of islands was set to 5 and the niche to 2. The hydrogen bond distance was set to 2.5 Å and the van der Waals linear cut-off to 4.0. Ligand flexibility options "flip pyramidal N", "flip amide bonds", and "flip ring corners" were allowed. Each best scored pose according to GOLD scoring function was re-scored by HINT. Owing to the huge dimension of the pocket, the molecules' positioning has been spatially restrained according to crystallographic pose of the inhibitory drug captopril.

The software HINT [33] was used as the re-scoring function on the basis of previous studies attesting the higher reliability of HINT scoring with respect to other scoring functions, as well as of its successful use in the search for ligands for other targets and in the estimation of ligand binding free energies. In more detail, the score provides the evaluation of thermodynamic benefits of protein-ligand interaction, and therefore low/negative scores indicate not appreciable protein-ligand interactions ([29–32,34]). GOLD uses a Lamarckian genetic algorithm and scores may slightly change from run to run. Therefore, in order to exclude a non-causative score assignment, we conducted simulations in quintuplicate and the mean values are reported.

2.8.4. Chemical Synthesis of Specific Peptides Predicted as Active

Peptides VPI and SPV were synthesized on solid phase according Fmoc/t-butyl strategy using a Syro I Fully Automated Peptide Synthesizer (Biotage, Uppsala, Sweden). The peptides were cleaved from the Wang resin using a TFA:TIS:H$_2$O (95:2.5:2.5) solution, precipitated with diethyl ether and desalted on Sep-Pack C18 cartridges (Waters Corporation, Milford, MA, USA). Characterization MH$^+$ (ESI–MS): 328.29 VPI, 302.15 SPV. Each peptide was tested as reported above for its ACE inhibitory activity, using 80 µL of NaBB solution as blank sample for ACEmax. In this case, the specific IC50 was also calculated. The IC50 value is defined as the inhibitor concentration that is able to decrease ACE activity by 50%. To determine IC50, different concentrations of peptides were prepared and their relative ACE inhibitory activity was evaluated. IC50 values were determined by plotting the percentage relative inhibition as a function of concentration of test compound.

3. Results and Discussion

The gastrointestinal digestion process has an influence not only on the hydrolysis of peptides still present in food, but also on the release of peptide sequences encrypted in food proteins. In order to have a complete picture of the fate of cocoa peptides during the digestion process, both those already present and those released by the proteins, different cocoa samples were analysed before and after simulated gastrointestinal digestion. In particular, fermented cocoa beans (currently used as healthy food or as ingredients in cocoa and bakery products), cocoa paste and dark chocolate were considered. The digestion procedure was adapted from Minekus et al. [22]: the method mimics the subsequent steps of the digestion process in terms of composition of the juices in the different compartments (simulation of salivary juice, gastric juices, duodenal juice and bile) as well as the relative residence times.

Typical MS chromatograms (full scan acquisition) of a cocoa bean sample of Congo origin are shown in Figure 1, showing the aqueous acidic extract containing peptides, obtained before digestion, compared with that of the digesta.

The quali-quantitative peptide profile of digesta from fermented cocoa beans is quite different with respect to that of the extract before digestion. The main peptides in the chromatograms were identified on the basis of data previously reported [8,9] on undigested cocoa beans, and some new peptides were further identified by HPLC/ESI-MS-MS. The complete list of peptides identified is reported in Table 1. In Figure 2 the amounts of total peptides in samples before and after digestion were compared, in order to better understand the formation/degradation of peptides in simulated digestion. Results showed that gastrointestinal digestion has different effects depending on the cocoa products: total peptide amount in fermented cocoa bean digested samples is generally reduced with respect to the corresponding non-digested samples (Figure 2), while in cocoa products (cocoa paste and chocolate) the amount of peptides released during digestion and resistant to the protease activities is higher respect the amount before digestion. This behaviour could be to the effect of thermal treatment occurring in cocoa beans to produce cocoa paste and chocolate. In the case of chocolate, it has to be taken into account also that other ingredients are added, such as sucrose, cocoa butter, lecithin etc. which probably have an effect on peptide bioaccessibility from proteases.

As far as the original peptides present in cocoa, many peptides seem to resist to simulated gastrointestinal digestion, even if to different extent. Peptides showing the higher decrease upon digestion are DVF, GDVF, IEF, PGDVF, SPGDVF, KDQPL, DEEGNFKIL, all having Phe or Leu as C-terminal. Some peptides are most abundant in digested samples respect to undigested, especially in cocoa-derived products (cocoa paste and chocolate), indicating that they are formed during gastrointestinal digestion. The peptides released are mainly dipeptides, e.g., II, LL, LI, IL, PI, PL, PI, VI, VL, AI. This behavior is in line with those of other food matrices, both animal and vegetal, as for example raw ham, cheese, soybean, which are recognized as a source of bioactive peptides mainly formed during digestion [35–38].

The pattern of resistant peptides after in vitro gastrointestinal digestion is of particular significance for the eventual bioactivity of a cocoa nitrogen fraction.

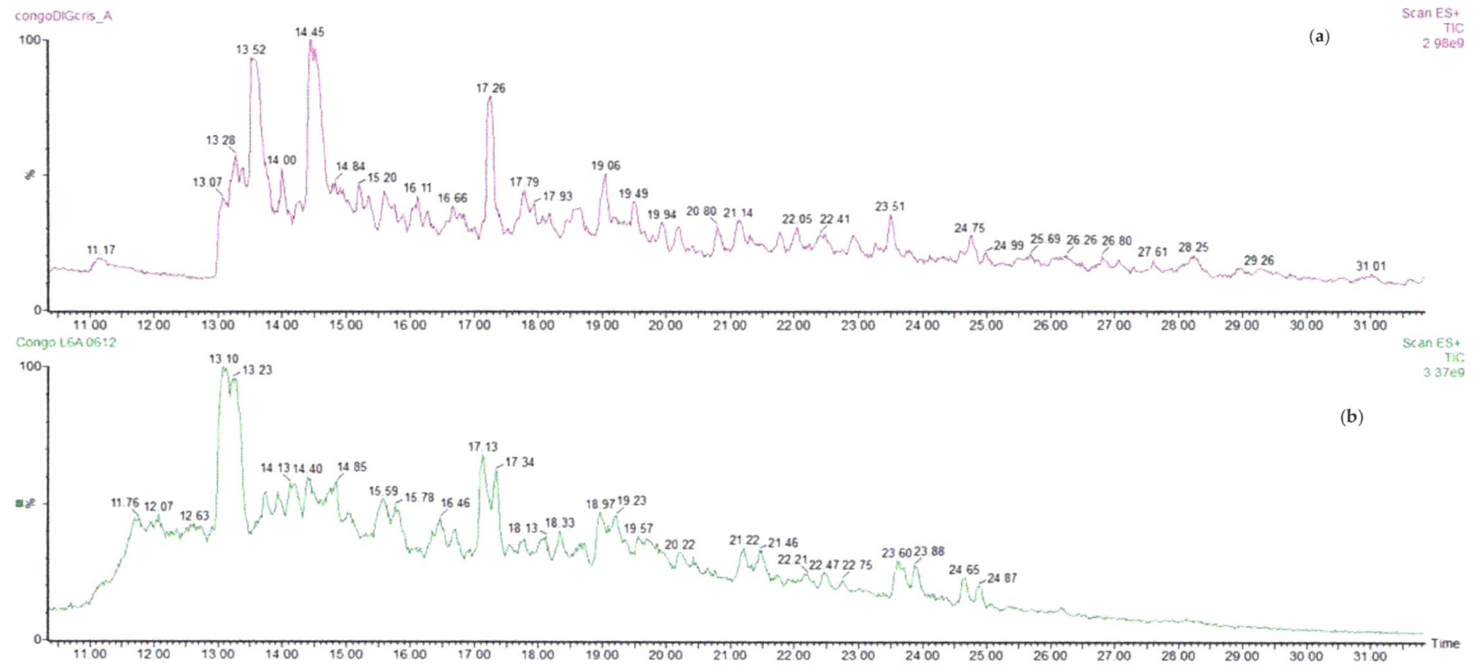

Figure 1. Ultra-high performance liquid chromatography/electrospray ionisation mass spectrometry (UPLC/ESI-MS) chromatogram of a digested sample of Congo fermented cocoa bean (**a**) compared to the peptide profile of the corresponding not digested, extracted sample (**b**).

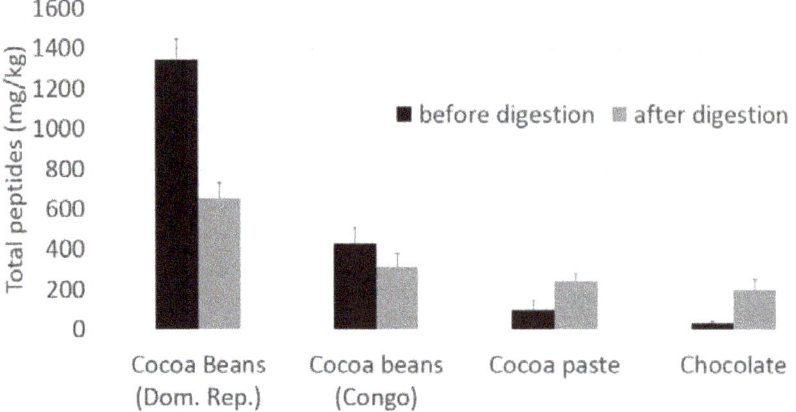

Figure 2. Peptides amount (mg/kg) in digested sample compared to the amount in the respective non-digested samples. Peptides were quantified as ratio of peptides area vs. internal standard area (phe-phe) and referred to the initial cocoa amount, making the raw assumption that the specific response factor for each peptides is equal to 1. The total peptides amount is referred to the sum of peptides listed in Table 1.

3.1. Potentially Bioactive Peptides

We identified for the first time the sequences of peptides derived from cocoa protein resistant to (or formed during) in vitro gastrointestinal digestion and, as a consequence, potentially bioaccessible, so it could be of interest to evaluate their potential bioactivity. It is known that in order to stimulate a biological response, the peptides must be bioavailable, i.e., following digestion, they must be able to cross the intestinal epithelial cells and enter the blood circulatory system, or produce local effects in the gastrointestinal tract [39]. The question of the absorption of oligopeptides is paramount to the science of food-derived bioactive peptides and it has been recently reviewed [40]. Although the complete mechanism of absorption and the bioavailability of the specific peptides were not investigated, there is some evidence that small food bioactive peptides are bioavailable and can be absorbed into the body [41]: whereas there is evidence for some uptake of intact di- and tripeptides from the human gastrointestinal tract, such uptake is not ubiquitous and there is little support for the uptake of tetrapeptides and larger peptides. Moreover, it is not completely clear what are the effects of cytosolic, vascular endothelial tissue peptidases and soluble plasma peptidases, and the half-life of many peptides in the plasma seems to be very short. Despite all these considerations on the limited evidence for the effective health effects of peptides in humans, the constant growthe of literature concerning bioactive peptides in food is a testament to the large interest in this field from the scientific community. Therefore, it has resulted in interest in predicting the potential bioactivity of new peptide sequences from new food sources, as in the case of cocoa.

As a first approach, we evaluated the bioactive potential of cocoa peptide formed during digestion using the BIOPEP database (http://www.uwm.edu.pl/biochemia/index.php/pl/biopep). Because no studies are present in literature reporting the sequences of bioactive peptides in cocoa, many of the peptides found in this work are not listed in the BIOPEP database. However, some cocoa peptide sequences, mainly dipeptides, found in digested samples, match sequences reported in the BIOPEP database, and their predicted activities are reported in Table 1. Cocoa samples showed high occurrence of sequences of angiotensin-converting enzyme-inhibitor peptides (ACE) as well as of peptides having glucose uptake stimulating activity and dipeptidyl peptidase IV inhibitor, suggesting some possible activities on blood pressure and diabetes, in agreement with the results of the previous works reporting the antioxidant, ACE inhibitory and hypoglycaemic activities of cocoa autolysates [18,19].

3.2. ACE Tests on Cocoa Digested Samples

Among the possible bioactivity of cocoa peptides, we focused on the angiotensin I converting enzyme (ACE) inhibitory activity, due to the beneficial effects of ACE inhibitory peptides on hypertension, representing one of the major risk in cardiovascular diseases [42]. ACE is an extracellular enzyme expressed in many types of endothelial cells, especially in the capillaries of the lung, as well as in epithelial cells in the kidney, small intestine and epididymis [43]. Bioactive peptides are currently viewed as a nutraceutical approach to moderately control the blood pressure level, even as prevention in healthy subjects, without bringing a sharp decrease of the pressure or other side effects.

The aqueous peptide fractions of digestate of cocoa bean samples and chocolate were tested for their ACE inhibitory activities. The digestates tested contained 50 mg/mL of cocoa sample, corresponding to an approximate range of peptide amount of 10–30 µg/mL, as resulted from the quantitative analysis of cocoa peptides in digestates reported in Table 1.

Because the aqueous fraction of cocoa digestate contained also significant amounts of other cocoa hydrosoluble compunds (mainly free amino acids, epicatechin and theobromine), separate solutions (model systems) of theobromine, epicatechin and a mixture of amino acids in the proportion found in cocoa proteins were prepared, with concentrations simulating those actually present in the digested fractions of cocoa products. These solutions were tested for their specific ACE inhibitory activity.

From the results of ACE inhibition (I%) reported in Table 2, it is possible to highlight that the activity of the ACE remains very high in the presence of the mixture of AA, theobromine and epicatechin, indicating that these components have a weak ACE inhibitory activity and do not significantly contribute to the activity that follows registered for cocoa samples.

Instead, the activity of ACE was strongly inhibited when cocoa beans, cocoa paste and chocolate were tested, indicating a strong ACE inhibitory activity especially for cocoa beans. This is coherent with the higher amount of peptides present in cocoa bean solution with respect to cocoa paste and chocolate, confirming that the peptide fraction is the most active against ACE. Consuming fermented beans is not so common, even if some products start to appear on the market (for example some cookies containing a part of cocoa beans), due to the higher polyphenol amount and, therefore, the better health properties. Anyway, the percentage of inhibition reached is higher than 50% (IC_{50}) for all the cocoa samples tested, comprising chocolate, representing a common cocoa product consumed all over the world. Considering that the 50% inhibition is obtained in this work digesting 2.5 g of cocoa beans/cocoa products, it is evident that it is possible to obtain a significant concentration of cocoa peptide by consuming a normal cocoa portions (30 g, as reported by LARN (Livelli di Assunzione Raccomandata di Nutrienti per la popolazione italiana, 2014 [44]). Therefore, one can hypothesize that physiologically relevant concentrations might be reached after dietary consumption of cocoa products, at least in the intestinal lumen.

Table 2. ACE inhibitory activity of cocoa samples digestates, reported as percentage of inhibition respect to ACE maximum activity (relative to digestion blank).

Solution	Concentration (mg/mL)	Esteemed Peptide Concentration (µg/mL)	I%
Theobromine	0.3	0	19 ± 5
Epicatechin	0.2	0	20 ± 5
Amino acid mixture	5	0	5 ± 4
Cocoa bean digestate	50	33	95 ± 4
Cocoa paste digestate	50	12	63 ± 10
Chocolate digestate	50	10	75 ± 8

3.3. In Silico Screening of Potential ACE Inhibitory Peptides

To go deeper inside into the molecular mechanisms of cocoa ACE inhibition, we proceeded to test the specific bioactivity of each resistant cocoa peptide identified in the digestates using a computational in silico approach. The principal advantage of using this approach is the reduction of the experimental efforts in the early stage of bioactive peptides identification, avoiding matrix isolation, chemical synthesis and an in vitro or in vivo bioactivity test of non-active peptides. Utilizing as a first screening the in silico approach used in this research, it is possible to assign a priority ranking of peptides in terms of ACE-inhibitory activity, with the final aim to guide the choice of peptides to be synthesized and then assessed for their inhibitory activity by using ACE in vitro assay.

ACE is a zinc-dependent carboxypeptidase organized in two distinct catalytic domains. Both hydrolyze substrates by removing one or more dipeptides from the C-terminal end. In particular, antihypertensive activity is based on the ability of several peptides to inhibit ACE, which in vivo converts the decapeptide angiotensin I into the octapeptide angiotensin II, the latter being able to induce vasoconstriction and to increase blood pressure Computational prediction of peptides ACE inhibitory activity is based on the assumption that the protein–ligand interaction is the condition for inhibitory activity. Consequently, in the case of ACE, a peptide is considered active if it is able to interact with at least one of the two domains. Therefore, the coupling of docking simulations and rescoring procedures by using the HINT scoring function—whose correlation with the free energy of binding was previously reported [34]—can be used effectively to predict the inhibitory activity.

Among the entire peptide profile identified after the in vitro physiological digestion of cocoa beans and derived products (Table 1), we focused only on low Mw peptides (di-, tri- and tetrapeptides) for the computational analysis. This inclusion criterion was based on the finding that short peptides may be easily adsorbed trough the intestinal epithelium [41], thus giving a greater physiological significance. The computational results obtained are reported in Table 3, where 19 peptides were predicted as active, 17 in both C-domain and N-domain of ACE, and 2 peptides that were active in only one domain (LLDR in N-domain and QLGN in C-domain).

Table 3. Computational in silico results obtained for the analyzed peptides sequences, presented as HINT (Hydrophatic INTeraction) scores and expected activity of peptides under analysis.

Compound	Cdominio (cutoff 260)		Ndominio (cutoff 540)		TOT
	HS	Interaction	HS	Interaction	
FE	2995	Positive	2443	positive	Active
FV	2494	Positive	2776	Positive	Active
II	2042	Positive	2380	Positive	Active
IL	1935	Positive	2262	Positive	Active
LI	2053	Positive	2045	Positive	Active
LL	1792	Positive	2062	Positive	Active
LY	3110	Positive	2709	Positive	Active
PI	1811	Positive	1717	positive	Active
PL	1992	Positive	1550	Positive	Active
QE	2091	Positive	1466	Positive	Active
SPV	1171	Positive	2075	Positive	Active
VEL	757	Positive	2160	Positive	Active
VF	2406	Positive	2366	Positive	Active
VI	1475	positive	2364	positive	Active
VL	1584	positive	2212	positive	Active
VPI	1258	positive	1707	positive	Active
VPL	1200	positive	1203	positive	Active
LLDR	−913	negative	1224	positive	Active
QLGN	799	positive	413	negative	Active

3.4. ACE in Vitro Test on Pure (Synthetized) Peptides

Two cocoa tripeptides predicted as active by computational procedure (VPI and SPV) were chemically synthetized by the Fmoc protocol on an automatic solid-phase peptide synthesizer (Syro I, Biotage) and tested for their effective in vitro ACE inhibitory activity. Cocoa dipeptides predicted as active were not synthetized and tested because their activity was in most cases still known in literature (BIOPEP database). ACE inhibitory activity of pure peptides VPI and SPV was determined by a UPLC/ESI-MS method, injecting the reaction products of ACE and hyppuryl-hystidyl-leucine as substrate, in presence and in absence of peptides. Both peptides predicted as ACE inhibitors by in silico approach, resulted inhibitors also in vitro. However, they are very weak inhibitors, because the IC50 is still not reached at concentrations higher than 1000 µM (Figure 3). These data suggest that the activity registered for cocoa samples is probably the effect of the synergic action of the whole peptide profile and, in particular, also the di-peptide pattern, could contribute significantly to the effect.

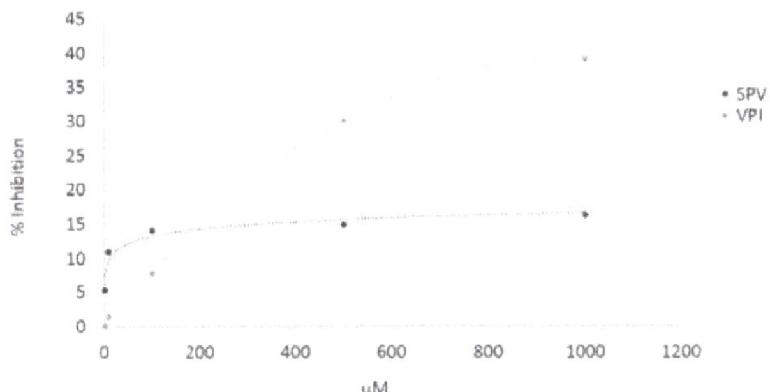

Figure 3. Sigmoid curves obtained by plotting the inhibitory activity of pure peptides SPV and VPI (synthetized by Fmoc protocol) against peptide concentration.

4. Conclusions

The research methodology utilized in this paper poses the bases for the determination of cocoa peptides' bioactivity, a topic largely under-investigated with respect to other food peptides and also with respect to other cocoa components (e.g., polyphenols). Cocoa peptide mixtures obtained after simulated gastrointestinal digestion are able to strongly reduce in vitro ACE activity. Moreover, the in silico approach resulted in an effective tool in prediction of cocoa peptides' bioactivity, giving the opportunity to reduce the experimental work aimed at the synthesis of peptides and bioactivity tests. However, the cocoa tri-peptides predicted as more active, synthetized and effectively tested for their in vitro bioactivity, resulted in being weak inhibitors, suggesting that probably the dipeptide pattern is the main one responsible for the activity registered for cocoa samples. A reasonable amount of the average chocolate portion (30 g) ensures a quantity of peptides after digestion that, assuming complete absorption, could reach almost a complete inhibition of ACE.

This study further supports the notion that cocoa could be an important component of a healthy diet by preventing chronic diseases. Taken as a whole, our findings are an important step forward to the in depth molecular characterization of such a relevant food product beyond its well-known nutritional properties.

Author Contributions: Conceptualization, S.S., P.C., G.G. and A.C.; Data curation, A.M., L.D., V.L. and P.C.; Investigation, A.M., L.D. and B.P.; Methodology, P.C., T.T. and A.C.; Project administration, A.C.; Software, L.D. and P.C.; Supervision, S.S., T.T., G.G. and A.C.; Writing—original draft, A.M. and A.C.; Writing—review & editing, S.S.

Funding: This research received no external funding.

References

1. Keen, C.L. Chocolate: Food as medicine/medicine as food. *J. Am. Coll. Nutr.* **2001**, *20*, 436S–439S. [CrossRef] [PubMed]
2. Katz, D.L.; Doughty, K.; Ali, A. Cocoa and chocolate in human health and disease. *Antioxid. Redox. Signal* **2011**, *15*, 2779–2811. [CrossRef] [PubMed]

3. AAVV. Scientific Opinion on the substantiation of a health claim related to cocoa flavanols and maintenance of normal endothelium-dependent vasodilation pursuant to Article 13 (5) of Regulation (EC) No 1924/2006. *EFSA J.* **2012**, *10*, 2809.

4. Rawel, H.M.; Huschek, G.; Sagu, S.T.; Homann, T. Cocoa bean proteins—Characterization, changes and modifications due to ripening and post-harvest processing. *Nutrients* **2019**, *11*, 428. [CrossRef]

5. Hebert, E.M.; Saavedra, L.; Ferranti, P. Bioactive peptides derived from casein and whey proteins. In *Biotechnology of Lactic Acid Bacteria: Novel Applications*; Mozzi, F., Raya, R., Vignolo, G., Eds.; Wiley-Blackwell: Ames, IA, USA, 2010; pp. 233–249.

6. Sánchez, A.; Vázquez, A. Bioactive peptides: A review. *Food Qual. Saf.* **2017**, *1*, 29–46. [CrossRef]

7. Saavedra, L.; Hebert, E.M.; Minahk, C.; Ferranti, P. An overview of "omic" analytical methods applied in bioactive peptide studies. *Food Res. Int.* **2013**, *54*, 925–934. [CrossRef]

8. Marseglia, A.; Sforza, S.; Faccini, A.; Bencivenni, M.; Palla, G.; Caligiani, A. Extraction, identification and semi-quantification of oligopeptides in cocoa beans. *Food Res. Int.* **2014**, *63*, 382–389. [CrossRef]

9. Caligiani, A.; Marseglia, A.; Prandi, B.; Palla, G.; Sforza, S. Influence of fermentation level and geographical origin on cocoa bean oligopeptide pattern. *Food Chem.* **2016**, *211*, 431–439. [CrossRef] [PubMed]

10. Kumari, N.; Grimbs, A.; D'Souza, R.N.; Verma, S.K.; Corno, M.; Kuhnert, N.; Ullrich, M.S. Origin and varietal based proteomic and peptidomic fingerprinting of Theobroma cacao in non-fermented and fermented cocoa beans. *Food Res. Int.* **2018**, *111*, 137–147. [CrossRef]

11. D'Souza, R.N.; Grimbs, A.; Grimbs, S.; Behrends, B.; Corno, M.; Ullrich, M.S.; Kuhnert, N. Degradation of cocoa proteins into oligopeptides during spontaneous fermentation of cocoa beans. *Food Res. Int.* **2018**, *109*, 506–516. [CrossRef]

12. Zak, D.L.; Keeney, P.G. Changes in cocoa proteins during ripening of fruit, fermentation, and further processing of cocoa beans. *J. Agric. Food Chem.* **1976**, *24*, 483–486. [CrossRef]

13. Voigt, J.; Janek, K.; Textoris-Taube, K.; Niewienda, A.; Woestemeyer, J. Partial purification and characterisation of the peptide precursors of the cocoa-specific aroma components. *Food Chem.* **2016**, *192*, 706–713. [CrossRef]

14. John, W.A.; Kumari, N.; Boettcher, N.L.; Koffi, K.J.; Grimbs, S.; Vrancken, G.; D'Souza, R.N.; Kuhnert, N.; Ullrich, M.S. Aseptic artificial fermentation of cocoa beans can be fashioned to replicate the peptide profile of commercial cocoa bean fermentations. *Food Res. Int.* **2016**, *89*, 764–772. [CrossRef]

15. Spencer, M.E.; Hodge, R. Cloning and sequencing of a cDNA encoding the major storage proteins of Theobroma cacao. *Planta* **1992**, *186*, 567–576. [CrossRef]

16. Kochhar, S.; Gartenmann, K.; Juillerat, M.A. Primary structure of the abundant seed albumin of theobroma cacao by mass spectrometry. *J. Agric. Food Chem.* **2000**, *48*, 5593–5599. [CrossRef]

17. Amin, I.; Jinap, S.; Jamilah, B.; Harikrisna, K.; Biehl, B. Oligopeptide patterns produced from Theobroma cacao L of various genetic origins. *J. Sci. Food Agric.* **2002**, *82*, 733–737. [CrossRef]

18. Sarmadi, B.; Amin, I.; Muhajir, H. Antioxidant and angiotensin converting enzyme (ACE) inhibitory activities of cocoa (*Theobroma cacao* L.) autolysates. *Food Res. Int.* **2011**, *44*, 290–296. [CrossRef]

19. Sarmadi, B.; Farhana, A.; Muhajir, H.; Nazamid, S.; Azizah, A.H.; Amin, I. Hypoglycemic effects of cocoa (*Theobroma cacao* L.) autolysates. *Food Chem.* **2012**, *134*, 905–911. [CrossRef]

20. Tovar-Perez, E.G.; Guerrero-Becerra, L.; Lugo-Cervantes, E. Antioxidant activity of hydrolysates and peptide fractions of glutelin from cocoa (*Theobroma cacao* L.) seed. *CyTA J. Food* **2017**, *15*, 489–496. [CrossRef]

21. Muguerza Marquinez, B.; Monzo Oltra, H.; Alepuz Rico, N.; Bataller Leiva, E.; Genoves Martinez, S.; Enrique Lopez, M.; Martorell Guerola, P.; Ramon Vidal, D. Bioactive Peptides with ACE- and PEP-Inhibiting Activity Obtained from Cocoa Extracts. PCT Int. Appl. WO 2010012845 A1. 4 February 2010.

22. Minekus, M.; Alminger, M.; Alvito, P.; Balance, S.; Bohn, T.; Bourlieu, C.; Carrière, F.; Boutrou, R.; Corredig, M.; Dupont, D.; et al. A standardised static in vitro digestion method suitable for food—An international consensus. *Food Funct.* **2014**, *5*, 1113–1124. [CrossRef]

23. Prandi, B.; Varani, M.; Faccini, A.; Lambertini, F.; Suman, M.; Leporati, A.; Tedeschi, T.; Sforza, S. Species specific marker peptides for meat authenticity assessment: A multispecies quantitative approach applied to Bolognese sauce. *Food Contr.* **2019**, *97*, 15–24. [CrossRef]

24. Sforza, S.; Galaverna, G.; Schivazappa, C.; Marchelli, R.; Dossena, A.; Virgili, R. Effect of extended aging of parma dry-cured ham on the content of oligopeptides and free amino acids. *J. Agric. Food Chem.* **2006**, *54*, 9422–9429. [CrossRef]

25. Cushman, D.W.; Cheung, H.S. Spectrophotometric assay properties of the angiotensin-converting enzyme of rabbit lung. *Biochem. Pharmacol.* **1971**, *20*, 1637–1648. [CrossRef]
26. Dellafiora, L.; Paolella, S.; Dall'Asta, C.; Dossena, A.; Cozzini, P.; Galaverna, G. Hybrid in silico/in vitro approach for the identification of angiotensin i converting enzyme inhibitory peptides from parma dry-cured ham. *J. Agric. Food Chem.* **2015**, *63*, 6366–6375. [CrossRef]
27. Baroni, M.; Cruciani, G.; Sciabola, S.; Perruccio, F.; Mason, J.S. A Common reference framework for analyzing/comparing proteins and ligands. fingerprints for ligands and proteins (FLAP): Theory and application. *J. Chem. Inform. Mod.* **2007**, *47*, 279–294. [CrossRef]
28. Amadasi, A.; Mozzarelli, A.; Meda, C.; Maggi, A.; Cozzini, P. Identification of xenoestrogens in food additives by an integrated in silico and in vitro approach. *Chem. Res. Toxicol.* **2009**, *22*, 52–63. [CrossRef]
29. Cozzini, P.; Dellafiora, L. In silico approach to evaluate molecular interaction between mycotoxins and the estrogen receptors ligand binding domain: a case study on zearalenone and its metabolites. *Toxicol. Lett.* **2012**, *214*, 81–85. [CrossRef]
30. Dellafiora, L.; Mena, P.; Cozzini, P.; Brighenti, F.; Del Rio, D. Modelling the possible bioactivity of ellagitannin-derived metabolites. In silico tools to evaluate their potential xenoestrogenic behavior. *Food Funct.* **2013**, *4*, 1442–1451. [CrossRef]
31. Dellafiora, L.; Mena, P.; Del Rio, D.; Cozzini, P. Modeling the effect of phase II conjugations on topoisomerase I poisoning: Pilot study with luteolin and quercetin. *J. Agric. Food Chem.* **2014**, *62*, 5881–5886. [CrossRef]
32. Salsi, E.; Bayden, A.S.; Spyrakis, F.; Amadasi, A.; Campanini, B.; Bettati, S.; Dodatko, T.; Cozzini, P.; Kellogg, G.E.; Cook, P.F.; et al. Design of O-acetylserine sulfhydrylase inhibitors by mimicking nature. *J. Med. Chem.* **2010**, *53*, 345–356. [CrossRef] [PubMed]
33. Kellogg, G.E.; Abraham, D.J. Hydrophobicity: Is LogP(o/w) more than the sum of its parts? *Eur. J. Med. Chem.* **2000**, *37*, 651–661. [CrossRef]
34. Cozzini, P.; Fornabaio, M.; Marabotti, A.; Abraham, D.J.; Kellogg, G.E.; Mozzarelli, A. Simple, intuitive calculations of free energy of binding for protein-ligand complexes. 1. Models without explicit constrained water. *J. Med. Chem.* **2002**, *45*, 2469–2483. [CrossRef] [PubMed]
35. Escudero, E.; Sentandreu, M.A.; Arihara, K.; Toldra, F. Angiotensin I-converting enzyme inhibitory peptides generated from in vitro gastrointestinal digestion of pork meat. *J. Agric. Food Chem.* **2010**, *58*, 2895–2901. [CrossRef]
36. Paolella, S.; Falavigna, C.; Faccini, A.; Virgili, R.; Sforza, S.; Dall'Asta, C.; Dossena, A.; Galaverna, G. Effect of dry-cured ham maturation time on simulated gastrointestinal digestion: Characterization of the released peptide fraction. *Food Res. Int.* **2015**, *67*, 136–144. [CrossRef]
37. Agudelo, R.A.; Gauthier, S.F.; Pouliot, Y.; Marin, J.; Savoie, L. Kinetics of peptide fraction release during in vitro digestion of casein. *J. Sci. Food Agric.* **2004**, *84*, 325–332. [CrossRef]
38. Roblet, C.; Amiot, J.; Lavigne, C.; Marette, A.; Lessard, M.M.; Jean, J.; Ramassamy, C.; Moresoli, C.; Bazinet, L. Screening of in vitro bioactivities of a soy protein hydrolysate separated by hollow fiber and spiral-wound ultrafiltration membranes. *Food Res. Int.* **2012**, *46*, 237–249. [CrossRef]
39. Korhonen, H.; Pihlanto, A. Food-derived bioactive peptides-Opportunities for designing future foods. *Curr. Pharm. Des.* **2003**, *9*, 1297–1308. [CrossRef]
40. Miner-Williams, W.M.; Stevens, B.R.; Moughan, P.J. Are intact peptides absorbed from the healthy gut in the adult human? *Nutr. Res. Rev.* **2014**, *27*, 308–329. [CrossRef]
41. Gonzalez de Mejia, E.; Martinez-Villaluenga, C.; Fernandez, D.; Urado, D.; Sato, K. Bioavailability and safety of food peptides. In *Food Proteins and Peptides: Chemistry, Functionality Interactions, and Commercialization*; Hettiarachchy, N.S., Sato, K., Kannan, A., Eds.; CRC Press: Boca Raton, FL, USA, 2012; Chapter 12; pp. 297–330.
42. Kannel, W.B.; Higgins, M. Smoking and hypertension as predictors of cardiovascular risk in population studies. *J. Hypertens. Suppl.* **1990**, *8*, S3–S8.
43. Riordan, J.F. Angiotensin-I-converting enzyme and its relatives. *Genome Biol.* **2003**, *4*, 225. [CrossRef]
44. Livelli di Assunzione Raccomandata di Nutrienti per la Popolazione Italiana. 2014. Available online: http://www.sinu.it/public/20141111_LARN_Porzioni.pdf (accessed on 29 April 2019).

Review

Cocoa Bean Proteins—Characterization, Changes and Modifications due to Ripening and Post-Harvest Processing

Harshadrai M. Rawel [1],[*] , Gerd Huschek [2], Sorel Tchewonpi Sagu [1] and Thomas Homann [1]

[1] Institute of Nutritional Science, University of Potsdam, Arthur-Scheunert-Allee 114-116, 14558 Nuthetal, Potsdam, Germany; sorelsagu@uni-potsdam.de (S.T.S.); homann@uni-potsdam.de (T.H.)
[2] IGV-Institut für Getreideverarbeitung GmbH, Arthur-Scheunert-Allee 40/41, D-14558 Nuthetal OT Bergholz-Rehbrücke, Germany; gerd.huschek@igv-gmbh.de
[*] Correspondence: rawel@uni-potsdam.de; Tel.:+49-33200-88-5525/5578

Received: 31 January 2019; Accepted: 15 February 2019; Published: 19 February 2019

Abstract: The protein fractions of cocoa have been implicated influencing both the bioactive potential and sensory properties of cocoa and cocoa products. The objective of the present review is to show the impact of different stages of cultivation and processing with regard to the changes induced in the protein fractions. Special focus has been laid on the major seed storage proteins throughout the different stages of processing. The study starts with classical introduction of the extraction and the characterization methods used, while addressing classification approaches of cocoa proteins evolved during the timeline. The changes in protein composition during ripening and maturation of cocoa seeds, together with the possible modifications during the post-harvest processing (fermentation, drying, and roasting), have been documented. Finally, the bioactive potential arising directly or indirectly from cocoa proteins has been elucidated. The "state of the art" suggests that exploration of other potentially bioactive components in cocoa needs to be undertaken, while considering the complexity of reaction products occurring during the roasting phase of the post-harvest processing. Finally, the utilization of partially processed cocoa beans (e.g., fermented, conciliatory thermal treatment) can be recommended, providing a large reservoir of bioactive potentials arising from the protein components that could be instrumented in functionalizing foods.

Keywords: cocoa processing; cocoa proteins; classification; extraction and characterization methods; fermentation-related enzymes; bioactive peptides; heath potentials; protein–phenol interactions

---

1. Introduction

Principal botanical varieties of *Theobroma cacao* L. are Forastero, Criollo and Trinitario. Forastero varieties are regarded as 'bulk cocoa in trade' and constitute almost 95% of the cocoa's total worldwide production [1]. Both the Trinitario and the Criollo varieties produce the 'fine flavor' cocoa beans, which account for less than 5% of the total cocoa's world production [1]. Cocoa protein constitutes 11–13% based on dry weight and may vary depending on geographical origin between 11.8% and 15.7% [2,3]. The average value for the amino acid-based protein content of cocoa bean cotyledons from different varieties was also investigated and lies at approx. 10.4% [4]; for Criollo it lies at 10%, for Trinitario it is between 8.8% and 10.7% and that for Forastero lies at 10.2–11.4% [4]. The value for crude protein (adjusted for alkaloids) is similar to that based on amino protein, although some of the latter values tend to be slightly lower [4]. The average protein content of roasted cotyledons (also termed "nibs") lies at around 12.5% [1]. Many factors affect not only the quality of proteins such as location (climate, soil, fertilizer, and stress) but also the considered botanical varieties (genomics). In the following, it is initially intended to encompass the extraction, characterization options and classification of cocoa been proteins. In the next step, we address the impact of different stages of cultivation and processing with regard to the induced changes in the protein fractions. Special focus is laid on the major seed storage proteins (vicilin and albumins) throughout the different stages of processing.

2. Extraction and Classification of Cocoa Proteins

Some of the early attempts to extract proteins from cocoa beans were conducted after the removal of lipids (soxhlet extraction with ethyl ether) and of phenolic compounds with methanol followed by extraction with buffering solutions containing different additives (acidic pH conditions using acetic acid, urea, hexadecyltrimethylammonium bromide, ascorbic acid, and sodium ethylenediaminetetraacetate (EDTA)), resulting in a maximum recovery of 25% of the protein nitrogen [5,6]. The extracted proteins are thereafter classified according to their solubility characteristics originating from the concept of T. B. Osborne (1859–1929) in the following manner: distilled water delivers the albumin or water-soluble proteins, a diluted salt solution to obtain a globulin fraction, extracted with 70% aqueous ethanol followed by 0.2% NaOH, yielding prolamine and glutelin fractions. Accordingly, 32–37% albumins, 19–25% globulins, 11–13% prolamines and 30–37% glutelins are allocated to non-pigmented cocoa bean varieties. Similarly, 51–71% albumins, 1–25% globulins, 12–20% prolamines and 8–12% glutelins are allocated to pigmented cocoa bean varieties, bearing in mind that only a partial protein recovery is determined [6]. The problems associated with discoloration and protein insolubility resulting in poor recovery are believed to be caused primarily by residual polyphenolic materials not removed by the preceding methanol extraction [6]. These protein–phenol interactions can be classified into two subgroups: non-covalent and covalent interactions [7]. Principally, three potential types of non-covalent interactions of phenolic compounds and proteins have been suggested: hydrogen, hydrophobic, and ionic bonding [7]. The phenolic compounds are also susceptible to both enzymatic and non-enzymatic oxidation in the presence of oxygen, leading to reactive and redox active o-quinones, an electrophilic species, capable of undergoing a nucleophilic addition to proteins [7]. This results in the derivatization of protein-bound amino acids invoking consequently also cross-linking reactions. Both of these two types of complexing interactions results in an increase in protein aggregation, insolubility and discoloration [7].

In the following years, the studies of Voigt et al. (1991–1997) were decisive in improving the protein extraction and characterization especially in the context of their role while analyzing the biochemical aspects of cocoa bean fermentation [8–17]. In the same decade, cDNAs encoding of the major albumin and globulin of cocoa seeds was achieved and the proteins were consequently cloned and sequenced [18–21]. The reported amino acid sequence of the albumin was homologous with the Kunitz protease and α-amylase inhibitor family [20,21]. Based on amino acid sequences, subunit compositions, and the processing of the corresponding polypeptide precursors, the globulins can be assigned to the vicilin-like globulins of storage proteins, previously found only in legumes and cotton [18,19]. In the cocoa beans, vicilin is synthesized (partly similar to that in cottonseed) as an approx. 70 kDa molecular weight precursor protein and then processed to 47 kDa and 31 kDa mature proteins [18]. A latter study documents that the entire cocoa vicilin is encoded by a single gene and that heterogeneity of the vicilin subunits may be attributed to statistical post-translational modifications [22]. In contrast to the major albumin and the vicilin class globulin, there are, however, at this stage no data available concerning prolamin and the glutelin which were also found in the seeds of *Theobroma cacao* [9].

Most of the cocoa seed proteins are solubilized thereafter while working with a dry polyphenol-free acetone powder and high-salt buffer systems. The separation of albumins is, for example, also achieved by the following desalting process while applying dialysis against a salt-free buffering solvent [22]. The major proteins of cocoa beans allocated to vicilin and albumin classes thereafter represent about 43% and 52% of the total cocoa seed proteins, respectively [9]. Other studies report that vicilin constitutes ca. 23% and the albumins constitute around 14% of the total soluble seed proteins [23]. The observed discrepancies in the values are most likely dependent on the extraction procedure and allocation method used. Both studies applied previously treated material with ice cold acetone to remove the interfering polyphenols. In the first study, the results are based on proteins which were extracted and fractionated into various solubility classes by using different buffering systems [9]. The protocol applies successive extractions with 10 mM Tris-HC1 (pH 7.5, containing 2 mM EDTA), 0.5 M NaCl (containing 2 mM EDTA and 10 mM Tris-HCl, pH 7.5), 70% (v/v) ethanol and 0.1 M NaOH, to obtain the albumin, globulin, prolamin and glutelin fractions, respectively [9]. The protein contents of these extracts deliver corresponding data reported. In the latter study [23], the proteins are extracted to be compatible with the high-resolution technique of immobilized pH gradient of the two-dimensional electrophoresis (2-DE). For this purpose, a solubilization solution containing different additives (3% (w/v) CHAPS (3-((3-cholamidopropyl) dimethylammonio)-1-propanesulfonate), 8.5 M urea, 0.15% (w/v) DTT and 3% (v/v) carrier ampholytes in the pH range of 3–10) can be recommended. The evaluation of 2-DE data deliver the 47, 31 and 15 kDa vicilin-type storage protein components [18,22] representing 23.1% of the soluble seed proteins for globulins, and 14.1% of the 21 kDa trypsin inhibitor for albumins, respectively [23]. These values are also lower than those reported (36% for globulin storage protein [24] and between 25% and 30% for the trypsin inhibitor [21]). The discrepancy was assumed by the authors [23] to be based on the electrophoretic methods used (estimation by SDS-PAGE being higher and since the polypeptide bands evaluated may contain different individual species). The most common gel-based technique used in a proteomic laboratory is 2-DE [25]. Protein quantities, as well as their profiles derived from two-dimensional gel electrophoresis, show striking differences for non-fermented cocoa beans, depending on their geographical origin [26]. Although 'Osborne fractionation' is still widely used, it is more usual today to classify seed proteins into three groups: storage proteins, structural and metabolic proteins, and protective proteins [27]. Figure 1 documents a tentative classification of the proteins based on different studies. Additional proteins in the cocoa powder after buffer extractions can be extracted with a solution containing chaotropic agents [28].

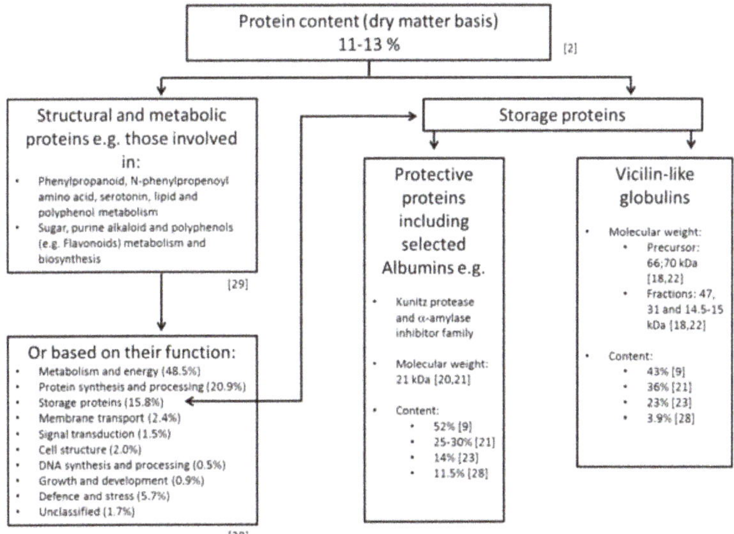

Figure 1. Contemporary classification of cocoa seed proteins [2,9,18,20–23,28,29].

During recent years, the area of proteomics has undergone rapid developments integrating high-resolution, fast, stable and accurate mass detection. Proteomic technologies have made an extensive development with the discovery of different protein ionization methods, notably the electrospray ionization (ESI) and matrix-assisted laser desorption/ionization (MALDI) techniques in mass spectrometry (MS), which enable proteins to be identified [25]. MALDI-time of flight/MS (MALDI-TOF/MS) is also employed to characterize the water-soluble portion of the proteomic seed extracts from different varieties (Forastero, Criollo, and Trinitario) of *Theobroma cacao*. Most of the proteins detected with this approach show molecular weights between 8 and 13 kDa, while a cluster at 21 kDa is attributed to albumin [30]. The development of gel-free proteomics (while using liquid chromatography (LC), and capillary electrophoresis) provides here an excellent alternative to more accurately quantitating protein and enabling deeper explorations of complex proteins [25,31]. Both the gel-based and gel-free methods integrate MS for protein profiling, protein identification (with prior tryptic digestion, and analysis of the digestion products), and quantification, as well as analysis of protein modifications and interactions [25]. Both gel-based and gel-free approaches have also been developed and utilized in a variety of combinations to separate proteins from tissue culture of cocoa beans prior to mass spectrometric analysis [32,33]. In this context, one of the first attempts to characterize the whole cocoa bean proteome by nano-LC-ESI MS/MS analysis using tryptic digests of cocoa bean protein extracts has recently been made indicating that more than 1000 proteins can be identified while applying a species-specific *Theobroma cacao* database [28]. Most of these are related to metabolism and energy household, protein synthesis and processing and response to different stress stimuli or connected with defense scenarios [28]. Vicilin and albumin are classified as storage proteins and show again the highest abundance among all detected proteins, although compared to the total protein amount their relative amounts are only 3.9% and 11.5%, respectively [28]. These levels are lower than the values discussed above of between 43% and 23% for vicilin, between 52% and 14% for albumin [9,23]. It is likely that again the discrepancies could arise from inefficiency of selective solubilization of proteins [9] and overestimation, due to detection insufficiency based on electrophoretic methods and eventually also due to incomplete tryptic digestion preceding the nano-LC-ESI MS/MS analysis. Figure 2 gives a summary of the currently applied methods for extraction, separation, protein allocation and identification.

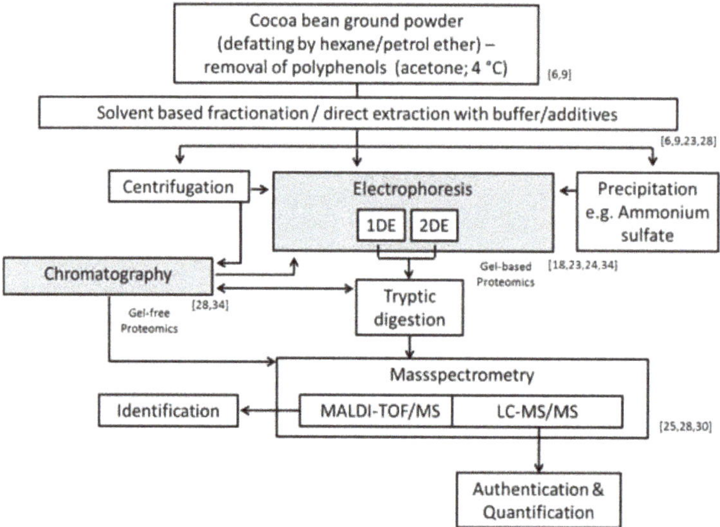

Figure 2. Compilation of the methods and characterization options applied for cocoa bean proteins denoted with corresponding relevant studies [6,9,18,23–25,28,30,34].

For the purpose of cocoa species verification, the extracted proteins and the individual peptides resulting from their tryptic digestion are initially controlled in silico (e.g., using software tools such as Skyline software [35]) with the BLAST algorithm (Basic Local Alignment Search Tool from the database; UniProt—the Universal Protein Resource: http://www.uniprot.org/) as described in [36,37]. The identification then proceeds with help of mass-spectra algorithms and search engines (e.g., Mascot, Matrix Science, London, UK; www.matrixscience.com) and available databases [28]. In order to assess whether searching different databases would yield a higher number of cocoa-specific protein hits, searches using different databases with taxonomy *Viridiplantae*, and custom databases containing only *Theobroma cacao* entries have been recommended [28]. The recently published cocoa genome sequences can also be useful to create a predicted proteolytic fragment database [32,38–40]. In this context, the Universal Protein Resource (UniProt) is the most comprehensive database/resource for specific protein sequences and annotation data (http://www.uniprot.org/). A recent search in this database confirmed at least 40,964 (compared to 40,941 according to [28]) *Theobroma cacao* protein entries (visited on 4th January, 2019), of which only six were reviewed representing records with information extracted from literature and curator-evaluated computational analysis. In comparison, based on this background and utilization of the available sophisticated MS tools, both gel-based and gel-free approaches have been more frequently applied for proteome analysis during zygotic and somatic embryo maturation, in order to identify alterations in protein abundance that correlate with maturation of cocoa embryos or with the intention to accelerate breeding programs and plant development [25,32,33,41–46].

In summary, it appears from the state of art that cocoa proteins need to be better accessed, especially while comparing the contents of different varieties of *Theobroma cacao*. The discrepancies in protein content observed in the reviewed literature need to be eliminated to obtain more consistent data. The methodical approach using bioinformatics algorithms, targeted peptide biomarkers and high-resolution MS can be recommended here for authentication of the analyzed proteins [36,37,47], especially while considering the different aspects of post-harvest processing.

3. Changes in Protein Composition during Ripening and Maturation of Cocoa Seeds

Cocoa pods need 4–5 months to grow to full size following pollination and the beans contained therein have reached the maximum development, thereafter they are allowed to ripen for approximately 4 weeks [5]. The major textural changes that cause softening of fruit during the ripening (and proceeding in the fermentation process) result from enzyme-mediated alterations in the structure and composition of the cell wall and partial or complete solubilization of cell wall polysaccharides, such as pectins [48]. During ripening, the mucilaginous pulp surrounding the beans undergoes changes critical to a successful fermentation, primarily resulting in an increase in fermentable carbohydrate components [5,49]. Data on changes in the amount of seed storage proteins during ripening and maturation of cocoa beans are sparse. During the ripening of cocoa fruit, both the amounts of the total and extractable protein content of seeds decreased by 25% or 19% respectively, but no consistent qualitative trends were not apparent [5]. Some differences in solubility trends between the investigated different ripening stages (135–160 days postpollination) afterwards were noted, although the amino acid analysis did not reveal any specific related differences [5]. One of the earliest studies to report on the accumulated globular storage protein during seed development/ripening was achieved while using antibody against a large subunit of the vicilin (7S) globulin, also capable of cross-reacting with all of the 7S globulin subunits [16]. The vicilin-like globulin of the cocoa seeds contains two prominent subunits with apparent molecular masses of 47 kDa and 31 kDa and three smaller polypeptides with apparent molecular masses of 15.5 kDa, 15.0 kDa, and 14.5 kDa arising from the proteolytic processing of a 66 kDa precursor [9,22]. Further data indicate that two 28 kDa and 16 kDa components also observed during characterization of vicilin are not intrinsic subunits of the cocoa vicilin but are generated by partial proteolysis during preparation of this particular globular storage protein, their content being strongly diminished while working with a sufficiently high concentration of pepstatin A, an inhibitor of aspartic endoproteinases [9,22]. These findings are derived from the combined results of the epitope mapping of the corresponding polyclonal antibody, sequence coverage as documented by the MALDI-TOF/MS mapping of the tryptic hydrolysates of the purified fractions and the amino acid profiling/kinetics of the release of amino acids by carboxypeptidase Y [22]. A comparison of two entries (60.8 kDa/525 amino acids reviewed and non-reviewed 65.5 kDa/566 amino acids version) from the database UniProt for cocoa vicilin showing similarities in the amino acid sequence are depicted in Figure 3a. In the same context, the results reported in [22] are related to the non-reviewed entry (A0A061EM85_THECC; 65.5 kDa/566 amino acids). Correspondingly, Figure 3b shows the predicted proteolytic processing of the vicilin precursor based on this particular sequence [22]. The localization of the 47 kDa vicilin based on the peptide fragments may differ as reported in the context of further studies [50]. Amino acid position 545 represents the C-terminal end-point of the 47 kDa subunit, as no peptide was obtained derived from the amino acid sequence beyond position 545 (Figure 3a,b). This result suggests that the absolute C-terminus of the 566-amino-acid-long vicilin precursor sequence undergoes a swift carboxypeptidase attack leaving no peptide(s) but just free amino acids or di- and tri-peptides behind [50]. Finally, the two-dimensional gel electrophoresis of these mature subunits of cocoa vicilin also revealed their heterogeneity, postulated to result from post-translational modifications of various amino acid side chains, e.g., due to the action of a protein deaminase during maturation [22]. A more recent study complements these previous findings and documents that the vicilin storage protein subunits may undergo phosphorylation (modification at positions 232 (Thr), 235 (Ser), 240 (Ser), and 518 (Ser)), or glycosylation (O-GLcNAc modifications at positions 193 (Thr), 235 (Ser), 338 (Thr) and 474 (Thr)); please also refer to Figure 3a [50]. Figure 4 shows, for the first time, the tentative modelling of the storage protein vicilin from *Theobroma cacao* (entry: Q43358; reviewed version from the database UniProt: http://www.uniprot.org/; 28th January, 2019), visualizing the accessibility of the amino acid residues for the post-translational modifications. The molecular modeling is based on the methods described in [51] (please also refer to Supplementary Figures S1–S3 and Table S1 provided). The model also shows the different theoretical sites for glycosylation (Figure 4a) or phosphorylation (Figure 4b). The homology modeling data further indicate that the phosphorylation and glycosylation sites as reported in [50] are buried within the molecule and are most probably accessible after the initial proteolytic processing of the 66 kDa precursor [9,22] as discussed above (please refer to supplementary information provided).

Finally, a major 43 kDa protein also present in the non-fermented bean sample could be identified belonging to protein kinase superfamily [50]. The most prominent protein with a molecular weight of 21 kDa is again allocated to albumin by tryptic digestion and MALDI-TOF/MS analysis [50].

Further studies address specific enzymes [52–56]. Acyl-thioesterase activity has been examined at two developmental stages (105/130 days postanthesis) in cocoa documenting low and high stearate productions [54]. A further study addresses a cysteine protease expressed during the process of the maturation of the cocoa seed [52]. This enzyme is part of the defense mechanism induced in response of the action of the parasite *Moniliophthora perniciosa*, a fungus that causes witches' broom, which is one of the diseases that most affects the production of cocoa, dramatically reducing crop yields.

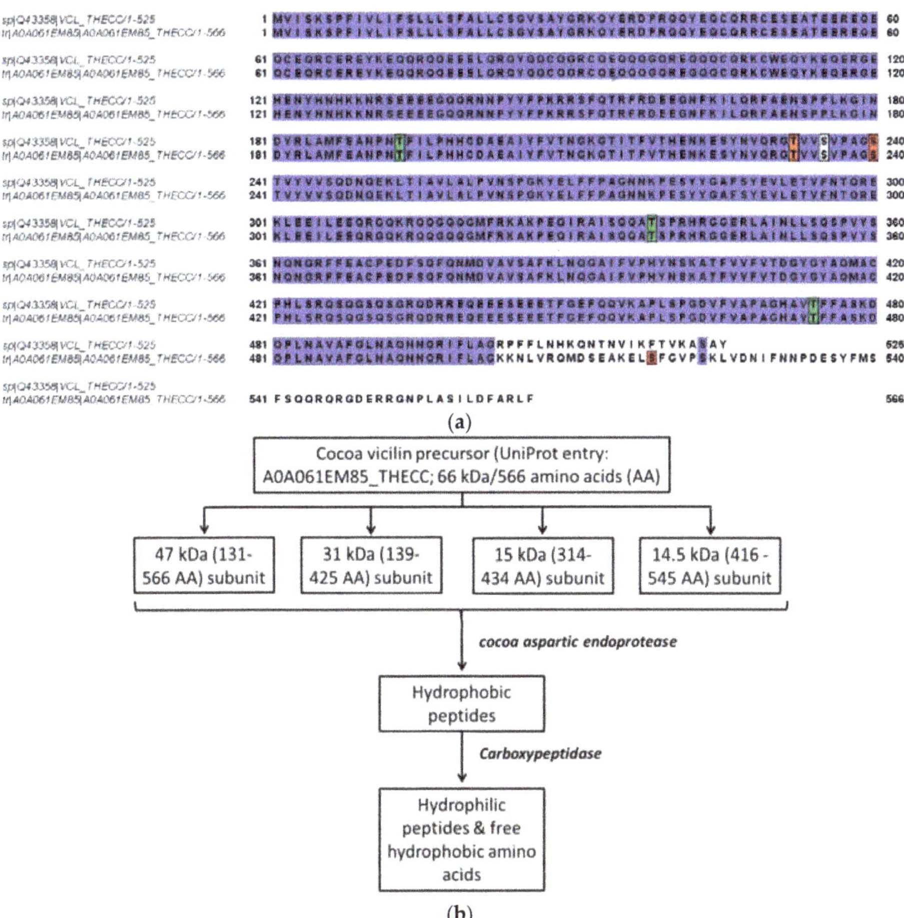

Figure 3. (**a**) Comparison of two entries of the storage protein vicilin from *Theobroma cacao* (a reviewed and a non-reviewed versions) from the database UniProt (http://www.uniprot.org/; 28th January, 2019) showing the similarities (blue) in the amino acid sequence. Positions for predicted post-translational modifications [50] are marked: phosphorylation (red; at positions 232 (Thr), 235 (Ser), 240 (Ser), and 518 (Ser)), or glycosylation (green; O-GLcNAc modifications at positions 193 (Thr), 235 (Ser), 338 (Thr) and 474 (Thr)). The common residue (white, 235 (Ser)) serves as a possible site for both modifications. (**b**) Postulated degradation of cocoa vicilin precursor during maturation by endogenous enzymes and fermentation adapted from [10–12,22,57]; the 47 kDa subunit may only encompass the sequence 131–545 amino acids [50]. Specific sites for phosphorylation or glycosylation of the subunits are given in Figure 3a [50]. AA represents the amino acid sequence of the subunits.

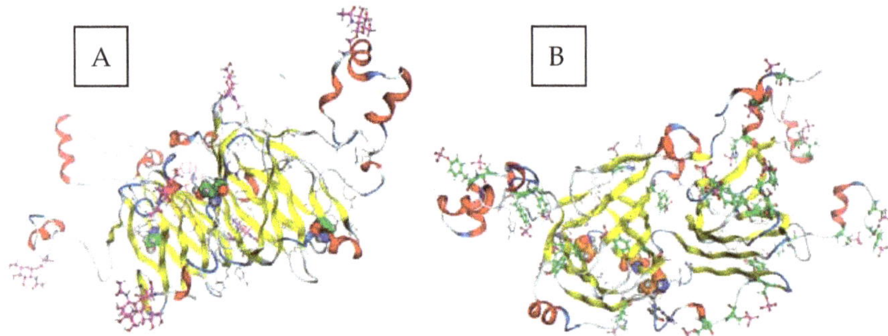

Figure 4. Homology modeling of the vicilin storage protein. (**A**) Postulated glycosylation (O-GLcNAc) and (**B**) phosphorylation sites (both in pink) in the storage protein vicilin from *Theobroma cacao* (entry: Q43358; UniProt: http://www.uniprot.org/; 28th January, 2019). The data indicate that the modifications (illustrated in green/orange spacefilling model) as reported in [50] are most probably occurring after the proteolytic processing of the 66 kDa precursor [9,22]. Please see supplementary information provided.

Cocoa beans also contain peroxidase, a heme-containing oxidoreductase, which efficiently oxidizes the phenolic molecules using H_2O_2 as a co-substrate. An increase of peroxidase activity in the seeds of cocoa during their ripening has been documented. The major cocoa isoperoxidase is an acidic enzyme with the isoelectric point (pI) of 4.7, together with two basic isoenzymes of the peroxidase with pIs of 8.6 and 9.0 detected, respectively, during the process of the fermentation [55]. The activity proceeds to increases further (about 10 times) during the fermentation and drying of the beans, again contributing to sensory perception of cocoa [55]. The role of these enzymes is important, since the oxidation products of phenolic compounds may not only modify amino acids, peptides and proteins [7], but also is determining in reducing the astringent and bitter taste, thus also contributing to flavor of cocoa [55]. In a similar context, a series of N-phenylpropenoyl amino acids categorized as multifunctional polyphenol derivatives (with aspartic acid amide of caffeic acid-(30,40-dihydroxy-(E)-cinnamoyl)-L-aspartic acid as the most abundant constituent) are identified as key contributors to the astringent taste of non-fermented cocoa beans [58,59] with their content decreasing during the cocoa seed development [29]. The results contradict with those reporting an increased accumulation in the advanced stage of seed development [60].

In *Theobroma cacao* seeds, an aspartic proteinase has been proposed to be a key enzyme involved in the formation of one group of cocoa aroma precursors [53]. At least two distinct aspartic proteinase genes (TcAP1 and TcAP2) are expressed during early cocoa seed development. Of these, the corresponding TcAP2 protein has been proposed to be primarily responsible for the majority of the industrially important protein hydrolysis that occurs during cocoa bean fermentation [53]. Aspartic endoproteinase activity increases rapidly during embryo expansion, reaching a maximal activity before final maturity, a prerequisite for the degradation of seed storage proteins during the following post-harvest processing [61]. Voigt et al., in 1995, showed that this enzyme accumulates with the vicilin-class globulin during bean ripening [16]. The enzyme consists of two peptides (29 and 13 kDa) that have been proposed to be derived from a single 42 kDa precursor zymogen, possibly by self-digestion [9,53]. The mature cocoa seed aspartic proteinase has been proposed to cleave protein substrates between hydrophobic amino acid residues producing oligopeptides with hydrophobic amino acid residues at their carboxyterminal ends [9,53]. The enzyme activity is shown to be optimal at pH 3.5 and is inhibited by pepstatin A. A recent check-up for aspartic proteinase in the protein database UniProt (http://www.uniprot.org/; 17.01.2019) delivers up to 107 *Theobroma cacao* entries, largely based on differing gene ontology (including the genes *AP1* and *AP2*) and representing different isoforms.

Finally, many reports encompass changes in protein expression during zygotic and somatic embryo maturation [32,43,44], where the seed storage protein is more strongly accumulated in cocoa zygotic embryos compared to that in their somatic counterpart [32]. The identified proteins represent an array of functional categories, including seed storage, stress response, photosynthesis and translation factors [32]. A system level analysis of cocoa seed ripening further revealed the accumulation of proteins and metabolites involved in biotic and abiotic stress resistance, leading to e.g., polyphenol accumulation [62], but also covered the interplay of different specific primary and secondary metabolism pathways important for the major compound classes (e.g., lipid and sugar metabolism as well as those of selected secondary bioactive plant metabolites such as alkaloids and polyphenolic compounds) involved in cocoa aroma and health benefits [29,62,63]. Accordingly, at the stage of reserve accumulation phase of cocoa seeds [29], most of the amino acids reach their lowest level similar to the trend observed for the reserve accumulation period of *Arabidopsis thaliana* seeds [64].

4. Changes in Proteins during Post-Harvest Processing

The complex composition of cocoa bean flavor has been discussed depending on bean genotype, post-harvest treatments such as pulp pre-conditioning, fermentation and drying, and industrial processes such as roasting as well as the type of soil and age of cocoa tree [65]. However, how the age of the cocoa tree and soil chemical compositions influence the formation of flavor precursors still remains unclear [65]. In this context, the impact of the farming system, the ripeness state of the pods, and the role of microbial interactions on the fermentation has also been evaluated [1]. Post-harvest processing on farms and plantations involves the following four main steps of pod opening and beans removal from the pod, beans fermentation, and drying. These steps are most likely also responsible for many post-transitional/post-processing modifications occurring to different proteins [1]. Consequently, the most effective and essentially critical step-hear appears to be that of fermentation which determines the development of flavor quality attributes of the commercial cocoa beans [1]. Most of the recent studies have been directed to enlighten the role of fermentation as a quality determining steps for the following roasting process [26,50,57,65–70]. Early studies of Voigt et al. in the 1990s confirm that cocoa seed storage proteins play an important role in flavor development since essential precursors of the cocoa-specific aroma components are formed from their degradation during the fermentation process [9–13,15]. Their experimental approach showed that these aroma precursors are released from seed proteins by the dual activity of the already mentioned aspartic endoprotease and a carboxypeptidase (Figure 3b) [10]. Their studies also document the proteolysis products obtained when these proteins are subjected to autolysis at pH 5.2, under conditions of optimal fermentation consisting of both hydrophilic peptides and hydrophobic free amino acids (Figure 3b), as also observed by other studies [53]. This specific mixture of hydrophilic peptides and hydrophobic free amino acids is capable of producing the typical cocoa flavor when roasted in the presence of reducing sugars and deodorized cocoa butter [10]. On the basis of sensory evaluation of the resulting aromas by sniffing analysis, it was further demonstrated that the fraction of hydrophilic peptides generated in vitro contains the essential cocoa-specific aroma precursors necessary for the following complex Maillard reaction under these conditions. In comparison, the patterns of free amino acids alone, specifically leucine, alanine, phenylalanine, and tyrosine found in fermented cocoa seeds, do not contribute towards the formation of this typical cocoa aroma under roasting conditions in vitro [10]. However, they are still likely to react with the reducing sugars fructose and glucose during the Maillard reaction as implicated in the following works [71].

To our present knowledge, the fermentation, which is essentially controlled by proteolytic activity within the cocoa bean, is also driven by changes in the presence of fermentation by-products as a result of microbial activity outside the bean [67]. This is so-called "post-mortem proteolysis" and therefore depends on the variable/desired pH value (4.0–5.5) in the seed, e.g., established by acetic acid absorbed from the fermenting pulp [72]. An aseptic artificial fermentation system, free from microbial activity, has also been reported, capable of simulating the proteolytic degradation of cocoa proteins as observed during commercial fermentation, where again acidification was the most crucial parameter for the protein degradation [67]. The changes occurring to vicilin-like storage protein at different stages of fermentation (from the non-fermented stage up to the dried cocoa beans) have been addressed in detail [50]. Analysis of vicilin breakdown peptide pool during fermentation by the UHPLC-ESI-MS/MS revealed an initial increase and subsequent decrease in the diversity of peptides with an increasing degree of fermentation [50]. The results also indicated that the sequences of free peptides are localized to distinct zones spread throughout the entire C-terminal vicilin sequence, except the N-terminal where no peptide hits could be found for the amino acid sequence 1–131 [50]. There are no detectable peptides found in the fully fermented and dried bean samples. This observation in turn shows the contribution to the pool of free amino acids or di- and tri-peptides as potential conjugative moieties for aroma compounds, such as Amadori or Strecker reaction products during the following roasting process [50,73]. The most abundant albumin storage protein is steadily and homogenously degraded without forming breakdown intermediates with a molecular weight larger than approximately 10 kDa [50].

The variability in the peptide pattern was observed among cocoa samples of different geographical origins, suggesting diversified proteolytic activities could be a relevant feature in this context [69]. The applied combination of proteomic and peptidomic fingerprinting enables a more comprehensive analysis of the attributes that characterize storage protein degradation in cocoa during microbial fermentation [26]. This study also confirms that the major differences in protein content of non-fermented cocoa beans are predominantly attributed to the geographic origin in terms of continental regions. The authors also attest that the formerly detected diversity of peptides could not be correlated to the geographical origin but rather to the degree of fermentation, depending on the fermentation method applied in the country of origin underlining again the role of diversified proteolytic activities [26]. In this context, more than 800 unique oligopeptides, excluding di- and tri-peptides, documenting the largest collection of cocoa oligopeptides, have been identified and relatively quantified by utilizing UHPLC-ESI-quadrupole-quadrupole-time-of-flight (Q-q-TOF) mass spectrometric analysis. In the same context, more than 800 fermentation peptides could recently also be unambiguously identified, providing unprecedented mechanistic details of cocoa fermentation [74]. The cocoa-specific aroma precursor fractions have also been characterized by MALDI-TOF and their amino acid sequences determined by ESI-MS/MS, allowing for a partial purification and a consequent detailed characterization of these peptides responsible for the generation of the cocoa-specific aroma components [66].

Polyphenols are also oxidized by polyphenol oxidase during fermentation and drying which reduce the astringency and bitterness of the beans, thus enhancing the flavor of cocoa beans [65]. The total amount of polyphenols in dried fresh cocoa beans may vary between 12% and 20% (*w/w*) and these are responsible for its high astringency, contributing to their bitterness as well [1]. Three main groups of polyphenols are present: anthocyanins, flavan-3-ol (catechins), and proanthocyanidins, corresponding to approximately 4%, 37%, and 58%, respectively [60,75–79]. An increase of peroxidase activity in the seeds of cocoa during their ripening and stronger during fermentation has been observed, implicating a possible oxidation of these compounds [55,65]. In a similar context, the polyphenols (depending on their structure) are also readily oxidized by polyphenol oxidase [7]. The combined effect of the interaction of the oxidized phenolic compounds with the degraded protein products (amino acids, and peptides; Figure 3b) during post-harvest processing may result in a further modification of the protein-based aroma precursors, although detailed studies in this respect, e.g., those similar to modification of coffee bean proteins [7,80,81], have not been addressed.

The vicilin-class globulins are quantitatively degraded during fermentation (88–90% of the initial content), providing the cocoa-specific aroma precursor fractions for the following Maillard reactions during drying and roasting [9,65,68,71,73,82,83]. The compounds produced in the complex interactions during the thermal mediated Maillard reaction comprise nitrogen and oxygen heterocyclic compounds, aldehydes and ketones, esters, alcohols, hydrocarbons, nitriles and sulphides, pyrazines, ethers, furans, thiazoles, pyrones, acids, phenols, imines, amines, oxazoles, and pyrroles [65]. Approximately

600 flavor compounds have been identified from cocoa beans and cocoa products [65,84]. The complex formation of cocoa and chocolate flavor is discussed in detail in [65,71,85]. It is influenced by different factors starting with the composition of the beans, post-harvest treatment (e.g., fermentation, and drying), processing (e.g., roasting encompassing the Maillard reactions and alkalization to change color) and eventually further fine-tuning during chocolate manufacture (conching). Depending on whether the proteolytic activity has been more or less intense during the fermentation, the result in the color development during the drying step will also be different. Similarly, depending on whether the fermentation will be more or less intense (more or less consumption of reducing sugars), the proportion of residual reducing sugar will also influence the development of the color from the resulting Maillard reaction products during drying and roasting. The relevant processes inducing the changes in the proteins fractions during post-harvest treatments are summarized in Table 1.

Table 1. Dominant reported and postulated changes in the protein fractions during the seed maturation and post-harvest processing.

Determinants	Changes in Protein Fraction	Mechanisms Involved	Relevant Studies	Remarks
Genetic predisposition Location/climate soil/fertilization Stress conditions Maturation	Content and composition/Variation Post-transitional modification	Protein expression and accumulation/ Phosphorylation/ Glycosylation/Oxidation/ Carbonylation	[28,29,32,43,44,62,63]	Reactive oxygen, carbonyl and nitrogen species that react with the proteins under stress conditions
Harvest/Storage, Pre-conditioning, Fermentation conditions (pH, temperature, method, and location/climate)	Degradation Post-transitional modification	Proteolytic processing/ Phosphorylation/ Glycosylation/Oxidation, Carbonylation, Deamidation, Decarboxylation Bound phenolics	[10–12,22,50,57]	Production of precursors for roasting—peptides/ amino acids/reducing sugars/lipid degradation/Phenol modification
Drying, Roasting conditions (temperature, time, method etc.)	Thermal modification/Reactions with other constituents	Maillard reaction Volatile compounds/Aroma and flavor development/ Browning/Covalent bound phenolics	[9,65,68,71,73,82,83]	600 flavor compounds [65,84] Melanoidin Fractions [86–88]
Further processing conditions/Alkalization	Interactions with phenolic compounds? [7,51]	Oxidation and polymerization of phenolic compounds/ Reduced acidity/ Improved sensory perception/quality	[65,71,85]	Browning/Melanoidin fractions [86–88]

5. Bioactive Potential Arising from Cocoa Proteins

The health benefits of cocoa and cocoa-based products have been reviewed, and its potential for prevention/treatment of allergies, cancers, oxidative injuries, inflammatory conditions, anxiety, hyperglycemia, and insulin resistance has been discussed in detail [89]. Most of these positive evaluated effects have been correlated with the high content of the different secondary plant metabolites present especially encompassing the group of polyphenolic compounds, where several mechanisms have been proposed that might confer cocoa's possible health benefit [90–93]. Exemplarily, being a rich source of flavonoids, cocoa represents a group of potent antioxidant and anti-inflammatory agents, documenting benefits for cardiovascular health [91,94–96]. Cocoa flavanols are also reported to have neuromodulatory, neuroprotective and antidiabetic actions in humans [45,97]. In this context, much recent work has been directed to elucidate the effect of processing on antioxidant properties of cocoa and cocoa products [98–101].

The bioactivity derived from intact cocoa proteins, or their degradation products and/or their derivatives, e.g., resulting from post-harvest processing, has been only sparsely addressed. A good example illustrating the biological activity of the intact proteins is that on cysteine protease of *Theobroma cacao* against witches' broom as already mentioned in the preceding section [52]. The utilization of such activities can be used for developing new products of commercial interest [52]. The formation of small peptides in fermented cocoa beans offers a large reservoir for other interesting aspects, since small peptides have been discussed as compounds imparting health benefits [26]. Food-derived bioactive peptides represent one such source of health-enhancing components, which can be released during gastrointestinal digestion or food processing. They can be physiologically active, either in the native protein state or as products of hydrolysis in vivo or prior to consumption [102]. Bioactive peptides usually contain 3–10 amino acid residues; their activity is based on their amino acid composition and sequence [103] which include several regulatory mechanisms related to nutrient uptake, immune defense, antioxidant, neuroprotective and antihypertensive properties [104–107]. Peptides which are not degraded in proteolysis can theoretically be absorbed intactly. It has been suggested that dipeptides and tripeptides are absorbed in the intestine [108,109]. Furthermore, it has also been reported that tripeptides containing a C-terminal proline–proline bond are usually resistant to human proteolytic enzymes [110,111].

In this context, a recent study reported the presence of a bioactive peptide (DNYDNSAGKWWVT) from a hydrolyzed cocoa by-product that was found to have antioxidant property, which could be used therapeutically for the prevention of age-related diseases [112]. The study documents that the peptide protects *Caenorhabditis elegans* from oxidative stress and is responsible for the modulation of synaptic and proteosomal functions. The peptide originates from the storage albumin fraction (21 kDa seed protein), a trypsin inhibitor from *Theobroma cacao* as confirmed by blast analysis in the protein database (https://www.uniprot.org; 28th January, 2019; entry: P32765). Similarly, antitumor activity was also observed in the albumin fraction, which inhibits the growth of cells in murine lymphoma, documenting one of the earliest reports on biological activity of semifermented-dry cocoa protein fractions [113]. The activity could be attributed to its hydrophobic and sulfur amino acids profile that confers antitumor and antioxidant potential. Free radical-scavenging capacity was also observed mainly in the albumin and glutelin fractions from cocoa [113,114]. Finally, the albumin fraction also shows antitumor activity in a mouse murine model of lymphoma L5178Y, indicating that it could be considered as a source of potential antitumor peptides [107]. Antioxidant and angiotensin-converting enzyme (ACE) inhibitory (anti-hypertensive) activities of cocoa autolysates after removing the partly interfering fat, alkaloids and polyphenols have also been elucidated, conferring again another attribute contributing to its health-promoting properties [115]. A recent review discusses on a possible antimicrobial potential of cocoa bean shell related to a diverse pool of bioactive compounds with antimicrobial properties including, beside other compounds, phenolics and bioactive peptides [116]. A further study suggests that cocoa products originating from different post-harvest processing steps may possess mild dipeptidyl peptidase-IV inhibitory activity, and that processing steps such as fermentation may actually enhance inhibition activity [117]. While considering the anti-obesity and anti-hyperglycemic effects, one potential mechanism relates to the inhibition of dipeptidyl peptidase-IV. Glucagon-like peptide-1 is a hormone, which is rapidly degraded by dipeptidyl peptidase-IV. The hormone also stimulates insulin release in response to glucose ingestion, increases satiety, and slows gastric emptying [117]. The compounds responsible for dipeptidyl peptidase-IV inhibition, according to the authors, may represent a previously uncharacterized pool of dietary bioactives beyond the flavanols and flavanol products and still remain to be elucidated [117]. While reducing native polyphenols, fermentation simultaneously produces complex polyphenol oxidation products, which may interact with proteins, peptides and amino acid components [7]. These non-native products may retain some polyphenol structure and activity and/or introduce potentially new activities [117].

Cocoa beans also contain a further interesting group of compounds arising from the enzymatic and/or (chemical) decarboxylation of amino acids representing bioactive amines, where mainly 2-phenylethylamine, tyramine, tryptamine, serotonin and dopamine are found [118–120]. Cocoa can also be a source of polyamines (spermidine and spermine), which may also contribute to cocoa's antioxidant activity [118]. The biogenic amines play relevant roles in plant development and human health and can be formed during fermentation [118]. At low levels, bioactive amines are positively correlated to human health; however, some amines, at high levels, may cause adverse effects to human health [118]. The changes in bioactive amines may be partly attributed through amino acid decarboxylation by microbial enzymes during the fermentation process of cocoa beans and their fate has been discussed in [118]. The roasting process also modifies significantly the profile and levels of biogenic amines [120].

Cocoa products undergo several steps of thermal treatment (drying/roasting) during processing where Maillard reaction products originating from the interactions of proteins, peptides and amino acids with reducing sugars, classified as early-stage "Amadori products" or advanced brown pigments termed "melanoidins", are formed. Melanoidins are brown, high-molecular-weight products of Maillard reaction [121] and may also contribute to the radical-scavenging potential [122]. Further, it has been suggested that complex polyphenol oxidation and condensation products may also be integrated in melanoidins [117]. Recent studies also discuss the loss of the naturally occurring antioxidants (flavonoids), while others, such as Maillard reaction products, are formed while considering the different stages of cocoa processing [99,123]. While comparing raw, pre-roasted and roasted cocoa samples, increased radical-scavenging activity and reduced growth of pathogenic bacteria in different molecular weight fractions (>30, 30–10, 10–5 and <5 kDa) of roasted cocoa were determined [122]. However, the study also documents that also beneficial bacteria are suppressed in their growth activity [122]. The structure and biological activities of melanoidins (antioxidant, antimicrobial, anticancer, antihypertensive, cytotoxic, genotoxic, and detoxifying activities) have recently been comprehensively reviewed, giving some implications for human health [121,124]. These compounds are only partially characterized and their activities are poorly understood, but provide a potential reservoir for novel and potent bioactivities, underlining the need for further research in this area [117,121,122,124]. Some very recent works not only underline the complexity of this compound class, but also reveal that the high-molecular-weight melanoidin fractions formed may integrate phenolic compounds during roasting, contributing correspondingly to their antioxidant activity [86–88]. Finally, the protein crosslinking via the Maillard reaction has been shown to alter the functional properties of several food proteins, but the potential to use this chemistry to alter the functional performance of cocoa proteins has yet to be fully explored [125]. In summary, Table 2 documents the biological activity of the most relevant protein and their modified fractions.

Table 2. Selected examples for bioactivity potentials connected with cocoa protein fraction.

Protein-Related Fractions	Documented Bioactivity	Relevant Studies
Intact proteins, e.g., cysteine protease Albumin and glutelin fractions	Response of the action of the parasite *Moniliophthora perniciosa*, a fungus that causes witches' broom	[56]
	Radical-scavenging capacity	[113,114]
Release of bioactive peptides during fermentation	Antioxidants—therapeutically interesting for the prevention of age-related diseases	[112]
	Antitumor activity in cell culture studies	[113]
	Anti-hypertensive activities	[115]
	Dipeptidyl peptidase IV inhibitory activity—anti-obesity and anti-hyperglycemic effects	[117]
Enzymatic and/or (chemical) decarboxylation of amino acids	Bioactive amines–contribution to antioxidant activity Depending on concentration—positive/adverse effects to human health	[118]
Thermal modification	Maillard reaction/Melanoidins—antioxidant, antimicrobial, anticancer, antihypertensive, cytotoxic, genotoxic, and detoxifying activities	[121,124]
	Interactions with phenolic compounds—antioxidant activity	[86–88]
Direct modification by phenolic compounds	Fermentation/Alkalization; retention of some of the polyphenol structure and activity and/or introduction of potentially new activities	[7,51,86–88,117]

6. Conclusions

In conclusion, the present study reviews the literature on the impact of different stages of cultivation and processing with focus on the changes induced in the protein fractions. It surely does not handle all the publications available in the field, but should prove helpful for these researchers needing a quick start in this particular field of research. Some of relevant research areas have been identified that need to be better accessed, especially while comparing the content and biological activities of different varieties of *Theobroma cacao*. The relevant "state of the art" also suggests that exploration of other potentially bioactive components in cocoa needs to be undertaken, while considering the complexity of reaction products occurring during the roasting phase of the post-harvest processing. In the same context, there is an increasing interest in two further involved compound classes (proteins and phenolics) from the chemical point of view, which is related directly or indirectly to their dual role as substrates for oxidative-monitored reactions, and integration of mass spectrometric methods may provide a valuable tool for their characterization. Finally, the utilization of partially processed cocoa beans (e.g., fermented, conciliatory thermal treatment) provides a large collection of bioactive potentials that could be included in the designing of functional foods, illuminating an alternative use of cocoa especially in the cocoa-producing countries to bolster the diets with corresponding positive impact on the health status of the local populations.

Supplementary Materials: References [126–129] are cited in the supplementary materials. The following are available online at http://www.mdpi.com/2072-6643/11/2/428/s1, Figure S1: Chronological order of the steps taken to produce the final model of the vicilin storage protein, Figure S2: Calculated and postulated phosphorylation sites (green) in the storage protein vicilin from *Theobroma cacao* (Entry: Q43358; 60.8 kDa/525 amino acids; reviewed version from the database UniProt-http://www.uniprot.org/; 28.01-2019). The data indicates that the reported modifications (orange) at positions 232 (Thr), 235 (Ser) and 240 (Ser), as reported in [50] are most probably occurring after the proteolytic processing of the 66-kDa precursor [9,22], Figure S3: Calculated and postulated glycosylation sites (O-GLcNAc; pink) in the storage protein vicilin from *Theobroma cacao* (Entry: Q43358; 60.8 kDa/525 amino acids; reviewed version from the database UniProt-http://www.uniprot.org/; 28.01-2019). The data indicates that the reported modifications at positions 193 (Thr), 235 (Ser), 338 (Thr) and 474 (Thr) as reported in [50] are most probably occurring after the proteolytic processing of the 66 kDa precursor [9,22], Table S1: Templates utilized for the homology model for vicilin from *Theobroma cacao*.

Author Contributions: Conceptualization, H.M.R., T.H., S.T.S. and G.H.; resources, S.T.S.; writing of original draft preparation, H.M.R.; writing of review and editing, T.H, G.H. and S.T.S.; visualization, T.H.

Funding: S.S. is supported by a postdoctoral scholarship (Georg Forster-Forschungsstipendium; Ref 3.4-CMR-1164093-GF-P) from the Alexander von Humboldt Foundation.

Conflicts of Interest: The authors declare no conflicts of interest.

References

1. Lima, L.J.R.; Almeida, M.H.; Nout, M.J.R.; Zwietering, M.H. *Theobroma cacao* L., "The Food of the Gods": Quality Determinants of Commercial Cocoa Beans, with Particular Reference to the Impact of Fermentation. *Crit. Rev. Food Sci.* **2011**, *51*, 731–761. [CrossRef] [PubMed]

2. Timbie, D.J.; Keeney, P.G. Extraction, fractionation, and amino acid composition of Brazilian comun cacao proteins. *J. Agric. Food Chem.* **1977**, *25*, 424–426. [CrossRef] [PubMed]

3. Bertazzo, A.; Comai, S.; Brunato, I.; Zancato, M.; Costa, C.V.L. The content of protein and non-protein (free and protein-bound) tryptophan in *Theobroma cacao* beans. *Food Chem.* **2011**, *124*, 93–96. [CrossRef]

4. Timbie, D.J.; Keeney, P.G. Comparison of Several Types of Cocoa Beans Relative to Fractionated Protein-Components. *J. Agr. Food Chem.* **1980**, *28*, 472–474. [CrossRef]

5. Zak, D.L.; Keeney, P.G. Changes in cocoa proteins during ripening of fruit, fermentation, and further processing of cocoa beans. *J. Agric. Food Chem.* **1976**, *24*, 483–486. [CrossRef] [PubMed]

6. Zak, D.L.; Keeney, P.G. Extraction and fractionation of cocoa proteins as applied to several varieties of cocoa beans. *J. Agric. Food Chem.* **1976**, *24*, 479–483. [CrossRef]

7. Rawel, H.M.; Rohn, S. Nature of hydroxycinnamate-protein interactions. *Phytochem Rev* **2010**, *9*, 93–109. [CrossRef]

8. Voigt, J.; Heinrichs, H.; Ziegelerberghausen, H.; Srivastava, S.; Xiong, Q.; Biehl, B. The Storage Proteins and Seed Proteases of *Theobroma cacao*. *Biol. Chem. H-S* **1991**, *372*, 772–773.

9. Voigt, J.; Biehl, B.; Wazir, S.K.S. The Major Seed Proteins of *Theobroma cacao* L. *Food Chem.* **1993**, *47*, 145–151. [CrossRef]

10. Voigt, J.; Biehl, B.; Heinrichs, H.; Kamaruddin, S.; Marsoner, G.G.; Hugi, A. In-Vitro Formation of Cocoa-Specific Aroma Precursors—Aroma-Related Peptides Generated from Cocoa-Seed Protein by Cooperation of an Aspartic Endoprotease and a Carboxypeptidase. *Food Chem.* **1994**, *49*, 173–180. [CrossRef]

11. Voigt, J.; Heinrichs, H.; Voigt, G.; Biehl, B. Cocoa-Specific Aroma Precursors Are Generated by Proteolytic Digestion of the Vicilin-Like Globulin of Cocoa Seeds. *Food Chem.* **1994**, *50*, 177–184. [CrossRef]

12. Voigt, J.; Voigt, G.; Heinrichs, H.; Wrann, D.; Biehl, B. In-Vitro Studies on the Proteolytic Formation of the Characteristic Aroma Precursors of Fermented Cocoa Seeds - the Significance of Endoprotease Specificity. *Food Chem.* **1994**, *51*, 7–14. [CrossRef]

13. Voigt, J.; Wrann, D.; Heinrichs, H.; Biehl, B. The Proteolytic Formation of Essential Cocoa-Specific Aroma Precursors Depends on Particular Chemical Structures of the Vicilin-Class Globulin of the Cocoa Seeds Lacking in the Globular Storage Proteins of Coconuts, Hazelnuts and Sunflower Seeds. *Food Chem.* **1994**, *51*, 197–205. [CrossRef]

14. Bytof, G.; Biehl, B.; Heinrichs, H.; Voigt, J. Specificity and Stability of the Carboxypeptidase Activity in Ripe, Ungerminated Seeds of *Theobroma cacao* L. *Food Chem.* **1995**, *54*, 15–21. [CrossRef]

15. Voigt, J.; Biehl, B. Precursors of the Cocoa-Specific Aroma Components Are Derived from the Vicilin-Class (7s) Globulin of the Cocoa Seeds by Proteolytic Processing. *Bot. Acta* **1995**, *108*, 283–289. [CrossRef]

16. Voigt, J.; Kamaruddin, S.; Heinrichs, H.; Wrann, D.; Senyuk, V.; Biehl, B. Developmental Stage-Dependent Variation of the Levels of Globular Storage Protein and Aspartic Endoprotease during Ripening and Germination of *Theobroma cacao* L. Seeds. *J. Plant Physiol.* **1995**, *145*, 299–307. [CrossRef]

17. Voigt, G.; Biehl, B.; Heinrichs, H.; Voigt, J. Aspartic proteinase levels in seeds of different angiosperms. *Phytochemistry* **1997**, *44*, 389–392. [CrossRef]

18. Spencer, M.E.; Hodge, R. Cloning and Sequencing of a Cdna-Encoding the Major Storage Proteins of *Theobroma cacao*—Identification of the Proteins as Members of the Vicilin Class of Storage Proteins. *Planta* **1992**, *186*, 567–576. [CrossRef]

19. McHenry, L.; Fritz, P.J. Comparison of the structure and nucleotide sequences of vicilin genes of cocoa and cotton raise questions about vicilin evolution. *Plant Mol. Bio.l* **1992**, *18*, 1173–1176. [CrossRef]

20. Tai, H.; McHenry, L.; Fritz, P.J.; Furtek, D.B. Nucleic acid sequence of a 21 kDa cocoa seed protein with homology to the soybean trypsin inhibitor (Kunitz) family of protease inhibitors. *Plant Mol. Biol.* **1991**, *16*, 913–915. [CrossRef]

21. Spencer, M.E.; Hodge, R. Cloning and sequencing of the cDNA encoding the major albumin of *Theobroma cacao*: Identification of the protein as a member of the Kunitz protease inhibitor family. *Planta* **1991**, *183*, 528–535. [CrossRef] [PubMed]

22. Kratzer, U.; Frank, R.; Kalbacher, H.; Biehl, B.; Wostemeyer, J.; Voigt, J. Subunit structure of the vicilin-like globular storage protein of cocoa seeds and the origin of cocoa- and chocolate-specific aroma precursors. *Food Chem.* **2009**, *113*, 903–913. [CrossRef]

23. Lerceteau, E.; Rogers, J.; Petiard, V.; Crouzillat, D. Evolution of cacao bean proteins during fermentation: A study by two-dimensional electrophoresis. *J. Sci. Food Agr.* **1999**, *79*, 619–625. [CrossRef]

24. Pettipher, G.L. The Extraction and Partial-Purification of Cocoa Storage Proteins. *Cafe Cacao The* **1990**, *34*, 23–26.

25. Chin, C.F.; Tan, H.S. The Use of Proteomic Tools to Address Challenges Faced in Clonal Propagation of Tropical Crops through Somatic Embryogenesis. *Proteomes* **2018**, *6*, 21. [CrossRef] [PubMed]

26. Kumari, N.; Grimbs, A.; D'Souza, R.N.; Verma, S.K.; Corno, M.; Kuhnert, N.; Ullrich, M.S. Origin and varietal based proteomic and peptidomic fingerprinting of *Theobroma cacao* in non-fermented and fermented cocoa beans. *Food Res. Int.* **2018**, *111*, 137–147. [CrossRef]

27. Shewry, P.R.; Halford, N.G. Cereal seed storage proteins: Structures, properties and role in grain utilization. *J. Exp. Bot.* **2002**, *53*, 947–958. [CrossRef]

28. Scollo, E.; Neville, D.; Oruna-Concha, M.J.; Trotin, M.; Cramer, R. Characterization of the Proteome of *Theobroma cacao* Beans by Nano-UHPLC-ESI MS/MS. *Proteomics* **2018**, *18*. [CrossRef]

29. Wang, L.; Nagele, T.; Doerfler, H.; Fragner, L.; Chaturvedi, P.; Nukarinen, E.; Bellaire, A.; Huber, W.; Weiszmann, J.; Engelmeier, D.; et al. System level analysis of cacao seed ripening reveals a sequential interplay of primary and secondary metabolism leading to polyphenol accumulation and preparation of stress resistance. *Plant J.* **2016**, *87*, 318–332. [CrossRef]

30. Bertazzo, A.; Agnolin, F.; Comai, S.; Zancato, M.; Costa, C.V.; Seraglia, R.; Traldi, P. The protein profile of *Theobroma cacao* L. seeds as obtained by matrix-assisted laser desorption/ionization mass spectrometry. *Rapid Commun. Mass Spectrom.* **2011**, *25*, 2035–2042. [CrossRef]

31. Scherp, P.; Ku, G.; Coleman, L.; Kheterpal, I. Gel-based and gel-free proteomic technologies. *Methods Mol. Biol.* **2011**, *702*, 163–190. [PubMed]

32. Niemenak, N.; Kaiser, E.; Maximova, S.N.; Laremore, T.; Guiltinan, M.J. Proteome analysis during pod, zygotic and somatic embryo maturation of *Theobroma cacao*. *J. Plant Physiol.* **2015**, *180*, 49–60. [CrossRef] [PubMed]

33. Noah, A.M.; Niemenak, N.; Sunderhaus, S.; Haase, C.; Omokolo, D.N.; Winkelmann, T.; Braun, H.P. Comparative proteomic analysis of early somatic and zygotic embryogenesis in *Theobroma cacao* L. *J. Proteomics* **2013**, *78*, 123–133. [CrossRef] [PubMed]

34. Zaini, N.M.; Awang, A.; Budiman, C.; Rodrigues, K.F. Single step purification of 2S albumin from *Theobroma cacao*. *Int. J. Adv. Appl. Sci.* **2017**, *4*, 57–61. [CrossRef]

35. MacLean, B.; Tomazela, D.M.; Shulman, N.; Chambers, M.; Finney, G.L.; Frewen, B.; Kern, R.; Tabb, D.L.; Liebler, D.C.; MacCoss, M.J. Skyline: An open source document editor for creating and analyzing targeted proteomics experiments. *Bioinformatics* **2010**, *26*, 966–968. [CrossRef] [PubMed]

36. Huschek, G.; Bonick, J.; Merkel, D.; Huschek, D.; Rawel, H. Authentication of leguminous-based products by targeted biomarkers using high resolution time of flight mass spectrometry. *LWT—Food Sci. Technol.* **2018**, *90*, 164–171. [CrossRef]

37. Bonick, J.; Huschek, G.; Rawel, H.M. Determination of wheat, rye and spelt authenticity in bread by targeted peptide biomarkers. *J. Food Compos. Anal.* **2017**, *58*, 82–91. [CrossRef]

38. Argout, X.; Salse, J.; Aury, J.M.; Guiltinan, M.J.; Droc, G.; Gouzy, J.; Allegre, M.; Chaparro, C.; Legavre, T.; Maximova, S.N.; et al. The genome of *Theobroma cacao*. *Nat. Genet.* **2011**, *43*, 101–108. [CrossRef] [PubMed]

39. Cornejo, O.E.; Yee, M.C.; Dominguez, V.; Andrews, M.; Sockell, A.; Strandberg, E.; Livingstone, D., 3rd; Stack, C.; Romero, A.; Umaharan, P.; et al. Population genomic analyses of the chocolate tree, *Theobroma cacao* L., provide insights into its domestication process. *Commun. Biol.* **2018**, *1*, 167. [CrossRef] [PubMed]

40. Argout, X.; Martin, G.; Droc, G.; Fouet, O.; Labadie, K.; Rivals, E.; Aury, J.M.; Lanaud, C. The cacao Criollo genome v2.0: An improved version of the genome for genetic and functional genomic studies. *BMC Genomics* **2017**, *18*, 730. [CrossRef]

41. Quinga, L.A.P.; Heringer, A.S.; Fraga, H.P.D.; Vieira, L.D.; Silveira, V.; Steinmacher, D.A.; Guerra, M.P. Insights into the conversion potential of *Theobroma cacao* L. somatic embryos using quantitative proteomic analysis. *Sci. Hortic.* **2018**, *229*, 65–76. [CrossRef]

42. Florez, S.L.; Erwin, R.L.; Maximova, S.N.; Guiltinan, M.J.; Curtis, W.R. Enhanced somatic embryogenesis in *Theobroma cacao* using the homologous BABY BOOM transcription factor. *BMC Plant Biol.* **2015**, *15*, 121. [CrossRef] [PubMed]

43. Zhang, Y.; Clemens, A.; Maximova, S.N.; Guiltinan, M.J. The *Theobroma cacao* B3 domain transcription factor TcLEC2 plays a duel role in control of embryo development and maturation. *BMC Plant Biol.* **2014**, *14*, 106. [CrossRef]

44. Maximova, S.N.; Florez, S.; Shen, X.; Niemenak, N.; Zhang, Y.; Curtis, W.; Guiltinan, M.J. Genome-wide analysis reveals divergent patterns of gene expression during zygotic and somatic embryo maturation of *Theobroma cacao* L., the chocolate tree. *BMC Plant Biol.* **2014**, *14*, 185. [CrossRef] [PubMed]

45. Sokolov, A.N.; Pavlova, M.A.; Klosterhalfen, S.; Enck, P. Chocolate and the brain: Neurobiological impact of cocoa flavanols on cognition and behavior. *Neurosci. Biobehav. Rev.* **2013**, *37*, 2445–2453. [CrossRef] [PubMed]

46. Kononowicz, H.; Janick, J. Changes in nucleus, nucleolus and cell size accompanying somatic embryogenesis of *Theobroma cacao* L. I. Relationship between DNA and total protein content and size of nucleus, nucleolus and cell. *Folia Histochem. Cytobiol.* **1988**, *26*, 237–247. [PubMed]

47. Orduna, A.R.; Husby, E.; Yang, C.T.; Ghosh, D.; Beaudry, F. Assessment of meat authenticity using bioinformatics, targeted peptide biomarkers and high-resolution mass spectrometry. *Food Addit. Contam. A* **2015**, *32*, 1709–1717. [CrossRef] [PubMed]

48. Prasanna, V.; Prabha, T.N.; Tharanathan, R.N. Fruit ripening phenomena—An overview. *Crit. Rev. Food Sci. Nutr.* **2007**, *47*, 1–19. [CrossRef] [PubMed]

49. Biehl, B.; Meyer, B.; Crone, G.; Pollmann, L.; Binsaid, M. Chemical and Physical Changes in the Pulp during Ripening and Post-Harvest Storage of Cocoa Pods. *J. Sci. Food Agr.* **1989**, *48*, 189–208. [CrossRef]

50. Kumari, N.; Kofi, K.J.; Grimbs, S.; D'Souza, R.N.; Kuhnert, N.; Vrancken, G.; Ullrich, M.S. Biochemical fate of vicilin storage protein during fermentation and drying of cocoa beans. *Food Res. Int.* **2016**, *90*, 53–65. [CrossRef]

51. Ali, M.; Homann, T.; Kreisel, J.; Khalil, M.; Puhlmann, R.; Kruse, H.P.; Rawel, H. Characterization and modeling of the interactions between coffee storage proteins and phenolic compounds. *J. Agric. Food Chem.* **2012**, *60*, 11601–11608. [CrossRef] [PubMed]

52. Andrade, D.V.G.d.; Góes-Neto, A.; Junior, M.C.; Taranto, A.G. Comparative modeling and QM/MM studies of cysteine protease mutant of *Theobroma cacao*. *Int. J. Quantum. Chem.* **2012**, *112*, 3164–3168. [CrossRef]

53. Laloi, M.; McCarthy, J.; Morandi, O.; Gysler, C.; Bucheli, P. Molecular and biochemical characterisation of two aspartic proteinases TcAP1 and TcAP2 from *Theobroma cacao* seeds. *Planta* **2002**, *215*, 754–762. [CrossRef] [PubMed]

54. Griffiths, G.; Walsh, M.C.; Harwood, J.L. Acyl-Thioesterase Activity in Developing Seeds of Cocoa. *Phytochemistry* **1993**, *32*, 1403–1405. [CrossRef]

55. Sakharov, I.Y.; Ardila, G.B. Variations of peroxidase activity in cocoa (*Theobroma cacao* L.) beans during their ripening, fermentation and drying. *Food Chem.* **1999**, *65*, 51–54. [CrossRef]

56. Hansen, C.E.; Manez, A.; Burri, C.; Bousbaine, A. Comparison of enzyme activities involved in flavour precursor formation in unfermented beans of different cocoa genotypes. *J Sci Food Agr.* **2000**, *80*, 1193–1198. [CrossRef]

57. Janek, K.; Niewienda, A.; Wostemeyer, J.; Voigt, J. The cleavage specificity of the aspartic protease of cocoa beans involved in the generation of the cocoa-specific aroma precursors. *Food Chem.* **2016**, *211*, 320–328. [CrossRef] [PubMed]

58. Stark, T.; Hofmann, T. Isolation, structure determination, synthesis, and sensory activity of N-phenylpropenoyl-L-amino acids from cocoa (*Theobroma cacao*). *J. Agric. Food Chem.* **2005**, *53*, 5419–5428. [CrossRef]

59. Lechtenberg, M.; Henschel, K.; Lieflander-Wulf, U.; Quandt, B.; Hensel, A. Fast determination of N-phenylpropenoyl-L-amino acids (NPA) in cocoa samples from different origins by ultra-performance liquid chromatography and capillary electrophoresis. *Food Chem.* **2012**, *135*, 1676–1684. [CrossRef]

60. Pereira-Caro, G.; Borges, G.; Nagai, C.; Jackson, M.C.; Yokota, T.; Crozier, A.; Ashihara, H. Profiles of phenolic compounds and purine alkaloids during the development of seeds of *Theobroma cacao* cv. Trinitario. *J. Agric. Food Chem.* **2013**, *61*, 427–434. [CrossRef]

61. Bucheli, P.; Rousseau, G.; Alvarez, M.; Laloi, M.; McCarthy, J. Developmental variation of sugars, carboxylic acids, purine alkaloids, fatty acids, and endoproteinase activity during maturation of *Theobroma cacao* L. seeds. *J. Agric. Food Chem.* **2001**, *49*, 5046–5051. [CrossRef] [PubMed]

62. Rojas, L.F.; Gallego, A.; Gil, A.; Londono, J.; Atehortua, L. Monitoring accumulation of bioactive compounds in seeds and cell culture of *Theobroma cacao* at different stages of development. *In Vitro Cell Dev-Pl* **2015**, *51*, 174–184. [CrossRef]

63. Dang, Y.K.T.; Nguyen, H.V.H. Effects of Maturity at Harvest and Fermentation Conditions on Bioactive Compounds of Cocoa Beans. *Plant Foods Hum. Nutr.* **2018**. [CrossRef] [PubMed]

64. Fait, A.; Angelovici, R.; Less, H.; Ohad, I.; Urbanczyk-Wochniak, E.; Fernie, A.R.; Galili, G. Arabidopsis seed development and germination is associated with temporally distinct metabolic switches. *Plant Physiol.* **2006**, *142*, 839–854. [CrossRef] [PubMed]

65. Kongor, J.E.; Hinneh, M.; de Walle, D.V.; Afoakwa, E.O.; Boeckx, P.; Dewettinck, K. Factors influencing quality variation in cocoa (*Theobroma cacao*) bean flavour profile—A review. *Food Res. Int.* **2016**, *82*, 44–52. [CrossRef]

66. Voigt, J.; Janek, K.; Textoris-Taube, K.; Niewienda, A.; Wostemeyer, J. Partial purification and characterisation of the peptide precursors of the cocoa-specific aroma components. *Food Chem.* **2016**, *192*, 706–713. [CrossRef] [PubMed]

67. John, W.A.; Kumari, N.; Bottcher, N.L.; Koffi, K.J.; Grimbs, S.; Vrancken, G.; D'Souza, R.N.; Kuhnert, N.; Ullrich, M.S. Aseptic artificial fermentation of cocoa beans can be fashioned to replicate the peptide profile of commercial cocoa bean fermentations. *Food Res. Int.* **2016**, *89*, 764–772. [CrossRef]

68. Hue, C.; Gunata, Z.; Breysse, A.; Davrieux, F.; Boulanger, R.; Sauvage, F.X. Impact of fermentation on nitrogenous compounds of cocoa beans (*Theobroma cacao* L.) from various origins. *Food Chem.* **2016**, *192*, 958–964. [CrossRef]

69. Caligiani, A.; Marseglia, A.; Prandi, B.; Palla, G.; Sforza, S. Influence of fermentation level and geographical origin on cocoa bean oligopeptide pattern. *Food Chem.* **2016**, *211*, 431–439. [CrossRef]

70. Voigt, J.; Lieberei, R. *Biochemistry of Cocoa Fermentation*; CRC Press: Boca Raton, FL, USA, 2015; pp. 193–225.

71. Afoakwa, E.O.; Paterson, A.; Fowler, M.; Ryan, A. Flavor formation and character in cocoa and chocolate: A critical review. *Crit. Rev. Food Sci. Nutr.* **2008**, *48*, 840–857. [CrossRef]

72. Biehl, B.; Heinrichs, H.; Ziegelerberghausen, H.; Srivastava, S.; Xiong, Q.; Passern, D.; Senyuk, V.I.; Hammoor, M. The Proteases of Ungerminated Cocoa Seeds and Their Role in the Fermentation Process. *Angew. Bot.* **1993**, *67*, 59–65.

73. Granvogl, M.; Bugan, S.; Schieberle, P. Formation of amines and aldehydes from parent amino acids during thermal processing of cocoa and model systems: New insights into pathways of the strecker reaction. *J. Agric. Food Chem.* **2006**, *54*, 1730–1739. [CrossRef] [PubMed]

74. D'Souza, R.N.; Grimbs, A.; Grimbs, S.; Behrends, B.; Corno, M.; Ullrich, M.S.; Kuhnert, N. Degradation of cocoa proteins into oligopeptides during spontaneous fermentation of cocoa beans. *Food Res. Int.* **2018**, *109*, 506–516. [CrossRef] [PubMed]

75. Wollgast, J.; Anklam, E. Review on polyphenols in *Theobroma cacao*: Changes in composition during the manufacture of chocolate and methodology for identification and quantification. *Food Res. Int.* **2000**, *33*, 423–447. [CrossRef]

76. Wollgast, J.; Anklam, E. Polyphenols in chocolate: Is there a contribution to human health? *Food Res. Int.* **2000**, *33*, 449–459. [CrossRef]

77. Kim, H.; Keeney, P.G. (-)-Epicatechin Content in Fermented and Unfermented Cocoa Beans. *J. Food Sci.* **1984**, *49*, 1090–1092. [CrossRef]

78. Sanchez-Rabaneda, F.; Jauregui, O.; Casals, I.; Andres-Lacueva, C.; Izquierdo-Pulido, M.; Lamuela-Raventos, R.M. Liquid chromatographic/electrospray ionization tandem mass spectrometric study of the phenolic composition of cocoa (*Theobroma cacao*). *J. Mass Spectrom.* **2003**, *38*, 35–42. [CrossRef] [PubMed]

79. Elwers, S.; Zambrano, A.; Rohsius, C.; Lieberei, R. Differences between the content of phenolic compounds in Criollo, Forastero and Trinitario cocoa seed (*Theobroma cacao* L.). *Eur. Food Res. Technol.* **2009**, *229*, 937–948. [CrossRef]

80. Montavon, P.; Mauron, A.F.; Duruz, E. Changes in green coffee protein profiles during roasting. *J. Agric. Food Chem.* **2003**, *51*, 2335–2343. [CrossRef]

81. Montavon, P.; Duruz, E.; Rumo, G.; Pratz, G. Evolution of green coffee protein profiles with maturation and relationship to coffee cup quality. *J. Agric. Food Chem.* **2003**, *51*, 2328–2334. [CrossRef]

82. Oliviero, T.; Capuano, E.; Cammerer, B.; Fogliano, V. Influence of roasting on the antioxidant activity and HMF formation of a cocoa bean model systems. *J. Agric. Food Chem.* **2009**, *57*, 147–152. [CrossRef] [PubMed]

83. Jinap, S.; Ikrawan, Y.; Bakar, J.; Saari, N.; Lioe, H.N. Aroma precursors and methylpyrazines in underfermented cocoa beans induced by endogenous carboxypeptidase. *J. Food Sci.* **2008**, *73*, H141–H147. [CrossRef]

84. Crafack, M.; Keul, H.; Eskildsen, C.E.; Petersen, M.A.; Saerens, S.; Blennow, A.; Skovmand-Larsen, M.; Swiegers, J.H.; Petersen, G.B.; Heimdal, H.; et al. Impact of starter cultures and fermentation techniques on the volatile aroma and sensory profile of chocolate. *Food Res. Int.* **2014**, *63*, 306–316. [CrossRef]

85. Afoakwa, E. *Cocoa Production and Processing Technology*; CRC Press: Boca Raton, FL, USA, 2014; pp. 1–329.

86. Oracz, J.; Nebesny, E.; Zyzelewicz, D. Identification and quantification of free and bound phenolic compounds contained in the high-molecular weight melanoidin fractions derived from two different types of cocoa beans by UHPLC-DAD-ESI-HR-MS(n). *Food Res. Int.* **2019**, *115*, 135–149. [CrossRef]

87. Oracz, J.; Nebesny, E. Effect of roasting parameters on the physicochemical characteristics of high-molecular-weight Maillard reaction products isolated from cocoa beans of different *Theobroma cacao* L. groups. *Eur. Food Res. Technol.* **2019**, *245*, 111–128. [CrossRef]

88. Quiroz-Reyes, C.N.; Fogliano, V. Design cocoa processing towards healthy cocoa products: The role of phenolics and melanoidins. *J. Funct. Foods* **2018**, *45*, 480–490. [CrossRef]

89. Latif, R. Health benefits of cocoa. *Curr. Opin. Clin. Nutr. Metab. Care* **2013**, *16*, 669–674. [CrossRef] [PubMed]

90. Koli, R.; Kohler, K.; Tonteri, E.; Peltonen, J.; Tikkanen, H.; Fogelholm, M. Dark chocolate and reduced snack consumption in mildly hypertensive adults: An intervention study. *Nutr. J.* **2015**, *14*, 84. [CrossRef]

91. Latham, L.S.; Hensen, Z.K.; Minor, D.S. Chocolate—Guilty pleasure or healthy supplement? *J. Clin. Hypertens. (Greenwich)* **2014**, *16*, 101–106. [CrossRef]

92. Gu, Y.; Yu, S.; Park, J.Y.; Harvatine, K.; Lambert, J.D. Dietary cocoa reduces metabolic endotoxemia and adipose tissue inflammation in high-fat fed mice. *J. Nutr. Biochem.* **2014**, *25*, 439–445. [CrossRef]

93. Gu, Y.; Yu, S.; Lambert, J.D. Dietary cocoa ameliorates obesity-related inflammation in high fat-fed mice. *Eur. J. Nutr.* **2014**, *53*, 149–158. [CrossRef] [PubMed]

94. Zyzelewicz, D.; Budryn, G.; Oracz, J.; Antolak, H.; Kregiel, D.; Kaczmarska, M. The effect on bioactive components and characteristics of chocolate by functionalization with raw cocoa beans. *Food Res. Int.* **2018**, *113*, 234–244. [CrossRef] [PubMed]

95. Zyzelewicz, D.; Zaklos-Szyda, M.; Juskiewicz, J.; Bojczuk, M.; Oracz, J.; Budryn, G.; Miskiewicz, K.; Krysiak, W.; Zdunczyk, Z.; Jurgonski, A. Cocoa bean (*Theobroma cacao* L.) phenolic extracts as PTP1B inhibitors, hepatic HepG2 and pancreatic beta-TC3 cell cytoprotective agents and their influence on oxidative stress in rats. *Food Res. Int.* **2016**, *89*, 946–957. [CrossRef]

96. Zeng, H.W.; Locatelli, M.; Bardelli, C.; Amoruso, A.; Coisson, J.D.; Travaglia, F.; Arlorio, M.; Brunelleschi, S. Anti-inflammatory Properties of Clovamide and *Theobroma cacao* Phenolic Extracts in Human Monocytes: Evaluation of Respiratory Burst, Cytokine Release, NF-kappa B Activation, and PPAR gamma Modulation. *J. Agr. Food Chem.* **2011**, *59*, 5342–5350. [CrossRef]

97. Martin, M.A.; Goya, L.; Ramos, S. Antidiabetic actions of cocoa flavanols. *Mol. Nutr. Food Res.* **2016**, *60*, 1756–1769. [CrossRef]

98. Salvador, I.; Massarioli, A.P.; Silva, A.P.S.; Malaguetta, H.; Melo, P.S.; Alencar, S.M. Can we conserve trans-resveratrol content and antioxidant activity during industrial production of chocolate? *J. Sci. Food Agr.* **2019**, *99*, 83–89. [CrossRef]

99. Di Mattia, C.D.; Sacchetti, G.; Mastrocola, D.; Serafini, M. From Cocoa to Chocolate: The impact of Processing on In Vitro Antioxidant Activity and the Effects of Chocolate on Antioxidant Markers in Vivo. *Front. Immunol.* **2017**, *8*. [CrossRef]

100. Hu, S.; Kim, B.Y.; Baik, M.Y. Physicochemical properties and antioxidant capacity of raw, roasted and puffed cacao beans. *Food Chem.* **2016**, *194*, 1089–1094. [CrossRef]

101. Schinella, G.; Mosca, S.; Cienfuegos-Jovellanos, E.; Pasamar, M.A.; Muguerza, B.; Ramon, D.; Rios, J.L. Antioxidant properties of polyphenol-rich cocoa products industrially processed. *Food Res. Int.* **2010**, *43*, 1614–1623. [CrossRef]

102. Erdmann, K.; Cheung, B.W.Y.; Schroder, H. The possible roles of food-derived bioactive peptides in reducing the risk of cardiovascular disease. *J. Nutr. Biochem.* **2008**, *19*, 643–654. [CrossRef]

103. Pihlanto-Leppala, A. Bioactive peptides derived from bovine whey proteins: Opioid and ace-inhibitory peptides. *Trends Food Sci. Tech.* **2000**, *11*, 347–356. [CrossRef]

104. Chen, J.R.; Suetsuna, K.; Yamauchi, F. Isolation and Characterization of Immunostimulative Peptides from Soybean. *J. Nutr. Biochem.* **1995**, *6*, 310–313. [CrossRef]

105. McLay, R.N.; Pan, W.H.; Kastin, A.J. Effects of peptides on animal and human behavior: A review of studies published in the first twenty years of the journal Peptides. *Peptides* **2001**, *22*, 2181–2255. [CrossRef]

106. Mendis, E.; Rajapakse, N.; Kim, S.K. Antioxidant properties of a radical-scavenging peptide purified from enzymatically prepared fish skin gelatin hydrolysate. *J. Agr. Food Chem.* **2005**, *53*, 581–587. [CrossRef] [PubMed]

107. Suetsuna, K.; Maekawa, K.; Chen, J.R. Antihypertensive effects of Undaria pinnatifida (wakame) peptide on blood pressure in spontaneously hypertensive rats. *J. Nutr. Biochem.* **2004**, *15*, 267–272. [CrossRef] [PubMed]

108. Satake, M.; Enjoh, M.; Nakamura, Y.; Takano, T.; Kawamura, Y.; Arai, S.; Shimizu, M. Transepithelial transport of the bioactive tripeptide, Val-Pro-Pro, in human intestinal Caco-2 cell monolayers. *Biosci. Biotech. Bioch.* **2002**, *66*, 378–384. [CrossRef]

109. Masuda, O.; Nakamura, Y.; Takano, T. Antihypertensive peptides are present in aorta after oral administration of sour milk containing these peptides to spontaneously hypertensive rats. *J. Nutr.* **1996**, *126*, 3063–3068. [CrossRef]

110. Ohsawa, K.; Satsu, H.; Ohki, K.; Enjoh, M.; Takano, T.; Shimizu, M. Producibility and digestibility of antihypertensive beta-casein tripeptides, Val-Pro-Pro and Ile-Pro-Pro, in the gastrointestinal tract: Analyses using an in vitro model of mammalian gastrointestinal digestion. *J. Agr. Food Chem.* **2008**, *56*, 854–858. [CrossRef]

111. Vanhoof, G.; Goossens, F.; Demeester, I.; Hendriks, D.; Scharpe, S. Proline Motifs in Peptides and Their Biological Processing. *Faseb. J.* **1995**, *9*, 736–744. [CrossRef]

112. Martorell, P.; Bataller, E.; Llopis, S.; Gonzalez, N.; Alvarez, B.; Monton, F.; Ortiz, P.; Ramon, D.; Genoves, S. A cocoa peptide protects *Caenorhabditis elegans* from oxidative stress and beta-amyloid peptide toxicity. *PLoS ONE* **2013**, *8*, e63283. [CrossRef]

113. Preza, A.M.; Jaramillo, M.E.; Puebla, A.M.; Mateos, J.C.; Hernandez, R.; Lugo, E. Antitumor activity against murine lymphoma L5178Y model of proteins from cacao (*Theobroma cacao* L.) seeds in relation with in vitro antioxidant activity. *BMC Complement. Altern. Med.* **2010**, *10*, 61. [CrossRef] [PubMed]

114. Tovar-Perez, E.G.; Guerrero-Becerra, L.; Lugo-Cervantes, E. Antioxidant activity of hydrolysates and peptide fractions of glutelin from cocoa (*Theobroma cacao* L.) seed. *Cyta-J. Food* **2017**, *15*, 489–496. [CrossRef]

115. Sarmadi, B.; Ismail, A.; Hamid, M. Antioxidant and angiotensin converting enzyme (ACE) inhibitory activities of cocoa (*Theobroma cacao* L.) autolysates. *Food. Res. Int.* **2011**, *44*, 290–296. [CrossRef]

116. Guil-Guerrero, J.L.; Ramos, L.; Moreno, C.; Zuniga-Paredes, J.C.; Carlosama-Yepez, M.; Ruales, P. Antimicrobial activity of plant-food by-products: A review focusing on the tropics. *Livest. Sci.* **2016**, *189*, 32–49. [CrossRef]

117. Ryan, C.M.; Khoo, W.; Stewart, A.C.; O'Keefe, S.F.; Lambert, J.D.; Neilson, A.P. Flavanol concentrations do not predict dipeptidyl peptidase-IV inhibitory activities of four cocoas with different processing histories. *Food Function* **2017**, *8*, 746–756. [CrossRef] [PubMed]

118. Brito, B.D.D.; Chiste, R.C.; Pena, R.D.; Gloria, M.B.A.; Lopes, A.S. Bioactive amines and phenolic compounds in cocoa beans are affected by fermentation. *Food Chem.* **2017**, *228*, 484–490. [CrossRef] [PubMed]

119. Baranowska, I.; Plonka, J. Simultaneous Determination of Biogenic Amines and Methylxanthines in Foodstuff-Sample Preparation with HPLC-DAD-FL Analysis. *Food Anal. Method.* **2015**, *8*, 963–972. [CrossRef]

120. Oracz, J.; Nebesny, E. Influence of roasting conditions on the biogenic amine content in cocoa beans of different *Theobroma cacao* cultivars. *Food Res. Int.* **2014**, *55*, 1–10. [CrossRef]

121. Langner, E.; Rzeski, W. Biological Properties of Melanoidins: A Review. *Int. J. Food Prop.* **2014**, *17*, 344–353. [CrossRef]

122. Summa, C.; McCourt, J.; Cammerer, B.; Fiala, A.; Probst, M.; Kun, S.; Anklam, E.; Wagner, K.H. Radical scavenging activity, anti-bacterial and mutagenic effects of Cocoa bean maillard reaction products with degree of roasting. *Mol. Nutr. Food Res.* **2008**, *52*, 342–351. [CrossRef]

123. Di Mattia, C.; Martuscelli, M.; Sacchetti, G.; Beheydt, B.; Mastrocola, D.; Pittia, P. Effect of different conching processes on procyanidin content and antioxidant properties of chocolate. *Food Res. Int.* **2014**, *63*, 367–372. [CrossRef]

124. Wang, H.Y.; Qian, H.; Yao, W.R. Melanoidins produced by the Mail lard reaction: Structure and biological activity. *Food Chem.* **2011**, *128*, 573–584. [CrossRef]

125. Jumnongpon, R.; Chaiseri, S.; Hongsprabhas, P.; Healy, J.P.; Meade, S.J.; Gerrard, J.A. Cocoa protein crosslinking using Maillard chemistry. *Food Chem.* **2012**, *134*, 375–380. [CrossRef]

126. Krieger, E.; Joo, K.; Lee, J.; Raman, S.; Thompson, J.; Tyka, M.; Baker, D.; Karplus, K. Improving physical realism, stereochemistry, and side-chain accuracy in homology modeling: Four approaches that performed well in CASP8. *Proteins* **2009**, *77*, 114–122. [CrossRef] [PubMed]

127. Krieger, E.; Koraimann, G.; Vriend, G. Increasing the precision of comparative models with YASARA NOVA—A self-parameterizing force field. *Proteins* **2002**, *47*, 393–402. [CrossRef] [PubMed]

128. Waterhouse, A.M.; Procter, J.B.; Martin, D.M.; Clamp, M.; Barton, G.J. Jalview Version 2—A multiple sequence alignment editor and analysis workbench. *Bioinformatics* **2009**, *25*, 1189–1191. [CrossRef]

129. Ali, M.; Homann, T.; Khalil, M.; Kruse, H.P.; Rawel, H. Milk whey protein modification by coffee-specific phenolics: Effect on structural and functional properties. *J. Agric. Food Chem.* **2013**, *61*, 6911–6920. [CrossRef] [PubMed]

Article

Changes in Antioxidants and Sensory Properties of Italian Chocolates and Related Ingredients Under Controlled Conditions During an Eighteen-Month Storage Period

Arianna Roda and Milena Lambri *

DiSTAS—Department for Sustainable Food Process, Università Cattolica del Sacro Cuore, 29122 Piacenza, Italy; arianna.roda@unicatt.it

* Correspondence: milena.lambri@unicatt.it; Tel.: +39-(0)523-599229; Fax: +39-(0)523-599232

Received: 24 August 2019; Accepted: 29 October 2019; Published: 9 November 2019

Abstract: Background: While there has been an increasing interest in the health properties of chocolate, limited research has looked into the changes of antioxidants occurring in the time span from production to the best before date, which was a period of 18 months in this study. Methods: Humidity, ash, pH, acidity, fiber, carotenoids, retinols, tocopherols, sugars, proteins, theobromine, caffeine, polyphenols, fats, the peroxide value, organic acids, and volatile compounds, along with the sensory profile, were monitored at 18-week intervals for 18 months under conditions simulating a factory warehouse or a point of sale. Results: At the end of the storage period, more polyphenols were lost (64% and 87%) than vitamin E (5% and 14%) in cocoa mass and cocoa powder, respectively. Conversely, a greater loss in vitamin E (34% and 86%) than in polyphenols (19% and 47%) was shown in the hazelnut paste and gianduja chocolate, respectively. The sensory profiling of cocoa mass, cocoa powder, and hazelnut paste revealed increases in grittiness and astringency, as well as decreases in melting, bitterness, and toasted aroma. Moreover, in the hazelnut paste and gianduja chocolate, oiliness increased with a toasted and caramel aroma. Furthermore, dark chocolate was more gritty, acidic, and bitter. Milk chocolate lost its nutty aroma but maintained its sweetness and creaminess. Conclusions: These results should contribute an important reference for companies and consumers, in order to preserve the antioxidants and understand how antioxidants and sensory properties change from the date of production until the best before date.

Keywords: Italian chocolate; quality; cocoa-based ingredients; monitoring; nutrition

1. Introduction

Many daily consumed foods, such as fruits, vegetables, wine, coffee, and chocolate, are attracting increasing attention due to their potential health effects thanks to their richness in polyphenols [1,2]. Recent studies have focused on cocoa products commonly consumed for pure pleasure, since they also generate tangible benefits for human health [3]. Indeed, cocoa is a complex product with over 300 constituents [4] and one of the richest sources of flavanols [5]. Cocoa belongs to the "nervine food group" because it contains xanthine alkaloids, with the most important being theobromine (2%) and caffeine [6]. When eaten in moderation as part of a balanced diet, it has been suggested that flavanols from cocoa products may exert beneficial effects on the cardiovascular risk via effects of lowering the blood pressure, anti-inflammation, antiplatelet function, higher HDL, and decreased LDL oxidation. This evidence has been supported by a systematic review of 136 publications [7] and by the European Food Safety Agency [8], which stated a health claim for dark chocolate with a high flavanol content due to its impact on "maintenance normal endothelium-dependent vasodilation which contributes to normal blood flow". In order to obtain the claimed effect, 200 mg of cocoa flavanols should be consumed daily. This amount could be provided by 2.5 g of high-flavanol cocoa powder or 10 g of high-flavanol dark chocolate.

On the other hand, the popularity of hedonic foods, like chocolate, mostly depends on its sensory properties, which in turn constitute the key for the acceptance of food products on the market. Consumers measure the quality and take raw information on the composition while evaluating the appearance, aroma, texture, taste, and flavor [9,10]. Additionally, experienced consumers have their own capability to search for defects leading to the rejection of chocolate and often caused by factors such as the temperature, relative humidity, and light [11], which in turn affect the stability of cocoa butter [9,12,13]. Indeed, the fat components are the main components responsible for oxidative deterioration which leads to the formation of off-flavors in chocolates [9,14].

Although there has been an increasing interest in the health properties of chocolate, mainly due to its polyphenol content, limited research has looked into the chocolate matrix to investigate the changes in polyphenols, vitamins, fats, and volatile compounds which also determine the quality [15–17]. While consumers are increasingly demanding high-quality food, they expect this quality to be maintained until the consumption time, but little is known about the evolution of chocolates and related raw materials in response to the storage conditions in a factory warehouse, as well as at a point of sale. The present study aims to monitor Italian chocolates and related ingredients for their polyphenol concentration, vitamin E, peroxide, and acidity values during storage for eighteen months under controlled conditions simulating a factory warehouse or a point of sale. In this study, sensory analysis was considered to outline the chemical changes which produce wider variations, so is more interesting from a commercial point of view.

2. Materials and Methods

2.1. Samples

Three types of chocolate (dark (D), milk (M), and gianduja (G)) and three ingredients (cocoa mass in solid form (CM), cocoa powder 22–24 (C), and hazelnut paste (HP)) were supplied by Venchi S.p.A. (Castelletto Stura, Cuneo, Italy). The recipes of the chocolates were as follows:

- D: cocoa 56% min., cocoa mass, sugar, cocoa butter, emulsifier: soy lecithin, and natural vanilla flavor;
- M: cocoa 31.8% min, milk: 23.5% min., sugar, whole milk powder, cocoa butter, cocoa mass, anhydrous milk fat, emulsifier: soy lecithin, and natural vanilla flavor;
- G: cocoa 21.2% min., Piedmont hazelnut pasta I.G.P. (33%), sugar, whole milk powder, cocoa butter, cocoa mass, anhydrous milk fat, emulsifier: soy lecithin, and natural vanilla flavor.

All the samples were collected immediately after production (t0) and stored in their own packages in an air-conditioned room at 21 ± 2 °C and 65% Relative Humidity (RH) for eighteen months, i.e., the best before date established by the company within the EU Regulation 2073:2005 [18]. An eighteen-week interval was decided for sampling considering both the number of samples made available by Venchi S.p.A and the storage time (t1, t2, t3, and t4).

2.2. Microbiological Analysis

The total microbial count (bacteria, fungi/molds, and yeasts) according to International Standard Organization (ISO) 4833-1:2013 [19], along with *Enterobacteriaceae* and coliforms as respectively described in ISO 21528-1:2017 [20] and ISO 4832:2006 [21], were detected at the production time (t0) and at the end of storage (t4). The number of colony-forming units (CFU) per gram of dry mass of the sample was provided.

2.3. Chemical Analysis

2.3.1. Humidity, Ash, pH, Acidity, and Fiber

Humidity and ash were determined according to Association of Official Analytical Chemists (AOAC) 931.04 [22], while food and raw fiber were analyzed using the AOAC 985.29 [23] methodology. Total acidity and pH were analyzed as reported in the official methods of the Office International du Cacao et du Chocolat et de la Confiserie [24]. For both analyses, a pH meter (CRISON MICRO TT 2050, Carpi, Modena, Italy) was used. The pH of a 10 g sample dissolved in 90 mL of boiling distilled water was measured. The acidity was dosed in the same solution with 20 mL of sodium hydroxide 0.1 N and then potentiometrically titrated with hydrochloric acid 0.1 N until pH 7.0. The result was finally expressed in mg equivalents of stearic acid in a 100 g sample.

2.3.2. Carotenoids, Retinol, Tocopherols, and Sugars

Carotenoids and retinol were determined by applying the AOAC 941.15 method [25], while tocopherols were dosed in compliance with Calvo et al. [26] and Belščak et al. [27]. The chromatographic determination of tocopherols was performed on an HPLC system, including a Perkin Elmer (Norwalk, CT, USA) 200 Series pump equipped with a Perkin-Elmer 650-10S fluorescence detector, Jasco LC-Net II/ADC (Oklahoma City, OK, USA) communication module, and ChromNAV Control Center software. A LiChrosorb Si60-5 C18 column 250 mm × 4.6 mm, 5 μm (Supelco, Bellefonte, PA, USA) was used, the mobile phase was hexane:isopropanol:ethanol (98.5:1:0.5) at a flow rate of 1.0 mL/min and the injection volume was 20 μL. The fluorescence detector was set at 290 nm excitation and 330 nm emission wavelengths. α-, γ-, and δ- tocopherols were identified by comparing the retention times with those of commercial standards. The results were expressed as mg of Vitamin E including α-, γ-, and δ- tocopherols contained in 100 g of dry sample. To analyze reducing and non-reducing sugars in cocoa liquor and hazelnut paste, the Luff-Schoorl volumetric analysis suggested by Balestrieri and Marini [6] was adopted.

2.3.3. Proteins

Protein determination was carried out following AOAC 939.02-1939 [28], as follows. Two grams of sample were dried in a stove for 24 h. Thirty milliliters of concentrated sulfuric acid (96%) were added to the sample and homogeneously dispersed, and the mixture was subsequently put into the digester (K-424 BUCHI) until complete digestion. Then, distillation was performed using a semi-automatic system (VELP SCIENTIFICA UDK R7), which adds 90 mL of sodium hydroxide (32%) and collects the distilled ammonia in 50 mL of boric acid (40 g/L), to which 0.5 mL of mixed indicator methyl red/bromocresol green was added. Finally, titration with 0.1 N sulfuric acid was carried out.

2.3.4. Theobromine, Caffeine, and Polyphenols

For determining theobromine, caffeine, and polyphenols, samples were defatted by extracting 50 g of each sample three times with 250 mL of *n*-hexane and drying the resulting powder under a nitrogen stream to remove the residual organic solvent [29]. For determinations of caffeine and theobromine, the powder was dispersed in methanol before spectrophotometric analysis, which was carried out by means of a Lambda Bio 40 UV-VIS spectrometer (Perkin Helmer).

The theobromine was determined following López-Martìnez et al.'s method [30]. Since the maximum theobromine absorption in our samples was observed at 240 nm, a calibration curve with methanolic solutions containing 5, 10, 15, 20, 25, 30, and 35 mg/L of theobromine was produced at 240 nm.

Caffeine was determined according to the procedure described by Hečimović et al. [31]. The absorbance was read at 274 nm and a calibration curve was built using solutions at 20, 40, 60, 80, and 100 mg/L of caffeine dissolved in methanol.

The total polyphenol content was determined for the defatted samples dispersed in distilled water according to Belšcak et al. [27], and the absorbance was recorded at 765 nm using a Shimadzu UV-1601 spectrophotometer (Shimadzu Europe, Duisburg, Germany). Gallic acid was used as a standard in a calibration curve with solutions of 20, 40, 80, 100, and 120 mg/L of gallic acid. The results were expressed as mg gallic acid equivalents (GAE) per 100 g of defatted sample.

2.3.5. Fat Content

To determine the fat content, the AOAC Official Method [25], modified as follows, was used. The samples initially underwent acid hydrolysis and extraction by means of a Soxhlet device. Different quantities of sample were weighed, including 3–4 g of cocoa liquor and hazelnut paste, and 4–5 g of cocoa powder. These quantities were mixed with 45 mL of distilled water at boiling point, after which 55 mL of hydrochloric acid (25% w/v) was added. The solutions boiled for about 30 min with a reflux condenser and were then filtered with a Whatman $n°$ 595 $\frac{1}{2}$ filter. The filter containing the hydrolyzed sample was thoroughly washed with distilled water until the chloride vanished, and it was then dried at 100 °C for 6 h. Finally, the fat matter was extracted by means of a Soxhlet device, where the dry filter underwent extraction with 50 mL of n-hexane for 4 h. The fatty acid and sterol profile were determined according to EC Regulation 2568/91 [32].

2.3.6. Peroxide Value

To separate the lipid fraction, 50 g of each sample was extracted three times with 250 mL of n-hexane [29]. The resulting mixture was centrifuged at 3000 rpm for 15 min (Varifuge 20 RS Hereaus Sepatech, Hanau, Germany), and the hexane was removed using Rotavapor (Büchi Rotavapor R-114, Flawil, Switzerland). The oil recovered was analyzed to determine the peroxide value using the method reported in the EC Regulation 2568/91 [32].

2.3.7. HPLC Analysis of Organic Acids

Twenty grams of sample were melted in 200 mL of hot distilled water, and then decolorized by means of a suitable quantity of carbon powder. The mix was centrifuged at 5000 rpm for 15 min at 5 °C (Centrifuge SL 16 R Thermo Scientific), with subsequent filtration by Whatman $n°$ 589/3 filters with a porosity of 150 μm. A collected volume of 30–35 mL was acidified with sulfuric acid (25% w/v) until reaching a pH 2.5. The acidified solution was centrifuged at 8000 rpm for 15 min at 15 °C (Varifuge 20 RS Hereaus Sepatech), in order to further purify the extract from any sediments, pigments, or turbidity. The supernatant obtained was filtered by means of 45 μm syringe filters. HPLC (Spectra system P4000) conditions were as follows: stationary phase Phenomenex Rezex column ROA-ORGANIC ACID H+ at 40°C; mobile phase 0.005 N sulfuric acid solution with a flow of 0.5 mL/min and a 20 μL injection volume; Spectra system UV 1000 detector at 210 nm.

2.3.8. GC Analysis of the Volatile Compounds

Chocolates and related ingredients' samples were first dispersed in MilliQ water heated at 80 °C, and then centrifuged at 5000 rpm for 15 min at 15 °C (Centrifuge SL 16 R Thermo Scientific) and filtered (Whatman $n°$ 595$\frac{1}{2}$). Before extraction, magnesium sulfate was added in order to minimize emulsions. Then, 1 mL of sample was substituted with the same volume of internal standard (1-heptanol) at a concentration of 50 ppm. Afterwards, 500 mL of sample was extracted for 6 h with the solvent mixture pentane:dichloromethane (2:1). During extraction, the sample was maintained at 60 °C. At the end it was purified with the addition of anhydrous sodium sulphate and then concentrated by means of Rotavapor (BUCHI Rotavapor R-114) up to a volume of 1 mL. For the determination of volatile compounds, gas chromatography (Autosystem XL Gas Chromatograph Perkin Helmer) with a Flame Ionization Detector (FID) was used, following the chromatographic conditions indicated by Bonvehì [33], by injecting 1 μL of extract. Standard solutions were prepared with pentane:dichloromethane (2:1) as the solvent.

2.4. Sensory Analysis

In order to allow a complete sensory description of products and to identify key sensory attributes of chocolates and related ingredients [11] the Quantitative Descriptive Analysis (QDA) [34,35] was applied as a suitable procedure for an assessment of the sensory quality of chocolates and related ingredients during the whole period of storage. No approval from the Human Ethics Committee was required by our institution to perform the sensory analysis in this research.

2.4.1. Determination of the Sensory Profile of Chocolates and Related Ingredients

Sensory properties of the chocolates and related ingredients were monitored at each control time (t0, t1, t2, t3, and t4) under conditions and procedures compliant to ISO standards [36–38]. The QDA was carried out using eight assessors with a broad experience in sensory evaluation, as well as interest and availability. The training was performed as reported by Donadini, Fumi, and Lambri [39]. The QDA was composed of the following stages [34,35]: (1) a lexicon generation process and (2) a set of sensory tests designed to quantify the intensity of the sensory descriptors established in the lexicon generation phase on a rating scale. The chocolate and ingredient samples were prepared and individually served to panelists. Three-digit random numbers were assigned to each sample for tracking purposes prior to service. The order of presentation was balanced and randomized across samples, panelists, and replicates, according to a rotated tasting plan [40].

2.4.2. Lexicon Generation Process

In the lexicon generation process, participants were preliminarily asked to name as many sensory characteristics as possible [41–43] which they considered important for the descriptive evaluation of chocolate [44–48]. Redundant terms were discussed openly, with the intervention of the panel leader as a moderator. Descriptors cited by at least 30% of the panel were retained and intensively discussed among panelists in an open session until agreement was reached on the final verbal definition. Selected sensory attributes as defined for each sample are reported in Table 1.

Table 1. Descriptors used for sensory profiling.

Attribute	Description	Range and References	Sample(*)
Brightness	Ability to reflect light; luminescence of color, with descriptions ranging from dull to shiny [42]	Low: Dull; dark chocolate 90% cocoa with fat blooming High: Shiny; dark chocolate 90% cocoa	D – M – G
Snap	The noise and force with which the sample breaks or fractures [39]	Low: Gianduja chocolate High: Dark chocolate 90% cocoa	D – M
Firmness	Force required for compressing the sample between molar teeth [45]	Low: Milk chocolate High: Dark chocolate 90% cocoa	MC – D – M
Crunchiness	Easily broken or ruptured Degree to which the sample fractures into pieces on the first bite with the molars [42]	Low: Dark chocolate 90% cocoa High: Milk chocolate	MC
Melting	Chocolate property of melting in mouth while chewing [45] till liquefaction [46]	Low: Dark chocolate 70% cocoa warmed in a microwave oven during 20 s. High: Dark chocolate 70% cocoa warmed in a microwave oven during 40 s.	MC – C – HP – D – M – G
Stickiness	The degree a sample sticks to the palate [42,47]		MC – HP – D – M – G

Table 1. *Cont.*

Attribute	Description	Range and References	Sample(*)
Chewiness	Length of time required to masticate the sample, at a constant rate of force application, to reduce it to a consistency suitable for swallowing [42]		G
Grittiness	Presence of perceptible particles in the oral cavity. The number of solid particles during mastication [42]		MC – C – HP – D – M – G
Astringency	Mouth drying and/or puckering effect which boosts the production of saliva; perceived between tongue and palate or at the back of the front teeth [42,47]	None: Milk chocolate High: Dark chocolate 90% cocoa	MC – C – HP – D
Oily	The amount of oil left on mouth surfaces [43]		HP – G
Fatness	Surface textural attributes relating to the perception of the quantity or quality of fat in a product [43]		HP – G
Creaminess	The mouth-feel related to the smoothness of the chocolate as related to fat [44]		D – M – G
Acidity	Citric acid (fruit), acetic acid (vinegar), lactic acid (sour milk), and mineral acid (metallic tasting) [42,46]		MC – C – HP – D – M
Bitterness	The taste on the tongue associated with substances such as caffeine and quinine [42]	None: Distilled water High: Caffeine solution at 0.5%	MC – C – HP – D
Sweetness	The taste on the tongue associated with sucrose and other sugars or sweeteners [42]	Low: Sugar solution at 1% High: Sugar solution at 10%	C – HP – D – M – G
Cocoa	The flavor associated with cocoa powder or cocoa beans [43]	Low: Powder cocoa solution at 0.5% High: Powder cocoa solution at 5.0%	MC – C – D – M – G
Toasted/Roasted	Flavor related to cocoa that is very toasted [46] The aroma associated with popcorn or roasted peanut [48]	Low: Dry cocoa seed without toasting. High: Cocoa seed toasted for 3 h	MC – C – HP – D – G
Coffee	The aroma associated with medium-high toasted coffee [39]		MC – C
Nutty	Delicate aroma of indistinguishable nuts without roast. Mixed raw nuts powder (hazelnut, walnut, peanut, and sunflower seeds) [48]		MC – HP – M – G
Caramel	The aroma associated with caramelized sugar [48]	Low: Dry sugar High: Sugar warmed at 120 °C until a brown color	HP – M – G
Vanillin	The aroma associated with vanillin [39]		D – M – G

(*) D=dark chocolate; M=milk chocolate; G=gianduja chocolate; CM=cocoa mass; C=cocoa powder 22–24; HP=hazelnut paste.

2.4.3. Sensory Tests of Chocolates and Ingredients to Rate Sensory Attributes

A set of sensory tests was designed to quantify the intensity of the sensory descriptors that were inserted into the score card on a 9-point scale (anchored at both extremes as "not perceived at all" and "extremely intense"). Panelists were provided with mineral water and unsalted breadsticks to cleanse their palates between samples.

2.5. Statistical Analysis

At each storage time starting from production, the sampling proceeded with two independent replications and each analysis was performed in triplicate. Data were subjected to Microsoft Excel 2017 and to Levene's test to point out the homogeneity of variance among sample subsets. Furthermore, data were analyzed by one-way analysis of variance (ANOVA) with Tukey's *t*-test at $p \leq 0.05$ to highlight the significance of the differences among the different storage times (t0, t1, t2, t3, and t4) for equal sample types (CM, C, HP, D, M, and G). The statistics were prepared using IBM SPSS Statistics 20 (IBM Corporation, New York, NY, USA).

3. Results and Discussion

3.1. Microbiological Analysis

The results of the microbiological analysis were compliant with the limits stated by EC Regulation 2073/2005 [18], as reported in Table 2. Different to other studies focused on microbial evolution in chocolate-based and confectionery products [49], in this study, no meaningful data were obtained. This outlined the microbiological quality of the samples under study, along with the strong hygienic conditions applied in production and storage.

Table 2. Results from microbiological analysis at the end of storage.

Parameter	CM	C	HP	D	M	G
Total microbial count	<5000	<5000	<5000	<5000	<5000	<5000
Enterobacteriaceae	n.d.	n.d.	n.d.	n.d.	n.d.	n.d.
Coliforms	n.d.	n.d.	n.d.	n.d.	n.d.	n.d.

n.d. = not detectable.

3.2. Nutritional Composition and Related Characterization

The sample characterization at t0 along with phenolic compounds, alkaloids, and vitamins is shown in Table 3, while determination of the fatty acid profile, fatty acid composition, and sterol composition is detailed in Table 4. Finally, the organic acids and the volatile compounds are reported in Tables 5 and 6, respectively.

Table 3. Characterization at t(0) along with phenolic compounds, alkaloids, and vitamins analyzed in cocoa mass (CM), cocoa 22-24 (C), hazelnut paste (HP), dark chocolate (D), milk chocolate (M), and gianduja chocolate (G). Data represent the mean ± SD ($n = 3$). Within each row, different superscript letters indicate statistically different values among samples according to a post-hoc comparison (Tukey's test) at $p \leq 0.05$.

Parameter	CM	C	HP	D	M	G
Humidity (%)	1.06 ± 0.03 [d]	2.11 ± 0.01 [a]	0.75 ± 0.01 [e]	0.70 ± 0.03 [f]	1.17 ± 0.09 [c]	1.37 ± 0.00 [b]
pH	5.51 ± 0.02 [e]	8.31 ± 0.10 [a]	5.71 ± 0.01 [d]	5.70 ± 0.05 [d]	6.29 ± 0.04 [b]	6.17 ± 0.03 [c]
Acidity (mg eq stearic acid/100 g)	361.00 ± 0.01 [a]	n.d.	159.00 ± 0.01 [c]	184.91 ± 2.84 [b]	110.95 ± 5.69 [e]	128.02 ± 2.84 [d]
Ash (%)	4.37 ± 0.97 [b]	11.70 ± 0.04 [a]	2.05 ± 0.01 [c]	1.67 ± 0.03 [e]	1.71 ± 0.01 [d]	1.65 ± 0.01 [f]
Protein (%)	14.11 ± 0.63 [b]	23.33 ± 0.08 [a]	3.90 ± 0.18 [f]	7.20 ± 0.01 [d]	6.40 ± 0.08 [e]	10.10 ± 0.41 [c]
Fat matter (%)	35.70 ± 0.04 [c]	17.90 ± 0.12 [f]	66.80 ± 1.75 [a]	23.20 ± 0.30 [e]	34.20 ± 1.61 [d]	37.50 ± 2.30 [b]
Total sugar (%)	0.80 ± 0.00 [e]	1.20 ± 0.03 [d]	1.60 ± 0.57 [d]	45.60 ± 0.31 [b]	54.70 ± 0.25 [a]	40.80 ± 1.39 [c]
Fiber (%) *	12.50 ± 0.06 [b]	28.03 ± 0.14 [a]	8.83 ± 0.04 [c]	7.56 ± 0.04 [d]	1.60 ± 0.01 [f]	5.45 ± 0.03 [e]
* raw fiber (%)	11.10	13.63	n.d.	n.d.	n.d.	n.d.
Phenols (mg Gallic Acid Equivalents /g)	2.06 ± 0.25 [b]	7.29 ± 0.98 [a]	0.31 ± 0.07 [e]	2.12 ± 0.18 [b]	0.64 ± 0.17 [d]	0.99 ± 0.09 [c]
Caffeine (mg/100 g)	65.67 ± 4.97 [b]	97.69 ± 5.46 [a]	3.40 ± 0.16 [e]	21.51 ± 1.04 [c]	2.06 ± 0.93 [f]	7.84 ± 2.27 [d]
Theobromine (mg/g)	6.77 ± 0.68 [b]	10.07 ± 0.10 [a]	2.04 ± 0.37 [d]	7.28 ± 0.59 [b]	3.97 ± 0.12 [c]	3.63 ± 0.25 [c,d]
13-Cis-β-Carotene (ppm)	<0.10 ± 15%	<0.10 ± 15%	<0.10 ± 15%	<0.10 ± 15%	<0.10 ± 15%	<0.10 ± 15%
9-Cis-β-Carotene (ppm)	<0.10 ± 15%	<0.10 ± 15%	<0.10 ± 15%	<0.10 ± 15%	<0.10 ± 15%	<0.10 ± 15%
All-Trans-α-Carotene (ppm)	<0.10 ± 15%	<0.10 ± 15%	<0.10 ± 15%	<0.10 ± 15%	<0.10 ± 15%	<0.10 ± 15%
All-Trans-β-Carotene (ppm)	0.34 ± 15%	<0.30 ± 15%	<0.10 ± 15%	<0.30 ± 15%	0.47 ± 15%	0.30 ± 15%
β-Cryptoxanthin (ppm)	<0.10 ± 15%	<0.10 ± 15%	<0.10 ± 15%	<0.10 ± 15%	<0.10 ± 15%	<0.10 ± 15%
Retinol (ppm)	<0.10 ± 15%	<0.10 ± 15%	<0.10 ± 15%	<5.00 ± 15%	82.00 ± 15%	5.00 ± 15%
Vitamin E (ppm)	6.77 ± 10% [d]	9.40 ± 10% [c]	304.00 ± 10% [a]	8.20 ± 10% [c]	6.63 ± 10% [d]	114.00 ± 10% [b]

Table 4. Fatty acid profile, fatty acid composition, and sterol composition expressed in % (as 100 of the total) in cocoa mass (CM), cocoa 22-24 (C), hazelnut paste (HP), dark chocolate (D), milk chocolate (M), and gianduja chocolate (G) at production time (t0).

Fatty acid / Sterol	CM	C	HP	D	M	G
Saturated fatty acids	63.90	61.29	9.49	63.87	65.23	37.67
Monounsaturated fatty acids	32.95	35.34	83.17	32.69	31.60	57.19
Polyunsaturated fatty acids	3.05	3.30	7.24	3.34	3.12	5.10
Trans-oleic fatty acids	<0.01	<0.01	<0.01	<0.01	0.64	0.21
Trans-linoleic fatty acids	<0.01	<0.01	<0.01	<0.01	<0.01	<0.01
Trans-linolenic fatty acids	<0.01	<0.01	<0.01	<0.01	<0.01	<0.01
Trans-palmitoleic fatty acids	<0.01	<0.01	<0.01	<0.01	<0.01	<0.01
C4:0 Butyric	<0.01	<0.01	<0.01	<0.01	0.69	0.23
C6:0 Capronic	<0.01	<0.01	<0.01	<0.01	0.43	0.13
C7:0 Enantiic	<0.01	<0.01	<0.01	<0.01	<0.01	<0.01
C8:0 Caprylic	<0.01	<0.01	<0.01	<0.01	0.28	0.10
C10:0 Capric	<0.01	<0.01	<0.01	<0.01	0.58	0.22
C10:1 Caproleic	<0.01	<0.01	<0.01	<0.01	0.07	<0.05
C12:0 Lauric	<0.01	<0.01	<0.01	<0.01	0.71	0.32
C12:1 Lauroleic	<0.01	<0.01	<0.01	<0.01	<0.05	<0.05
C13:0 Tridecanoic	<0.01	<0.01	<0.01	<0.01	<0.01	<0.01
C13:1 Tridecenoic	<0.01	<0.01	<0.01	<0.01	<0.01	<0.01
C14:0 Myristic	0.10	0.11	<0.05	0.21	2.34	0.98
C14:1 Miristoleic	0.01	<0.01	<0.01	<0.01	0.20	0.08
C15:0 Pentadecanoic	<0.05	<0.01	<0.01	0.05	0.26	0.12
C15:1 Pentadecenoic	<0.01	<0.01	<0.01	<0.01	<0.01	<0.01
C16:0 Palmitic	26.66	27.83	6.61	25.99	26.96	16.57
C16:1 Palmitoleic	0.27	0.28	0.27	0.24	0.51	0.35
C17.0 Eptadecanoic	0.22	0.23	<0.05	0.23	0.38	0.16
C17:1 Eptadecenoic	<0.05	<0.05	0.08	<0.05	0.09	0.08
C18:0 Stearic	35.67	32.08	2.68	36.15	31.53	18.10
C18:1 Oleic	32.68	35.06	82.68	32.40	30.73	56.60
C18:2 Linoleic	2.87	3.10	7.16	3.11	2.83	4.94
C18:3 Linolenic	0.18	0.20	0.08	0.23	0.29	0,16
C20:0 Arachic	1.01	0.90	0.15	1.03	0.90	0,57
C:20:1 Eicosenoic	<0.05	<0.05	0.14	0.05	<0.05	0.08
C22:0 Behenic	0.17	0.14	0.05	0.17	0.16	0.11
C22:1 Erucic	<0.01	<0.01	<0.01	<0.01	<0.01	<0.01
C22:0 Lignoceric	0.07	<0.01	<0.05	0.09	0.08	0.06
Cholesterol	1.00	1.00	0.30	1.30	26.80	14.70
Brassicasterol	<0.10	<0.10	<0.10	<0.10	<0.10	<0.10
2,4-methylene cholesterol	0.30	0.50	0.10	0.20	0.20	0.20
Campesterol	9.00	9.40	4.10	9.60	7.20	6.60
Campestanol	0.20	0.30	0.40	0.20	0.10	0.20
Stigmasterol	25.80	26.10	1.20	24.70	17.80	13.90
Delta-7-campesterol	<0.10	<0.10	<0.10	<0.10	<0.10	<0.10
Delta-5,23-stigmastadienol	<0.10	<0.10	<0.10	<0.10	<0.10	<0.10
Clerosterol	0.80	0.70	0.70	0.70	0.50	0.90
Beta-sitosterol	58.70	56.30	83.40	58.40	44.20	56.70
Sitostanol	0.70	0.70	1.40	0.70	0.40	1.00
Delta-5-avenasterol	2.50	2.50	6.20	2.40	2.00	3.00
Delta-7,9(11)-stigmastadienol	0.20	<0.10	<0.10	<0.10	<0.10	<0.10
Delta-5,24-stigmastanediol	0.40	0.20	0.80	0.80	0.20	0.80
Delta-7-stigmastenol	0.40	0.90	0.70	0.60	0.20	1.00
Delta-7-avenasterol	0.20	0.70	0.80	0.30	0.10	0.50

Table 5. Organic acids (mg/kg) in cocoa mass (CM), cocoa 22–24 (C), hazelnut paste (HP), dark chocolate (D), milk chocolate (M), and gianduja chocolate (G) analysed at production time (t0). Data represent the mean ± SD (n = 3). Within each row, different superscript letters indicate statistically different values among samples according to a post-hoc comparison (Tukey's test) at $p \leq 0.05$.

Organic Acids	CM	C	HP	D	M	G
Ossalic acid	889 ± 27.3 [b]	1824 ± 33.2 [a]	195 ± 4.6 [d]	421 ± 9.7 [c]	37 ± 0.4 [f]	79 ± 12.7 [e]
Citric acid	1765 ± 69.7 [b]	3640 ± 51.2 [a]	547 ± 36.1 [d]	1163 ± 98.5 [c]	1681 ± 78.5 [b]	1253 ± 34.7 [c]
Acetic acid	902 ± 13.4 [b]	2750 ± 5.8 [a]	610 ± 7.9 [c]	109 ± 0.3 [e]	547 ± 34.6 [d]	454 ± 78.2 [d,e]
L-malic acid	444 ± 15.9 [b]	n.d.	1524 ± 167.4 [a]	399 ± 2.5 [c]	194 ± 12.8 [d]	1388 ± 54.9 [a]
Lactic acid	570 ± 3.3 [b]	n.d.	17 ± 0.7 [e]	194 ± 0.8 [d]	874 ± 56.8 [a]	434 ± 23.8 [c]
Formic acid	n.d.	n.d.	208 ± 16.1 [b]	843 ± 6.4 [a]	229 ± 34.1 [b]	202 ± 5.8 [b]
Tartaric acid	235 ± 15.7	n.d.	n.d.	n.d.	n.d.	n.d.
Succinic acid	48 ± 0.7 [b]	n.d.	n.d.	n.d.	54 ± 4.7 [a]	n.d.

n.d. = not detectable.

Table 6. Volatile compounds (mg/kg) in cocoa mass (CM), cocoa 22–24 (C), hazelnut paste (HP), dark chocolate (D), milk chocolate (M), and gianduja chocolate (G) analysed at production time (t0). Data represent the mean ± SD (n = 3). Within each row, different superscript letters indicate statistically different values among samples according to a post-hoc comparison (Tukey's test) at $p \leq 0.05$.

Volatile Compounds	CM	C	HP	D	M	G	References	Descriptors
2,5-Dimethylpyrazine	n.d.	22.69 ± 2.20 [a]	5.21 ± 1.54 [b]	n.d.	0.46 ± 0.07[d]	0.95 ± 0.02 [c]	0.23–1.69	cocoa, roast nuts
2,6-Dimethylpyrazine	3.32 ± 0.03 [b]	2.36 ± 0.47 [c]	12.67 ± 2.58 [a]	15.94 ± 3.19 [a]	n.d.	14.10 ± 1.19 [a]	0.11–0.39	nutty, coffee, green
2,3,5-Trimethylpyrazine	41.75 ± 7.39 [a]	19.18 ± 3.57 [b]	3.63 ± 0.45 [d]	7.38 ± 2.69 [c]	0.12 ± 0.02 [f]	0.33 ± 0.17 [e]	0.21–1.71	cocoa, roast nuts, peanut
2,3,5,6-Tetramethylpyrazine	6.50 ± 1.51 [a]	4.50 ± 2.11a [b]	2.03 ± 1.26 [c]	8.15 ± 1.89 [a]	1.58 ± 0.11 [c]	4.53 ± 0.57 [a,b]	0.52–8.28	chocolate, cocoa, coffee
Benzaldehyde	0.59 ± 0.31 [d]	3.86 ± 0.12[b]	0.48 ± 0.04 [e]	7.52 ± 3.50 [a]	1.80 ± 0.60 [c]	0.73 ± 0.29 [d]	0.5–1.89	bitter
2-Acetyl-5-methylfuran	n.d.	5.06 ± 0.92 [a]	1.43 ± 0.16 [b]	1.76 ± 0.45 [b]	0.50 ± 0.14 [c]	0.43 ± 0.01 [c]		
2-Phenylacetaldehyde	0.55 ± 0.07 [c]	2.15 ± 0.75 [b]	4.02 ± 0.35 [a]	1.47 ± 1.12b [c]	2.59 ± 0.54 [b]	1.55 ± 0.07b [c]	2–8.90	berry, nutty
α-Terpenilformato	0.16 ± 0.01 [c]	n.d.	0.72 ± 0.11 [b]	n.d.	0.71 ± 0.22 [b]	2.89 ± 0.34 [a]	0–0.38	herbaceous, citrus
Benzyl acetate	0.37 ± 0.29 [d]	n.d.	1.92 ± 0.01 [a]	0.59 ± 0.12 [c]	0.45 ± 0.04 [d]	1.32 ± 0.13 [b]	0–0.033	floral, jasmine
Octanoic acid	0.40 ± 0.29 [b]	1.93 ± 0.96 [a]	1.12 ± 0.09 [a]	1.62 ± 0.02 [a]	0.61 ± 0.23 [b]	0.85 ± 0.01 [a]	0.021–0.37	unpleasant, oily, fatty
2-Acetyl pyrrole	0.18 ± 0.03 [c]	2.74 ± 1.71 [a]	1.52 ± 0.31 [a]	n.d.	0.36 ± 0.01 [b]	1.58 ± 1.11 [a]	0.021–0.38	bread, walnut, licorice
3-Hydroxy-2-methylpyridine	n.d.	1.63 ± 0.06 [b]	1.84 ± 0.32 [b]	4.90 ± 1.94 [a]	0.59 ± 0.04 [d]	0.94 ± 0.06 [c]	0.14–0.38	wizened
2,3-Dihydro-3,5-dihydro-6-methyl-4-pyrone	n.d.	n.d.	n.d.	n.d.	0.50 ± 0.14	n.d.	0.28–1.87	roasted
3,5-Hydroxy-6-methyl-4-pyrone	n.d.	n.d.	n.d.	n.d.	2.59 ± 0.54	n.d.	0.02–0.37	roasted

n.d. = not detectable.

3.2.1. Chocolates

For D and M chocolates, the content of ash, proteins, and fats (Table 3) confirmed CREA [50] and other authors' findings [19]. The ash levels were similar between samples and minor at 2%, while proteins were higher in the D sample (at 7.2%) than in the M sample with a content of 6.4% (Table 3). Since D is obtained from the processing of cocoa mass (CM), its results (Table 4) outlined a composition of mainly oleic acid (C18:1), stearic acid (C18:0), palmitic acid (C16:0), and small quantities of linoleic acid (C18:2), confirming what has been reported in the literature [51,52]. As observed by Çakmak et al. [52], oleic acid (C18:1), stearic acid (C18:0), palmitic acid (C16:0), and linoleic acid (C18:2) prevailed in the cocoa mass (CM).

The theobromine (Table 3) at a concentration of about 7.3 mg/g for D and almost 4 mg/g for M agreed with what has been described by Belščak-Cvitanović et al. [53]. Besides, the concentration in D was compliant with Meng et al. [54], who indicated a range of 237–519 mg of theobromine per 50 g portion of dark chocolate. The total soluble polyphenols, with a value of about 2.1 mg GAE/g, and caffeine, with more than 21 mg/100 g in D (Table 3), were similarly remarked upon in a previous study [53].

Citric, oxalic, acetic, malic, lactic, and formic acids were present in all the samples (Table 5), with formic acid in the D sample having a value of 843 ppm, which is greater than in other research [10,55]. The formic acid did not seem to come from the raw materials, but rather from the transformations that occurred during the process: despite its volatility, it was maintained at medium-high levels in the finished chocolates (Table 5). Tartaric acid was completely absent in the chocolates (Table 5), while succinic acid was only found in the M sample with almost a 55-ppm concentration, probably deriving from the added milk.

Finally, considering the volatile compounds detected in the D, M, and G samples, Table 6 shows that D chocolate was characterized by a higher level of pyrazines and benzaldehyde, reaching a value of about 31.5 and 7.5 ppm, respectively, in line with the results of other studies [48,56]. As a matter of fact, pyrazines contribute to 40% of the volatile compounds of roasted cocoa aroma with multiple descriptors. In the study of Liu et al. [48], dark chocolate was characterized by a cocoa flavor with malty, nutty, and toasted notes. Furthermore, Aprotosoaie et al. [57] stated that different cocoa types may exhibit various and specific flavors since the concentration and sensory characters of these compounds vary significantly, depending on the nature and origin of the identified molecules. The literature has stressed a possible microbial fermentation derivation of *Bacillus subtilis* and *Bacillus megatrium*; however, most cocoa and chocolate pyrazines originate in Strecker degradation and Maillard reactions, and need heat and precursors such as aminocinetones, acetoin, and diacetyl [56]. This allows us to conclude that the roasting of both cocoa and hazelnuts has a considerable impact on the aroma profile of the chocolates and related ingredients.

3.2.2. Ingredients

Polyphenols exceeded 0.30 mg GAE/g (Table 3) in HP, as also reported in other research [58,59]. The highest concentrations were found in the C sample, with more than 7 mg GAE/g of polyphenols, as in Belščak et al. [27], although the literature mentions very variable polyphenol contents, depending on the geographical origin of the beans, degree of maturation, processing, and packaging [7,9,11,29,46]. The content in theobromine and caffeine of the C sample under study (Table 2) was coherent with what was reported by Jalil and Ismail [60], whilst the pH value (Table 3) matched Miller et al.'s results [61], since this sample may be considered as quite alkalized cocoa, showing a pH value greater than 8.00. The alkalization of cocoa powder causes a decrease in the total polyphenols and an increase in pH up to 8.0. Natural cocoa powder has a pH of about 5.3–5.8, while alkalized cocoa powder may be classified according to treatment: light treatment (pH 6.5–7.2), medium treatment (pH 7.2–7.6), and heavy treatment (pH 7.6 and above) [61].

The overall fat content (Table 3) was lower than what was reported by CREA [50], with the most in HP and the least in the C sample. The lipid content is influenced by the geographical origin of cocoa, ranging from 16% to 22% for commercial typologies [6]. The fatty acid composition of C (Table 4) matched what was stated by Elkhori et al. [51]. As for the triglycerides of cocoa butter, our results (Table 4) outlined mainly oleic acid (C18:1), stearic acid (C18:0), palmitic acid (C16:0), and small amounts of linoleic acid (C18:2), as presented in the literature [51], while the Vitamin E content found in C (Table 3), i.e., more than 9%, complied with Lipp et al.'s findings [62]. The saturated, monounsaturated, and polyunsaturated fatty acid content (Table 4) of CM agreed with the values presented by Çakmak et al. [52]. Oleic acid exceeded 30% in all of the samples, reaching almost 83% in the HP sample (Table 3). Moreover, there were other fatty acids (lauric, myristic, myristoleic, pentadecanoic, pentadecenoic, palmitoleic, heptadecanoic, heptadecenoic, linolenic, arachic, eicosadienoic and eicosatrienoic, behenic, and lignoceric) ranging between 0.01% and 1% (Table 5), as found by Bignami et al. [63]. Finally, vitamin E (Table 2) ranged around 300 ppm, confirming what was previously reported [63].

Despite various works being carried out on the lipid fraction of hazelnuts, few data are currently available on the organic acids of HP [63–65], which, in our study (Table 5), showed malic as the most abundant acid. Citric, oxalic, and acetic were the only acids detected in the C sample (Table 5). Tartaric and succinic acid were observed in the CM (Table 5), but they were completely absent in the other samples, while formic acid was only present in the HP (Table 4). Succinic acid contained in CM at a concentration of 48 ppm was probably a derived fermentation product. D-L lactic and L-malic acid were present in both CM and HP (Table 5).

The analysis of the aroma compounds led to the identification of the only octanoic acid as off-flavor, which showed the highest amounts in C (Table 6). Among the main contributors to the overall profile of cocoa producing, the greatest impacts to chocolate aroma [48,56], tetramethylpyrazine and trimethylpyrazine, responsible for milk coffee-mocha roasted, nutty, and earthy aromas, respectively, were present, especially in CM (Table 5), whilst phenylacetaldehyde (rosy-like aroma) showed the highest value in HP (Table 5).

3.3. Eighteen-Month Evolution of Chocolates and Ingredients: Nutritional and Sensory Changes

As chocolate is a continuous lipid phase, the structural changes in its fat matter may alter volatile release, thus changing the flavor profile of the chocolate [11,14,17,47,48,56,57,66]. As a consequence, chocolates and the related ingredients under study were chemically checked during the 18 months of storage for peroxide and acidity values, polyphenols, and vitamin E [67]. This chemical-physical monitoring was applied with an eighteen-weekly check frequency and outlined for both chocolates and ingredients, and the results are reported in Figure 1.

On the other hand, the sensory analysis was performed at t0, t1, t2, t3, and t4 by means of QDA profiling that allowed identifying significant trends of increase or decrease in the visual, auditory, mechanical, and flavor perceptions of the samples [11,68]. In the QDA sensory sheet, panelists could indicate descriptors related to the sensory flaws: this was useful for displaying any "anomalies" that arose during storage. The evolution of the sensory attributes of each sample during the eighteen months of conservation is represented in Figures 2 and 3 for ingredients and chocolates, respectively.

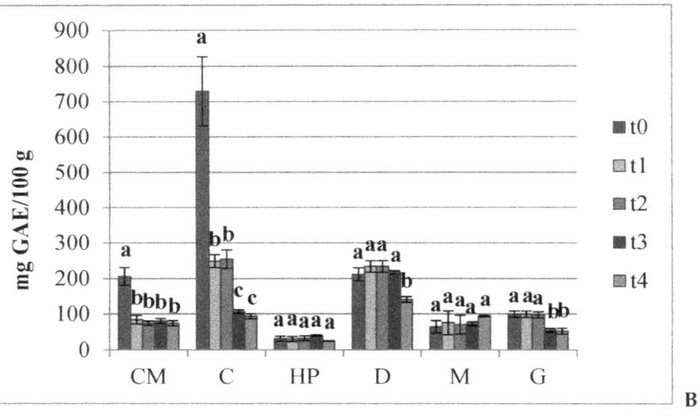
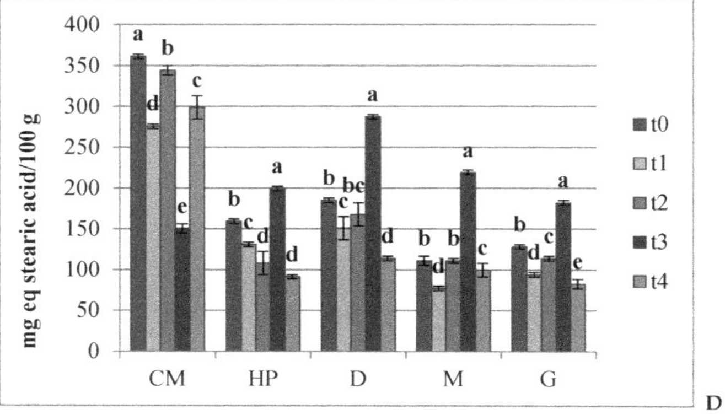
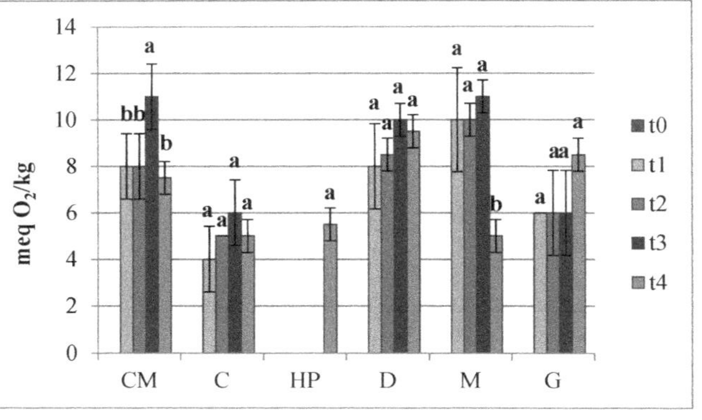
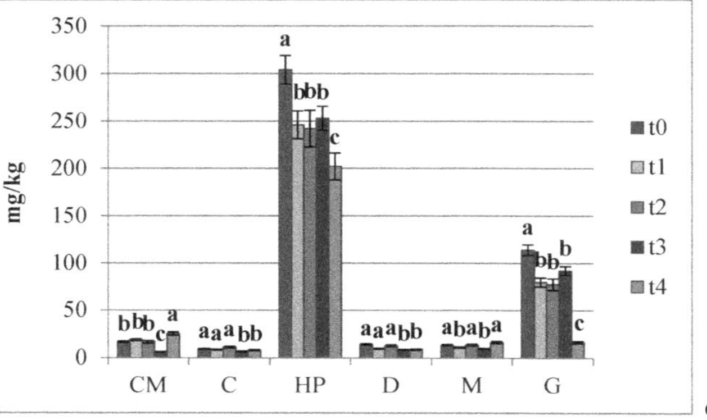

Figure 1. Evolution of the lipid and phenolic matrix at t0, t1, t2, t3, and t4 of (**A**) Peroxide values (expressed as meq of O_2/kg); (**B**) Unoxidized polyphenol concentration (expressed as mg of GAE per 100 g); (**C**) Vitamin E concentration (mg/kg); (**D**) Acidity values (expressed as mg of stearic acid equivalents per 100 g). Data represent the mean ± SD ($n = 3$). Within each sample, different letters indicate statistically different values among times according to a post-hoc comparison (Tukey's test) at $p \leq 0.05$.

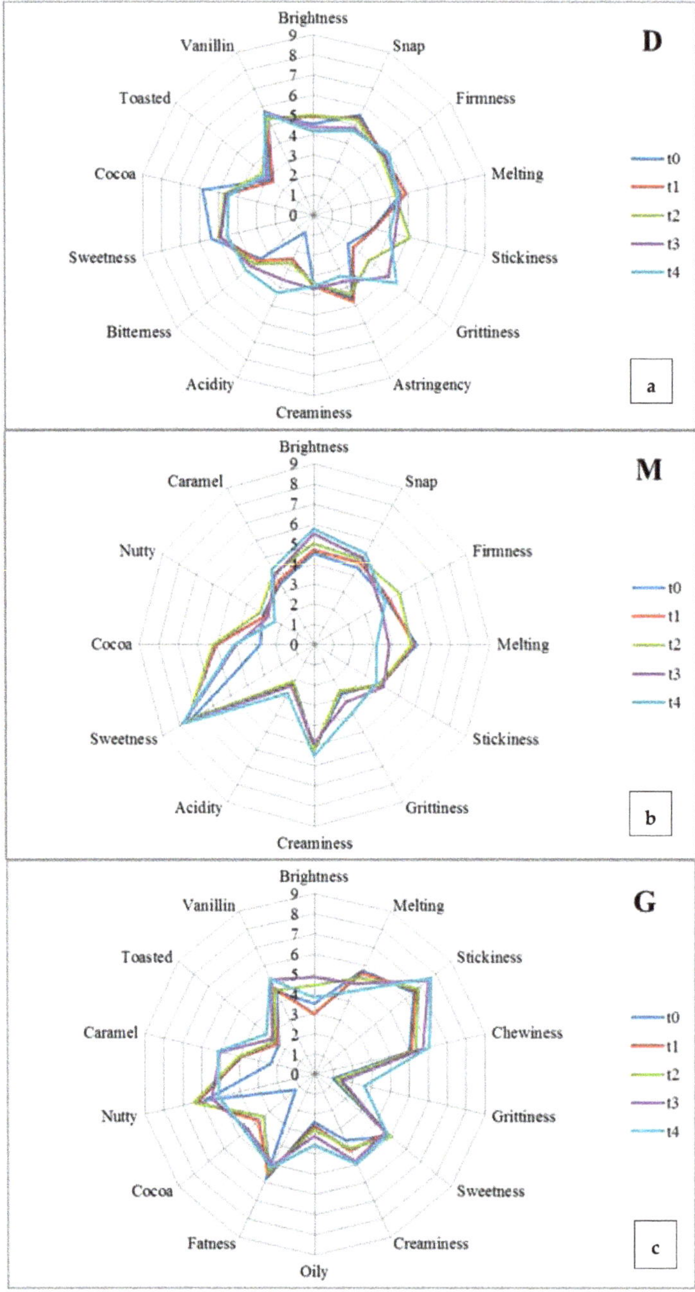

Figure 2. Sensory profile at t0, t1, t2, t3, and t4 of the dark chocolate (**D, subfigure a**), milk chocolate (**M, subfigure b**), and gianduja chocolate (**G, subfigure c**).

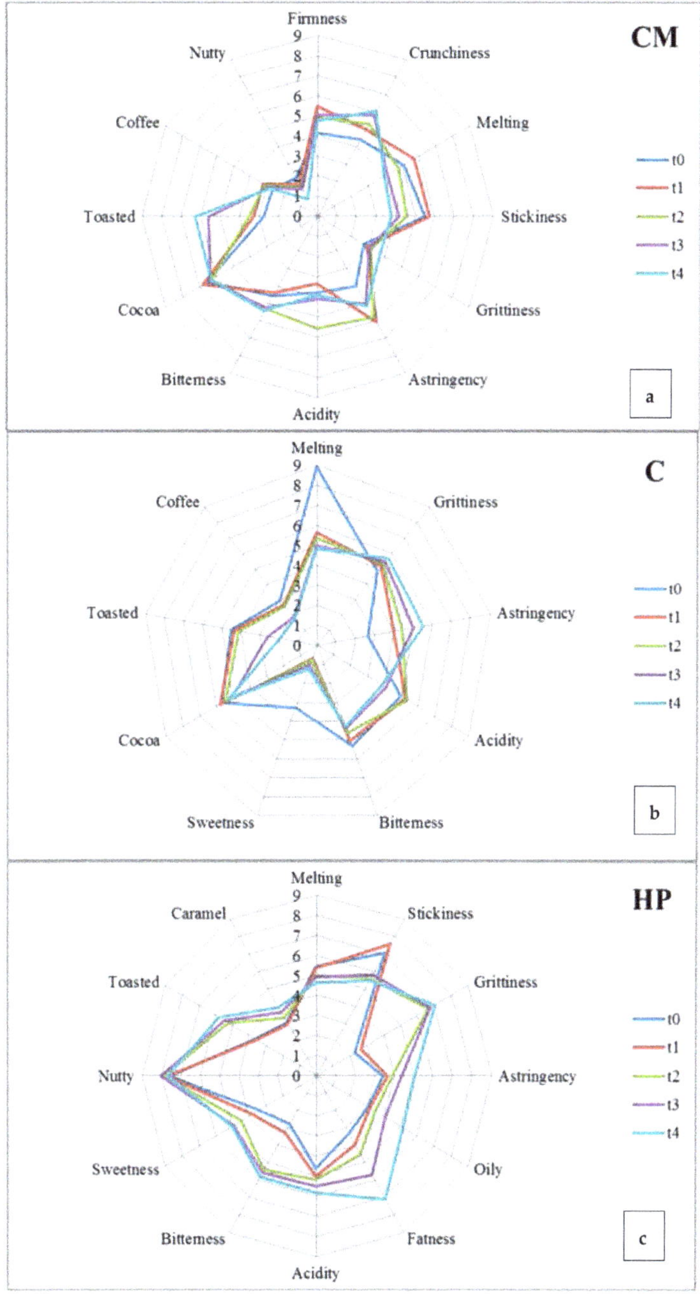

Figure 3. Sensory profile at t0, t1, t2, t3, and t4 of the cocoa mass (**CM, subfigure a**), cocoa 22–24 (**C, subfigure b**), and hazelnut paste (**HP, subfigure c**).

3.3.1. Chemo-Sensory Evolution of Chocolates

In the milk chocolate (M sample), brightness and snap increased over time (Figure 2), whereas the intensity of brightness remained constant in D, where the snap became less intense from t0 to t4 (Figure 2a). This was in accordance with Machálková et al. [66], who found a slight deterioration of some mechanical descriptors in the chocolate samples stored at 20 °C. The firmness did not change significantly in D and M samples (Figure 2a,b), while the melting dropped in G and M chocolate (Figure 2c,b). In this regard, Thamke et al. [68] concluded that chocolate with a lower cocoa content was characterized by the greatest melting and creaminess, while the product with the highest cocoa content was characterized as dry dough. This was confirmed by the results from Figure 2c, which show a lower value of creaminess in D chocolate than in M and G samples.

It is interesting to compare the behavior of fat matter with the modification in structural properties observed in chocolate samples and previously described, since the physical state of triglycerides is known to affect the firmness, creaminess, and melting of a chocolate-based product [67]. The alteration of the physical state of fat matter, which is widely observed over time in the storage of fatty food products, is mainly due to rancidity phenomena starting with the hydrolysis of triglycerides giving rise to free fatty acids, which in turn undergo oxidation [69]. This emphasized the role acquired by the monitoring of both acidity and peroxide values that, together with sensory outcomes, in terms of perceived acidity, may help to get quick and consistent information about the fat matter evolution of chocolate-based products during their storage.

Our results evidenced a reduction in the acidity in the first eighteen weeks (t1) (Figure 1D), followed by a fast and similar increase in all the samples after t2 with values of almost 290, 220, and 185 mg eq. stearic acid/100 g in D, M, and G samples, respectively. Afterwards, it fell again until the eighteenth month of storage (t4). Considering the M chocolate (Figure 2b), acidity increased, this was only slightly with respect to the D sample (Figure 2a). Regarding peroxides, the three chocolates maintained a value ranging between 5 and 11 meq of O_2/kg (Figure 1A). In particular, the results showed no detectable peroxides at the production time (t0), while at the t1, there was a significant increase, especially in M and D chocolate. The fact that in G, the peroxide value remained lower than the other two chocolates, confirms the observations of other authors [70,71], who have attributed this behavior to the phytochemical compounds present in dried fruit, in this case, hazelnuts. As observed for M samples after 18 months (Figure 1A), the peroxide value decreases due to them changing into short-chain aldehydes, or in secondary products deriving from their decomposition [69,72]. Therefore, despite chocolates having a high fat content (Table 3), the lipid oxidation at the end of storage (Figure 1A) was very slow. Both the lyophobic and the lyophilic antioxidants were supposed to perform a continuously protective activity towards fats [14,69] due to their well-known biological effects [3,16,17,27,53,60,61,73,74].

At the end of the eighteen months, the residual content in polyphenols of the chocolates ranged from 50 to 217 mg of GAE per 100 g, highlighting a significant difference between the different types of chocolate. Phenolic substances are involved in the chocolate's flavor and in the primary sensory characteristics as bitterness and astringency. The analysis (Figure 1B) showed a slight increase in D and M from time t0 to time t1, while in G, the concentration remained constant until t2, and then decreased significantly up to the end of storage (t4), reaching the value of 54 mg GAE/100 g (Figure 1B). Although D chocolate showed an important loss of polyphenols (Figure 1B), it maintained the highest content at the end of storage (t4) when the astringency was perceived to be at a medium level, contrary to M and G chocolate (Figure 2b,c). Studies [11–13,66,67] have already observed the depletion of polyphenols in cocoa-based products during storage, correlating this loss with their oxidation in the corresponding quinones, which might lead to increases in bitterness, as outlined in D chocolate (Figure 2a). In the same sample, starting from t2, some panelists marked descriptors related to oxidation as "pungent", "closed", "cork", and "dried fig". This was in line with Subramaniam [72], who reported that dark chocolate loses some of its chocolate flavor and develops a "stale note" over time. Additionally, she reported that the chocolate develops a stale, "cardboardy" flavor due to oxidative rancidity prior to the onset of fat bloom.

In this regard, results from vitamin E showing a protective antioxidant effect with a reduction potential of 500 mV, comparable to that of epigallocatechin gallate (430 mV) [75], were particularly interesting in the sample derived from hazelnut, i.e., in G chocolate, where a moderate decrease was recorded at t1 and t4 control times (Figure 1C). The concentration in vitamin E detected in G supported its non-significant lipid oxidation (Figure 1A), even if the G polyphenol content was lower than in D and M chocolates (Figure 1B). The G sample (Figure 2c) showed a growth in stickiness, chewiness, creaminess, and grittiness. A drop in the intensity of positive perceptions was also reported by Bomba [76], who noted that the addition of nuts to chocolate shortens its shelf-life, even though the antioxidants in chocolate may be of some benefit to the oil in the nuts. In this regard, the oily attribute continuously rises, even if the fatness perception slightly decreases (Figure 2c). In the M chocolate (Figure 2b), the attributes of "liquorice" and "aged" have also been reported, although at medium levels. Additionally, milk chocolate flavors tend to blend with aging, but extended aging may result in undesirable fruity notes [77]. Liu et al. [48] reported that the typical flavors of milk chocolate are milky, nutty, and caramel with coconut notes, as observed in the present study (Table 5).

3.3.2. Chemo-Sensory Evolution of Ingredients

As regards the evolution of ingredients, the three semi-finished products maintained a peroxide value (Figure 1A) corresponding to a good state of conservation [46], ranging between 4 and 11 meq of O_2/kg, and even if from t2, the parameter was significantly elevated in the CM. On the contrary, in the HP, the data is not detectable in the first three control times and only at t4 was a value of almost 6 meq of O_2/kg registered (Figure 1A). This result confirmed the observations of other authors [70], who attributed this behavior to the phytochemical compounds present in the hazelnuts. Indeed, in vitro studies have shown that incubating cells with nut extracts very rich in polyphenols can inhibit oxidative susceptibility [64]. It should be noted that, as observed for C and CM, after 18 months (Figure 1A), the peroxide value may decrease when the hydro-peroxides formed evolve into short-chain aldehydes, i.e., in secondary products deriving from their decomposition. Conversely, the acidity was risen in a linear way only in the HP sample (Figure 3c), while in CM and C, the trend of this attribute was quite fluctuating (Figure 3a,b). In detail, the acidity of CM and HP decreased at t1, whereas C had undetectable values at each sampling point (Figure 1D). After the t2 control, the acidity of CM increased, while in HP, it continued to decrease, as also reported by Fardelli [70], who analyzed roasted and natural nuts for nine months of storage.

As well as in chocolates and other studies [12–15,76], a decrease in the total polyphenol content due to their oxidation was especially observed in CM and C from time t0 to time t1 (Figure 1B) and then from the t2 to t3 for the C sample till the lowest value of 93 mg GAE/100 g. On the other hand, the unoxidized polyphenols remained almost constant for HP throughout the full storage period, ranging between 25 and 39 mg GAE/100 g (Figure 1B). This was quite different from what was observed by Fardelli [70], who reported decreases in polyphenols during nine-month storage of natural and roasted nuts.

The results of vitamin E indicated no significant change during the whole storage time in the HP under examination (Figure 1C), with the exception of a decrease at times t1 and t4. The retained level of vitamin E explains why the HP did not demonstrate detectable lipid oxidation (Figure 1A), even though it had polyphenol content levels that were lower than other samples (Figure 1B). The antioxidant properties of tocopherols come from their ability to donate their phenolic hydrogen to lipid free radicals and to retard the autocatalytic lipid peroxidation processes [78]. Moreover, they show a good stability: after six months of maintenance at room temperature, a 13% decrease in antioxidant activity was already observed [79].

Finally, regarding flavors, an emphasis should be placed on the toasted aroma, whose intensity increased during storage for CM and HP, while it decreased drastically for C. During the eighteen months of storage, in the CM (Figure 3a), the nutty aroma decreased, while cocoa and coffee remained stable. In C, the coffee aroma dropped significantly in the last months of storage, while the cocoa intensity remained constant (Figure 3b). On the contrary, in HP, the caramel aroma increased, in addition to oily and fattiness perception (Figure 3c).

4. Conclusions

In the present work, two main groups of cocoa-based foods (chocolates and related ingredients) were investigated by monitoring the evolution of some nutritional components during eighteen months of storage under conditions simulating a point of sale or factory warehouse.

Although the matrices under study contained variable amounts of cocoa butter and hazelnut oil, the evolution of their fats revealed only slow oxidation phenomena. Due to the known effects of catechins, flavonols, and proanthocyanins against lipid peroxidation, the polyphenols measured in chocolates and related ingredients continued to exhibit their protective activity towards fats, even though a general loss in unoxidized polyphenols was observed over time. Finally, despite some alteration of mechanical and structural properties, as well as some losses in aroma, the samples exhibited good sensory scores for the descriptive analysis till the eighteenth month of storage.

Since chocolate, as part of a balanced diet, is becoming a commodity with health and nutritional benefits, these outcomes are really useful for confectionery companies in order to gain detailed information about the state of chocolates and related ingredients which is perceptible by humans and associated with their nutritional composition. Enhancing this knowledge represents a stimulus to people involved in this kind of production, processing, and consumption.

Author Contributions: Conceptualization and methodology, M.L.; Writing—review and editing, A.R.

Funding: This research was funded by the Venchi S.p.A. through grant from the Ministry of Economic Development for the "Industrial innovation project: New Technologies for the Made in Italy" (D.M. 22 December 2010; G.U.R.I. no. 53 of 5 March 2011).

Acknowledgments: The authors are grateful to Massimiliano Cavicchioli for technical support.

Conflicts of Interest: The authors declare no conflict of interest.

References

1. Lin, X.; Zhang, I.; Li, A.; Manson, J.E.; Sesso, H.D.; Wang, L.; Liu, S. Cocoa flavanol intake and biomarkers for cardiometabolic health: A systematic review and meta-analysis of randomized controlled trials. *J. Nutr.* **2016**, *146*, 2325–2333. [CrossRef]
2. Ludovici, V.; Barthelmes, J.; Nägele, M.P.; Enseleit, F.; Ferri, C.; Flammer, A.J.; Ruschitzka, F.; Sudano, I. Cocoa, blood pressure, and vascular function. *Front. Nutr.* **2017**, *4*, 36. [CrossRef]
3. Kim, J.; Kim, J.; Shim, J.; Lee, C.Y.; Lee, K.W.; Lee, H.J. Cocoa phytochemicals: Recent advances in molecular mechanisms on health. *Crit. Rev. Food Sci. Nutr.* **2014**, *54*, 1458–1472. [CrossRef]
4. Fung, T. *Healthy Eating: A Guide to the New Nutrition*; Harvard School of Public Health, Nutrition Department: Boston, MA, USA, 2011; p. 48.
5. Manach, C.; Scalbert, A.; Morand, C.; Remesy, C.; Jimenez, L. Polyphenols: Food sources and bioavailability. *Am. J. Clin. Nutr.* **2004**, *79*, 727–747. [CrossRef]
6. Balestieri, F.; Marini, D. *Metodi di Analisi Chimica dei Prodotti Alimentari*; Monolite: Roma, Italy, 1996; ISBN 8873310052.
7. Badrie, N.; Bekele, F.; Sikora, E.; Sikora, M. Cocoa agronomy, quality, nutritional, and health aspects. *Crit. Rev. Food Sci. Nutr.* **2015**, *55*, 620–659. [CrossRef]
8. EFSA Panel on Dietetic Products, Nutrition and Allergies (NDA). Scientific Opinion on the substantiation of a health claim related to cocoa flavanols and maintenance of normal endothelium-dependent vasodilation pursuant to Article 13(5) of Regulation (EC) No 1924/2006. *EFSA J.* **2012**, *10*, 2809.
9. Afoakwa, E.O. *Chocolate Science and Technology*; Wiley-Blackwell: Singapore, 2010.
10. Januszewska, R.; Viaene, J. Acceptance of chocolate by preference cluster mapping across Belgium and Poland. *J. Euromark.* **2002**, *11*, 61–86. [CrossRef]

11. Popov-Raljić, J.V.; Laličić-Petronijević, J.G. Sensory properties and color measurements of dietary chocolates with different compositions during storage for up to 360 days. *Sensors* **2009**, *9*, 1996–2016. [CrossRef]

12. Kumara, B.; Jinap, S.; Che Man, Y.B.; Yusoff, M.S.A. Comparison of colour techniques to measure chocolate fat bloom. *Food Sci. Technol. Int.* **2003**, *9*, 295–299. [CrossRef]

13. Pastor, C.; Santamaria, J.; Chiralt, A.; Aguilera, J.M. Gloss and colour of dark chocolate during storage. *Food Sci. Technol. Int.* **2007**, *13*, 27–30. [CrossRef]

14. Simic, M.G.; Jovanovic, S.V.; Niki, E. Mechanisms of lipid oxidative processes and their inhibition. In *Lipid Oxidation in Food*; St. Angelo, A.J., Ed.; American Chemical Society: New York, NY, USA, 1992; pp. 14–32.

15. Jolic, S.M.; Redovnikovic, I.R.; Markovic, K.; Sipusic, D.I.; Delonga, K. Changes of phenolic compounds and antioxidant capacity in cocoa beans processing. *Int. J. Food Sci. Technol.* **2011**, *46*, 1793–1800. [CrossRef]

16. McShea, A.; Ramiro-Puig, E.; Munro, S.B.; Casadesus, G.; Castell, M.; Smith, M.A. Clinical benefit and preservation of flavonols in dark chocolate manufacturing. *Nutr. Rev.* **2008**, *66*, 630–641. [CrossRef]

17. Wollgast, J.; Anklam, A. Review on polyphenols in Theobroma cacao: Changes in composition during the manufacture of chocolate and methodology for identification and quantification. *Food Res. Int.* **2000**, *33*, 423–447. [CrossRef]

18. Commission Regulation (EC) No 2073/2005 of 15 November 2005 on microbiological criteria for foodstuffs. *Off. J. Eur. Union* **2005**, *OJL 338*, 1–26.

19. ISO. *ISO 4833-1:2013. Microbiology of Food and Animal Feeding Stuffs—Horizontal Method for the Enumeration of Microorganisms—Colony-Count Technique at 30 °C*; ISO: Geneva, Switzerland, 2013.

20. ISO. *ISO 21528-1:2017. Microbiology of Food and Animal Feeding Stuffs—Horizontal Methods for the Detection and Enumeration of Enterobacteriaceae—Part 1: Detection and Enumeration by MPN Technique with Pre-Enrichment*; ISO: Geneva, Switzerland, 2017.

21. ISO. *ISO 4832:2006. Microbiology of Food and Animal Feeding Stuffs—Horizontal Method for the Enumeration of Coliforms—Colony-Count Technique*; ISO: Geneva, Switzerland, 2006.

22. AOAC International. AOAC International. AOAC 931.04, Moisture in Cocoa Products. In *Official Methods of Analysis of AOAC International*, 17th ed.; AOAC International: Gaithersburg, MD, USA, 2013.

23. AOAC International. Total dietary fiber in foods, enzymatic-gravimetric method. In *Official Methods of Analysis of AOAC International*, 17th ed.; AOAC International: Gaithersburg, MD, USA, 2003.

24. Office International du Cacao et du Chocolat et de la Confiserie (OICCC). *Analytical Method*; OICCC: Brussels, Belgium, 1972.

25. AOAC Authors. *Official Methods of Analysis Lipids, Fats and Oils Analysis Total Carotenoids—Item 22*, 17th ed.; Association of Analytical Communities: Gaithersburg, MD, USA, 2006.

26. Calvo, P.; Castaño, Á.L.; Hernández, M.T.; González-Gómez, D. Effects of microcapsule constitution on the quality of microencapsulated walnut oil. *Eur. J. Lipid Sci. Technol.* **2011**, *113*, 1273–1280. [CrossRef]

27. Belšcak, A.; Komes, D.; Horzic, D.; Ganic, K.K.; Karlovic, D. Comparative study of commercially available cocoa products in terms of their bioactive composition. *Food Res. Int.* **2009**, *42*, 707–716. [CrossRef]

28. AOAC International. AOAC 939.02-1939, Protein (milk) in milk chocolate. Kjeldahl method. In *Official Methods of Analysis of AOAC International*, 17th ed.; AOAC International: Gaithersburg, MD, USA, 2000.

29. Adamson, G.E.; Lazarus, A.; Mitchell, A.E.; Prior, R.L.; Cao, G.; Jacobs, P.H.; Kremers, B.G.; Hammerstone, J.F.; Rucker, R.B.; Ritter, K.A.; et al. HPLC method for the quantification of procyanidins in cocoa and chocolate samples and correlation to total antioxidant capacity. *J. Agric. Food Chem.* **1999**, *47*, 4184–4188. [CrossRef]

30. López-Martìnez, L.; López-de-Alba, P.L.; Garcìa-Campos, R.; De León-Rodrìguez, L.M. Simultaneous determination of methylxanthines in coffees and teas by UV-Vis spectrophotometry and partial least squares. *Anal. Chim. Acta* **2003**, *493*, 83–94. [CrossRef]

31. Hečimović, I.; Belšcak-Cvitanović, A.; Horžić, D.; Komes, D. Comparative study of polyphenols and caffeine in different coffee varieties affected by the degree of roasting. *Food Chem.* **2011**, *129*, 991–1000. [CrossRef]

32. EC Commission Regulation no. 2568/91, annex III, on the characteristics of olive oil and olive-residue oil and on the relevant methods of analysis. *Off. J. Eur. Union* **1991**, *OJL 248*, 1.

33. Bonvehì, J.S. Investigation of aromatic compounds in roasted cocoa powder. *Eur. Food Res. Technol.* **2005**, *221*, 19–29. [CrossRef]

34. Murray, J.M.; Delahunty, C.M.; Baxter, I.A. Descriptive Sensory Analysis: Past, Present, and Future. *Food Res. Int.* **2001**, *34*, 461–471. [CrossRef]

35. Stone, H.; Sidel, J. Sensory Evaluation Practices. In *Food and Science Technology Series*, 2nd ed.; Taylor, S., Ed.; Academic Press: Cambridge, MA, USA, 1992; ISBN 9780323139762.

36. ISO. *ISO 8589:2007. Sensory Analysis—General Guidance for the Design of Test Rooms*; ISO: Geneva, Switzerland, 2007.

37. ISO. *ISO 5492:2008. Sensory Analysis—Vocabulary*; ISO: Geneva, Switzerland, 2008.

38. ISO. *ISO 8586:2012. Sensory Analysis—General Guidelines for the Selection, Training and Monitoring of Selected Assessors And Expert Sensory Assessors*; ISO: Geneva, Switzerland, 2012.

39. Donadini, G.; Fumi, M.D.; Lambri, M. The hedonic response to chocolate and beverage pairing: A preliminary study. *Food Res. Int.* **2012**, *48*, 703–711. [CrossRef]

40. MacFie, H.J.; Bratchell, N.; Greenhoff, K.; Vallis, L.V. Designs to balance the effect of order of presentation and first-order carry-over effect in halls tests. *J. Sens. Stud.* **1989**, *4*, 129–148. [CrossRef]

41. Kemp, S.E.; Hollowood, T.; Hort, J. *Sensory Evaluation: A Practical Handbook*; Wiley-Blackwell: Singapore, 2009. [CrossRef]

42. Meilgaard, M.; Civille, G.V.; Carr, B.T. *Sensory Evaluation Techniques*, 4th ed.; CRC Press: Boca Raton, FL, USA, 2007.

43. Civille, G.V.; Lyon, B.G. *Aroma and Flavor Lexicon for Sensory Evaluation: Terms, Definitions, References, and Examples*; ASTM Data Series Publication DS 66; American Society for Testing and Materials (ASTM) International: West Conshohocken, PA, USA, 1996.

44. Thompson, J.L.; Drake, M.A.; Lopetcharat, K.; Yates, M.D. Preference mapping of commercial chocolate milks. *J. Food Sci.* **2004**, *69*, S406–S413. [CrossRef]

45. Ferreira, J.M.M.; Marcacini Azevedo, B.; Luccas, V.; Bolini, H.M.A. Sensory profile and consumer acceptability of prebiotic white chocolate with sucrose substitutes and the addition of goji berry (*Lycium barbarum*). *J. Food Sci.* **2017**, *82*, 818–824. [CrossRef]

46. Leite, P.B.; Da Silva Bispo, E.; Randomille De Santana, L.R. Sensory profiles of chocolates produced from cocoa cultivars resistant to *Moniliophtora Perniciosa*. *Rev. Bras. Frutic.* **2013**, *35*, 594–602. [CrossRef]

47. Gills, L.A.; Resurreccion, A.V.A. Sensory and physical properties of peanut butter treated with palm oil and hydrogenated vegetable oil to prevent oil separation. *J. Food Sci.* **2000**, *65*, 173–180. [CrossRef]

48. Liu, J.; Liu, M.; He, C.; Song, H.; Guo, J.; Wang, Y.; Yang, H.; Su, X. A comparative study of aroma-active compounds between dark and milk chocolate: Relationship to sensory perception. *J. Sci. Food Agric.* **2015**, *95*, 1362–1372. [CrossRef]

49. Marvig, C.L.; Kristiansen, R.M.; Madsen, M.G.; Nielsen, D.S. Identification and characterization of organisms associated with chocolate pralines and sugar syrups used for their production. *Int. J. Food Microbiol.* **2014**, *185*, 167–176. [CrossRef]

50. CREA (Il Centro di Ricerca per gli Alimenti e la Nutrizione). Available online: https://www.crea.gov.it/web/alimenti-e-nutrizione/banche-dati/ (accessed on 31 October 2019).

51. Elkhori, S.; Parè, J.R.J.; Bélanger, J.M.R.; Pérez, E. The microwave-assisted process (MAPTM1): Extraction and determination of fat from cocoa powder and cocoa nibs. *J. Food Eng.* **2007**, *79*, 1110–1114. [CrossRef]

52. Çakmak, Y.S.; Güler, G.O.; Aktümsek, A. Trans fatty acid contents in chocolates and chocolate wafers in Turkey. *Czech J. Food Sci.* **2010**, *28*, 177–184. [CrossRef]

53. Belščak-Cvitanović, A.; Durgo, K.; Gačina, T.; Horžić, D.; Franekić, J.; Komes, D. Comparative study of cytotoxic and cytoprotective activities of cocoa products affected by their cocoa solids content and bioactive composition. *Eur. Food Res. Technol.* **2012**, *234*, 173–186. [CrossRef]

54. Meng, C.C.; Jalil, A.M.; Ismail, A. Phenolic and theobromine contents of commercial dark, milk and white chocolates on the Malaysian market. *Molecules* **2009**, *14*, 200–209. [CrossRef] [PubMed]

55. Copetti, M.V.; Iamanaka, B.T.; Mororó, R.C.; Pereira, J.L.; Frisvad, J.C.; Taniwaki, M.H. The effect of cocoa fermentation and weak organic acids on growth and ochratoxin A production by Aspergillus species. *Int. J. Food Microbiol.* **2012**, *155*, 158–164. [CrossRef] [PubMed]

56. Afoakwa, E.O.; Paterson, A.; Fowler, M.; Ryan, A. Flavor formation and character in cocoa and chocolate: A critical Review. *Crit. Rev. Food Sci. Nutr.* **2008**, *48*, 840–857. [CrossRef] [PubMed]

57. Aprotosoaie, A.C.; Luca, S.V.; Miron, A. Flavor chemistry of cocoa and cocoa products—An Overview. *Compr. Rev. Food Sci. Food Saf.* **2016**, *15*, 73–91. [CrossRef]

58. Abe, L.T.; Lajolo, F.M.; Genovese, M.I. Comparison of phenol content and antioxidant capacity of nuts. *Ciênc. Tecnol. Aliment.* **2010**, *30*, 254–259. [CrossRef]

59. Parcerisa, J.; Codony, R.; Boatella, J.; Rafecas, M. Triacylglycerol and phospholipid composition of hazelnut (*Corylus avellana* L.) lipid fraction during fruit development. *J. Agric. Food Chem.* **1999**, *47*, 1410–1415. [CrossRef]

60. Jalil, A.M.; Ismail, A. Polyphenols in cocoa and cocoa products: Is there a link between antioxidant properties and health? *Molecules* **2008**, *13*, 2190–2219. [CrossRef]

61. Miller, K.B.; Hurst, W.J.; Payne, M.J.; Stuart, D.A.; Apgar, J.; Sweigart, D.S.; Ou, B. Impact of alkalization on the antioxidant and flavanol content of commercial cocoa powders. *J. Agric. Food Chem.* **2008**, *56*, 8527–8533. [CrossRef] [PubMed]

62. Lipp, M.; Simoneau, C.; Ulberth, F.; Anklam, E.; Crews, C.; Brereton, P.; de Greyt, W.; Schwack, W.; Wiedmaier, C. Composition of genuine cocoa butter and cocoa butter equivalents. *J. Food Compos. Anal.* **2001**, *14*, 399–408. [CrossRef]

63. Bignami, C.; Cristofori, V.; Troso, D.; Bertazza, G. Kernel quality and composition of hazelnut (*Corylus avellana* L.) cultivars. *Acta Hortic.* **2005**, *686*, 477–484. [CrossRef]

64. Ebrahem, K.S.; Richardson, D.G.; Tetley, R.M.; Mehlenbacher, S.A. Oil content, fatty acid composition and vitamin E concentration of hazelnut varieties, compared to other types of nuts and oilseeds. *Acta Hortic.* **1994**, *351*, 685–692. [CrossRef]

65. Botta, R.; Gianotti, C.; Richardson, D.; Suwanagul, A.; Carlos, L.S. Hazelnut variety organic acids, sugars, and total lipid fatty acids. *Acta Hortic.* **1994**, *351*, 693–699. [CrossRef]

66. Machálková, L.; Hřivna, L.; Nedomová, S.; Jůzl, M. The effect of storage temperature on the quality and formation of blooming defects in chocolate confectionery. *Potravinarstvo* **2015**, *9*, 39–47. [CrossRef]

67. Thamke, I.; Dürrschmid, K.; Rohm, H. Sensory description of dark chocolate by consumers. *LWT Food Sci. Technol.* **2009**, *42*, 534–539. [CrossRef]

68. Yadav, P.; Pandey, J.P.; Garc, S.K. Biochemical changes during storage of chocolates. *Int. Res. J. Biochem. Bioinform.* **2011**, *1*, 242–247.

69. Gordon, M.H. Chapter 2—The development of oxidative randicity in foods. In *Antioxidants in Food-Practical Applications*; Pokorni, J., Yanishlieva, N., Gordon, M., Eds.; Woodhead Publishing Ltd.: Shaston, UK, 2001; pp. 7–21.

70. Fardelli, A. Effect of Storage Conditions on Hazelnuts (*Corylus avellana* L.) Quality. Ph.D. Thesis, Università degli Studi della Tuscia, Via Santa Maria in Gradi, Italy, 2008.

71. Lopez-Uriarte, P.; Bulló, M.; Casas-Agustench, P.; Babio, N.; Salas-Salvadó, J. Nuts and oxidation: A systematic review. *Nutr. Rev.* **2009**, *67*, 497–508. [CrossRef]

72. Subramaniam, P.J. Confectionery products. In *The Stability and Shelf-Life of Food*; Llcast, D., Subramaniam, P.J., Eds.; CRC Press: Boca Raton, FL, USA, 2000; pp. 221–248.

73. Lamuela-Raventòs, R.M.; Romero-Pérez, A.I.; Andrés-Lacueva, C.; Tornero, A. Review: Health Effects of Cocoa Flavonoids. *Food Sci. Technol. Int.* **2005**, *11*, 159–176. [CrossRef]

74. Othman, A.; Ismail, A.; Ghani, A.N.; Adenan, I. Antioxidant capacity and phenolic content of cocoa beans. *Food Chem.* **2007**, *100*, 1523–1530. [CrossRef]

75. Kornsteiner, M.; Wagner, K.H.; Elmadfa, I. Tocopherols and total phenolics in 10 different nut types. *Food Chem.* **2006**, *98*, 381–387. [CrossRef]

76. Bomba, P.C. Shelf life of chocolate confectionery products. In *Shelf Life Studies of Foods and Beverages*; Charalambous, G., Ed.; Elsevier: Amsterdam, The Netherlands, 1993; pp. 341–351.

77. Stauffer, M. The flavor of milk chocolate—Changes caused by processing. *Manuf. Confect.* **2000**, *80*, 113–118.

78. Seppanen, C.M.; Song, Q.; Csallany, A.S. The antioxidant functions of tocopherol and tocotrienol homologues in oils, fats, and food systems. *J. Am. Oil Chem. Soc.* **2010**, *87*, 469–481. [CrossRef]

79. Baccelloni, S. Recupero di Antiossidanti Naturali dai Sottoprodotti di Trasformazione Dell'industria Agroalimentare: Noccioli di Oliva (*Olea europaea* L.) Gusci e Perisperma di Nocciole (*Corylus avellana* L.). Ph.D. Thesis, Università degli Studi della Tuscia, Via Santa Maria in Gradi, Italy, 2006.

Article

Effects of Particle Size and Extraction Methods on Cocoa Bean Shell Functional Beverage

Olga Rojo-Poveda [1,2], Letricia Barbosa-Pereira [1,3,*], Lívia Mateus-Reguengo [1], Marta Bertolino [1], Caroline Stévigny [2] and Giuseppe Zeppa [1,*]

[1] Department of Agriculture, Forestry and Food Sciences (DISAFA), University of Turin, 10095 Grugliasco, Italy; olgapaloma.rojopoveda@unito.it (O.R.-P.); liviareguengo@gmail.com (L.M.-R.); marta.bertolino@unito.it (M.B.)

[2] RD3 Department-Unit of Pharmacognosy, Bioanalysis and Drug Discovery, Faculty of Pharmacy, Université Libre de Bruxelles, 1050 Brussels, Belgium; Caroline.Stevigny@ulb.be

[3] Department of Analytical Chemistry, Nutrition and Food Science, Faculty of Pharmacy, University of Santiago de Compostela, 15782 Santiago de Compostela, Spain

* Correspondence: letricia.barbosa.pereira@usc.es (L.B.-P.); giuseppe.zeppa@unito.it (G.Z.); Tel.: +39-011-670-8705 (L.B.-P. & G.Z.)

Received: 15 March 2019; Accepted: 15 April 2019; Published: 17 April 2019

Abstract: One of the main by-products in cocoa industry is the cocoa bean shell (CBS), which represents approximately 12–20% of the bean. This product has been suggested as a food ingredient because of its aroma and high dietary fiber and polyphenol contents. The purpose of this work was to evaluate the effects of the CBS particle size and extraction methods on the chemical composition and consumer acceptance of a functional beverage, in order to find the best combination of technological parameters and health benefits. Five particle sizes of CBS powder and six home techniques were used for beverage preparation. The influence of these factors on the physico-chemical characteristics, methylxanthine and polyphenolic contents, antioxidant and antidiabetic properties, and consumer acceptance was evaluated. Total phenolic content values up to 1803.83 mg GAE/L were obtained for the beverages. Phenolic compounds and methylxanthines were identified and quantified by HPLC-PDA. These compounds may be related to the high antioxidant capacity (up to 7.29 mmol TE/L) and antidiabetic properties (up to 52.0% of α-glucosidase inhibition) observed. Furthermore, the consumer acceptance results indicated that CBS may represent an interesting ingredient for new functional beverages with potential health benefits, reducing the environmental and economic impact of by-product disposal.

Keywords: cocoa by-product; functional food; polyphenols; α-glucosidase inhibition; antidiabetic capacity; antioxidant capacity; methylxanthines

---

1. Introduction

According to the International Cocoa Organization (ICCO), each year, more than 4000 tons of cocoa beans are processed and consumed worldwide [1]. Considering that the main products of cocoa are obtained from its roasted bean, which represents only 10% of the total weight of the fruit, cocoa processing produces a large amount of vegetal residue. Besides being expensive, the disposal of these by-products can be harmful for the environment because they contain potentially phytotoxic polyphenols [2] and high concentrations of theobromine, which may be toxic for non-human mammals [3]. Such underutilization of residual biomass can be overcome by the development of an added-value foodstuff based on cocoa by-products, particularly the cocoa bean shell (CBS), which represents 12% to 20% of the cocoa bean [4]. CBS has been reported to be a considerable source of proteins and dietary fiber, with low fat content in comparison with cocoa beans [5] but with a similar profile of volatiles [6]. Considering that CBS is a final-stage by-product from cocoa processing, it appears to be an economical, organoleptic, and nutritionally rewarding substance for the transformation of cocoa bean industries. Most of the research into CBS utilization is related to animal feeding and, despite the presence of theobromine, CBS has positive effects on fortified diets for ruminants, pigs, and poultry [3], among other animals. Besides, CBS has been applied as an additive in organic fertilizer [7], as biomass for biogas production [8], or as a pectin source [5], among other applications. The application of this cocoa by-product to food has attracted some attention due to its nutritional characteristics and high concentration of phenolic compounds, mainly flavonoids [9]. In recent years, some studies in the food research field were published, suggesting CBS as a food ingredient [4]. This research interest can be linked to the sustainability and bioeconomy framework of the modern food and agricultural industries, leading to the valorization of functional foods developed from by-products, as an opportunity to make healthier foods. Functional foods are similar to conventional ones in that they are part of a standard diet and consumed on a regular basis and in regular amounts. A functional food is claimed to have proven benefits for the maintenance or promotion of a state of well-being or health or a reduction in the risk of a pathological process or disease [10]. Although this market niche is not well defined, by influencing the data on global sales, functional foods irrefutably represent a top trend in the food industry.

Because of their antioxidant capacity, phenolic compounds may be capable of protecting cell components from oxidative damage, thus limiting the risk of several diseases associated with oxidative stress, for instance, diabetes [11]. The relevance of polyphenols to the management of blood glucose is mainly due to their inhibition of digestive enzymes involved in the metabolism of carbohydrates (α-glucosidase and α-amylase) [9]. Even if these phenolic compounds are poorly absorbed, they can still act on membrane-bound enzymes located in the intestinal epithelium. Moreover, other mechanisms of action of polyphenols on glucose uptake, after ingestion of carbohydrate-rich meals, are being studied [12]. The presence of bioactive components in CBS and its sustainable character may arouse interest in such a product as a potential ingredient in the functional-beverage industry. Despite their resource-rich matrix, the effective extraction of plant bioactive compounds and their concentrations in the final product can be influenced by processing or preparation methods.

Unground CBS has already been utilized as one of the ingredients for commercialized herbal infusion bags, such as the "ChocoTea" infusion bags from Valberbe® or the "Choco" tisane bags from YogiTea®. Nonetheless, other preparations destined for the use of this by-product in home-based beverage-making techniques have not been proposed yet. In this paper, several CBS preparations were developed to be employed in six diffused techniques for coffee home-preparation available at a consumer level, such as the Moka, Neapolitan flip, American, Espresso, Capsule, and French press coffee makers. Instead of coffee powder, CBS at different grinding degrees (GDs) was employed to find the optimal GD for each extraction technique. The aim of this study was to find the best combination of the CBS GD and the beverage preparation technique in order to obtain a new functional beverage with the optimal chemical composition, biological effects (antioxidant and antidiabetic properties), and sensory characteristics.

2. Materials and Methods

2.1. Chemicals

Folin & Ciocalteu's phenol reagent, sodium carbonate ($\geq$99.5%), 2,2'-diphenyl-1-picrylhydrazyl (95%) (DPPH), 6-hydroxy-2,5,7,8-tetramethylchroman-2-carboxylic acid (97%; trolox), vanillin (99%), (+)-catechin hydrate (>98%), methanol ($\geq$99.9%), hydrochloric acid (fuming 37%), aluminum chloride (99%), sodium nitrite ($\geq$99%), α-glucosidase from intestinal acetone powders from rat, p-nitrophenyl-α-D-glucopyranoside ($\geq$99%; p-PNG), acarbose ($\geq$95%), potassium phosphate monobasic ($\geq$99%), formic acid ($\geq$98%), quercetin-3-O-glucoside ($\geq$90%; Q-3-G), theobromine ($\geq$98.5%), caffeine ($\geq$98.5%), and quercetin ($\geq$98.5%) were provided by Sigma-Aldrich (Milan, Italy). Potassium phosphate dibasic ($\geq$ 98%) was acquired from Carlo Erba (Milan, Italy). Gallic acid, ethanol ($\geq$99.9%), sodium hydroxide (1 M), (-)-epicatechin (>90%), procyanidin B1 (≥98.5%; PCB1), procyanidin B2 (≥98.5%; PCB2), protocatechuic acid (>97%), caffeic acid (≥95%), and vanillic acid (≥99%) were supplied by Fluka (Milan, Italy). Ultrapure water was prepared in a Milli-Q filter system (Millipore, Milan, Italy).

2.2. Samples

CBS from São Tomé cocoa beans (Forastero variety) was kindly supplied by Pastiglie Leone S.r.l. (Turin, Italy). The CBS was divided into different grain sizes using a BA200N vibrating sieve (CISA, Barcelona, Spain). Five GDs were obtained: above 4000 µm (GD1), 2000–4000 µm (GD2), 1000–2000 µm (GD3), 500–1000 µm (GD4), and 250–500 µm (GD5).

The chemical and nutritional characterization of CBS was carried out according to Bertolino et al. [13].

2.3. Preparation of Beverages

The beverages were prepared by six techniques available to consumers, i.e., Moka Express 6 cups (Bialetti, Brescia, Italy), Neapolitan flip coffee pot (Ilsa, Turin, Italy), American coffee maker Cucina HD 7502 (Phillips, Milan, Italy), Espresso Saeco HD 8423/11 (Phillips, Milan, Italy), Capsule LM 3100 (AEG, Milan, Italy), and French press Kaffe (Ikea, Collegno, Italy). Still mineral water (Valmora, Luserna San Giovanni, Italy) was used for the beverage production. Each of the GDs was tested with all the coffee makers, resulting in 30 beverages. Three production batches were generated for each beverage.

The water quantities employed varied for each technique following the established rules for the use of the different machines. For Espresso and Capsule techniques, the initial volumes of water are unknown because both machines operate with a continuous inlet of water and therefore it is possible to define only the final volume of the obtained beverage. In addition, the quantities of CBS powders used for the beverage preparations were adapted to each technique and GD to obtain a technologically viable formulation.

All the beverages were centrifuged on a MPW-260R centrifuge (MPW, Warsaw, Poland) at 3075 ×*g* for 10 min and then passed through a 0.45 µm cellulose acetate filter (Carlo Erba, Milan, Italy) before analyses.

2.4. Analytical Procedures

2.4.1. Physicochemical Analysis

A pH meter MICROpH 2002 (CRISON, Carpi, Italy) served for pH measurement.

Determination of total acidity (expressed in grams of acetic acid per liter of a beverage) was performed by potentiometric titration of 5 mL of a beverage (diluted to 50 mL with distilled water) by means of 0.01N NaOH up to pH 8.2.

The dry extract of each beverage was analyzed gravimetrically for a 5 mL sample of each beverage dried in an oven at 110 °C until constant weight.

The color analysis was conducted in transmittance mode on a CM-5 spectrophotometer (Konica Minolta, Tokyo, Japan). L^*, a^*, and b^* CIELab parameters were used to measure the color, where L^* is a coefficient of lightness ranging from 0 (black) to 100 (white), a^* indicates the colors red-purple (when positive a^*) and bluish-green (when negative a^*), and b^* denotes the colors yellow (when positive b^*) and blue (negative b^*). The ΔE parameter, which represents the difference between two colors [14] and its perceptibility by the human eye when ΔE > 2.5, was calculated according to the equation

$$\Delta E = \sqrt{\left(L_{*a} - L_{*b}\right)^2 + \left(a_{*a} - a_{*b}\right)^2 + \left(b_{*a} - b_{*b}\right)^2}$$

2.4.2. Total Phenolic, Tannin, and Flavonoid Contents

The total phenolic (TPC), total flavonoid (TFC), and total tannin (TTC) contents were determined according to the methods described by Barbosa-Pereira et al. [15], in 96-well microplates, using a BioTek Synergy HT spectrophotometric multi-detection microplate reader (BioTek Instruments, Milan, Italy). All the measurements were performed in triplicate. For the TPC analysis, a calibration curve of gallic acid (20–100 mg/L) was constructed to quantify the concentration, which was expressed in milligrams of gallic acid equivalents per liter of a beverage (mg GAE/L). The quantification of both TFC and TTC was performed based on a standard curve of catechin (5–500 mg/L), and the concentrations were expressed in milligrams of catechin equivalents per liter of a beverage (mg CE/L).

2.4.3. Antioxidant Capacity

The antioxidant capacity of the beverages was assessed by the 2,2′-diphenyl-1-picrylhydrazyl (DPPH•) radical–scavenging method described by Barbosa-Pereira et al. [15]. All the assays were conducted in triplicate in 96-well microplates with the BioTek Synergy HT spectrophotometric multi-detection microplate reader (BioTek Instruments). Antioxidant capacity was calculated as the inhibition percentage (IP) of the DPPH radical as

$$\mathbf{IP}\ (\%) = \frac{(\mathbf{A}_0 - \mathbf{A}_{30})}{\mathbf{A}_0} \times 100$$

where A_0 is absorbance at the initial time point, and A_{30} is the absorbance after 30 min.

A standard curve of trolox was constructed (12.5–300 μM) for assessment of the radical-scavenging activity values, which were expressed as millimoles of trolox equivalents per liter of a beverage (mmol TE/L).

2.4.4. Antidiabetic Capacity

The antidiabetic effects of the beverages were determined by the α-glucosidase colorimetric assay adapted from the method described by Kwon et al. [16].

An aliquot (50 μL) of the sample was mixed with 100 μL of α-glucosidase (10 mg/mL), prepared in 0.1 M phosphate buffer pH 6.9, and incubated for 5 min. After that, 50 μL of substrate p-PNG at 4 mM (prepared in the phosphate buffer) was added, and the solution was mixed. The solution was incubated for 30 min at 37 °C, and then absorbance was measured at 405 nm against a blank control. Acarbose at 0.5 mM (half-maximal inhibitory concentration, IC_{50}) served as a positive control, and the antidiabetic capacity was expressed as the α-glucosidase inhibition percentage. All of the measurements were conducted in triplicate in 96-well microplates, using the BioTek Synergy HT spectrophotometric multi-detection microplate reader (BioTek Instruments).

2.4.5. RP-HPLC-PDA Analysis

Characterization of the polyphenols contained in the beverages was performed by means of reversed-phase high-pressure liquid chromatography with a photodiode array detector (RP-HPLC-PDA) Thermo-Finnigan Spectra System (Thermo-Finnigan, Waltham, MA, USA). The instrument was equipped with a P2000 binary gradient pump, SCM 1000 degasser, AS 3000 automatic injector, and Finnigan Surveyor PDA Plus detector. Instrument control, data collection, and data processing were performed using the ChromQuest software, version 5.0 (Thermo-Finnigan, Waltham, MA, USA).

For separation of compounds, a reverse-phase Kinetex Phenyl-Hexyl C18 column (150 × 4.6 mm internal diameter and 5 μm particle size; Phenomenex, Castel Maggiore, Italy) was utilized at 35 °C.

Two solvents served as a mobile phase: water containing formic acid at 0.1% *v/v* (solvent A) and 100% methanol (solvent B). The sample injection volume was 10 μL. To separate the different compounds, gradient elution at a flow rate of 1 mL/min was conducted during 45 min as follows: minutes 0–2, 90% A and 10% B; minutes 2–18, a linear gradient from 10% to 50% B; minutes 18–40, a linear gradient from 50% to 80% B; minutes 40–42, a linear gradient from 80% to 90% B; and minutes 42–45, a linear gradient until 90% A and 10% B were reached.

Detection was carried out via continuous scanning of wavelengths between 200 and 400 nm. Methylxanthines (theobromine and caffeine) were quantified at 272 nm, protocatechuic acid at 293 nm, caffeic acid at 325 nm, flavan-3-ols (catechin, epicatechin, and catechin-3-O-glucoside), and procyanidins B (type B procyanidin and procyanidin B2) were quantified at 280 nm, and flavonols (quercetin, quercetin-3-O-glucoside, and quercetin-3-O-rhamnoside) at 365 nm.

The quantification was performed based on external linear calibration curves analyzed under the same conditions and the following correlation coefficients were obtained: $R^2 = 0.9995$ for theobromine, $R^2 = 0.9996$ for caffeine, $R^2 = 0.9999$ for catechin, $R^2 = 0.9998$ for epicatechin, $R^2 = 0.9997$ for protocatechuic acid, $R^2 = 0.9999$ for caffeic acid, $R^2 = 0.9998$ for procyanidin B1 (PB1), $R^2 = 0.9999$ for procyanidin B2 (PB2), $R^2 = 0.9996$ for quercetin-3-O-glucoside (Q-3-G), and $R^2 = 0.9988$ for quercetin. For catechin-3-O-glucoside, type B procyanidin, and quercetin-3-O-rhamnoside, concentrations were expressed as catechin, procyanidin B1, and quercetin-3-O-glucoside equivalents, respectively.

2.5. Consumer Acceptance Evaluation

For each beverage, a consumer test was carried out with 20 tasters where appearance, odor, taste, flavor, texture, overall liking, and purchase predisposition were evaluated on a nine-point hedonic scale (1 = extremely dislike, 9 = extremely like) [17]. The tests were performed in an air-conditioned room with white light at approximately 21°C.

2.6. Statistical Analysis

All the obtained results were subjected to analysis of variance (ANOVA) with Duncan's post hoc test at 95% confidence level and to linear regression analysis in the Windows software called STATISTICA, version 13.3 (StatSoft Inc., Tulsa, OK, USA).

Values obtained by the consumer test were analyzed by the Kruskal–Wallis test (test H).

3. Results and Discussion

3.1. Cocoa Bean Shell—Chemical and Nutritional Composition

The chemical and nutritional composition of the cocoa bean shell employed in beverage preparation, expressed for 100 g of dried product, was as follows: protein: 20.9 g, fat: 2.3 g, carbohydrates: 7.85 g, dietary fiber: 55.1 g (42.3 g of insoluble fiber and 12.8 g of soluble fiber), water: 5.9 g, and ash: 7.9 g.

3.2. Beverage Yield

Thirty formulations were developed, resulting in 30 beverages with different yields, mostly depending on the CBS GD (Table 1). Yields ranged from 72.0% to 93.3% when GDs above 500 µm were used, and a notable substantial decrease in the recovery percentage was observed with a reduction in the GD, thereby leading to such values as 29.6% for the beverage prepared with the Moka and GD5. This decrease could be mostly due to the water-holding capacity of the insoluble fiber present in the CBS; this fiber became more available when the surface-to-volume ratio of the CBS powders increased. In some cases, also the larger CBS powder quantities used in order to obtain technologically realistic preparations influenced the water-holding capacity. Nevertheless, this influence may not affect the beverages obtained with other coffee makers such as the American pot or French press where the CBS quantities remain the same at all the GDs, and the reduction in the yield is due to the lower GD only.

For the lowest GD (GD5, 250–500 µm), the beverage yields decreased considerably, and this parameter, in general, lost the repeatability observed for larger GDs, mostly owing to the technological problems during beverage preparation such as machine blockage in cases where the CBS absorbed too much water or with extremely long preparation periods. Due to the aforementioned problems, the beverages obtained by the Moka, Neapolitan, Espresso, and Capsule techniques with GD5 were assumed to be not technologically viable and it was not possible to proceed with further analyses. Only the beverages obtained with the French press and American techniques were considered for GD5.

Table 1. Amounts of cocoa bean shell (CBS) powder and water utilized for the beverage preparations, volume of the obtained beverages, and process yield registered for each production technique and GD. Analysis of variance (ANOVA) was performed on the process yields among GDs and extraction techniques.

		Moka	Neapolitan	American	Espresso	Capsule	French press	Sig.
>4000 μm	Water (mL)	230.00	400.00	200	n/a	n/a	100.00	
	CBS powder (g)	10.00	14.00	8.00	7.00	2.00	6.00	***
	Beverage (mL)	188.00 ± 3.46	373.33 ± 2.89	172.00 ± 1.73	123.67 ± 1.53	59.33 ± 0.58	82.67 ± 0.58	
	Yield (%)	81.74 ± 1.51 [bC]	93.33 ± 0.72 [aA]	86.00 ± 0.87 [aB]	n/a	n/a	82.67 ± 0.58 [aC]	
2000–4000 μm	Water (mL)	230.00	400.00	200.00	n/a	n/a	100.00	
	CBS powder (g)	18.00	17.00	8.00	13.00	2.80	6.00	***
	Beverage (mL)	190.00 ± 2.00	363.33 ± 2.89	159.67 ± 6.81	125.67 ± 2.08	60.33 ± 0.58	83.00 ± 1.00	
	Yield (%)	82.61 ± >0.87 [abBC]	90.83 ± 0.72 [aA]	79.83 ± 3.40 [bC]	n/a	n/a	83.00 ± 1.00 [aB]	
1000–2000 μm	Water (mL)	230.00	400.00	200.00	n/a	n/a	100.00	
	CBS powder (g)	22.00	30.00	8.00	16.00	6.50	6.00	***
	Beverage (mL)	188.33 ± 1.53	329.66 ± 1.53	152.67 ± 1.15	122.67 ± 0.29	60.33 ± 0.58	78.67 ± 1.15	
	Yield (%)	81.88 ± 0.66 [bA]	82.42 ± 0.38 [aA]	76.33 ± 0.58 [cC]	n/a	n/a	78.67 ± 1.15 [bB]	
500–1000 μm	Water (mL)	230.00	400.00	200.00	n/a	n/a	100.00	
	CBS powder (g)	26.00	30.00	8.00	18.00	8.00	6.00	***
	Beverage (mL)	186.67 ± 2.31	316.67 ± 5.77	148.67 ± 4.16	119.67 ± 0.58	60.33 ± 0.58	72.00 ± 1.73	
	Yield (%)	81.16 ± 1.00 [bA]	79.17 ± 1.44 [aA]	74.33 ± 2.08 [cB]	n/a	n/a	72.00 ± 1.73 [cC]	
250–500 μm	Water (mL)	230.00	400.00	200.00	n/a	n/a	100.00	
	CBS powder (g)	26.00	30.00	8.00	18.00	8.00	6.00	***
	Beverage (mL)	68.00 ± 13.86	200.00 ± 81.85	146.67 ± 2.31	71.00 ± 9.64	36.35 ± 0.28	71.33 ± 1.53	
	Yield (%)	29.57 ± 6.02 [cC]	50.00 ± 20.46 [bB]	73.33 ± 1.15 [cA]	n/a	n/a	71.33 ± 1.53 [cA]	
Sig.		***	**	***	n/a	n/a	***	

n/a, not applicable. Means followed by different lower case superindexes within the same column (different grinding degrees) and by upper case superindexes within the same row (different techniques) are significantly different at $p < 0.05$. Significance: ** $p < 0.01$; *** $p < 0.001$. Data are expressed as mean values ($n = 3$) ± standard deviation.

3.3. Physico-Chemical Characterization

3.3.1. Acidity, Dry Matter, and Color

The pH and the titratable acidity results obtained for the functional beverages are shown in Table 2. The beverages had pH and titratable acidity ranging from 4.84 to 5.19 and from 0.12 to 1.64 g of acetic acid equivalents per liter of a beverage, respectively. In general, lower pH and higher acidity were observed in beverages produced with a lower GD, except for the beverages produced with the Moka and the Neapolitan techniques, which showed nonsignificant differences in pH when the GD was varied. Nevertheless, they still showed the same tendency for acidity, which increased with a decrease in particle size, probably owing to a major acid extraction when the CBS surface-to-volume ratio was increased. This tendency was not observed for the beverage produced with the French press technique where pH and acidity were found to be independent of the CBS particle size.

For all the beverages obtained by percolation techniques (Moka, Neapolitan, American, Espresso, and Capsule), the quantity of dry matter increased when the GD was reduced (Table 2) because the extraction seemed to be more effective at low GD. For French press, the quantity of dry matter did not correlate with the GD of CBS, suggesting that the extraction by this maceration technique was barely affected by the particle size of the CBS powder.

Table 2. pH, titratable acidity, dry matter, and CIELab values of the beverages obtained by each extraction technique and grinding degrees (GD) of CBS, and ANOVA among GDs for each technique.

Technique	Grinding Degree (μm)	pH	Titratable acidity (g acetic acid eq/L)	Dry weight (%)	L^*	a^*	b^*
Moka	>4000	4.91 ± 0.05 [a]	0.40 ± 0.05 [d]	0.52 ± 0.06 [d]	82.73 ± 1.51 [a]	5.69 ± 1.36 [d]	50.84 ± 3.99 [c]
	2000–4000	4.88 ± 0.01 [a]	0.61 ± 0.05 [c]	0.84 ± 0.11 [c]	75.91 ± 0.64 [b]	11.65 ± 1.06 [c]	64.77 ± 2.83 [b]
	1000–2000	4.89 ± 0.03 [a]	0.95 ± 0.06 [b]	1.53 ± 0.06 [b]	61.20 ± 2.83 [c]	25.76 ± 1.85 [b]	80.92 ± 0.54 [a]
	500–1000	4.93 ± 0.01 [a]	1.64 ± 0.09 [a]	2.70 ± 0.17 [a]	33.21 ± 1.96 [d]	37.75 ± 1.12 [a]	55.08 ± 3.30 [c]
	250–500	n/a	n/a	n/a	n/a	n/a	n/a
	Significance	ns	***	***	***	***	***
Neapolitan	>4000	4.92 ± 0.03 [a]	0.29 ± 0.03 [c]	0.41 ± 0.01 [c]	82.76 ± 0.70 [a]	5.52 ± 0.50 [c]	48.74 ± 1.33 [c]
	2000–4000	4.91 ± 0.04 [a]	0.37 ± 0.04 [c]	0.49 ± 0.06 [c]	80.56 ± 2.80 [a]	7.33 ± 2.19 [c]	54.14 ± 5.51 [c]
	1000–2000	4.88 ± 0.03 [a]	0.95 ± 0.08 [b]	1.50 ± 0.15 [b]	56.56 ± 4.34 [b]	28.45 ± 2.89 [b]	79.31 ± 0.68 [a]
	500–1000	4.93 ± 0.01 [a]	1.23 ± 0.04 [a]	2.00 ± 0.13 [a]	39.04 ± 2.66 [c]	37.24 ± 0.70 [a]	63.97 ± 4.02 [b]
	250–500	n/a	n/a	n/a	n/a	n/a	n/a
	Significance	ns	***	***	***	***	***
American	>4000	4.96 ± 0.01 [b]	0.26 ± 0.01 [d]	0.36 ± 0.01 [c]	87.98 ± 0.53 [a]	1.98 ± 0.18 [e]	38.34 ± 0.68 [e]
	2000–4000	5.07 ± 0.04 [a]	0.37 ± 0.02 [c]	0.51 ± 0.04 [bc]	85.17 ± 0.64 [b]	4.02 ± 0.51 [d]	48.39 ± 1.86 [d]
	1000–2000	5.07 ± 0.01 [a]	0.40 ± 0.01 [b]	0.69 ± 0.01 [b]	80.12 ± 0.34 [c]	8.35 ± 0.26 [c]	59.79 ± 0.60 [c]
	500–1000	4.99 ± 0.02 [b]	0.55 ± 0.01 [a]	1.01 ± 0.26 [a]	73.99 ± 1.33 [d]	14.54 ± 1.37 [b]	71.05 ± 1.92 [b]
	250–500	4.87 ± 0.01 [c]	0.57 ± 0.01 [a]	1.00 ± 0.04 [a]	64.30 ± 0.74 [e]	23.80 ± 0.72 [a]	80.12 ± 0.31 [a]
	Significance	***	***	***	***	***	***
Espresso	>4000	5.09 ± 0.07 [a]	0.20 ± 0.02 [d]	0.28 ± 0.04 [d]	91.14 ± 1.37 [a]	0.58 ± 0.69 [d]	29.45 ± 3.67 [d]
	2000–4000	4.98 ± 0.04 [b]	0.37 ± 0.04 [c]	0.51 ± 0.05 [c]	86.15 ± 1.81 [b]	3.32 ± 1.29 [c]	44.39 ± 4.08 [c]
	1000–2000	4.98 ± 0.01 [b]	0.61 ± 0.05 [b]	0.89 ± 0.07 [b]	77.63 ± 2.16 [c]	10.61 ± 2.13 [b]	64.10 ± 3.74 [b]
	500–1000	4.87 ± 0.02 [c]	1.08 ± 0.02 [a]	1.62 ± 0.05 [a]	65.87 ± 0.52 [d]	22.10 ± 0.51 [a]	79.52 ± 0.69 [a]
	250–500	n/a	n/a	n/a	n/a	n/a	n/a
	Significance	**	***	***	***	***	***
Capsule	>4000	5.19 ± 0.04 [a]	0.12 ± 0.01 [d]	0.15 ± 0.00 [d]	94.73 ± 0.12 [a]	−0.47 ± 0.05 [c]	18.15 ± 0.56 [d]
	2000–4000	5.15 ± 0.04 [a]	0.17 ± 0.01 [c]	0.24 ± 0.00 [c]	92.73 ± 0.21 [a]	−0.08 ± 0.06 [c]	25.28 ± 0.90 [c]
	1000–2000	5.15 ± 0.02 [a]	0.43 ± 0.00 [b]	0.66 ± 0.01 [b]	82.41 ± 0.67 [b]	6.22 ± 0.58 [b]	53.72 ± 1.56 [b]
	500–1000	4.84 ± 0.01 [b]	0.93 ± 0.03 [a]	1.33 ± 0.06 [a]	68.71 ± 1.92 [c]	19.06 ± 1.73 [a]	75.91 ± 1.66 [a]
	250–500	n/a	n/a	n/a	n/a	n/a	n/a
	Significance	***	***	***	***	***	***
French press	>4000	4.91 ± 0.05 [ab]	0.62 ± 0.00 [a]	0.91 ± 0.06 [a]	77.73 ± 1.53 [b]	10.80 ± 1.47 [a]	67.09 ± 2.30 [a]
	2000–4000	4.94 ± 0.02 [ab]	0.65 ± 0.05 [a]	0.91 ± 0.05 [a]	76.48 ± 2.49 [b]	11.97 ± 2.55 [a]	68.31 ± 4.37 [a]
	1000–2000	4.94 ± 0.01 [ab]	0.50 ± 0.04 [b]	0.86 ± 0.10 [a]	77.42 ± 2.71 [b]	10.96 ± 2.65 [a]	65.83 ± 4.68 [a]
	500–1000	4.96 ± 0.01 [a]	0.40 ± 0.02 [c]	0.66 ± 0.05 [b]	82.23 ± 1.42 [a]	6.53 ± 1.27 [b]	56.03 ± 3.50 [b]
	250–500	4.90 ± 0.02 [b]	0.49 ± 0.03 [b]	0.83 ± 0.05 [a]	75.22 ± 0.36 [b]	12.75 ± 1.40 [a]	67.87 ± 3.18 [a]
	Significance	ns	***	**	*	*	*

n/a, not applicable. Means followed by different letters are significantly different at $p < 0.05$. Significance: * $p < 0.05$; ** $p < 0.01$; *** $p < 0.001$; ns = not significant. Data are expressed as mean values ($n = 3$) ± standard deviation.

Regarding the chromatic parameters (Table 2), generally, the brightness parameter decreased with the decreasing GD due to the increase in the surface-to-volume ratio of the CBS powder, thereby allowing for better extraction of color pigments. The beverages having the lowest values of L^* and therefore, darker beverages, were those obtained at the GD4 with Moka and Neapolitan techniques. Generally, both parameters a^* and b^* rose with the decreasing GD, as the beverages became browner. Always following a similar trend, various beverages showed significant differences when the GD was changed except for the beverages produced with the French press. This fact can be numerically explained by the ΔE parameter (data not shown), which determines whether two colors can be distinguished by the human eye ($\Delta E > 2.5$). Considering the different CBS GDs within each technique, the beverages showed values of ΔE higher than 2.5 except for some beverages obtained with the French press technique, where $\Delta E_{GD1\text{-}GD2} = 2.10$, $\Delta E_{GD1\text{-}GD3} = 1.31$, and $\Delta E_{GD2\text{-}GD5} = 1.55$, which means that these beverages had colors indistinguishable for the human eye.

3.3.2. Polyphenolic Content

TPC, TFC, and TTC data are presented in Figure 1 (Figure 1a, Figure 1b, and Figure 1c, respectively).

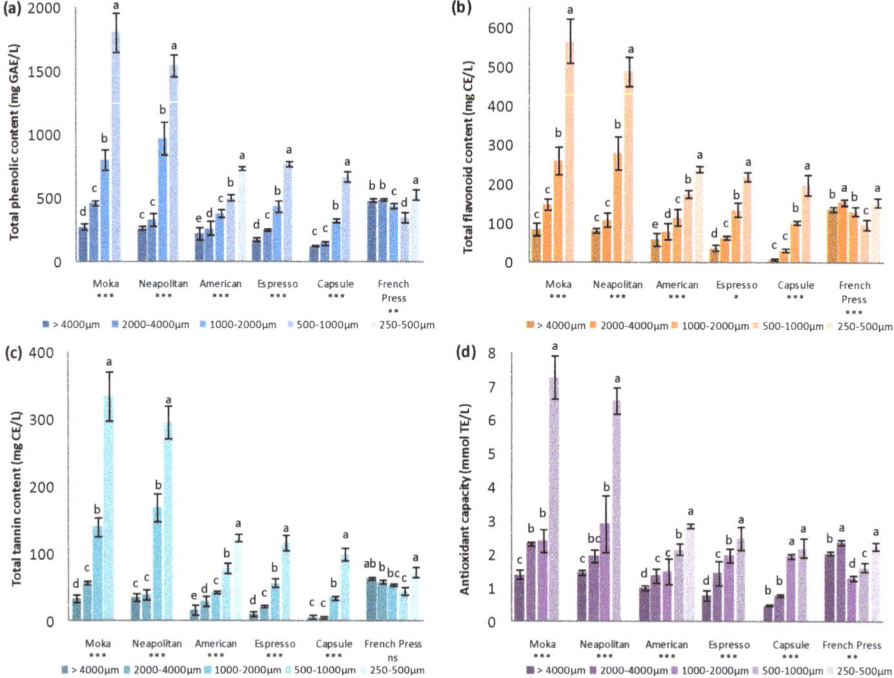

Figure 1. Total phenolic content (TPC), (**a**), total flavonoid content (TFC) (**b**), and total tannin content TTC (**c**), and antioxidant capacity (**d**) for the beverages produced by the six techniques at different CBS GDs; ANOVA among GDs for each technique. GAE = gallic acid equivalent, CE = catechin equivalent, and TE = trolox equivalent. Different letters indicate significant differences at $p < 0.05$. Significance: * $p < 0.05$; ** $p < 0.01$; *** $p < 0.001$; ns = not significant.

TPC varied considerably among the beverages obtained by different techniques and at varied GDs, with values that ranged from 126.86 mg GAE/L for the beverage obtained by the Capsule technique at GD1 to 1803.83 mg GAE/L for the beverage obtained by the Moka technique using GD4. Except for the beverage obtained with the French press, large significant differences were detected for all the other beverages when the GD of the CBS was varied. For the beverages obtained by percolation techniques,

TPC increased with a reduction in the GD, whereas for the beverage produced by the maceration technique (French press), TPC values were not influenced by the GD. The highest phenolic content was seen in beverages prepared by the Moka and Neapolitan techniques, where the results ranged from 276 mg GAE/L at GD1 to 1803.83 mg GAE/L at GD4 and from 263.67 mg GAE/L at GD1 to 1545.87 mg GAE/L at GD4, respectively. The beverage obtained by the Capsule technique manifested the lowest values of this parameter, ranging from 126.86 mg GAE/L at GD1 to 671.68 mg GAE/L at GD4. On the other hand, considering the intake, one cup (200 mL) of the beverage obtained by the French press technique at any GDs or by the American technique using CBS at GDs 3–5, provided the same quantity of polyphenols as one cup (60 mL) of the beverages obtained by the Moka or the Neapolitan techniques with CBS at GD4. The different values of TPC obtained for the beverages ranged between various values presented in previous studies, where beverages thought to have 'high-polyphenol content' were evaluated. The values of TPC in the present study were in most cases even higher than those found by Zujko & Witkowska [18] for drinking chocolate (600 mg GAE/L) or hot cocoa (300 mg GAE/L), but also higher in some cases than those of different tea types such as white tea (1040 mg GAE/L), green tea (850 mg GAE/L), black tea (720 mg GAE/L), and red tea (380 mg GAE/L). Reported TPC values for red wine (2410 mg GAE/L) and white wine (260 mg GAE/L) [18] were also within the range of the values obtained for the beverages studied in the present work. Regarding the values reported for some fruit juices by Gardner et al. [19] such as orange (755 mg GAE/L), apple (339 mg GAE/L), or pineapple juice (358 mg GAE/L), the levels of total phenols in CBS beverages were always between these values or even higher.

Flavonoids were the main compounds that contributed to TPC, constituting from 20.8% to 34.7% of this value, depending on the preparation technique but these contributions remained constant at different GDs within each technique. TFC and TPC were highly correlated ($r = 0.9965$), and thus the former followed the same tendencies of abundance depending on the technique and GD. In this way, the highest value of TFC was seen in the beverages obtained by the Moka technique at GD4 (566.42 mg CE/L), followed by the beverage obtained by the Neapolitan technique with GD4 (489.43 mg CE/L). Beverages obtained by the two techniques had higher values at each GD compared with the other techniques except for the French press, for which the beverages showed slight differences with the variation of the GD. The TFC values of the beverages produced by the French press technique ranged between 97.98 and 155.63 mg CE/L and therefore had the highest values when larger CBS particle sizes were chosen. Again, the lowest values were observed for the Capsule beverage, which had TFC between 8.78 and 198.47 mg CE/L, which increased with a reduction in the CBS particle size.

Values of TTC accounted for 3.4% to 19.2% of TPC with a significantly high correlation ($r = 0.9968$). Contrary to what was observed between TPC and TFC, the percentage of TTC's contribution to TPC increased with a reduction in the GD within the values obtained for each technique, except for the beverages produced by the maceration technique, which maintained the contribution of the tannin content (12.0% to 13.9%) to TPC values independently of the GD. This fact can be noticed as the differences between big and small CBS particle sizes in TTC values become higher than those of TPC and TFC. CBS particle size could have an influence on selective extraction of some polyphenol groups for the percolation techniques, as could be the case for tannins. Tannins are normally larger molecules than flavonoids and therefore, a decrease in the particle size could facilitate their extraction compared to that of flavonoids, which were already extracted using the CBS powder with high particle sizes. In this way, greater increases were observed for TTC within the same technique when the particle size was reduced in comparison with those observed for TPC and TFC. Nevertheless, the highest concentrations of tannins were again detected in the Moka and Neapolitan beverages at GD4 (334.64 and 296.06 mg CE/L, respectively) and the lowest values of TTC were observed in the Capsule beverage, which ranged from 5.18 to 99.86 mg CE/L.

More than 30 polyphenolic compounds that may contribute to the above values were detected and quantified by HPLC analysis. Only the concentrations determined for the main cocoa marker phenolic compounds and those showing the highest concentrations are given in Table 3. Belonging to the group of phenolic acids, protocatechuic acid, and caffeic acid were found in CBS beverages, with the former being the most abundant. Both phenolic acids have already been quantified in cocoa bean and chocolate samples [20,21]. Nonetheless even if the caffeic acid was present at lower concentrations than protocatechuic acid, in the CBS beverages it showed higher levels (with respect to protocatechuic acid) than those found in chocolate 100% cocoa made from Sao Tome cocoa beans (Forastero variety) studied by Rodríguez-Carrasco et al. [21]. These results indicate that the proportions of these two components diverge in the CBS with respect to the cocoa bean. For protocatechuic acid, the concentrations depending on the technique and GD used to produce the beverages followed a trend similar to the one already observed. The highest concentration was obtained in the beverage produced by the Moka technique using the minimum GD (18.14 mg/L) while the lowest values were found for the Capsule beverage (ranging from 1.32 to 10.80 mg/L) and the beverage produced by the American technique (ranging from 2.91 to 7.23 mg/L). Similar data were obtained for the French press beverage regardless of the particle size (ranging from 5.06 to 7.70 mg/L). As for caffeic acid, the lowest concentrations were seen in the beverage produced by the American technique, with values ranging from 0.15 to 0.33 mg/L. A slight decrease in the caffeic acid concentration was observed in the beverages prepared by the French press technique in comparison with the other techniques.

As for flavan-3-ols, we detected and quantified catechin, epicatechin, and catechin-3-O-glucoside in all the beverages; these are three characteristic flavan-3-ols for cocoa beans and chocolate already reported in CBS [15]. They can be found as free monomers or forming condensed tannins as monomeric constituents [22]. Variable concentrations were noted for these compounds, with catechin-3-O-glucoside being the most abundant, followed by epicatechin and catechin. Nevertheless, the concentrations obtained for both catechin and epicatechin were lower than those obtained with other types of extraction procedures where organic solvents were used as in the work of Hérnandez-Hérnandez et al. [23]. Catechin and epicatechin showed low solubility in water, even when pressure and high temperatures were applied in the present work, whereas catechin-3-O-glucoside was present at higher concentrations, probably because of the water solubility ensured by the glycoside group. As a flavonoid compound, epicatechin significantly correlated with the results obtained for both TPC and TFC ($r = 0.9628$ and $r = 0.9574$, respectively).

Two procyanidins of type B were detected and quantified by HPLC analysis. Procyanidins are flavan-3,4-diols, generally forming condensation compounds with epicatechin at 4–8 or 4–6 bonds [22]. Type B procyanidin was found to be the polyphenolic compound at the highest concentration among those quantified in the studied samples. This compound was present at the highest concentrations in both Moka and Neapolitan beverages with the smallest CBS particle size (36.30 and 27.10 mg procyanidin B1 eq/L, respectively) and with the lowest values for the beverages produced by the Capsule and American techniques; the concentration increased with a decrease in the GD. As epicatechin, this compound highly correlated with both TPC and TFC ($r = 0.9621$ and $r = 0.9572$, respectively). Procyanidin B2 was also found at its highest concentrations in the Moka and Neapolitan beverages, and at lower concentrations in the Capsule and American ones. These results followed the trend of an increasing concentration when the GD was decreased for almost all the beverages obtained by the percolation techniques; and an intermediate constant concentration was observed for the maceration technique, independently of the GD.

Finally, three flavonols were quantified in the CBS beverages, quercetin and two of its glycoside derivates: quercetin-3-O-glucoside and quercetin-3-O-rhamnoside. Quercetin and its derivates are part of the more abundant and recurrent flavonoids in foods known for their bitter flavor [22].

All these polyphenolic compounds, which possess antioxidant properties, could be beneficial in terms of prevention of diseases related to oxidative stress. Therefore, it is highly important for humans to consume them with nutrition. The new beverages (based on CBS) developed in this study may be an interesting source of these bioactive compounds with potential health benefits.

Table 3. Content (mg/L) of methylxanthines (theobromine and caffeine) and polyphenols (phenolic acids, flavan-3-ols, procyanidins B, and flavonols) evaluated by HPLC for the beverages obtained via the different techniques and GDs of the CBS and ANOVA among GDs for each technique.

Technique	Grinding Degree (μm)	Methylxanthines		Phenolic acids			Flavan-3-ol		Procyanidins B		Flavonols		
		Theobromine	Caffeine	Protocatechuic acid	Caffeic acid	Catechin-3-O-glucoside	Catechin	Epicatechin	Type B procyanidin	Procyanidin B2	Quercetin-3-O-glucoside	Quercetin-3-O-rhamnoside	Quercetin
Moka	>4000	147.33 ± 12.08 [d]	25.44 ± 2.85 [d]	5.25 ± 0.66 [c]	0.17 ± 0.01 [c]	6.43 ± 0.76 [c]	1.01 ± 0.06 [b]	0.51 ± 0.03 [d]	8.64 ± 1.52 [c]	0.79 ± 0.04 [d]	0.19 ± 0.01 [c]	0.24 ± 0.02 [d]	0.88 ± 0.00 [d]
	2000–4000	261.85 ± 12.05 [c]	37.75 ± 2.38 [c]	5.98 ± 0.95 [c]	0.35 ± 0.05 [b]	6.16 ± 0.19 [c]	0.41 ± 0.02 [b]	1.77 ± 0.02 [c]	10.50 ± 1.80 [c]	1.43 ± 0.17 [c]	0.44 ± 0.02 [b]	0.39 ± 0.07 [c]	1.76 ± 0.01 [c]
	1000–2000	384.78 ± 13.15 [b]	68.07 ± 5.73 [b]	10.29 ± 1.00 [b]	0.41 ± 0.08 [b]	13.49 ± 0.43 [b]	0.81 ± 0.09 [b]	3.20 ± 0.18 [b]	20.38 ± 3.03 [b]	1.81 ± 0.32 [b]	0.50 ± 0.04 [b]	0.51 ± 0.02 [b]	1.80 ± 0.01 [b]
	500–1000	703.79 ± 52.64 [a]	124.84 ± 15.91 [a]	18.14 ± 1.38 [a]	0.81 ± 0.05 [a]	21.73 ± 1.98 [a]	1.92 ± 0.33 [a]	6.38 ± 0.10 [a]	36.30 ± 5.29 [a]	3.30 ± 0.13 [a]	0.93 ± 0.13 [a]	0.66 ± 0.04 [a]	3.54 ± 0.02 [a]
	250–500	n/a	n/a	n/a	n/a	n/a	n/a	n/a	n/a	n/a	n/a	n/a	n/a
	Sig.	***	***	***	***	***	***	***	***	***	***	***	***
Neapolitan	>4000	127.84 ± 3.97 [d]	19.17 ± 0.80 [c]	3.44 ± 0.15 [c]	0.16 ± 0.00 [d]	4.21 ± 0.06 [b]	0.59 ± 0.08 [bc]	0.45 ± 0.01 [d]	5.56 ± 0.06 [c]	0.74 ± 0.03 [d]	0.17 ± 0.01 [d]	0.17 ± 0.01 [d]	0.88 ± 0.01 [d]
	2000–4000	185.78 ± 12.03 [c]	26.11 ± 2.15 [c]	4.10 ± 0.38 [c]	0.31 ± 0.01 [c]	4.61 ± 0.27 [b]	0.36 ± 0.02 [c]	1.55 ± 0.04 [c]	7.29 ± 0.46 [c]	1.30 ± 0.04 [c]	0.36 ± 0.03 [c]	0.32 ± 0.00 [c]	1.75 ± 0.00 [c]
	1000–2000	418.97 ± 24.11 [b]	80.49 ± 8.83 [b]	10.94 ± 1.55 [b]	0.40 ± 0.04 [b]	13.67 ± 0.67 [a]	0.83 ± 0.20 [b]	3.61 ± 0.04 [b]	20.69 ± 2.15 [b]	1.59 ± 0.10 [b]	0.51 ± 0.04 [b]	0.52 ± 0.04 [b]	1.78 ± 0.00 [b]
	500–1000	583.12 ± 29.74 [a]	102.98 ± 7.30 [a]	13.23 ± 0.88 [a]	0.67 ± 0.01 [a]	13.58 ± 1.13 [a]	1.57 ± 0.20 [b]	5.24 ± 0.16 [a]	27.10 ± 1.90 [a]	2.88 ± 0.08 [a]	0.98 ± 0.07 [a]	0.75 ± 0.01 [a]	3.54 ± 0.02 [a]
	250–500	n/a	n/a	n/a	n/a	n/a	n/a	n/a	n/a	n/a	n/a	n/a	n/a
	Sig.	***	***	***	***	***	***	***	***	***	***	***	***
American	>4000	115.00 ± 3.29 [e]	14.57 ± 1.42 [e]	2.91 ± 0.24 [e]	0.15 ± 0.00 [c]	3.89 ± 0.28 [c]	0.36 ± 0.08 [c]	0.25 ± 0.03 [b]	5.71 ± 0.14 [c]	0.72 ± 0.03 [c]	0.16 ± 0.00 [d]	0.15 ± 0.00 [b]	0.87 ± 0.00 [b]
	2000–4000	145.57 ± 5.42 [d]	21.24 ± 1.26 [d]	4.20 ± 0.23 [d]	0.17 ± 0.01 [c]	5.95 ± 0.23 [b]	0.52 ± 0.04 [bc]	0.43 ± 0.07 [b]	7.84 ± 0.21 [b]	0.77 ± 0.06 [c]	0.19 ± 0.00 [c]	0.15 ± 0.01 [b]	0.87 ± 0.00 [b]
	1000–2000	233.01 ± 3.92 [a]	34.38 ± 0.97 [c]	5.54 ± 0.07 [c]	0.33 ± 0.03 [a]	6.17 ± 0.26 [b]	0.39 ± 0.03 [bc]	1.30 ± 0.06 [a]	10.54 ± 0.86 [a]	1.37 ± 0.04 [a]	0.37 ± 0.01 [a]	0.35 ± 0.03 [a]	1.75 ± 0.01 [a]
	500–1000	180.89 ± 1.11 [c]	37.09 ± 0.72 [b]	6.76 ± 0.30 [b]	0.22 ± 0.02 [b]	8.65 ± 0.26 [a]	1.08 ± 0.05 [a]	1.54 ± 0.43 [a]	12.19 ± 0.58 [a]	0.99 ± 0.09 [b]	0.20 ± 0.02 [c]	0.17 ± 0.01 [b]	0.88 ± 0.00 [b]
	250–500	193.03 ± 0.84 [b]	45.69 ± 0.91 [a]	7.23 ± 0.27 [a]	0.21 ± 0.01 [b]	8.44 ± 0.56 [a]	0.55 ± 0.16 [b]	1.65 ± 0.13 [a]	12.33 ± 2.33 [a]	0.97 ± 0.06 [b]	0.24 ± 0.02 [b]	0.16 ± 0.01 [b]	0.87 ± 0.00 [b]
	Sig.	***	***	***	***	***	***	***	***	***	***	***	***
Espresso	>4000	92.34 ± 9.99 [d]	10.63 ± 0.81 [d]	2.10 ± 0.13 [d]	0.16 ± 0.01 [c]	2.67 ± 0.21 [d]	0.24 ± 0.05 [c]	0.04 ± 0.01 [c]	4.16 ± 0.27 [d]	0.65 ± 0.02 [c]	0.17 ± 0.00 [c]	0.17 ± 0.01 [d]	0.87 ± 0.00 [c]
	2000–4000	195.25 ± 20.65 [c]	24.91 ± 3.64 [c]	4.27 ± 0.54 [c]	0.32 ± 0.01 [b]	4.76 ± 0.54 [c]	0.28 ± 0.03 [c]	1.42 ± 0.10 [b]	7.25 ± 1.23 [c]	1.29 ± 0.02 [b]	0.36 ± 0.02 [b]	0.34 ± 0.02 [c]	1.74 ± 0.00 [b]
	1000–2000	292.82 ± 21.28 [b]	42.90 ± 4.16 [b]	7.16 ± 0.48 [b]	0.32 ± 0.00 [b]	9.43 ± 0.53 [b]	0.53 ± 0.01 [a]	1.84 ± 0.16 [b]	13.97 ± 0.97 [b]	1.34 ± 0.13 [b]	0.38 ± 0.05 [b]	0.45 ± 0.00 [b]	1.76 ± 0.01 [a]
	500–1000	392.36 ± 4.40 [a]	66.81 ± 1.55 [a]	12.59 ± 0.18 [a]	0.43 ± 0.00 [a]	13.98 ± 0.22 [a]	0.45 ± 0.04 [b]	2.87 ± 0.10 [a]	20.72 ± 0.66 [a]	1.60 ± 0.08 [a]	0.45 ± 0.03 [a]	0.56 ± 0.05 [a]	1.77 ± 0.00 [a]
	250–500	n/a	n/a	n/a	n/a	n/a	n/a	n/a	n/a	n/a	n/a	n/a	n/a
	Sig.	***	***	***	***	***	***	***	***	***	***	***	***
Capsule	>4000	59.82 ± 2.66 [d]	6.09 ± 0.50 [d]	1.32 ± 0.05 [d]	0.14 ± 0.00 [b]	1.71 ± 0.17 [d]	0.12 ± 0.04 [c]	0.00 ± 0.00 [c]	2.77 ± 0.11 [d]	0.60 ± 0.02 [b]	0.15 ± 0.01 [c]	0.17 ± 0.00 [c]	0.86 ± 0.00 [b]
	2000–4000	88.78 ± 3.27 [c]	9.75 ± 0.59 [c]	2.13 ± 0.11 [c]	0.15 ± 0.00 [b]	2.61 ± 0.03 [c]	0.22 ± 0.03 [bc]	0.14 ± 0.02 [c]	3.96 ± 0.31 [c]	0.68 ± 0.04 [b]	0.15 ± 0.01 [c]	0.13 ± 0.01 [c]	0.87 ± 0.00 [b]
	1000–2000	230.27 ± 9.07 [b]	33.46 ± 0.15 [b]	5.49 ± 0.36 [b]	0.40 ± 0.07 [a]	6.26 ± 0.33 [b]	0.31 ± 0.09 [ab]	1.27 ± 0.01 [b]	9.49 ± 0.25 [b]	1.31 ± 0.08 [a]	0.35 ± 0.01 [b]	0.39 ± 0.02 [b]	1.76 ± 0.00 [a]
	500–1000	361.79 ± 5.64 [a]	59.60 ± 1.99 [a]	10.80 ± 0.33 [a]	0.38 ± 0.01 [a]	12.21 ± 0.59 [a]	0.37 ± 0.03 [a]	2.53 ± 0.11 [a]	17.99 ± 1.03 [a]	1.39 ± 0.04 [a]	0.44 ± 0.02 [a]	0.49 ± 0.04 [a]	1.76 ± 0.01 [a]
	250–500	n/a	n/a	n/a	n/a	n/a	n/a	n/a	n/a	n/a	n/a	n/a	n/a
	Sig.	***	***	***	***	***	**	***	***	***	***	***	***
French press	>4000	289.14 ± 10.27 [a]	43.07 ± 3.46 [ab]	7.51 ± 0.28 [a]	0.33 ± 0.01 [b]	8.40 ± 0.57 [a]	1.00 ± 0.03 [a]	2.35 ± 0.23 [a]	14.05 ± 0.79 [a]	1.55 ± 0.05 [a]	0.41 ± 0.02 [a]	0.44 ± 0.04 [a]	1.76 ± 0.01 [a]
	2000–4000	272.30 ± 11.82 [a]	40.47 ± 2.09 [b]	7.34 ± 0.24 [a]	0.34 ± 0.03 [b]	8.16 ± 0.59 [a]	0.54 ± 0.02 [b]	1.77 ± 0.18 [b]	13.58 ± 1.35 [a]	1.61 ± 0.10 [a]	0.41 ± 0.07 [a]	0.39 ± 0.04 [a]	1.76 ± 0.01 [a]
	1000–2000	172.48 ± 9.12 [b]	31.03 ± 4.31 [c]	6.54 ± 0.63 [b]	0.20 ± 0.00 [c]	8.82 ± 1.09 [a]	0.94 ± 0.08 [a]	0.88 ± 0.11 [c]	12.99 ± 1.82 [a]	0.87 ± 0.02 [b]	0.18 ± 0.02 [b]	0.20 ± 0.03 [b]	0.88 ± 0.00 [b]
	500–1000	154.85 ± 22.81 [b]	26.73 ± 2.38 [c]	5.06 ± 0.22 [c]	0.21 ± 0.01 [c]	5.68 ± 0.19 [b]	0.52 ± 0.05 [b]	0.63 ± 0.03 [c]	8.68 ± 0.74 [b]	1.02 ± 0.21 [b]	0.18 ± 0.01 [b]	0.18 ± 0.02 [b]	0.88 ± 0.01 [b]
	250–500	286.68 ± 16.56 [a]	46.65 ± 3.02 [a]	7.70 ± 0.33 [a]	0.36 ± 0.00 [a]	7.80 ± 0.65 [a]	0.57 ± 0.07 [b]	2.27 ± 0.09 [a]	12.59 ± 1.26 [a]	1.44 ± 0.08 [a]	0.42 ± 0.05 [a]	0.39 ± 0.02 [a]	1.76 ± 0.01 [a]
	Sig.	***	***	***	***	**	*	*	**	***	***	***	***

n/a, not applicable. Means followed by different letters are significantly different at $p < 0.05$. Significance: * $p < 0.05$; ** $p < 0.01$; *** $p < 0.001$. Data are expressed as mean values ($n = 3$) ± standard deviation.

3.3.3. Methylxanthines

The concentrations of theobromine and caffeine evaluated by HPLC are given in Table 3. The amounts of theobromine were approximately 5–7-fold higher than those of caffeine. The concentrations of both methylxanthines significantly increased with a reduction in the GD for all the beverages obtained by percolation techniques. In general, the beverages that manifested the highest concentrations were those obtained by the Moka and Neapolitan methods, ranging from 147.33 to 703.79 mg theobromine/L and 25.44 to 124.84 mg caffeine/L for the former, and from 127.84 to 583.12 mg theobromine/L and 19.17 to 102.98 mg caffeine/L for the latter. The observed caffeine contents were lower than those observed for other kinds of beverages such as coffee (567 mg/L), mate (520 mg/L) [24], matcha (300 mg/L), loose leaf teas (99.21–296.86 mg/L), or bagged teas (151.73–246.71 mg/L) [25]. Nonetheless, the theobromine contents of CBS beverages are significantly higher than those observed for the same beverages, showing such amounts as 12.18 mg theobromine/L for matcha, 7.55–86.18 mg/L for loose leaf teas, and 21.58–66.91 mg/L for bagged teas [25].

Nevertheless, considering the expected intake of each type of beverage (60 mL for Moka, Neapolitan, Espresso, and Capsule and 200 mL for French press and American), the largest amounts of theobromine and caffeine would be consumed with the beverage produced by the French press technique, for which up to 60 mg of theobromine and 9 mg of caffeine would be consumed with each expected dose.

Cocoa is known for stimulating the brain due to the presence of theobromine and caffeine. As mentioned above, these two methylxanthines are also present in the CBS, the former being notably more abundant than the latter. They both influence alertness and mood in a positive way, acting on the central nervous system, and thus, may partly account for cocoa acceptance by consumers. Besides, several beneficial biological activities are linked to these methylxanthines such as the anticarcinogenic, antiobesity, antioxidant, antitumor, diuretic, or energizer effects of caffeine or the cAMP-inhibitory (with $IC_{50} = 0.06$ mg/mL), cAMP-phosphodiesterase-inhibitory, diuretic (when consumed at 300–600 mg/day), stimulant, or myorelaxant activities of theobromine [24]. Theobromine is also linked to the beneficial effects of cocoa consumption because of various other benefits associated with it, all of them without some of the unwanted effects of caffeine [26]. Usmani et al. [27] observed that a single 1000 mg dose of theobromine has a significantly greater antitussive effect on humans than a single 60 mg dose of codeine, an opioid drug whose clinical use is limited due to its unacceptable side effects, which could be avoided with an alternative theobromine treatment. Neufingerl et al. [28] reported that 850 mg of theobromine per day increases serum high-density lipoprotein cholesterol concentrations by 0.16 mmol/L. Another example of theobromine consumption benefits has been demonstrated by Kargul et al. [29], who showed that application of theobromine at concentrations of 100 and 200 mg/L to teeth could significantly protect enamel surface via a cariostatic effect, thus being an alternative to fluoride treatments. According to these data, CBS is likely to represent the proper combination of both methylxanthines in order to exert all the aforementioned beneficial actions without the secondary effects of big doses of caffeine, such as tachycardia or increased blood pressure if consumed at doses over 250 mg [30]. Regarding the above-mentioned examples of theobromine's benefits, one single expected dose of the CBS beverage would not yield the needed levels to observe the effects on high-density lipoprotein cholesterol level or the antitussive properties but it would clearly contribute to these benefits. Moreover, further optimizations or extract concentrations of the prepared beverages could be proposed to improve their functionality. Additionally, harmful levels of caffeine will not be an issue when expected doses of the beverages are consumed.

3.4. Biofunctional Characteristics

Polyphenols, as antioxidants, chelators of divalent cations, or inhibitors of enzymatic activities, have been reported to have several possible beneficial effects, e.g., anti-carcinogenic, anti-ulcer, anti-thrombotic, anti-inflammatory, anti-allergenic, immunomodulating, antimicrobial, vasodilatory, analgesic, or antidiabetic effects [16,22].

The beverages studied in the present work were found to contain considerably large quantities of various polyphenols and thus could exert a particular functional effect on the human body through the different properties of their polyphenols. Amongst these features, antioxidant and antidiabetic properties were studied here to evaluate the potential bioactivity of the beverages.

3.4.1. Antioxidant Capacity

The development of some chronic diseases such as cancer, cardiovascular diseases, and diabetes is tightly related to oxidative stress. Therefore, new chemopreventive approaches have been developed for preventing the damaging effects of free radicals and oxidants, mostly based on acquisition of radical scavengers and antioxidants from the diet. It has been already demonstrated that polyphenols from cocoa products (mostly flavanols) can interfere with these harmful processes and thus prevent the pathogenesis of the aforementioned diseases [31].

Results showing the radical scavenging or antioxidant capacity of the functional beverages expressed in mmol TE/L are presented in Figure 1d. The highest antioxidant capacity values were found in the beverages produced by the Moka and the Neapolitan methods at GD4 (7.29 and 6.58 mmol TE/L, respectively). These values were at least twice higher than those observed in the other beverages. A general trend was observed with a proportional increase in the antioxidant capacity with the decreasing CBS particle size for all the beverages produced by percolation techniques, whereas those obtained by the French press manifested no dependence on the GD, having the highest level of antioxidant capacity at the bigger CBS particle sizes, as reported about other parameters above. Antioxidant capacity showed significantly high correlations with the obtained values of TPC, TFC, and TTC ($r = 0.9656$, $r = 0.9716$, and $r = 0.9649$, respectively), even though lower correlations were seen between the antioxidant capacity and the single compounds detected by HPLC (ranging from $r = 0.7713$ for quercetin-3-O-rhamnoside to $r = 0.9348$ for procyanidin B2).

3.4.2. Antidiabetic Capacity

Cocoa polyphenols, in particular flavanols, have been reported to possess several antidiabetic bioactivities such as the improvement of insulin secretion by protecting β-pancreatic cells and the improvement of insulin sensitivity by protecting insulin-sensitive tissues from oxidative damage, among other reasons [9]. However, the main and more extended antidiabetic property of polyphenols have been reported to be the inhibition of key enzymes involved in glucose metabolism and absorption such as α-glucosidase or α-amylase [16]. This effect could be achieved by means of some drugs such as acarbose, a potent α-glucosidase inhibitor that is in disuse because of its considerable side effects that affect quality of life, e.g., abdominal distension, flatulence, meteorism, and possibly diarrhea [16]. For all these reasons, there is great interest among researchers in the search for healthy and sustainable alternatives such as the CBS preparations for beverages presented in this work. We carried out the study of the antidiabetic capacity due to the α-glucosidase inhibition of the beverages and the results are depicted in Figure 2.

As expected, according to TPC, TFC, and TTC results and taking into account the potential influence of these compounds on the antidiabetic capacity, the beverage exerting the higher percentage of α-glucosidase inhibition was the one obtained with Moka at the smallest CBS particle size GD4 (52.0%), followed by the Neapolitan (36.8%), Espresso (32.0%), and Capsule (26.2%) beverages at the same GD. The smallest percentage of α-glucosidase inhibition was observed for the beverage produced by the Capsule technique at the biggest GD of the CBS (4.7% with GD1). In general, beverages produced with big particle sizes of the CBS are those exerting the smallest α-glucosidase inhibition, especially for the techniques that employ pressure for extraction, where the contact time between water and the CBS decreases, and so does extraction performance. The French press beverages instead showed an intermediate value of α-glucosidase inhibition independently of the CBS particle size. It is important to note that the contribution to the antidiabetic effect manifested by the new functional beverages is close to that of 0.5 mM acarbose serving as a control sample corresponding to the IC_{50} concentration for this drug. The α-glucosidase inhibition parameters showed a significantly high correlation with TPC (r = 0.9537) and some of the detected polyphenolic compounds such as protocatechuic acid (r = 0.9826), type B procyanidin (r = 0.9870), and, as expected, both flavan-3-ols catechin-3-O-glucoside and epicatechin (r = 0.9803 and r = 0.9500, respectively).

This is the first study where the α-glucosidase inhibition due to polyphenolic content is reported for CBS. Nonetheless, the hypoglycemic effects of this by-product due to its fiber content have already been described by Nsor-Atindana et al. [32]. We did not evaluate this effect.

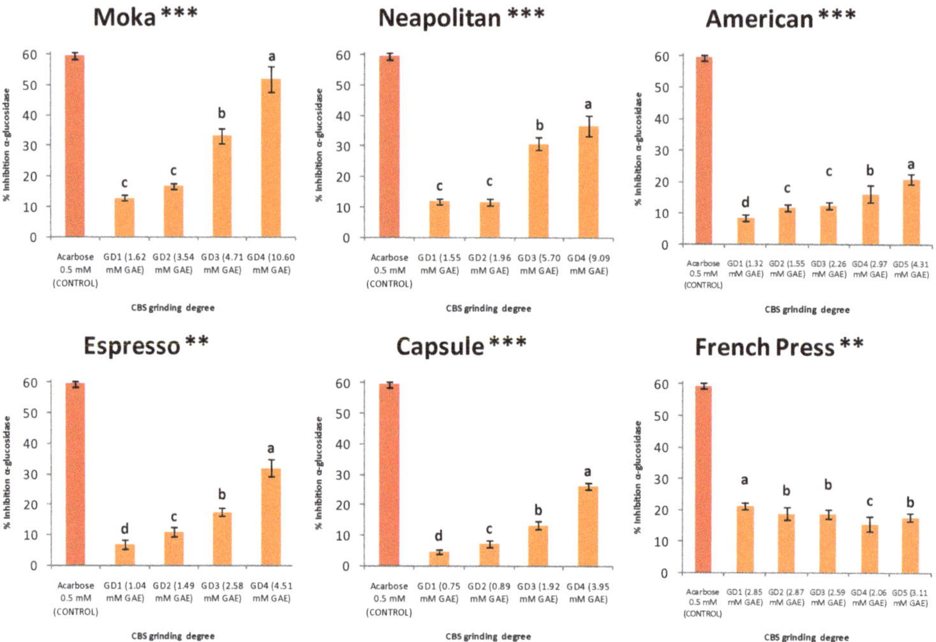

Figure 2. α-Glucosidase inhibition by the beverages produced via the six techniques at various CBS GDs related to the total phenolic content expressed in mM GAE (gallic acid equivalents); ANOVA among GDs for each technique. Acarbose at 0.5 mM (IC_{50}) served as a control. Different letters indicate significant differences among the GDs at $p < 0.05$. Significance: ** $p < 0.01$; *** $p < 0.001$.

3.5. Consumer Acceptance Evaluation

Table 4 shows the sums of ranks values calculated for the consumer evaluation parameters for each beverage preparation technique and CBS particle size. These values were subjected to the Kruskal–Wallis test to highlight the differences in acceptance for the different beverages prepared by the same technique, comparing the GDs of the CBS.

As far as the aspect was concerned, slight differences were evidenced, and a general preference for the beverages obtained with smaller GDs was observed. Taking into account the results obtained for the beverages' color, where in general, the brightness decreased while the GD decreased and both parameters $a*$ and $b*$ rose when the CBS GD decreased, it could be assumed that darker and browner beverages were preferred over lighter ones. Regarding the odor, the beverages obtained by the Neapolitan and the Moka techniques were generally the most appreciated while those obtained by the Capsule and the French press techniques were the least appreciated. In the case of taste, the beverage obtained by the Capsule technique showed high levels of consumer acceptance for almost all of the GDs. In some cases, even the beverages obtained by the Neapolitan and American methods were highly liked. The situation was again ambiguous for the flavor where the liking seems to be influenced by the technique–GD interaction, though in general, the beverages obtained by the Neapolitan or American technique were among the most appreciated. It was observed that overall, the beverages that had high values of titratable acidity were less liked in terms of taste and flavor. The overall rating obviously reflects all the previous variability for the various parameters of consumer acceptance, with the Neapolitan and the American techniques yielding the most liked beverages at the highest GDs, whereas the Capsule and Espresso methods seem to be the ones preferred at small GDs. Again, this pattern is reflected in the purchase predisposition, with a greater preference for Neapolitan technique beverages with large CBS particle sizes and for those generated by Capsule and Espresso at smaller GDs.

It should be remarked that the most active beverages (Moka and Neapolitan with small GDs) are between those most appreciated as far as the appearance or the odor are concerned. But, on the contrary, the appreciation of these beverages decreases considerably when evaluating taste and flavor, and so do the overall liking and the purchase predisposition. This fact could be related to the unpleasant, bitter and, in some cases, astringent flavor related to the polyphenols and the methylxantines which are present in high amounts in these beverages.

Table 4. Consumer evaluation of the beverages and results of the Kruskal–Wallis test. Data are expressed as the sum of ranks of the results obtained from 20 tasters who filled out a nine-point hedonic scale (1 = extremely dislike, 9 = extremely like).

Production Technique	Grinding Degree	Appearance	Odor	Taste	Flavor	Overall liking	Purchase predisposition
Moka	>4000 μm	309.5 [ab]	234.0 [b]	482.5 [a]	473.0 [a]	486.5 [a]	395.0 [a]
	2000–4000 μm	188.0 [b]	220.0 [b]	335.0 [ab]	283.5 [b]	312.5 [b]	265.5 [a]
	1000–2000 μm	288.5 [ab]	311.5 [ab]	206.0 [bc]	188.0 [b]	177.5 [b]	279.0 [a]
	500–1000 μm	390.0 [a]	410.5 [a]	152.5 [c]	231.5 [b]	199.5 [b]	236.5 [a]
	250–500 μm	n/a	n/a	n/a	n/a	n/a	n/a
	Significance	*	*	***	***	***	ns
Neapolitan	>4000 μm	372.5 [a]	291.0 [a]	449.5 [a]	362.0 [ab]	433.5 [a]	439.5 [a]
	2000–4000 μm	230.5 [a]	344.5 [a]	361.0 [ab]	382.5 [a]	406.0 [a]	376.5 [a]
	1000–2000 μm	271.0 [a]	231.5 [a]	245.5 [bc]	202.5 [b]	181.5 [b]	205.0 [b]
	500–1000 μm	302.0 [a]	309.0 [a]	120.0 [c]	229.0 [ab]	155.0 [b]	155.0 [b]
	250–500 μm	n/a	n/a	n/a	n/a	n/a	n/a
	Significance	ns	ns	***	**	***	***
American	>4000 μm	138.5 [b]	338.0 [a]	400.5 [ab]	446.0 [a]	436.0 [ab]	467.5 [a]
	2000–4000 μm	510.5 [a]	444.0 [a]	595.0 [a]	557.5 [a]	569.5 [a]	403.0 [a]
	1000–2000 μm	380.5 [a]	288.0 [a]	328.0 [bc]	342.0 [a]	309.0 [bc]	318.0 [a]
	500–1000 μm	461.0 [a]	310.0 [a]	363.0 [bc]	396.0 [a]	392.0 [ab]	372.0 [a]
	250–500 μm	339.5 [ab]	450.0 [a]	143.5 [c]	88.5 [b]	123.5 [c]	269.5 [a]
	Significance	***	ns	***	***	***	ns
Espresso	>4000 μm	120.0 [b]	138.0 [b]	277.0 [ab]	247.0 [ab]	172.5 [b]	214.0 [bc]
	2000–4000 μm	316.0 [a]	337.0 [a]	413.0 [a]	369.0 [a]	402.0 [a]	394.5 [a]
	1000–2000 μm	356.5 [a]	290.0 [ab]	366.5 [a]	391.5 [a]	351.5 [a]	368.5 [ab]
	500–1000 μm	383.5 [a]	410.0 [a]	119.5 [b]	168.0 [b]	250.0 [ab]	199.0 [c]
	250–500 μm	n/a	n/a	n/a	n/a	n/a	n/a
	Significance	***	***	***	**	**	**
Capsule	>4000 μm	125.5 [b]	151.0 [b]	431.0 [b]	292.5 [ab]	235.5 [b]	201.0 [b]
	2000–4000 μm	185.0 [b]	248.0 [b]	321.0 [b]	188.0 [b]	195.5 [b]	206.0 [b]
	1000–2000 μm	389.0 [a]	310.0 [ab]	172.0 [ab]	283.5 [ab]	300.0 [ab]	339.5 [ab]
	500–1000 μm	476.0 [a]	467.0 [a]	252.0 [a]	412.0 [a]	445.0 [a]	429.5 [a]
	250–500 μm	n/a	n/a	n/a	n/a	n/a	n/a
	Significance	***	***	***	**	***	***
French press	>4000 μm	464.5 [a]	526.5 [a]	192.0 [c]	472.0 [a]	436.0 [a]	319.0 [ab]
	2000–4000 μm	312.0 [a]	172.0 [b]	237.5 [bc]	119.0 [b]	132.5 [b]	224.5 [b]
	1000–2000 μm	366.0 [a]	140.5 [b]	367.5 [abc]	290.0 [ab]	285.5 [ab]	493.0 [a]
	500–1000 μm	265.5 [a]	494.0 [a]	580.0 [a]	503.0 [a]	488.0 [a]	461.5 [a]
	250–500 μm	422.0 [a]	497.0 [a]	453.0 [ab]	446.0 [a]	488.0 [a]	332.0 [ab]
	Significance	ns	***	***	***	***	**

n/a, not applicable. Means followed by different letters are significantly different at $p < 0.05$. Significance: * $p < 0.05$; ** $p < 0.01$; *** $p < 0.001$; ns = not significant. Data are expressed as sums of ranks.

3.6. Technological Efficiency of Polyphenol Extraction

To identify the technique allowing higher extraction yield of polyphenols, normalization of the TPC values was performed according to the different CBS powder amounts and water quantities employed. The new TPC values, expressed in milligrams of gallic acid equivalents per gram of CBS ranged between 2.72 and 16.32 mg GAE/g (Table 5). A significant increase in TPC was observed for the beverages obtained by the forced percolation techniques (Moka, Espresso, and Capsule) at GDs below 1000 μm and for the beverages produced by the Neapolitan method (natural percolation) at GDs below 2000 μm. For the beverage produced by the American technique (natural percolation), TPC increased progressively with the decreasing GDs. On the contrary, for the beverage produced with the French press, no influence of the GDs was observed, thereby leading to the conclusion that the difference in GDs had only a minor impact on the polyphenol extraction by this technique based on a maceration process. Neapolitan was always the most effective technique when total phenol extraction was compared among the beverages obtained by all the techniques at the same GD, followed by those produced by the American, French press, or Moka techniques. The lowest TPC values belong to the beverages produced by the Capsule and Espresso. On the other hand, these techniques usually afforded higher extraction of essential oils because they produce more aromatic coffee. The highest TPC observed (16.32 mg GAE/g for the beverage obtained with Neapolitan using GD4) was significantly higher than those reported for other types of CBS extraction in previous works. Hernández-Hernández et al. [23] obtained a TPC value of 3 mg GAE/g when extracting 1 g of CBS at 500 μm in 6 mL of water at 70 °C. Manzano et al. [33] observed a TPC value of 6.04 mg GAE/g using 2 g of CBS screened at 75 μm, which was processed in 50 mL of water with 5-min reflux extraction. Nevertheless, in other studies in which assisted extraction was carried out, higher values of TPC were obtained as compared to those obtained for the beverages in the present work, as expected. Nsor-Atindana et al. [32] reported a value of 17.21 mg GAE/g after the extraction of 2 g of CBS ground up at 250 μm in 50 mL of water using microwaves. In any case, it is important to note that the solid/liquid ratio, which was different in all studies, could also have some influence on these results.

Table 5. Total phenolic content for the beverages after normalization considering CBS and water quantities. ANOVA among GDs and extraction techniques. Values are expressed in milligrams of gallic acid equivalents for each gram of CBS powder employed for the beverage preparation (mg GAE/g CBS).

	Moka	Neapolitan	American	Espresso	Capsule	French Press	*Sig.*
>4000 μm	5.21 ± 0.54 [cB]	7.03 ± 0.47 [cA]	4.84 ± 0.28 [dB]	3.12 ± 0.30 [bC]	3.76 ± 0.25 [bC]	6.68 ± 0.64 [aA]	***
2000–4000 μm	6.37 ± 2.25 [bcAB]	7.14 ± 1.06 [cA]	5.26 ± 0.09 [dB]	2.72 ± 0.39 [bC]	3.25 ± 0.30 [cC]	6.74 ± 0.72 [aA]	***
1000–2000 μm	6.92 ± 0.72 [bB]	10.00 ± 1.33 [bA]	7.32 ± 0.34 [cB]	3.41 ± 0.33 [bD]	3.03 ± 0.11 [cD]	5.78 ± 0.30 [bC]	***
500–1000 μm	12.94 ± 0.96 [aB]	16.32 ± 1.04 [aA]	9.39 ± 0.92 [bC]	5.10 ± 0.15 [aD]	5.06 ± 0.25 [aD]	4.10 ± 0.31 [cE]	***
250–500 μm	n/a	n/a	13.45 ± 0.86 [aA]	n/a	n/a	6.29 ± 0.26 [aB]	***
Sig.	***	***	***	***	***	***	***

n/a, not applicable. Means followed by different lower case superindexes within the same column (different grinding degrees) and by upper case superindexes within the same row (different techniques) are significantly different at $p < 0.05$. Significance: *** $p < 0.001$. Data are expressed as mean values ($n = 3$) ± standard deviation.

4. Conclusions

The various extraction techniques used for CBS ground to different degrees allowed us to obtain beverages with different chemical characteristics and consumer-related parameters. Several compounds were identified and quantified by HPLC (phenolic acids, flavan-3-ols, quercetin-glycosides, catechin-glycosides, and procyanidins), which may underlie the high radical scavenging capacity and significant α-glucosidase inhibition results shown by the beverages. The GD was optimized for each extraction technique; the smallest GDs allowed us to obtain the most functional beverages when using percolation techniques, whereas the maceration technique (French press) in general, showed no dependence on the CBS particle size. This finding may be of great interest as with the French press technique, no further CBS grinding treatments will be needed to obtain a beverage having high-potential biological activities. In terms of consumer acceptance, it was found that, in general, the most active beverages were the least appreciated as far as taste and flavor are concerned, probably because of the bigger presence of polyphenols and methylxanhines. This fact could open a possibility for further research in order to optimize these beverages, aiming at a higher consumer acceptance. Such optimization could be achieved by bioactive compounds encapsulation or by adding new pleasant ingredients, among other options.

For the first time, it was demonstrated that the Moka and Neapolitan techniques may be the most effective methods for polyphenol extraction, affording the highest radical scavenging activity and α-glucosidase inhibition capacity, whereas the beverage produced by the Capsule technique showed the poorest extraction. Therefore, this work indicates that CBS may be an optimal ingredient for home-made functional beverages with potential health benefits for consumers, thereby reducing the environmental and economic impact of by-product disposal.

Author Contributions: Writing—original Draft Preparation, O.R.-P.; Investigation, O.R.-P. and L.M.-R.; Review and Editing, L.B.-P., G.Z., C.S., and M.B.; Conceptualization, L.B.-P., G.Z., and M.B.; Supervision, L.B.-P., G.Z., and C.S.; Project Administration, L.B.-P. and G.Z.; Funding acquisition, L.B.-P. and G.Z.

Funding: The present work has been supported by COVALFOOD "Valorisation of high added-value compounds from cocoa industry by-products as food ingredients and additives". This project has received funding from the European Union's Seventh Framework program for research and innovation under the Marie Skłodowska-Curie grant agreement No. 609402-2020 researchers: Train to Move (T2M).

Acknowledgments: O. Rojo-Poveda gratefully acknowledges the University of Turin for the award of a PhD student fellowship. L. Barbosa-Pereira is grateful to the European Union's Seventh Framework program for the Marie Skłodowska-Curie grant.

Conflicts of Interest: The authors declare no conflict of interest.

Abbreviations

CBS, cocoa bean shell; GD, grinding degree; TPC, total phenolic content; TFC, total flavonoid content; TTC, total tannin content, RSA, radical scavenging activity; GAE, gallic acid equivalents; CE, catechin equivalents; TE, trolox equivalents; RP-HLPC-PDA, reversed-phase high-pressure liquid chromatography equipped with photodiode array detector; PB1, procyanidin B1; PB2, procyanidin B2; Q-3-G, quercetin-3-O-glucoside; ANOVA, analysis of variance.

References

1. ICCO. *Quarterly Bulletin of Cocoa Statistics*; International Cocoa Organization: Abidjan, Ivory Coast, 2018; Volume XLIV.
2. Kofink, M.; Papagiannopoulos, M.; Galensa, R. (−)-Catechin in cocoa and chocolate: Occurence and analysis of an atypical flavan-3-ol enantiomer. *Molecules* **2007**, *12*, 1274–1288. [CrossRef] [PubMed]
3. Adamafio, N. Theobromine toxicity and remediation of cocoa by-products: An overview. *J. Boil. Sci.* **2013**, *13*, 570–576. [CrossRef]
4. Okiyama, D.C.G.; Navarro, S.L.B.; Rodrigues, C.E.C. Cocoa shell and its compounds: Applications in the food industry. *Trends Food Sci. Technol.* **2017**, *63*, 103–112. [CrossRef]
5. Arlorio, M.; Coisson, J.; Restani, P.; Martelli, A. Characterization of pectins and some secondary compounds from Theobroma cacao hulls. *J. Food Sci.* **2001**, *66*, 653–656. [CrossRef]
6. Grillo, G.; Boffa, L.; Binello, A.; Mantegna, S.; Cravotto, G.; Chemat, F.; Dizhbite, T.; Lauberte, L.; Telysheva, G. Analytical dataset of Ecuadorian cocoa shells and beans. *Data Brief* **2019**, *22*, 56–64. [CrossRef]
7. Prasad, M.; Simmons, P.; Maher, M. Release characteristics of organic fertilizers. *Acta Hortic.* **2004**, 163–170. [CrossRef]
8. Mancini, G.; Papirio, S.; Lens, P.N.L.; Esposito, G. Effect of *N*-methylmorpholine-*N*-oxide Pretreatment on Biogas Production from Rice Straw, Cocoa Shell, and Hazelnut Skin. *Environ. Eng. Sci.* **2016**, *33*, 843–850. [CrossRef]
9. Martin, M.Á.; Goya, L.; Ramos, S. Antidiabetic actions of cocoa flavanols. *Mol. Nutr. Food Res.* **2016**, *60*, 1756–1769. [CrossRef]
10. Tur, J.; Bibiloni, M. Functional foods. Reference Module in Food Science. In *Encyclopedia of Food and Health*; Elsevier B.V.: Kidlington, UK, 2016; pp. 157–161.

11. Maritim, A.; Sanders, A.; Watkins, J., 3rd. Diabetes, oxidative stress, and antioxidants: A review. *J. Biochem. Mol. Toxicol.* **2003**, *17*, 24–38. [CrossRef] [PubMed]

12. Vinayagam, R.; Xu, B. Antidiabetic properties of dietary flavonoids: A cellular mechanism review. *Nutr. Metab.* **2015**, *12*, 60. [CrossRef]

13. Bertolino, M.; Belviso, S.; Dal Bello, B.; Ghirardello, D.; Giordano, M.; Rolle, L.; Gerbi, V.; Zeppa, G. Influence of the addition of different hazelnut skins on the physicochemical, antioxidant, polyphenol and sensory properties of yogurt. *LWT-Food Sci. Technol.* **2015**, *63*, 1145–1154. [CrossRef]

14. Hill, B.; Roger, T.; Vorhagen, F.W. Comparative analysis of the quantization of color spaces on the basis of the CIELAB color-difference formula. *ACM Trans. Graph.* **1997**, *16*, 109–154. [CrossRef]

15. Barbosa-Pereira, L.; Guglielmetti, A.; Zeppa, G. Pulsed Electric Field Assisted Extraction of Bioactive Compounds from Cocoa Bean Shell and Coffee Silverskin. *Food Bioprocess Technol.* **2018**, *11*, 818–835. [CrossRef]

16. Kwon, Y.I.; Apostolidis, E.; Shetty, K. In vitro studies of eggplant (*Solanum melongena*) phenolics as inhibitors of key enzymes relevant for type 2 diabetes and hypertension. *Bioresour. Technol.* **2008**, *99*, 2981–2988. [CrossRef] [PubMed]

17. Torri, L.; Piochi, M.; Marchiani, R.; Zeppa, G.; Dinnella, C.; Monteleone, E. A sensory- and consumer-based approach to optimize cheese enrichment with grape skin powders. *J. Dairy Sci.* **2016**, *99*, 194–204. [CrossRef] [PubMed]

18. Zujko, M.E.; Witkowska, A.M. Antioxidant potential and polyphenol content of beverages, chocolates, nuts, and seeds. *Int. J. Food Prop.* **2014**, *17*, 86–92. [CrossRef]

19. Gardner, P.T.; White, T.A.; McPhail, D.B.; Duthie, G.G. The relative contributions of vitamin C, carotenoids and phenolics to the antioxidant potential of fruit juices. *Food Chem.* **2000**, *68*, 471–474. [CrossRef]

20. Damm, I.; Enger, E.; Chrubasik-Hausmann, S.; Schieber, A.; Zimmermann, B.F. Fast and comprehensive analysis of secondary metabolites in cocoa products using ultra high-performance liquid chromatography directly after pressurized liquid extraction. *J. Sep. Sci.* **2016**, *39*, 3113–3122. [CrossRef]

21. Rodríguez-Carrasco, Y.; Gaspari, A.; Graziani, G.; Sandini, A.; Ritieni, A. Fast analysis of polyphenols and alkaloids in cocoa-based products by ultra-high performance liquid chromatography and Orbitrap high resolution mass spectrometry (UHPLC-Q-Orbitrap-MS/MS). *Food Res. Int.* **2018**, *111*, 229–236. [CrossRef]

22. Wollgast, J.; Anklam, E. Review on polyphenols in Theobroma cacao: Changes in composition during the manufacture of chocolate and methodology for identification and quantification. *Food Res. Int.* **2000**, *33*, 423–447. [CrossRef]

23. Hernández-Hernández, C.; Morales Sillero, A.; Fernández-Bolaños, J.; Bermúdez Oria, A.; Azpeitia Morales, A.; Rodríguez-Gutiérrez, G. Cocoa bean husk: Industrial source of antioxidant phenolic extract. *J. Sci. Food Agric.* **2018**, *99*, 325–333. [CrossRef] [PubMed]

24. Heck, C.I.; De Mejia, E.G. Yerba Mate Tea (Ilex paraguariensis): A comprehensive review on chemistry, health implications, and technological considerations. *J. Food Sci.* **2007**, *72*, R138–R151. [CrossRef] [PubMed]

25. Komes, D.; Horžić, D.; Belščak, A.; Ganić, K.K.; Vulić, I. Green tea preparation and its influence on the content of bioactive compounds. *Food Res. Int.* **2010**, *43*, 167–176. [CrossRef]

26. Martínez-Pinilla, E.; Oñatibia-Astibia, A.; Franco, R. The relevance of theobromine for the beneficial effects of cocoa consumption. *Front. Pharmacol.* **2015**, *6*, 30. [CrossRef]

27. Usmani, O.S.; Belvisi, M.G.; Patel, H.J.; Crispino, N.; Birrell, M.A.; Korbonits, M.; Korbonits, D.; Barnes, P.J. Theobromine inhibits sensory nerve activation and cough. *Faseb J.* **2005**, *19*, 231–233. [CrossRef]

28. Neufingerl, N.; Zebregs, Y.E.; Schuring, E.A.; Trautwein, E.A. Effect of cocoa and theobromine consumption on serum HDL-cholesterol concentrations: A randomized controlled trial–. *Am. J. Clin. Nutr.* **2013**, *97*, 1201–1209. [CrossRef]

29. Kargul, B.; Özcan, M.; Peker, S.; Nakamoto, T.; Simmons, W.B.; Falster, A.U. Evaluation of human enamel surfaces treated with theobromine: A pilot study. *Oral Health Prev. Dent.* **2012**, *10*, 275.

30. Pelchovitz, D.J.; Goldberger, J.J. Caffeine and cardiac arrhythmias: A review of the evidence. *Am. J. Med.* **2011**, *124*, 284–289. [CrossRef]

31. Martín, M.A.; Ramos, S. Cocoa polyphenols in oxidative stress: Potential health implications. *J. Funct. Foods* **2016**, *27*, 570–588. [CrossRef]

32. Nsor-Atindana, J.; Zhong, F.; Mothibe, K.J.; Bangoura, M.L.; Lagnika, C. Quantification of total polyphenolic content and antimicrobial activity of cocoa (Theobroma cacao L.) Bean Shells. *Pak. J. Nutr.* **2012**, *11*, 574.
33. Manzano, P.; Hernández, J.; Quijano-Avilés, M.; Barragán, A.; Chóez-Guaranda, I.; Viteri, R.; Valle, O. Polyphenols extracted from Theobroma cacao waste and its utility as antioxidant. *Emir. J. Food Agric.* **2017**, *29*, 45. [CrossRef]